MCAT® PHYSICS AND MATH REVIEW 2022–2023

ONLINE + BOOK

Edited by Alexander Stone Macnow, MD

ACKNOWLEDGMENTS

Editor-in-Chief, 2022–2023 Edition
Lauren K. White

Contributing Editors, 2022–2023 Edition
Brandon Deason, MD, Christopher Durland, Tyler Fara

Prior Edition Editorial Staff: Brandon Deason, Christopher Durland, M. Dominic Eggert, Tyler Fara, Elizabeth Flagge, Adam Grey, Lauren K. White

Published by Kaplan Publishing, a division of Kaplan, Inc.
750 Third Avenue
New York, NY 10017

ISBN: 978-1-5062-7673-1

10 9 8 7 6 5 4 3 2 1

Kaplan Publishing print books are available at special quantity discounts to use for sales promotions, employee premiums, or educational purposes. For more information or to purchase books, please call the Simon & Schuster special sales department at 866-506-1949.

TABLE OF CONTENTS

kaptest.com/booksonline

The *Kaplan MCAT Review* Team vi

Getting Started Checklist vii

Preface viii

About *Scientific American* ix

About the MCAT x

How This Book Was Created xxii

Using This Book xxiii

Studying for the MCAT xxvii

CHAPTER 1: KINEMATICS AND DYNAMICS 1

1.1 Units 8

1.2 Vectors and Scalars 10

1.3 Displacement and Velocity 16

1.4 Forces and Acceleration 19

1.5 Newton's Laws 24

1.6 Motion with Constant Acceleration 26

1.7 Mechanical Equilibrium 33

CHAPTER 2: WORK AND ENERGY 49

2.1 Energy 56

High-Yield 2.2 Work 62

2.3 Mechanical Advantage 67

CHAPTER 3: THERMODYNAMICS 85

3.1 Zeroth Law of Thermodynamics 92

3.2 Systems 96

High-Yield 3.3 First Law of Thermodynamics 98

3.4 Second Law of Thermodynamics and Entropy 106

CHAPTER 4: FLUIDS 119
4.1 Characteristics of Fluids and Solids 128
4.2 Hydrostatics 133
High-Yield 4.3 Fluid Dynamics 139
4.4 Fluids in Physiology 148

CHAPTER 5: ELECTROSTATICS AND MAGNETISM 159
5.1 Charges 168
5.2 Coulomb's Law 170
5.3 Electric Potential Energy 174
5.4 Electric Potential 176
5.5 Special Cases in Electrostatics 178
5.6 Magnetism 184

CHAPTER 6: CIRCUITS 203
6.1 Current 212
High-Yield 6.2 Resistance 216
6.3 Capacitance and Capacitors 225
6.4 Meters 231

CHAPTER 7: WAVES AND SOUND 243
7.1 General Wave Characteristics 250
High-Yield 7.2 Sound 257

CHAPTER 8: LIGHT AND OPTICS 279
High-Yield 8.1 Electromagnetic Spectrum 286
High-Yield 8.2 Geometrical Optics 288
8.3 Diffraction 306
8.4 Polarization 311

CHAPTER 9: ATOMIC AND NUCLEAR PHENOMENA 323
9.1 The Photoelectric Effect 330
9.2 Absorption and Emission of Light 334
9.3 Nuclear Binding Energy and Mass Defect 336
High-Yield 9.4 Nuclear Reactions 339

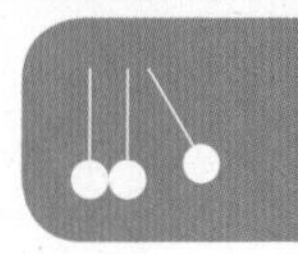

CHAPTER 10: MATHEMATICS 357

10.1 Arithmetic and Significant Figures 364

10.2 Exponents and Logarithms 368

10.3 Trigonometry 373

10.4 Problem-Solving 376

CHAPTER 11: REASONING ABOUT THE DESIGN AND EXECUTION OF RESEARCH 391

11.1 The Scientific Method 400

High-Yield 11.2 Basic Science Research 403

11.3 Human Subjects Research 408

11.4 Ethics 412

11.5 Research in the Real World 416

CHAPTER 12: DATA-BASED AND STATISTICAL REASONING 427

12.1 Measures of Central Tendency 436

12.2 Distributions 439

12.3 Measures of Distribution 442

12.4 Probability 447

12.5 Statistical Testing 449

High-Yield 12.6 Charts, Graphs, and Tables 451

12.7 Applying Data 460

GLOSSARY 471

INDEX 481

ART CREDITS 493

THE *KAPLAN MCAT REVIEW* TEAM

Alexander Stone Macnow, MD
Editor-in-Chief

Áine Lorié, PhD
Editor

Kristen L. Russell, ME
Editor

Derek Rusnak, MA
Editor

Pamela Willingham, MSW
Editor

Mikhail Alexeeff
Kaplan MCAT Faculty

Melinda Contreras, MS
Kaplan MCAT Faculty

Laura L. Ambler
Kaplan MCAT Faculty

Samantha Fallon
Kaplan MCAT Faculty

Krista L. Buckley, MD
Kaplan MCAT Faculty

Jason R. Selzer
Kaplan MCAT Faculty

Faculty Reviewers and Editors: Elmar R. Aliyev; James Burns; Jonathan Cornfield; Alisha Maureen Crowley; Nikolai Dorofeev, MD; Benjamin Downer, MS; Colin Doyle; Christopher Durland; M. Dominic Eggert; Marilyn Engle; Eleni M. Eren; Raef Ali Fadel; Elizabeth Flagge; Adam Grey; Rohit Gupta, Jonathan Habermacher; Tyra Hall-Pogar, PhD; Justine Harkness, PhD; Scott Huff; Samer T. Ismail; Ae-Ri Kim, PhD; Elizabeth A. Kudlaty; Kelly Kyker-Snowman, MS; Ning-fei Li; John P. Mahon; Brandon McKenzie; Matthew A. Meier; Nainika Nanda; Caroline Nkemdilim Opene; Kaitlyn E. Prenger; Uneeb Qureshi; Jason Selzer; Allison St. Clair; Bela G. Starkman, PhD; Chris Sun; Michael Paul Tomani, MS; Bonnie Wang; Ethan Weber; Lauren K. White; Nicholas M. White; Allison Ann Wilkes, MS; Kerranna Williamson, MBA; and Tony Yu

Thanks to Kim Bowers; Eric Chiu; Tim Eich; Tyler Fara; Owen Farcy; Dan Frey; Robin Garmise; Rita Garthaffner; Joanna Graham; Allison Harm; Beth Hoffberg; Aaron Lemon-Strauss; Keith Lubeley; Diane McGarvey; Petros Minasi; Camellia Mukherjee; John Polstein; Deeangelee Pooran-Kublall, MD, MPH; Rochelle Rothstein, MD; Larry Rudman; Sylvia Tidwell Scheuring; Carly Schnur; Karin Tucker; Lee Weiss; and the countless others who made this project possible.

GETTING STARTED CHECKLIST

- [ ] Register for your free online assets—including full-length tests, Science Review Videos, and additional practice materials—at **www.kaptest.com/moreonline**.
- [ ] Create a study calendar that ensures you complete content review and sufficient practice by Test Day!
- [ ] As you finish a chapter and the online practice for that chapter, check it off on the table of contents.
- [ ] Register to take the MCAT at **www.aamc.org/mcat**.
- [ ] Set aside time during your prep to make sure the rest of your application—personal statement, recommendations, and other materials—is ready to go!
- [ ] Take a moment to admire your completed checklist, then get back to the business of prepping for this exam!

PREFACE

And now it starts: your long, yet fruitful journey toward wearing a white coat. Proudly wearing that white coat, though, is hopefully only part of your motivation. You are reading this book because you want to be a healer.

If you're serious about going to medical school, then you are likely already familiar with the importance of the MCAT in medical school admissions. While the holistic review process puts additional weight on your experiences, extracurricular activities, and personal attributes, the fact remains: along with your GPA, your MCAT score remains one of the two most important components of your application portfolio—at least early in the admissions process. Each additional point you score on the MCAT pushes you in front of thousands of other students and makes you an even more attractive applicant. But the MCAT is not simply an obstacle to overcome; it is an opportunity to show schools that you will be a strong student and a future leader in medicine.

We at Kaplan take our jobs very seriously and aim to help students see success not only on the MCAT, but as future physicians. We work with our learning science experts to ensure that we're using the most up-to-date teaching techniques in our resources. Multiple members of our team hold advanced degrees in medicine or associated biomedical sciences, and are committed to the highest level of medical education. Kaplan has been working with the MCAT for over 50 years and our commitment to premed students is unflagging; in fact, Stanley Kaplan created this company when he had difficulty being accepted to medical school due to unfair quota systems that existed at the time.

We stand now at the beginning of a new era in medical education. As citizens of this 21st-century world of healthcare, we are charged with creating a patient-oriented, culturally competent, cost-conscious, universally available, technically advanced, and research-focused healthcare system, run by compassionate providers. Suffice it to say, this is no easy task. Problem-based learning, integrated curricula, and classes in interpersonal skills are some of the responses to this demand for an excellent workforce—a workforce of which you'll soon be a part.

We're thrilled that you've chosen us to help you on this journey. Please reach out to us to share your challenges, concerns, and successes. Together, we will shape the future of medicine in the United States and abroad; we look forward to helping you become the doctor you deserve to be.

Good luck!

Alexander Stone Macnow, MD
Editor-in-Chief
Department of Pathology and Laboratory Medicine
Hospital of the University of Pennsylvania

BA, Musicology—Boston University, 2008
MD—Perelman School of Medicine at the University of Pennsylvania, 2013

ABOUT *SCIENTIFIC AMERICAN*

As the world's premier science and technology magazine, and the oldest continuously published magazine in the United States, *Scientific American* is committed to bringing the most important developments in modern science, medicine, and technology to our worldwide audience in an understandable, credible, and provocative format.

Founded in 1845 and on the "cutting edge" ever since, *Scientific American* boasts over 200 Nobel laureate authors including Albert Einstein, Francis Crick, Stanley Prusiner, and Richard Axel. *Scientific American* is a forum where scientific theories and discoveries are explained to a broader audience.

Scientific American published its first foreign edition in 1890, and in 1979 was the first Western magazine published in the People's Republic of China. Today, *Scientific American* is published in 14 foreign language editions. *Scientific American* is also a leading online destination (**www.ScientificAmerican.com**), providing the latest science news and exclusive features to more than 10 million visitors each month.

The knowledge that fills our pages has the power to spark new ideas, paradigms, and visions for the future. As science races forward, *Scientific American* continues to cover the promising strides, inevitable setbacks and challenges, and new medical discoveries as they unfold.

ABOUT THE MCAT

Anatomy of the MCAT

Here is a general overview of the structure of Test Day:

Section	Number of Questions	Time Allotted
Test-Day Certification		4 minutes
Tutorial (optional)		10 minutes
Chemical and Physical Foundations of Biological Systems	59	95 minutes
Break (optional)		10 minutes
Critical Analysis and Reasoning Skills (CARS)	53	90 minutes
Lunch Break (optional)		30 minutes
Biological and Biochemical Foundations of Living Systems	59	95 minutes
Break (optional)		10 minutes
Psychological, Social, and Biological Foundations of Behavior	59	95 minutes
Void Question		3 minutes
Satisfaction Survey (optional)		5 minutes

The structure of the four sections of the MCAT is shown below.

Chemical and Physical Foundations of Biological Systems	
Time	95 minutes
Format	• 59 questions • 10 passages • 44 questions are passage-based, and 15 are discrete (stand-alone) questions. • Score between 118 and 132
What It Tests	• Biochemistry: 25% • Biology: 5% • General Chemistry: 30% • Organic Chemistry: 15% • Physics: 25%

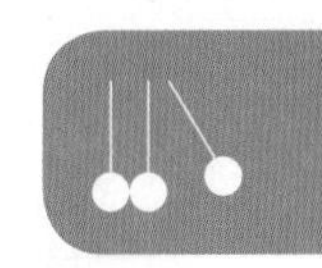

Critical Analysis and Reasoning Skills (CARS)	
Time	90 minutes
Format	• 53 questions • 9 passages • All questions are passage-based. There are no discrete (stand-alone) questions. • Score between 118 and 132
What It Tests	Disciplines: • Humanities: 50% • Social Sciences: 50% Skills: • *Foundations of Comprehension*: 30% • *Reasoning Within the Text*: 30% • *Reasoning Beyond the Text*: 40%
Biological and Biochemical Foundations of Living Systems	
Time	95 minutes
Format	• 59 questions • 10 passages • 44 questions are passage-based, and 15 are discrete (stand-alone) questions. • Score between 118 and 132
What It Tests	• Biochemistry: 25% • Biology: 65% • General Chemistry: 5% • Organic Chemistry: 5%
Psychological, Social, and Biological Foundations of Behavior	
Time	95 minutes
Format	• 59 questions • 10 passages • 44 questions are passage-based, and 15 are discrete (stand-alone) questions. • Score between 118 and 132
What It Tests	• Biology: 5% • Psychology: 65% • Sociology: 30%
Total	
Testing Time	375 minutes (6 hours, 15 minutes)
Total Seat Time	447 minutes (7 hours, 27 minutes)
Questions	230
Score	472 to 528

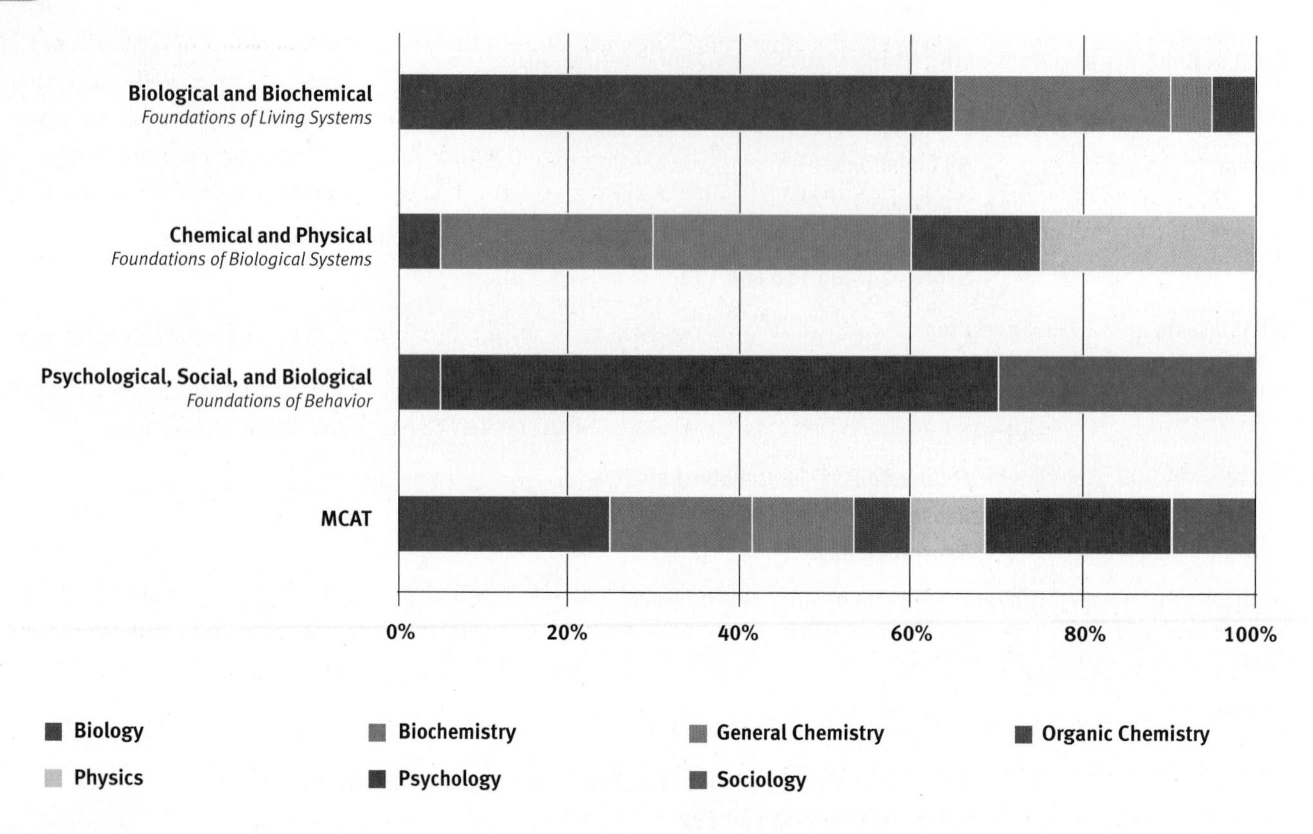

Scientific Inquiry and Reasoning Skills (SIRS)

The AAMC has defined four *Scientific Inquiry and Reasoning Skills* (SIRS) that will be tested in the three science sections of the MCAT:

1. *Knowledge of Scientific Concepts and Principles* (35% of questions)
2. *Scientific Reasoning and Problem-Solving* (45% of questions)
3. *Reasoning About the Design and Execution of Research* (10% of questions)
4. *Data-Based and Statistical Reasoning* (10% of questions)

Let's see how each one breaks down into more specific Test Day behaviors. Note that the bullet points of specific objectives for each of the SIRS are taken directly from the *Official Guide to the MCAT Exam*; the descriptions of what these behaviors mean and sample question stems, however, are written by Kaplan.

Skill 1: *Knowledge of Scientific Concepts and Principles*

This is probably the least surprising of the four SIRS; the testing of science knowledge is, after all, one of the signature qualities of the MCAT. Skill 1 questions will require you to do the following:

- Recognize correct scientific principles
- Identify the relationships among closely related concepts
- Identify the relationships between different representations of concepts (verbal, symbolic, graphic)
- Identify examples of observations that illustrate scientific principles
- Use mathematical equations to solve problems

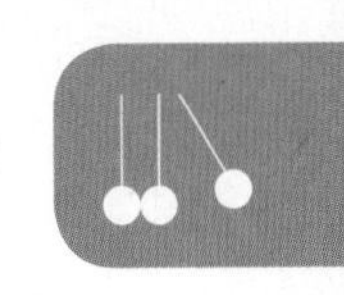

At Kaplan, we simply call these Science Knowledge or Skill 1 questions. Another way to think of Skill 1 questions is as "one-step" problems. The single step is either to realize which scientific concept the question stem is suggesting or to take the concept stated in the question stem and identify which answer choice is an accurate application of it. Skill 1 questions are particularly prominent among discrete questions (those not associated with a passage). These questions are an opportunity to gain quick points on Test Day—if you know the science concept attached to the question, then that's it! On Test Day, 35% of the questions in each science section will be Skill 1 questions.

Here are some sample Skill 1 question stems:

- How would a proponent of the James–Lange theory of emotion interpret the findings of the study cited in the passage?
- Which of the following most accurately describes the function of FSH in the human female menstrual cycle?
- If the products of Reaction 1 and Reaction 2 were combined in solution, the resulting reaction would form:
- Ionic bonds are maintained by which of the following forces?

Skill 2: *Scientific Reasoning and Problem-Solving*

The MCAT science sections do, of course, move beyond testing straightforward science knowledge; Skill 2 questions are the most common way in which it does so. At Kaplan, we also call these Critical Thinking questions. Skill 2 questions will require you to do the following:

- Reason about scientific principles, theories, and models
- Analyze and evaluate scientific explanations and predictions
- Evaluate arguments about causes and consequences
- Bring together theory, observations, and evidence to draw conclusions
- Recognize scientific findings that challenge or invalidate a scientific theory or model
- Determine and use scientific formulas to solve problems

Just as Skill 1 questions can be thought of as "one-step" problems, many Skill 2 questions are "two-step" problems, and more difficult Skill 2 questions may require three or more steps. These questions can require a wide spectrum of reasoning skills, including integration of multiple facts from a passage, combination of multiple science content areas, and prediction of an experiment's results. Skill 2 questions also tend to ask about science content without actually mentioning it by name. For example, a question might describe the results of one experiment and ask you to predict the results of a second experiment without actually telling you what underlying scientific principles are at work—part of the question's difficulty will be figuring out which principles to apply in order to get the correct answer. On Test Day, 45% of the questions in each science section will be Skill 2 questions.

Here are some sample Skill 2 question stems:

- Which of the following experimental conditions would most likely yield results similar to those in Figure 2?
- All of the following conclusions are supported by the information in the passage EXCEPT:
- The most likely cause of the anomalous results found by the experimenter is:
- An impact to a man's chest quickly reduces the volume of one of his lungs to 70% of its initial value while not allowing any air to escape from the man's mouth. By what percentage is the force of outward air pressure increased on a 2 cm^2 portion of the inner surface of the compressed lung?

Skill 3: *Reasoning About the Design and Execution of Research*

The MCAT is interested in your ability to critically appraise and analyze research, as this is an important day-to-day task of a physician. We call these questions Skill 3 or Experimental and Research Design questions for short. Skill 3 questions will require you to do the following:

- Identify the role of theory, past findings, and observations in scientific questioning
- Identify testable research questions and hypotheses
- Distinguish between samples and populations and distinguish results that support generalizations about populations
- Identify independent and dependent variables
- Reason about the features of research studies that suggest associations between variables or causal relationships between them (such as temporality and random assignment)
- Identify conclusions that are supported by research results
- Determine the implications of results for real-world situations
- Reason about ethical issues in scientific research

Over the years, the AAMC has received input from medical schools to require more practical research skills of MCAT test takers, and Skill 3 questions are the response to these demands. This skill is unique in that the outside knowledge you need to answer Skill 3 questions is not taught in any one undergraduate course; instead, the research design principles needed to answer these questions are learned gradually throughout your science classes and especially through any laboratory work you have completed. It should be noted that Skill 3 comprises 10% of the questions in each science section on Test Day.

Here are some sample Skill 3 question stems:

- What is the dependent variable in the study described in the passage?
- The major flaw in the method used to measure disease susceptibility in Experiment 1 is:
- Which of the following procedures is most important for the experimenters to follow in order for their study to maintain a proper, randomized sample of research subjects?
- A researcher would like to test the hypothesis that individuals who move to an urban area during adulthood are more likely to own a car than are those who have lived in an urban area since birth. Which of the following studies would best test this hypothesis?

Skill 4: *Data-Based and Statistical Reasoning*

Lastly, the science sections of the MCAT test your ability to analyze the visual and numerical results of experiments and studies. We call these Data and Statistical Analysis questions. Skill 4 questions will require you to do the following:

- Use, analyze, and interpret data in figures, graphs, and tables
- Evaluate whether representations make sense for particular scientific observations and data
- Use measures of central tendency (mean, median, and mode) and measures of dispersion (range, interquartile range, and standard deviation) to describe data
- Reason about random and systematic error

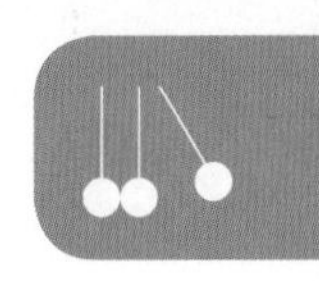

- Reason about statistical significance and uncertainty (interpreting statistical significance levels and interpreting a confidence interval)
- Use data to explain relationships between variables or make predictions
- Use data to answer research questions and draw conclusions

Skill 4 is included in the MCAT because physicians and researchers spend much of their time examining the results of their own studies and the studies of others, and it's very important for them to make legitimate conclusions and sound judgments based on that data. The MCAT tests Skill 4 on all three science sections with graphical representations of data (charts and bar graphs), as well as numerical ones (tables, lists, and results summarized in sentence or paragraph form). On Test Day, 10% of the questions in each science section will be Skill 4 questions.

Here are some sample Skill 4 question stems:

- According to the information in the passage, there is an inverse correlation between:
- What conclusion is best supported by the findings displayed in Figure 2?
- A medical test for a rare type of heavy metal poisoning returns a positive result for 98% of affected individuals and 13% of unaffected individuals. Which of the following types of error is most prevalent in this test?
- If a fourth trial of Experiment 1 was run and yielded a result of 54% compliance, which of the following would be true?

SIRS Summary

Discussing the SIRS tested on the MCAT is a daunting prospect given that the very nature of the skills tends to make the conversation rather abstract. Nevertheless, with enough practice, you'll be able to identify each of the four skills quickly, and you'll also be able to apply the proper strategies to solve those problems on Test Day. If you need a quick reference to remind you of the four SIRS, these guidelines may help:

Skill 1 (**Science Knowledge**) questions ask:

- Do you remember this science content?

Skill 2 (**Critical Thinking**) questions ask:

- Do you remember this science content? And if you do, could you please apply it to this novel situation?
- Could you answer this question that cleverly combines multiple content areas at the same time?

Skill 3 (**Experimental and Research Design**) questions ask:

- Let's forget about the science content for a while. Could you give some insight into the experimental or research methods involved in this situation?

Skill 4 (**Data and Statistical Analysis**) questions ask:

- Let's forget about the science content for a while. Could you accurately read some graphs and tables for a moment? Could you make some conclusions or extrapolations based on the information presented?

Critical Analysis and Reasoning Skills (CARS)

The *Critical Analysis and Reasoning Skills* (CARS) section of the MCAT tests three discrete families of textual reasoning skills; each of these families requires a higher level of reasoning than the last. Those three skills are as follows:

1. *Foundations of Comprehension* (30% of questions)
2. *Reasoning Within the Text* (30% of questions)
3. *Reasoning Beyond the Text* (40% of questions)

These three skills are tested through nine humanities- and social sciences–themed passages, with approximately 5 to 7 questions per passage. Let's take a more in-depth look into these three skills. Again, the bullet points of specific objectives for each of the CARS are taken directly from the *Official Guide to the MCAT Exam*; the descriptions of what these behaviors mean and sample question stems, however, are written by Kaplan.

Foundations of Comprehension

Questions in this skill will ask for basic facts and simple inferences about the passage; the questions themselves will be similar to those seen on reading comprehension sections of other standardized exams like the SAT® and ACT®. *Foundations of Comprehension* questions will require you to do the following:

- Understand the basic components of the text
- Infer meaning from rhetorical devices, word choice, and text structure

This admittedly covers a wide range of potential question types including Main Idea, Detail, Inference, and Definition-in-Context questions, but finding the correct answer to all *Foundations of Comprehension* questions will follow from a basic understanding of the passage and the point of view of its author (and occasionally that of other voices in the passage).

Here are some sample *Foundations of Comprehension* question stems:

- **Main Idea**—The author's primary purpose in this passage is:
- **Detail**—Based on the information in the second paragraph, which of the following is the most accurate summary of the opinion held by Schubert's critics?
- **(Scattered) Detail**—According to the passage, which of the following is FALSE about literary reviews in the 1920s?
- **Inference (Implication)**—Which of the following phrases, as used in the passage, is most suggestive that the author has a personal bias toward narrative records of history?
- **Inference (Assumption)**—In putting together her argument in the passage, the author most likely assumes:
- **Definition-in-Context**—The word "obscure" (paragraph 3), when used in reference to the historian's actions, most nearly means:

Reasoning Within the Text

While *Foundations of Comprehension* questions will usually depend on interpreting a single piece of information in the passage or understanding the passage as a whole, *Reasoning Within the Text* questions require more thought because they will ask you to identify the purpose of a particular piece of information in the context of the passage, or ask how one piece of information relates to another. *Reasoning Within the Text* questions will require you to:

- Integrate different components of the text to draw relevant conclusions

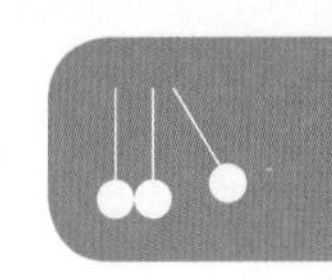

The CARS section will also ask you to judge certain parts of the passage or even judge the author. These questions, which fall under the *Reasoning Within the Text* skill, can ask you to identify authorial bias, evaluate the credibility of cited sources, determine the logical soundness of an argument, identify the importance of a particular fact or statement in the context of the passage, or search for relevant evidence in the passage to support a given conclusion. In all, this category includes Function and Strengthen–Weaken (Within the Passage) questions, as well as a smattering of related—but rare—question types.

Here are some sample *Reasoning Within the Text* question stems:

- **Function**—The author's discussion of the effect of socioeconomic status on social mobility primarily serves which of the following functions?
- **Strengthen–Weaken (Within the Passage)**—Which of the following facts is used in the passage as the most prominent piece of evidence in favor of the author's conclusions?
- **Strengthen–Weaken (Within the Passage)**—Based on the role it plays in the author's argument, *The Possessed* can be considered:

Reasoning Beyond the Text

The distinguishing factor of *Reasoning Beyond the Text* questions is in the title of the skill: the word *Beyond*. Questions that test this skill, which make up a larger share of the CARS section than questions from either of the other two skills, will always introduce a completely new situation that was not present in the passage itself; these questions will ask you to determine how one influences the other. *Reasoning Beyond the Text* questions will require you to:

- Apply or extrapolate ideas from the passage to new contexts
- Assess the impact of introducing new factors, information, or conditions to ideas from the passage

The *Reasoning Beyond the Text* skill is further divided into Apply and Strengthen–Weaken (Beyond the Passage) questions, and a few other rarely appearing question types.

Here are some sample *Reasoning Beyond the Text* question stems:

- **Apply**—If a document were located that demonstrated Berlioz intended to include a chorus of at least 700 in his *Grande Messe des Morts*, how would the author likely respond?
- **Apply**—Which of the following is the best example of a "virtuous rebellion," as it is defined in the passage?
- **Strengthen–Weaken (Beyond the Text)**—Suppose Jane Austen had written in a letter to her sister, "My strongest characters were those forced by circumstance to confront basic questions about the society in which they lived." What relevance would this have to the passage?
- **Strengthen–Weaken (Beyond the Text)**—Which of the following sentences, if added to the end of the passage, would most WEAKEN the author's conclusions in the last paragraph?

CARS Summary

Through the *Foundations of Comprehension* skill, the CARS section tests many of the reading skills you have been building on since grade school, albeit in the context of very challenging doctorate-level passages. But through the two other skills (*Reasoning Within the Text* and *Reasoning Beyond the Text*), the MCAT demands that you understand the deep structure of passages and the arguments within them at a very advanced level. And, of course, all of this is tested under very tight timing restrictions: only 102 seconds per question—and that doesn't even include the time spent reading the passages.

Here's a quick reference guide to the three CARS skills:

Foundations of Comprehension questions ask:

- Did you understand the passage and its main ideas?
- What does the passage have to say about this particular detail?
- What must be true that the author did not say?

Reasoning Within the Text questions ask:

- What's the logical relationship between these two ideas from the passage?
- How well argued is the author's thesis?

Reasoning Beyond the Text questions ask:

- How does this principle from the passage apply to this new situation?
- How does this new piece of information influence the arguments in the passage?

Scoring

Each of the four sections of the MCAT is scored between 118 and 132, with the median at 125. This means the total score ranges from 472 to 528, with the median at 500. Why such peculiar numbers? The AAMC stresses that this scale emphasizes the importance of the central portion of the score distribution, where most students score (around 125 per section, or 500 total), rather than putting undue focus on the high end of the scale.

Note that there is no wrong answer penalty on the MCAT, so you should select an answer for every question—even if it is only a guess.

The AAMC has released the 2020–2021 correlation between scaled score and percentile, as shown on the following page. It should be noted that the percentile scale is adjusted and renormalized over time and thus can shift slightly from year to year.

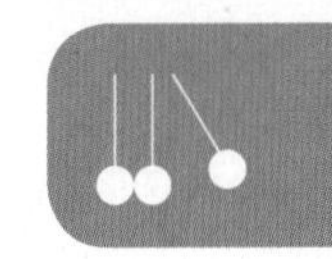

Total Score	Percentile	Total Score	Percentile
528	100	499	43
527	100	498	40
526	100	497	37
525	100	496	33
524	100	495	30
523	99	494	28
522	99	493	25
521	98	492	22
520	98	491	20
519	97	490	17
518	96	489	15
517	94	488	13
516	93	487	11
515	91	486	10
514	89	485	8
513	87	484	7
512	85	483	6
511	82	482	5
510	80	481	4
509	77	480	3
508	74	479	2
507	70	478	2
506	67	477	1
505	64	476	1
504	60	475	<1
503	57	474	<1
502	53	473	<1
501	50	472	<1
500	46		

Source: AAMC. 2020. *Summary of MCAT Total and Section Scores.* Accessed November 2020. **https://students-residents.aamc.org/advisors/article/percentile-ranks-for-the-mcat-exam/**

Further information on score reporting is included at the end of the next section (see *After Your Test*).

MCAT Policies and Procedures

We strongly encourage you to download the latest copy of *MCAT® Essentials*, available on the AAMC's website, to ensure that you have the latest information about registration and Test Day policies and procedures; this document is updated annually. A brief summary of some of the most important rules is provided here.

MCAT Registration

The only way to register for the MCAT is online. You can access AAMC's registration system at **www.aamc.org/mcat**.

You will be able to access the site approximately six months before Test Day. The AAMC designates three registration "Zones"—Gold, Silver, and Bronze. Registering during the Gold Zone (from the opening of registration until approximately one month before Test Day) provides the most flexibility and lowest test fees. The Silver Zone runs until approximately two to three weeks before Test Day and has less flexibility and higher fees; the Bronze Zone runs until approximately one to two weeks before Test Day and has the least flexibility and highest fees.

Fees and the Fee Assistance Program (FAP)

Payment for test registration must be made by MasterCard or VISA. As described earlier, the fees for registering for the MCAT—as well as rescheduling the exam or changing your testing center—increase as one approaches Test Day. In addition, it is not uncommon for test centers to fill up well in advance of the registration deadline. For these reasons, we recommend identifying your preferred Test Day as soon as possible and registering. There are ancillary benefits to having a set Test Day, as well: when you know the date you're working toward, you'll study harder and are less likely to keep pushing back the exam. The AAMC offers a Fee Assistance Program (FAP) for students with financial hardship to help reduce the cost of taking the MCAT, as well as for the American Medical College Application Service (AMCAS®) application. Further information on the FAP can be found at **www.aamc.org/students/applying/fap**.

Testing Security

On Test Day, you will be required to present a qualifying form of ID. Generally, a current driver's license or United States passport will be sufficient (consult the AAMC website for the full list of qualifying criteria). When registering, take care to spell your first and last names (middle names, suffixes, and prefixes are not required and will not be verified on Test Day) precisely the same as they appear on this ID; failure to provide this ID at the test center or differences in spelling between your registration and ID will be considered a "no-show," and you will not receive a refund for the exam.

During Test Day registration, other identity data collected may include: a digital palm vein scan, a Test Day photo, a digitization of your valid ID, and signatures. Some testing centers may use a metal detection wand to ensure that no prohibited items are brought into the testing room. Prohibited items include all electronic devices, including watches and timers, calculators, cell phones, and any and all forms of recording equipment; food, drinks (including water), and cigarettes or other smoking paraphernalia; hats and scarves (except for religious purposes); and books, notes, or other study materials. If you require a medical device, such as an insulin pump or pacemaker, you must apply for accommodated testing. During breaks, you are allowed to access food and drink, but not to electronic devices, including cell phones.

Testing centers are under video surveillance and the AAMC does not take potential violations of testing security lightly. The bottom line: *know the rules and don't break them.*

Accommodations

Students with disabilities or medical conditions can apply for accommodated testing. Documentation of the disability or condition is required, and requests may take two months—or more—to be approved. For this reason, it is recommended that you begin the process of applying for accommodated testing as early as possible. More information on applying for accommodated testing can be found at **www.aamc.org/students/applying/mcat/accommodations**.

After Your Test

When your MCAT is all over, no matter how you feel you did, be good to yourself when you leave the test center. Celebrate! Take a nap. Watch a movie. Ride your bike. Plan a trip. Call up all of your neglected friends or stalk them on Facebook. Totally consume a cheesesteak and drink dirty martinis at night (assuming you're over 21). Whatever you do, make sure that it has absolutely nothing to do with thinking too hard—you deserve some rest and relaxation.

Perhaps most importantly, do not discuss specific details about the test with anyone. For one, it is important to let go of the stress of Test Day, and reliving your exam only inhibits you from being able to do so. But more significantly, the Examinee Agreement you sign at the beginning of your exam specifically prohibits you from discussing or disclosing exam content. The AAMC is known to seek out individuals who violate this agreement and retains the right to prosecute these individuals at their discretion. This means that you should not, under any circumstances, discuss the exam in person or over the phone with other individuals—including us at Kaplan—or post information or questions about exam content to Facebook, Student Doctor Network, or other online social media. You are permitted to comment on your "general exam experience," including how you felt about the exam overall or an individual section, but this is a fine line. In summary: *if you're not certain whether you can discuss an aspect of the test or not, just don't do it!* Do not let a silly Facebook post stop you from becoming the doctor you deserve to be.

Scores are typically released approximately one month after Test Day. The release is staggered during the afternoon and evening, ending at 5 p.m. Eastern Standard Time. This means that not all examinees receive their scores at exactly the same time. Your score report will include a scaled score for each section between 118 and 132, as well as your total combined score between 472 and 528. These scores are given as confidence intervals. For each section, the confidence interval is approximately the given score ± 1; for the total score, it is approximately the given score ± 2. You will also be given the corresponding percentile rank for each of these section scores and the total score.

AAMC Contact Information

For further questions, contact the MCAT team at the Association of American Medical Colleges:

MCAT Resource Center
Association of American Medical Colleges
www.aamc.org/mcat
(202) 828-0600
www.aamc.org/contactmcat

HOW THIS BOOK WAS CREATED

The *Kaplan MCAT Review* project began shortly after the release of the *Preview Guide for the MCAT 2015 Exam*, 2nd edition. Through thorough analysis by our staff psychometricians, we were able to analyze the relative yield of the different topics on the MCAT, and we began constructing tables of contents for the books of the *Kaplan MCAT Review* series. A dedicated staff of 30 writers, 7 editors, and 32 proofreaders worked over 5,000 combined hours to produce these books. The format of the books was heavily influenced by weekly meetings with Kaplan's learning science team.

In the years since this book was created, a number of opportunities for expansion and improvement have occurred. The current edition represents the culmination of the wisdom accumulated during that time frame, and it also includes several new features designed to improve the reading and learning experience in these texts.

These books were submitted for publication in April 2021. For any updates after this date, please visit www.kaptest.com/retail-book-corrections-and-updates.

If you have any questions about the content presented here, email KaplanMCATfeedback@kaplan.com. For other questions not related to content, email booksupport@kaplan.com.

Each book has been vetted through at least ten rounds of review. To that end, the information presented in these books is true and accurate to the best of our knowledge. Still, your feedback helps us improve our prep materials. Please notify us of any inaccuracies or errors in the books by sending an email to KaplanMCATfeedback@kaplan.com.

USING THIS BOOK

Kaplan MCAT Physics and Math Review, and the other six books in the *Kaplan MCAT Review* series, bring the Kaplan classroom experience to you—right in your home, at your convenience. This book offers the same Kaplan content review, strategies, and practice that make Kaplan the #1 choice for MCAT prep.

This book is designed to help you review the physics and math topics covered on the MCAT. Please understand that content review—no matter how thorough—is not sufficient preparation for the MCAT! The MCAT tests not only your science knowledge but also your critical reading, reasoning, and problem-solving skills. Do not assume that simply memorizing the contents of this book will earn you high scores on Test Day; to maximize your scores, you must also improve your reading and test-taking skills through MCAT-style questions and practice tests.

Learning Objectives

At the beginning of each section, you'll find a short list of objectives describing the skills covered within that section. Learning objectives for these texts were developed in conjunction with Kaplan's learning science team, and have been designed specifically to focus your attention on tasks and concepts that are likely to show up on your MCAT. These learning objectives will function as a means to guide your study, and indicate what information and relationships you should be focused on within each section. Before starting each section, read these learning objectives carefully. They will not only allow you to assess your existing familiarity with the content, but also provide a goal-oriented focus for your studying experience of the section.

MCAT Concept Checks

At the end of each section, you'll find a few open-ended questions that you can use to assess your mastery of the material. These MCAT Concept Checks were introduced after numerous conversations with Kaplan's learning science team. Research has demonstrated repeatedly that introspection and self-analysis improve mastery, retention, and recall of material. Complete these MCAT Concept Checks to ensure that you've got the key points from each section before moving on!

Science Mastery Assessments

At the beginning of each chapter, you'll find 15 MCAT-style practice questions. These are designed to help you assess your understanding of the chapter before you begin reading the chapter. Using the guidance provided with the assessment, you can determine the best way to review each chapter based on your personal strengths and weaknesses. Most of the questions in the Science Mastery Assessments focus on the first of the Scientific Inquiry and Reasoning Skills (*Knowledge of Scientific Concepts and Principles*), although there are occasional questions that fall into the second or fourth SIRS (*Scientific Reasoning and Problem-Solving and Data-Based and Statistical Reasoning*, respectively). In addition, in your online resources you'll find a test-like passage set covering the same content you just studied to ensure you can also apply your knowledge the way the MCAT will expect you to!

Sidebars

The following is a guide to the five types of sidebars you'll find in *Kaplan MCAT Physics and Math Review:*

- **Bridge:** These sidebars create connections between science topics that appear in multiple chapters throughout the *Kaplan MCAT Review* series.
- **Key Concept:** These sidebars draw attention to the most important takeaways in a given topic, and they sometimes offer synopses or overviews of complex information. If you understand nothing else, make sure you grasp the Key Concepts for any given subject.
- **MCAT Expertise:** These sidebars point out how information may be tested on the MCAT or offer key strategy points and test-taking tips that you should apply on Test Day.
- **Mnemonic:** These sidebars present memory devices to help recall certain facts.
- **Real World:** These sidebars illustrate how a concept in the text relates to the practice of medicine or the world at large. While this is not information you need to know for Test Day, many of the topics in Real World sidebars are excellent examples of how a concept may appear in a passage or discrete (stand-alone) question on the MCAT.

What This Book Covers

The information presented in the *Kaplan MCAT Review* series covers everything listed on the official MCAT content lists. Every topic in these lists is covered in the same level of detail as is common to the undergraduate and postbaccalaureate classes that are considered prerequisites for the MCAT. Note that your premedical classes may include topics not discussed in these books, or they may go into more depth than these books do. Additional exposure to science content is never a bad thing, but all of the content knowledge you are expected to have walking in on Test Day is covered in these books.

Chapter profiles, on the first page of each chapter, represent a holistic look at the content within the chapter, and will include a pie chart as well as text information. The pie chart analysis is based directly on data released by the AAMC, and will give a rough estimate of the importance of the chapter in relation to the book as a whole. Further, the text portion of the Chapter Profiles includes which AAMC content categories are covered within the chapter. These are referenced directly from the AAMC MCAT exam content listing, available on the test maker's website.

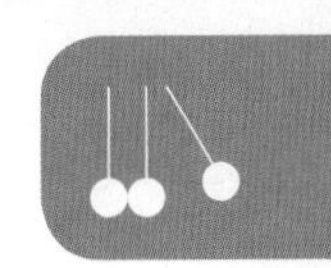

You'll also see new High-Yield badges scattered throughout the sections of this book:

In This Chapter

1.1 **Amino Acids Found in Proteins** HY
- A Note on Terminology 4
- Stereochemistry of Amino Acids 5
- Structures of the Amino Acids 5
- Hydrophobic and Hydrophilic Amino Acids 9
- Amino Acid Abbreviations 9

1.2 **Acid–Base Chemistry of Amino Acids**
- Protonation and Deprotonation...... 12
- Titration of Amino Acids............ 14

1.3 **Peptide Bond Formation and Hydrolysis** HY
- Peptide Bond Formation 17
- Peptide Bond Hydrolysis 18

1.4 **Primary and Secondary Protein Structure** HY
- Primary Structure 19
- Secondary Structure................ 20

1.5 **Tertiary and Quaternary Protein Structure** HY
- Tertiary Structure 22
- Folding and the Solvation Layer 23
- Quaternary Structure 24
- Conjugated Proteins................ 25

1.6 **Denaturation** 28

Concept Summary 30

1.1 Amino Acids Found in Proteins

High-Yield

LEARNING OBJECTIVES

After Chapter 1.1, you will be able to:

These badges represent the top 100 topics most tested by the AAMC. In other words, according to the testmaker and all our experience with their resources, a High-Yield badge means more questions on Test Day.

This book also contains a thorough glossary and index for easy navigation of the text.

In the end, this is your book, so write in the margins, draw diagrams, highlight the key points—do whatever is necessary to help you get that higher score. We look forward to working with you as you achieve your dreams and become the doctor you deserve to be!

Studying With This Book

In addition to providing you with the best practice questions and test strategies, Kaplan's team of learning scientists are dedicated to researching and testing the best methods for getting the most out of your study time. Here are their top four tips for improving retention:

Review multiple topics in one study session. This may seem counterintuitive—we're used to practicing one skill at a time in order to improve each skill. But research shows that weaving topics together leads to increased learning. Beyond that consideration, the MCAT often includes more than one topic in a single question. Studying in an integrated manner is the most effective way to prepare for this test.

Customize the content. Drawing attention to difficult or critical content can ensure you don't overlook it as you read and re-read sections. The best way to do this is to make it more visual—highlight, make tabs, use stickies, whatever works. We recommend highlighting only the most important or difficult sections of text. Selective highlighting of up to about 10 percent of text in a given chapter is great for emphasizing parts of the text, but over-highlighting can have the opposite effect.

Repeat topics over time. Many people try to memorize concepts by repeating them over and over again in succession. Our research shows that retention is improved by spacing out the repeats over time and mixing up the order in which you study content. For example, try reading chapters in a different order the second (or third!) time around. Revisit practice questions that you answered incorrectly in a new sequence. Perhaps information you reviewed more recently will help you better understand those questions and solutions you struggled with in the past.

Take a moment to reflect. When you finish reading a section for the first time, stop and think about what you just read. Jot down a few thoughts in the margins or in your notes about why the content is important or what topics came to mind when you read it. Associating learning with a memory is a fantastic way to retain information! This also works when answering questions. After answering a question, take a moment to think through each step you took to arrive at a solution. What led you to the answer you chose? Understanding the steps you took will help you make good decisions when answering future questions.

Online Resources

In addition to the resources located within this text, you also have additional online resources awaiting you at **www.kaptest.com/booksonline**. Make sure to log on and take advantage of free practice and other resources!

Please note that access to the online resources is limited to the original owner of this book.

STUDYING FOR THE MCAT

The first year of medical school is a frenzied experience for most students. To meet the requirements of a rigorous work schedule, students either learn to prioritize their time or else fall hopelessly behind. It's no surprise, then, that the MCAT, the test specifically designed to predict success in medical school, is a high-speed, time-intensive test. The MCAT demands excellent time-management skills, endurance, as well as grace under pressure both during the test as well as while preparing for it. Having a solid plan of attack and sticking with it are key to giving you the confidence and structure you need to succeed.

Creating a Study Plan

The best time to create a study plan is at the beginning of your MCAT preparation. If you don't already use a calendar, you will want to start. You can purchase a planner, print out a free calendar from the Internet, use a built-in calendar or app on one of your smart devices, or keep track using an interactive online calendar. Pick the option that is most practical for you and that you are most likely to use consistently.

Once you have a calendar, you'll be able to start planning your study schedule with the following steps:

1. **Fill in your obligations and choose a day off.**

   Write in all your school, extracurricular, and work obligations first: class sessions, work shifts, and meetings that you must attend. Then add in your personal obligations: appointments, lunch dates, family and social time, etc. Making an appointment in your calendar for hanging out with friends or going to the movies may seem strange at first, but planning social activities in advance will help you achieve a balance between personal and professional obligations even as life gets busy. Having a happy balance allows you to be more focused and productive when it comes time to study, so stay well-rounded and don't neglect anything that is important to you.

   In addition to scheduling your personal and professional obligations, you should also plan your time off. Taking some time off is just as important as studying. Kaplan recommends taking at least one full day off per week, ideally from all your study obligations but at minimum from studying for the MCAT.

2. **Add in study blocks around your obligations.**

   Once you have established your calendar's framework, add in study blocks around your obligations, keeping your study schedule as consistent as possible across days and across weeks. Studying at the same time of day as your official test is ideal for promoting recall, but if that's not possible, then fit in study blocks wherever you can.

   To make your studying as efficient as possible, block out short, frequent periods of study time throughout the week. From a learning perspective, studying one hour per day for six days per week is much more valuable than studying for six hours all at once one day per week. Specifically, Kaplan recommends studying for no longer than three hours in one sitting. Within those three-hour blocks, also plan to take ten-minute breaks every hour. Use these breaks to get up from your seat, do some quick stretches, get a snack and drink, and clear your mind. Although ten minutes of break for every 50 minutes of studying may sound like a lot, these breaks will allow you to deal with distractions and rest your brain so that, during the 50-minute study blocks, you can remain fully engaged and completely focused.

3. **Add in your full-length practice tests.**

   Next, you'll want to add in full-length practice tests. You'll want to take one test very early in your prep and then spread your remaining full-length practice tests evenly between now and your test date. Staggering tests in this way allows you to form a baseline for comparison and to determine which areas to focus on right away, while also providing realistic feedback throughout your prep as to how you will perform on Test Day.

When planning your calendar, aim to finish your full-length practice tests and the majority of your studying by one week before Test Day, which will allow you to spend that final week completing a final review of what you already know. In your online resources, you'll find sample study calendars for several different Test Day timelines to use as a starting point. The sample calendars may include more focus than you need in some areas, and less in others, and it may not fit your timeline to Test Day. You will need to customize your study calendar to your needs using the steps above.

The total amount of time you spend studying each week will depend on your schedule, your personal prep needs, and your time to Test Day, but it is recommended that you spend somewhere in the range of 300–350 hours preparing before taking the official MCAT. One way you could break this down is to study for three hours per day, six days per week, for four months, but this is just one approach. You might study six days per week for more than three hours per day. You might study over a longer period of time if you don't have much time to study each week. No matter what your plan is, ensure you complete enough practice to feel completely comfortable with the MCAT and its content. A good sign you're ready for Test Day is when you begin to earn your goal score consistently in practice.

How to Study

The MCAT covers a large amount of material, so studying for Test Day can initially seem daunting. To combat this, we have some tips for how to take control of your studying and make the most of your time.

Goal Setting

To take control of the amount of content and practice required to do well on the MCAT, break the content down into specific goals for each week instead of attempting to approach the test as a whole. A goal of "I want to increase my overall score by 5 points" is too big, abstract, and difficult to measure on the small scale. More reasonable goals are "I will read two chapters each day this week." Goals like this are much less overwhelming and help break studying into manageable pieces.

Active Reading

As you go through this book, much of the information will be familiar to you. After all, you have probably seen most of the content before. However, be very careful: Familiarity with a subject does not necessarily translate to knowledge or mastery of that subject. Do not assume that if you recognize a concept you actually know it and can apply it quickly at an appropriate level. Don't just passively read this book. Instead, read actively: Use the free margin space to jot down important ideas, draw diagrams, and make charts as you read. Highlighting can be an excellent tool, but use it sparingly: highlighting every sentence isn't active reading, it's coloring. Frequently stop and ask yourself questions while you read (e.g., *What is the main point? How does this fit into the overall scheme of things? Could I thoroughly explain this to someone else?*). By making connections and focusing on the grander scheme, not only will you ensure you know the essential content, but you also prepare yourself for the level of critical thinking required by the MCAT.

Focus on Areas of Greatest Opportunity

If you are limited by only having a minimal amount of time to prepare before Test Day, focus on your biggest areas of opportunity first. Areas of opportunity are topic areas that are highly tested and that you have not yet mastered. You likely won't have time to take detailed notes for every page of these books; instead, use your results from practice materials to determine which areas are your biggest opportunities and seek those out. After you've taken a full-length test, make sure you are using

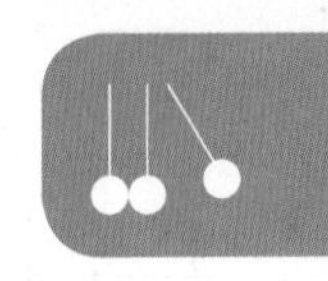

your performance report to best identify areas of opportunity. Skim over content matter for which you are already demonstrating proficiency, pausing to read more thoroughly when something looks unfamiliar or particularly difficult. Begin with the Science Mastery Assessment at the beginning of each chapter. If you can get all of those questions correct within a reasonable amount of time, you may be able to quickly skim through that chapter, but if the questions prove to be more difficult, then you may need to spend time reading the chapter or certain subsections of the chapter more thoroughly.

Practice, Review, and Tracking

Leave time to review your practice questions and full-length tests. You may be tempted, after practicing, to push ahead and cover new material as quickly as possible, but failing to schedule ample time for review will actually throw away your greatest opportunity to improve your performance. The brain rarely remembers anything it sees or does only once. When you carefully review the questions you've solved (and the explanations for them), the process of retrieving that information reopens and reinforces the connections you've built in your brain. This builds long-term retention and repeatable skill sets—exactly what you need to beat the MCAT!

While reviewing, take notes about the specific reasons why you missed questions you got wrong or had to guess on, perhaps by using a spreadsheet like the one below. Keep adding to the same How I'll Fix It Sheet (HIFIS) as you complete more practice questions, and periodically review your HIFIS to identify any patterns you see, such as consistently missing questions in certain content areas or falling for the same test maker traps. As you move through your MCAT prep, adjust your study plan based on your available study time and the results of your review. Your strengths and weaknesses are likely to change over the course of your prep. Keep addressing the areas that are most important to your score, shifting your focus as those areas change. For more help with reviewing and making the most of your full-length tests, including a How I'll Fix It Sheet template, make sure to check out the videos and resources in your online syllabus.

Section	Q #	Type/Topic	Wrong answer chosen	How I'll Fix It
Chem/Phys	*42*	*Nuclear chem.*	*Opposite*	*Flashcard electron absorption and emission*
Chem/Phys	*47*	K_{eq}	*Miscalculation*	*Practice* K_{eq} *problems*
CARS	*2*	*Detail*	*Opposite*	*Didn't read "not" in answer choice; slow down and rephrase the question!*
CARS	*4*	*Inference*	*Out of Scope*	*Go back to passage and make a prediction*

Where to Study

One often-overlooked aspect of studying is the environment where the learning actually occurs. Although studying at home is many students' first choice, several problems can arise in this environment, chief of which are distractions. Studying can be a mentally draining process, so as time passes, these distractions become ever more tempting as escape routes. Although you may have considerable willpower, there's no reason to make staying focused harder than it needs to be. Instead of studying at home, head to a library, quiet coffee shop, or another new location whenever possible. This will eliminate many of the usual distractions and also promote efficient studying; instead of studying off and on at home over the course of an entire day, you can stay at the library for three hours of effective studying and enjoy the rest of the day off from the MCAT.

No matter where you study, make your practice as much like Test Day as possible. Just as is required during the official test, don't have snacks or chew gum during your study blocks. Turn off your music, television, and phone. Practice on the computer with your online resources to simulate the computer-based test environment. When completing practice questions, do your work on scratch paper or noteboard sheets rather than writing directly on any printed materials since you won't have that option on Test Day. Because memory is tied to all of your senses, the more test-like you can make your studying environment, the easier it will be on Test Day to recall the information you're putting in so much work to learn.

CHAPTER 1

Kinematics and Dynamics

SCIENCE MASTERY ASSESSMENT

Every pre-med knows this feeling: there is so much content I have to know for the MCAT! How do I know what to do first or what's important?

While the high-yield badges throughout this book will help you identify the most important topics, this Science Mastery Assessment is another tool in your MCAT prep arsenal. This quiz (which can also be taken in your online resources) and the guidance below will help ensure that you are spending the appropriate amount of time on this chapter based on your personal strengths and weaknesses. Don't worry though—skipping something now does not mean you'll never study it. Later on in your prep, as you complete full-length tests, you'll uncover specific pieces of content that you need to review and can come back to these chapters as appropriate.

How to Use This Assessment

If you answer 0–7 questions correctly:

Spend about 1 hour to read this chapter in full and take limited notes throughout. Follow up by reviewing **all** quiz questions to ensure that you now understand how to solve each one.

If you answer 8–11 questions correctly:

Spend 20–40 minutes reviewing the quiz questions. Beginning with the questions you missed, read and take notes on the corresponding subchapters. For questions you answered correctly, ensure your thinking matches that of the explanation and you understand why each choice was correct or incorrect.

If you answer 12–15 questions correctly:

Spend less than 20 minutes reviewing all questions from the quiz. If you missed any, then include a quick read-through of the corresponding subchapters, or even just the relevant content within a subchapter, as part of your question review. For questions you got correct, ensure your thinking matches that of the explanation and review the Concept Summary at the end of the chapter.

1. A man walks 30 m east and then 40 m north. What is the difference between his traveled distance and his displacement?
   - **A.** 0 m
   - **B.** 20 m
   - **C.** 50 m
   - **D.** 70 m

2. A 1000 kg rocket ship, travelling at 100 $\frac{m}{s}$, is acted upon by an average force of 20 kN applied in the direction of its motion for 8 s. What is the change in velocity of the rocket?
   - **A.** 160 $\frac{m}{s}$
   - **B.** 260 $\frac{m}{s}$
   - **C.** 160,000 $\frac{m}{s}$
   - **D.** 260,000 $\frac{m}{s}$

3. A car is traveling at 40 $\frac{km}{hr}$ and the driver puts on the brakes, bringing the car to rest in a time of 6 s. What is the magnitude of the average acceleration of the car?
   - **A.** 240 $\frac{km}{hr^2}$
   - **B.** 12,000 $\frac{km}{hr^2}$
   - **C.** 24,000 $\frac{km}{hr^2}$
   - **D.** 30,000 $\frac{km}{hr^2}$

4. An elevator is designed to carry a maximum weight of 9800 N (including its own weight), and to move upward at a speed of 5 $\frac{m}{s}$ after an initial period of acceleration. What is the relationship between the maximum tension in the elevator cable and the maximum weight of the elevator while the elevator is accelerating upward?
   - **A.** The tension is greater than 9800 N.
   - **B.** The tension is less than 9800 N.
   - **C.** The tension equals 9800 N.
   - **D.** It cannot be determined from the information given.

5. A student must lift a mass of 4 kg a distance of 0.5 m. The ambient temperature is 298 K and the student must lift the mass in 30 seconds. Which of the following values is NOT necessary to calculate power?
   - **A.** Mass
   - **B.** Distance
   - **C.** Temperature
   - **D.** Time

6. A firefighter jumps horizontally from a burning building with an initial speed of 1.5 $\frac{m}{s}$. At what time is the angle between his velocity and acceleration vectors the greatest?
   - **A.** The instant he jumps
   - **B.** When he reaches terminal velocity
   - **C.** Halfway through his fall
   - **D.** Right before he lands on the ground

7. A 10 kg wagon rests on an inclined plane. The plane makes an angle of 30° with the horizontal. Approximately how large is the force required to keep the wagon from sliding down the plane (Note: sin 30° = 0.5, cos 30° = 0.866)?
   - **A.** 10 N
   - **B.** 49 N
   - **C.** 85 N
   - **D.** 98 N

8. Which of the following expressions correctly illustrates the SI base units for each of the variables in the formula below?

$$m\Delta\mathbf{v} = \mathbf{F}\Delta t$$

   - **A.** $\text{lb} \times \text{mph} = \text{ft} \times \text{lb} \times \text{s}$
   - **B.** $\text{kg} \times \frac{\text{m}}{\text{s}} = \text{N} \times \text{s}$
   - **C.** $\text{kg} \times \frac{\text{m}}{\text{s}} = \frac{\text{kg}\cdot\text{m}}{\text{s}^2} \times \text{s}$
   - **D.** $\text{g} \times \frac{\text{m}}{\text{s}} = \frac{\text{g}\cdot\text{m}}{\text{s}^2} \times \text{s}$

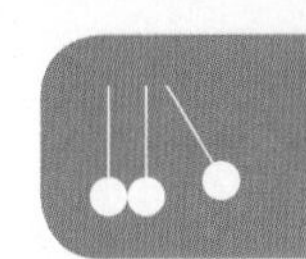

9. The diagram below shows two vectors.

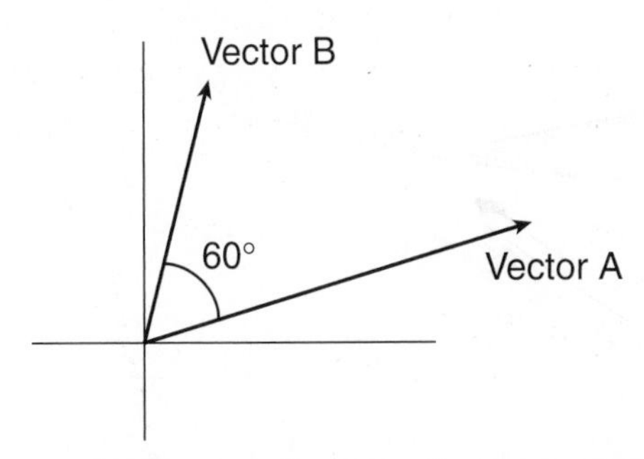

If a student uses the formula $ab \cdot \sin(60°)$, where a and b are the magnitudes of the respective vectors A and B, then which of the following best describes the computed value?

- **A.** Positive scalar
- **B.** Negative scalar
- **C.** Vector into the page
- **D.** Vector out of the page

10. Which of the following quantities is NOT a vector?

- **A.** Velocity
- **B.** Force
- **C.** Displacement
- **D.** Distance

11. A 30 kg girl sits on a seesaw at a distance of 2 m from the fulcrum. Where must her father sit to balance the seesaw if he has a mass of 90 kg?

- **A.** 67 cm from the girl
- **B.** 67 cm from the fulcrum
- **C.** 133 cm from the girl
- **D.** 267 cm from the fulcrum

12. A physics major builds a potato gun and tests it in an open field. He launches a potato with a velocity of 12 m/s at an angle of 30°. The potato is found 60 meters from the launch site. Which of the following represents the maximum height achieved by the potato? (Note: $\cos(60°) = 0.86$)

- **A.** 0.3 m
- **B.** 1.8 m
- **C.** 5 m
- **D.** 18 m

13. A rock ($m = 2$ kg) is shot up vertically at the same time that a ball ($m = 0.5$ kg) is projected horizontally. If both start from the same height:

- **A.** the rock and ball will reach the ground at the same time.
- **B.** the rock will reach the ground first.
- **C.** the ball will reach the ground first.
- **D.** the rock and ball will collide in the air before reaching the ground.

14. Centrifugal force is an apparent outward force during circular motion. It has been described as a reaction force according to Newton's third law. Which of the following statements is most likely to be correct regarding centrifugal force?

- **A.** Centrifugal force exists only for uniform circular motion, not nonuniform circular motion.
- **B.** Centrifugal force exists only when tension or a normal force provides centripetal acceleration.
- **C.** Centrifugal force always acts antiparallel to the centripetal force vector.
- **D.** Centrifugal force is result of repulsive electrostatic interactions.

15. Which of the following statements is true of movement on a plane with friction?

- **I.** Acceleration is a function of applied force only.
- **II.** More force is needed to accelerate a stationary object than an identical moving object.
- **III.** The force of friction is independent of the mass of objects.

- **A.** I only
- **B.** II only
- **C.** I and II only
- **D.** I and III only

Answer Key

1. **B** (Ch. 1.3)
2. **A** (Ch. 1.3)
3. **C** (Ch. 1.4)
4. **A** (Ch. 1.5)
5. **C** (Ch. 1.1)
6. **A** (Ch. 1.4)
7. **B** (Ch. 1.5)
8. **C** (Ch. 1.1)
9. **D** (Ch. 1.2)
10. **D** (Ch. 1.2)
11. **B** (Ch. 1.7)
12. **B** (Ch. 1.6)
13. **C** (Ch. 1.6)
14. **C** (Ch. 1.7)
15. **B** (Ch. 1.4)

CHAPTER 1

Kinematics and Dynamics

In This Chapter

1.1 **Units**
- Fundamental Measurements 8

1.2 **Vectors and Scalars**
- Vector Addition 10
- Vector Subtraction 13
- Multiplying Vectors by Scalars 13
- Multiplying Vectors by Other Vectors 14

1.3 **Displacement and Velocity**
- Displacement 17
- Velocity 17

1.4 **Forces and Acceleration**
- Forces 19
- Mass and Weight 21
- Acceleration 23

1.5 **Newton's Laws**
- First Law 24
- Second Law 25
- Third Law 25

1.6 **Motion with Constant Acceleration**
- Linear Motion 26
- Projectile Motion 28
- Inclined Planes 29
- Circular Motion 31

1.7 **Mechanical Equilibrium**
- Free Body Diagrams 33
- Translational Equilibrium 35
- Rotational Equilibrium 36

Concept Summary 39

CHAPTER PROFILE

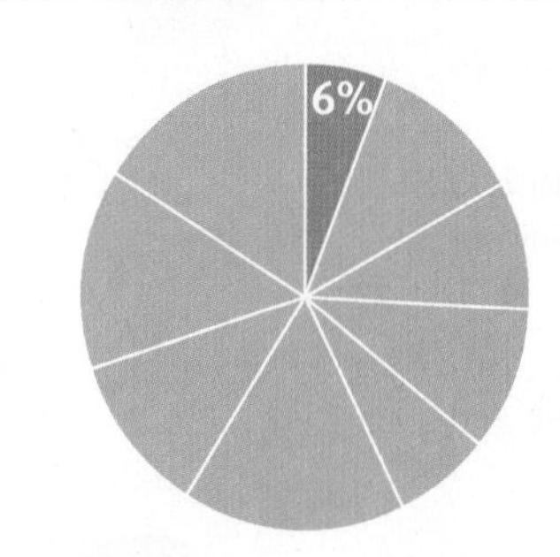

The content in this chapter should be relevant to about 6% of all questions about physics on the MCAT.

This chapter covers material from the following AAMC content category:

4A: Translational motion, forces, work, energy, and equilibrium in living systems

Introduction

A professor once said: *Biology is chemistry. Chemistry is physics. Physics is life.* Not surprisingly, this was the claim of a physics professor.

Walking into MCAT preparation, many students think of physics as the least applicable science to medicine, reflecting on calculus-heavy premedical classes. But even in the medical field, physics is all around us. When we treat patients at a rehab hospital, we often talk about motion, forces, and bone strength. An ophthalmologist may draw diagrams to help students better understand myopia and hyperopia. When we talk about mitochondria functioning as the batteries of the cell, we mean that fairly literally.

This first chapter reviews the three systems of units encountered on the MCAT: MKS (meter–kilogram–second), CGS (centimeter–gram–second), and SI (International System of Units). We'll take a few moments to review the geometry of physics questions, especially vector mathematics. Next, we'll move into true physics content as we consider kinematics—the equations that deal with the motion of objects—and Newtonian mechanics and dynamics—the study of forces and their effects.

REAL WORLD

Natural phenomena occur on many scales, as shown in Figure 1.1. We often assume that the fine details have little bearing on the larger scale of the universe, but the rapid inflation of the universe allows the infinitesimally small to affect the astronomically big.

10^{26} meters: Observable universe

10^{21} meters: Milky Way galaxy

10^{13} meters: Solar system

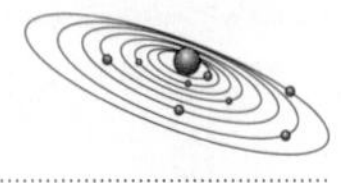

10^{7} meters: Earth

10^{-2} meters: Insect

10^{-10} meters: Atom

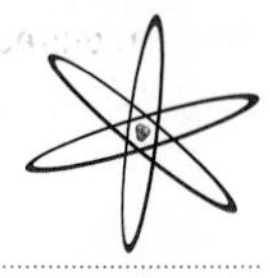

10^{-15} meters: Atomic nucleus

10^{-18} meters: Smallest distance probed by particle accelerators

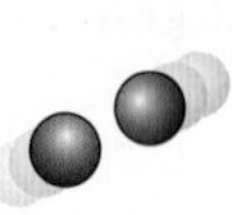

10^{-18} to 10^{-35} meters: Typical size of fundamental strings and of extra dimensions

10^{-35} meters: Minimum meaningful length in nature

Figure 1.1 Size of Natural Phenomena

1.1 Units

LEARNING OBJECTIVES

After Chapter 1.1, you will be able to:

- Recall the fundamental measurements and their units
- Order a given set of units that measure the same type of quantity from smallest to largest

Before we begin our discussion of motion, we must define a consistent vocabulary for our discussion of physics throughout this book. Physics relies on the language of mathematics to convey important descriptions and explanations of the world around us. Yet those numbers would themselves be meaningless—or vague at best—without the labels of units.

Fundamental Measurements

Over the years, various systems of units have been developed for specific purposes. Some of these systems are commonly used in everyday life but rarely in science. The **British** or **Imperial system** (foot–pound–second or **FPS**) is used commonly in the United States but virtually nowhere else—not even in Britain. Basic units for length, weight, and time are the **foot** (**ft**), the **pound** (**lb**), and the **second** (**s**), respectively. Because weight, and not mass, is used, the British system later derived the **slug** as a unit of mass. The MCAT rarely—if ever—utilizes FPS in passages or questions.

The most common system of units is the metric system, which is the basis for the SI units used on the MCAT. Depending on the context of a passage or question, the **metric system** may be given in meters, kilograms, and seconds (**MKS**) or centimeters, grams, and seconds (**CGS**). **SI units** include the MKS system as well as four other base units, as shown in Table 1.1.

Quantity	Unit	Symbol
Length	meter	m
Mass (*not* weight)	kilogram	kg
Time	second	s
Current	ampère (coulomb/second)	A
Amount of Substance	mole	mol
Temperature	kelvin	K
Luminous Intensity	candela	cd

Table 1.1 SI Units

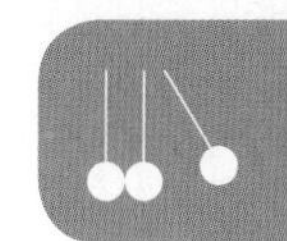

In each measurement system, there are base units and derived units. **Base units** are the standard units around which the system itself is designed. **Derived units,** as the name implies, are created by associating base units with each other. For example, the newton—a unit of force—is derived from kilograms, meters, and seconds: $1\text{ N} = 1\ \frac{\text{kg}\cdot\text{m}}{\text{s}^2}$. Table 1.2 contains examples of important derived units from each of the systems described above. Prefixes for metric units and conversions between metric and Imperial units are discussed in Chapter 10 of *MCAT Physics and Math Review.*

Quantity	FPS	CGS	MKS (SI)
Mass	slug $\left(\frac{\text{lb}\cdot\text{s}^2}{\text{ft}}\right)$ or blob $\left(\frac{\text{lb}\cdot\text{s}^2}{\text{in}}\right)$		
Force		dyne $\left(\frac{\text{g}\cdot\text{cm}}{\text{s}^2}\right)$	newton $\left(\frac{\text{kg}\cdot\text{m}}{\text{s}^2}\right)$
Work and Energy	foot–pound (ft • lb)	erg $\left(\frac{\text{g}\cdot\text{cm}^2}{\text{s}^2}\right)$	joule $\left(\frac{\text{kg}\cdot\text{m}^2}{\text{s}^2}\right)$
Power	foot–pound per second	erg per second	watt $\left(\frac{\text{kg}\cdot\text{m}^2}{\text{s}^3}\right)$

Table 1.2 Derived Units in Various Systems of Measurement

At the molecular, atomic, or subatomic level, different units may be given that are easier to work with at such a small scale. For example, length may be given in **ångströms** ($1\ \text{Å} = 10^{-10}\ \text{m}$) or nanometers ($1\ \text{nm} = 10^{-9}\ \text{m}$). Energy on the atomic scale can be expressed in **electron-volts** ($1\ \text{eV} = 1.6 \times 10^{-19}\ \text{J}$), which represent the amount of energy gained by an electron accelerating through a potential difference of one volt.

MCAT EXPERTISE

While it is good to be aware of the various systems of measurement, the only system that you are required to memorize for the MCAT is the SI system.

BRIDGE

Solutions to concept checks for a given chapter in *MCAT Physics and Math Review* can be found near the end of the chapter in which the concept check is located, following the Concept Summary for that chapter.

MCAT CONCEPT CHECK 1.1

Before you move on, assess your understanding of the material with these questions.

1. If the newton is the product of kilograms and meters/second^2, what units comprise the pound?

____

2. Order the following units from smallest to largest: centimeter, angstrom, inch, mile, foot.

______ < ______ < ______ < ______ < ______

1.2 Vectors and Scalars

LEARNING OBJECTIVES

After Chapter 1.2, you will be able to:

- Explain the importance of order when performing vector calculations
- Calculate a scalar or a vector as a product of two vectors, using the right-hand rule when applicable:

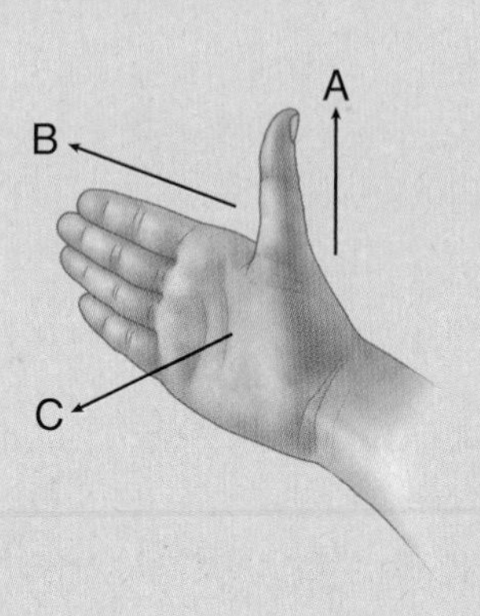

Vectors are numbers that have magnitude and direction. Vector quantities include displacement, velocity, acceleration, and force. **Scalars** are numbers that have magnitude only and no direction. Scalar quantities include distance, speed, energy, pressure, and mass. The difference between a vector and scalar quantity can be quite pronounced when there is a nonlinear path involved. For example, in the course of a year, the Earth travels a distance of roughly 940 million kilometers. However, because this is a circular path, the displacement of the Earth in one year is zero kilometers. This difference between distance and displacement can be further illustrated with vector representations.

Vectors may be represented by arrows; the direction of the arrow indicates the direction of the vector. The length of the arrow is usually proportional to the magnitude of the vector quantity. Common notations for a vector quantity are either an arrow or boldface. For example, the straight-line path from *here* to *there* might be represented by a vector identified as $\vec{\mathbf{A}}$ or **A**. The magnitude of the displacement between the two positions can be represented as $|\vec{\mathbf{A}}|$, $|\mathbf{A}|$, or A. Scalar quantities are generally represented with italic type: the distance between two points could be represented by d.

In this book (and all books of the *Kaplan MCAT Review* series), we will consistently use **boldface** to represent a vector quantity and *italic* to represent the magnitude of a vector or a scalar quantity.

Vector Addition

KEY CONCEPT

When adding vectors, always add tip-to-tail!

The sum or difference of two or more vectors is called the **resultant** of the vectors. One way to find the sum or resultant of two vectors **A** and **B** is to place the tail of **B** at the tip of **A** without changing either the length or the direction of either arrow. In this **tip-to-tail method**, the lengths of the arrows must be proportional to the

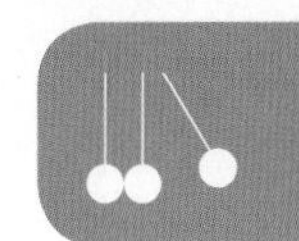

magnitudes of the vectors. The vector sum **A** + **B** is the vector joining the tail of **A** to the tip of **B** and pointing toward the tip of **B**. Vector addition is demonstrated in Figure 1.2 below.

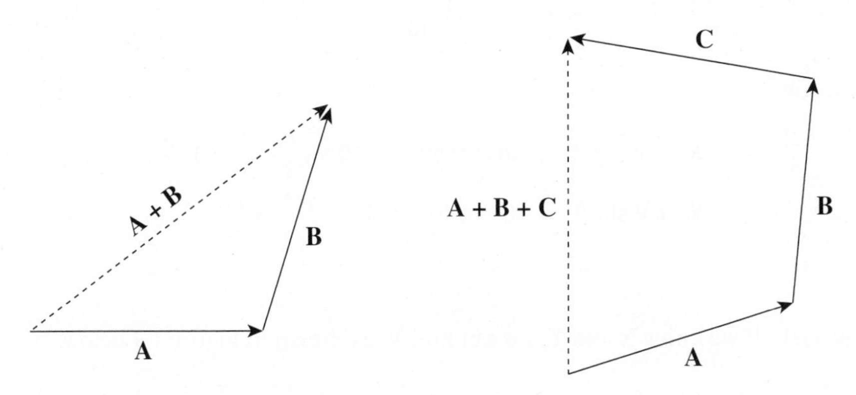

Figure 1.2 The Tip-to-Tail Method of Vector Addition
*(left) Vectors **A** and **B** with resultant **A** + **B**; (right) Vectors **A**, **B**, and **C** with resultant **A** + **B** + **C***

Another method for finding the resultant of several vectors involves breaking each vector into perpendicular **components**. In most cases, these components are horizontal and vertical (***x***- and ***y***-components, respectively); however, in some instances—such as inclined planes—it may make more sense to define the components as parallel and perpendicular ($||$ and $\perp$, respectively) to some other surface.

Given any vector **V**, we can find the *x*- and *y*-components (**X** and **Y**) by drawing a right triangle with **V** as the hypotenuse, as shown in Figure 1.3.

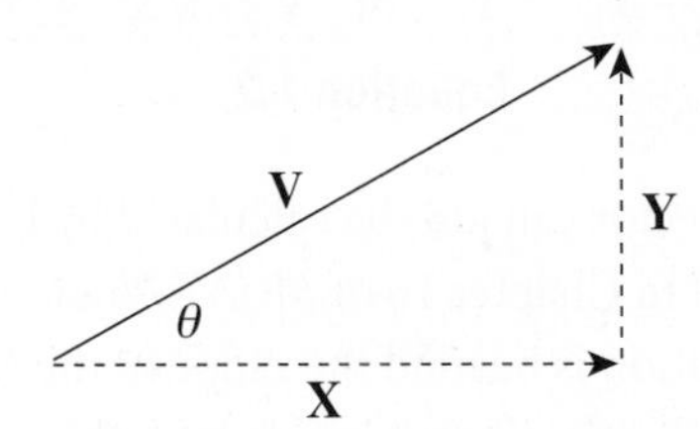

Figure 1.3 Splitting a Vector into Components

If θ is the angle between **V** and the *x*-component, then $\cos\theta = \frac{\mathbf{X}}{\mathbf{V}}$ and $\sin\theta = \frac{\mathbf{Y}}{\mathbf{V}}$. In other words:

$$\mathbf{X} = \mathbf{V}\cos\theta$$
$$\mathbf{Y} = \mathbf{V}\sin\theta$$

Equation 1.1

Example: Find the *x*- and *y*-components of the following vector

$$\mathbf{V} = 10\ \frac{\text{m}}{\text{s}}$$
$$\theta = 30^\circ$$

Solution:

$$\mathbf{X} = \mathbf{V}\cos(\theta) = 10\cos(30^\circ) = 10 \times \frac{\sqrt{3}}{2} = 5\sqrt{3}\ \frac{\text{m}}{\text{s}}$$
$$\mathbf{Y} = \mathbf{V}\sin(\theta) = 10\sin(30^\circ) = 10 \times \frac{1}{2} = 5\ \frac{\text{m}}{\text{s}}$$

Conversely, if we know **X** and **Y**, we can find **V**, as shown in Figure 1.4 below.

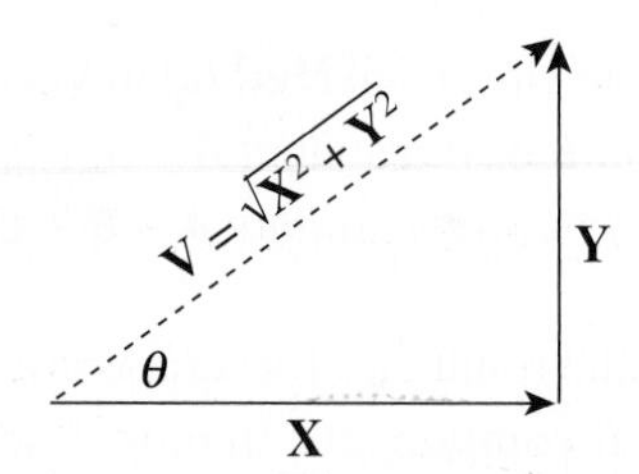

Figure 1.4 Using the Pythagorean Theorem to Determine the Magnitude of the Resultant Vector

Calculating the magnitude of **V** requires use of the **Pythagorean theorem**:

$$\mathbf{X}^2 + \mathbf{Y}^2 = \mathbf{V}^2 \quad \text{or} \quad \mathbf{V} = \sqrt{\mathbf{X}^2 + \mathbf{Y}^2}$$

Equation 1.2

No calculator! More logic than math.

The angle of the resultant vector can also be calculated by knowing inverse trigonometric functions, discussed in Chapter 10 of *MCAT Physics and Math Review* (Note: This inverse tangent calculation is beyond the scope of the MCAT):

$$\theta = \tan^{-1}\frac{\mathbf{Y}}{\mathbf{X}}$$

Equation 1.3

Example: What is the magnitude of the vector with the following components?

$$\mathbf{X} = 3\ \frac{\text{m}}{\text{s}}$$
$$\mathbf{Y} = 4\ \frac{\text{m}}{\text{s}}$$

Solution:

$$\mathbf{V} = \sqrt{3^2 + 4^2} = \sqrt{25} = 5\ \frac{\text{m}}{\text{s}}$$

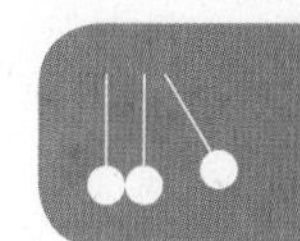

The x-component of a resultant vector is simply the sum of the x-components of the vectors being added. Similarly, the y-component of a resultant vector is simply the sum of the y-components of the vectors being added. This is illustrated in Figure 1.5.

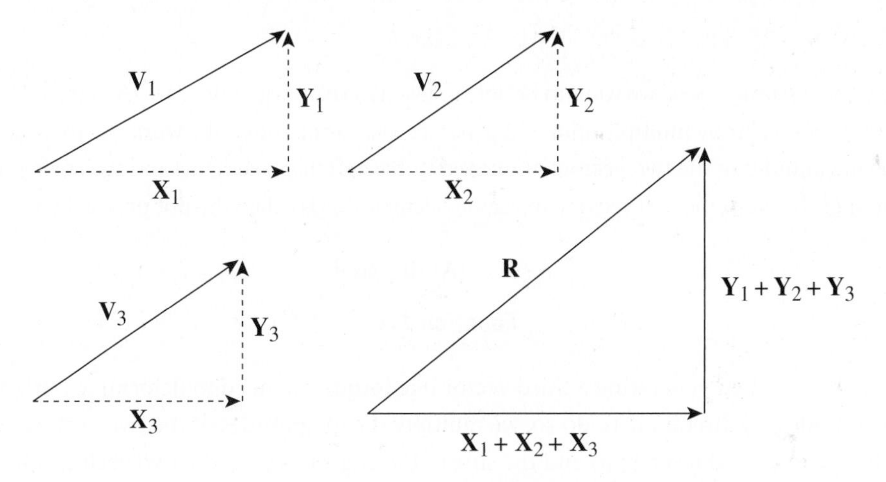

Figure 1.5 Finding the Resultant (R) of $V_1 + V_2 + V_3$

To find the resultant (**R**) using the components method, follow these steps:

1. Resolve the vectors to be added into their x- and y-components.
2. Add the x-components to get the x-component of the resultant ($\mathbf{R}_x$). Add the y-components to get the y-component of the resultant ($\mathbf{R}_y$).
3. Find the magnitude of the resultant by using the Pythagorean theorem. If $\mathbf{R}_x$ and $\mathbf{R}_y$ are the components of the resultant, then $R = \sqrt{R_x^2 + R_y^2}$.
4. Find the direction (θ) of the resultant by using the relationship $\theta = \tan^{-1} \frac{R_y}{R_x}$.

Vector Subtraction

Subtracting one vector from another can be accomplished by adding a vector with equal magnitude—but opposite direction—to the first vector. This can be expressed mathematically as $\mathbf{A} - \mathbf{B} = \mathbf{A} + (-\mathbf{B})$, where $-\mathbf{B}$ represents a vector with the same magnitude as **B**, but pointing in the opposite direction. Vector subtraction may also be performed on the component vectors first and then combined to create a final vector. As with vector addition, the x-component of the resultant vector is the difference of the x-components of the vectors being subtracted. Similarly, the y-component of the resultant vector is the difference of the y-components of the vectors being subtracted.

KEY CONCEPT

Notice that when you subtract vectors, you are simply flipping the direction of the vector being subtracted and then following the same rules as normal: adding tip-to-tail.

Multiplying Vectors by Scalars

When a vector is multiplied by a scalar, its magnitude will change. Its direction will be either parallel or antiparallel to its original direction. If a vector **A** is multiplied by the scalar value n, a new vector, **B**, is created such that $\mathbf{B} = n\mathbf{A}$. To find the magnitude of the new vector, **B**, simply multiply the magnitude of **A** by $|n|$, the absolute value of n. To determine the direction of the vector **B**, we must look at the sign on n. If n is a positive number, then **B** and **A** are in the same direction. However, if n is a negative number, then **B** and **A** point in opposite directions. For example, if vector **A** is multiplied by the scalar +3, then the new vector **B** is three times as long as **A**, and points in the same direction. If vector **A** is multiplied by the scalar −3, then **B** would still be three times as long as **A** but would now point in the opposite direction.

Multiplying Vectors by Other Vectors

Dot product, multiply to vectors to get a scalar!

A • B = |A||B| cosθ

In some circumstances, we want to be able to use two vector quantities to generate a third vector or a scalar by multiplication. To generate a scalar quantity like work, we multiply the magnitudes of the two vectors of interest (force and displacement) and the cosine of the angle between the two vectors. In vector calculus, this is called the **dot product** ($\mathbf{A} \cdot \mathbf{B}$):

$$\mathbf{A} \cdot \mathbf{B} = |\mathbf{A}|\,|\mathbf{B}| \cos \theta$$

Equation 1.4

Cross product, multiply vectors to get a vector;

A × B = |A||B| sinθ

use right hand rule!

In contrast, when generating a third vector like torque, we need to determine both its magnitude and direction. To do so, we multiply the magnitudes of the two vectors of interest (force and lever arm) and the sine of the angle between the two vectors. Once we have the magnitude, we use the **right-hand rule** to determine its direction. In vector calculus, this is called the **cross product** ($\mathbf{A} \times \mathbf{B}$):

$$\mathbf{A} \times \mathbf{B} = |\mathbf{A}|\,|\mathbf{B}| \sin \theta$$

Equation 1.5

The resultant of a cross product will always be perpendicular to the plane created by the two vectors. Because the MCAT is a two-dimensional test, this usually means that the vector of interest will be going into or out of the page (or screen).

There are multiple versions of the right-hand rule that can be used to determine the direction of a cross product resultant vector. Figure 1.6 shows one method considering a resultant **C** where $\mathbf{C} = \mathbf{A} \times \mathbf{B}$:

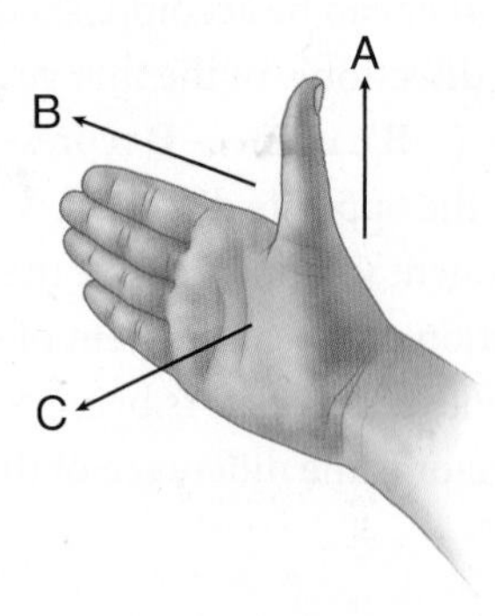

Figure 1.6 Applying the Right-Hand Rule

1. Start by pointing your thumb in the direction of vector **A**.
2. Extend your fingers in the direction of vector **B**. You may need to rotate your wrist to get the correct configuration of thumb and fingers.
3. Your palm establishes the plane between the two vectors. The direction your palm points is the direction of the resultant **C**.

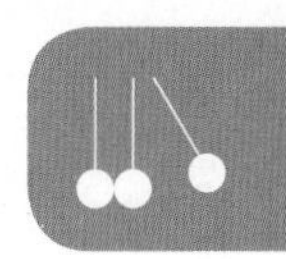

Note that you may have learned a version of the right-hand rule that is different from what is described here. For example, some students learn to point the right index finger in the direction of **A** and the right middle finger in the direction of **B**; when one holds the thumb perpendicular to these two fingers, it points in the direction of **C**. It makes no difference which version of the right-hand rule you use, as long as you are comfortable with it and are skilled in its proper use.

MCAT EXPERTISE

There are several different methods for determining the direction of a cross product resultant vector. Choose whichever method you prefer and stick with it—it's more important to be skilled in using one method than to be only somewhat familiar with multiple methods.

Example: What are the magnitudes and directions of the resultant vectors from the following cross products: $\mathbf{C} = \mathbf{A} \times \mathbf{B}$ and $\mathbf{D} = \mathbf{B} \times \mathbf{A}$?

$$\mathbf{A}: \mathbf{X} = -3\text{ N}, \mathbf{Y} = 0$$

$$\mathbf{B}: \mathbf{X} = 0, \mathbf{Y} = +4\text{ m}$$

Solution: The magnitude of the resultant vector is simply the product of the magnitudes of the factor vectors and the sine of the angle between them. In this case, because one is oriented in the *x*-direction and the other is in the *y*-direction, the angle between them is 90°.

$$|\mathbf{A}| \times |\mathbf{B}| \times \sin 90° = 3\text{ N} \times 4\text{ m} \times 1 = 12\text{ N}\cdot\text{m}$$

The magnitude is therefore 12 N•m.

Now, to determine the direction of **C**, start by pointing your right thumb toward the left (negative *x*-direction). Your fingers will point toward the top of the page (positive *y*-direction). Your palm is therefore pointing into the page.

Now, to determine the direction of **D**, start by pointing your right thumb toward the top of the page (positive *y*-direction). Your fingers will point toward the left (negative *x*-direction). Your palm is therefore pointing out of the page.

Therefore, **C** is 12 N•m [⊗ (into the page)] and **D** is 12 N•m [⊙ (out of the page)].

KEY CONCEPT

For cross products and the right-hand rule, order matters! Unlike scalar multiplication, which is commutative ($3 \times 4 = 4 \times 3$), vector multiplication is not commutative ($\mathbf{A} \times \mathbf{B} \neq \mathbf{B} \times \mathbf{A}$)!

MCAT CONCEPT CHECK 1.2

Before you move on, assess your understanding of the material with these questions.

1. When calculating the sum of vectors **A** and **B** (**A** + **B**), we put the tail of **B** at the tip of **A**. What would be the effect of reversing this order (**B** + **A**)?

___

2. When calculating the difference of vectors **A** and **B** (**A** − **B**), we invert **B** and put the tail of this new vector at the tip of **A**. What would be the effect of reversing this order (**B** − **A**)?

___

3. How is a scalar calculated from the product of two vectors? How is a vector calculated?

   - Scalar:

___

   - Vector:

___

4. True or False: If **C** = **A** × **B**, where **A** is directed toward the right side of the page and **B** is directed to the top of the page, then **C** is directed midway between **A** and **B** at a 45° angle.

1.3 Displacement and Velocity

LEARNING OBJECTIVES

After Chapter 1.3, you will be able to:

- Describe the relationship between the average and instantaneous versions of velocity and speed
- Distinguish between total distance and total displacement
- Connect displacement and velocity with an equation

Now that we've covered the basic geometry that serves as the foundation of physics, we can examine the related physical quantities. The basic quantities that relate to kinematics are displacement, velocity, and acceleration.

Displacement

An object in motion may experience a change in its position in space, known as **displacement** (**x** or **d**). This is a vector quantity and, as such, has both magnitude and direction. The displacement vector connects (in a straight line) the object's initial position and its final position. Understand that displacement does not account for the actual pathway taken between the initial and the final positions—only the net change in position from initial to final. **Distance** (*d*) traveled, on the other hand, considers the pathway taken and is a scalar quantity.

Example: What is the displacement of a man who walks 2 km east, then 2 km north, then 2 km west, and then 2 km south?

Solution: While his total distance traveled is 8 km, his displacement is a vector quantity that represents the change in position. In this case, his displacement is zero because the man ends up the same place he started, as shown below.

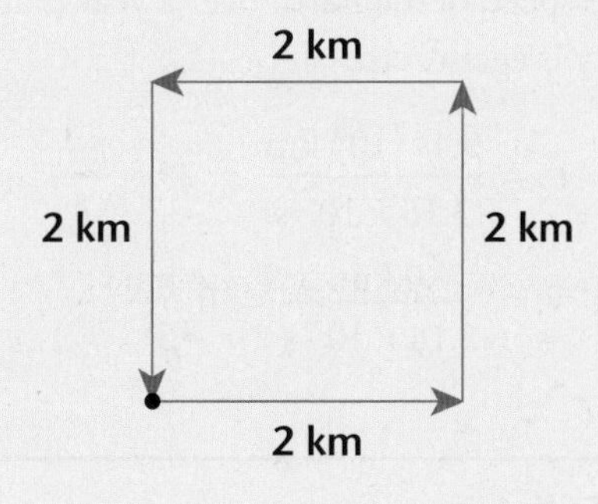

Velocity

As was mentioned earlier, **velocity** (**v**) is a vector. Its magnitude is measured as the rate of change of displacement in a given unit of time, and its SI units are meters per second. The direction of the velocity vector is necessarily the same as the direction of the displacement vector. **Speed** (*v*) is the rate of actual distance traveled in a given unit of time.

*speed = scalar
velocity = vector

The distinction is subtle, so let's examine this a little more carefully. The **instantaneous speed** of an object will always be equal to the magnitude of the object's **instantaneous velocity**, which is a measure of the average velocity as the change in time (Δt) approaches zero:

$$\mathbf{v} = \lim_{\Delta t \to 0} \frac{\Delta \mathbf{x}}{\Delta t}$$

Equation 1.6

where **v** is the instantaneous velocity, $\Delta\mathbf{x}$ is the change in position, and Δt is the change in time. As a measure of speed, instantaneous speed is a scalar number. Average speed will not necessarily always be equal to the magnitude of the average velocity. This is because average velocity is the ratio of the displacement vector over the change in time (and is a vector), whereas average speed (which is scalar) is the ratio of the total distance traveled over the change in time. Average speed accounts for actual distance traveled, whereas average velocity does not:

$$\bar{\mathbf{v}} = \frac{\Delta \mathbf{x}}{\Delta t}$$

Equation 1.7

where $\bar{\mathbf{v}}$ is the average velocity, $\Delta \mathbf{x}$ is the change in position, and Δt is the change in time.

Consider the example given earlier regarding the Earth's orbit. In one year, the Earth travels roughly 940 million kilometers, but its displacement is zero:

$$d = 9.4 \times 10^8 \text{ km}$$
$$\mathbf{x} = 0 \text{ km}$$

The average speed is a measure of distance traveled in a given period of time; the average velocity is a measure of the displacement of an object over a given period of time. While the average speed of the Earth over a year is about 30 kilometers per second, its average velocity is again zero:

$$v = \frac{9.4 \times 10^8 \text{ km}}{3.16 \times 10^7 \text{ s}} = 29.8 \frac{\text{km}}{\text{s}}$$
$$\bar{\mathbf{v}} = \frac{0 \text{ km}}{3.16 \times 10^7 \text{ s}} = 0 \frac{\text{km}}{\text{s}}$$

MCAT CONCEPT CHECK 1.3

Before you move on, assess your understanding of the material with these questions.

1. What is the relationship between instantaneous velocity and instantaneous speed? Between average velocity and average speed?

____

2. True or False: Total distance traveled can never be less than the total displacement.

3. Provide a definition for displacement or velocity in terms of the other variable.

____

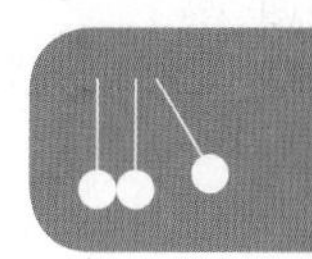

1.4 Forces and Acceleration

LEARNING OBJECTIVES

After Chapter 1.4, you will be able to:

- Calculate a frictional force and predict its direction
- Describe the relationship between force, velocity, and acceleration
- Identify common forces, including frictional and gravitational forces
- Explain the concept of normal forces and how they impact a system

Every change in velocity is motivated by a push or a pull—a **force**. In this section, we'll examine how forces interact with one another, as well as how acceleration results from those forces.

all velocity Δ's related to forces

Forces

Force (F) is a vector quantity that is experienced as pushing or pulling on objects. Forces can exist between objects that are not even touching. While it is common for forces to be exerted by one object pushing on another, there are even more instances in which forces exist between objects nowhere near each other, such as gravity or electrostatic forces between point charges. The SI unit for force is the **newton (N)**, which is equivalent to one $\frac{\text{kg}\cdot\text{m}}{\text{s}^2}$.

Gravity

When Newton observed apples falling out of trees, he was struck by the fact that they always fell perpendicularly to the ground, rather than sideways or even away from the ground. Furthermore, Newton began to wonder about the farthest reaches of gravity. If the apple feels this attractive pull toward the Earth, then what about the Moon? This force is what Newton would later term "universal gravitation."

Gravity is an attractive force that is felt by all forms of matter. We usually think of gravity as acting on us to keep us from floating off of the Earth's surface, or for holding the planets of our solar system in orbit. However, all objects exert gravitational forces on each other; there is a small (but measurable) force of gravity between you and this *MCAT Physics and Math Review* book, the chair you're sitting on, and all the objects around you. Gravitational forces usually do not have much significance on a small scale because other forces tend to be much larger in magnitude. Only on the planetary level do gravitational forces really take on a significant value.

The magnitude of the **gravitational force** between two objects is

$$F_g = \frac{Gm_1m_2}{r^2}$$

Equation 1.8

MCAT EXPERTISE

Acceleration due to gravity, g, decreases with height above the Earth and increases the closer one gets to the Earth's center of mass. Near the Earth's surface, use $g = 10\ \frac{\text{m}}{\text{s}^2}$.

REAL WORLD

Newton's third law states that the force of gravity on m_1 from m_2 is equal and opposite to the force of gravity on m_2 from m_1. This means that the force of gravity on you from the Earth is equal and opposite to the force of gravity from you on the Earth. Because the forces are equal but the masses are very different, the accelerations must also be very different, from **F** = *m***a** (discussed later in this chapter). Because your mass is very small compared to the Earth, you experience a large acceleration from it. In contrast, because the Earth is massive, it experiences a tiny acceleration from the same magnitude of force.

where G is the universal gravitational constant $\left(6.67 \times 10^{-11} \frac{\text{N}\cdot\text{m}^2}{\text{kg}^2}\right)$, m_1 and m_2 are the masses of the two objects, and r is the distance between their centers of mass. This equation is commonly tested in the context of proportionalities. For instance, the magnitude of the gravitational force is inversely related to the square of the distance (that is, if r is halved, then F_g will quadruple). The magnitude of the gravitational force is also directly related to the masses of the objects (that is, if m_1 is tripled, then F_g will triple).

Example: Find the gravitational force between an electron and a proton that are 10^{-11} m apart. (Note: Mass of a proton $= 1.67 \times 10^{-27}$ kg; mass of an electron $= 9.11 \times 10^{-31}$ kg)

Solution: Use Newton's law of gravitation:

$$F_g = \frac{Gm_1m_2}{r^2} = \frac{\left(6.67 \times 10^{-11} \frac{\text{N}\cdot\text{m}^2}{\text{kg}^2}\right)\left(1.67 \times 10^{-27}\text{ kg}\right)\left(9.11 \times 10^{-31}\text{ kg}\right)}{\left(10^{-11}\text{ m}\right)^2}$$

$$F_g \approx \frac{\left(\frac{20}{3} \times 10^{-11}\right)\left(\frac{5}{3} \times 10^{-27}\right)\left(9 \times 10^{-31}\right)}{10^{-22}} = \frac{\frac{100 \times 9}{9} \times 10^{-69}}{10^{-22}} = \frac{10^{-67}}{10^{-22}}$$

$$= 10^{-45}\text{ N}\left(\text{actual} = 1.02 \times 10^{-45}\text{ N}\right)$$

Friction

Friction is a type of force that opposes the movement of objects. Unlike other kinds of forces, such as gravity or electromagnetic force, which can cause objects either to speed up or slow down, friction forces always oppose an object's motion and cause it to slow down or become stationary. There are two types of friction: static and kinetic.

Static friction (f_s) exists between a stationary object and the surface upon which it rests. The inequality that describes the magnitude of static friction is

$$0 \leq f_s \leq \mu_s N$$

Equation 1.9

where μ_s is the coefficient of static friction and N is the magnitude of the normal force. The **coefficient of static friction** is a unitless quantity that is dependent on the two materials in contact. The **normal force** is the component of the force between two objects in contact that is perpendicular to the plane of contact between the object and the surface upon which it rests.

Note the less-than-or-equal-to signs in the equation. These signify that there is a range of possible values for static friction. The minimum, of course, is zero. This would be the case if an object were resting on a surface with no applied forces.

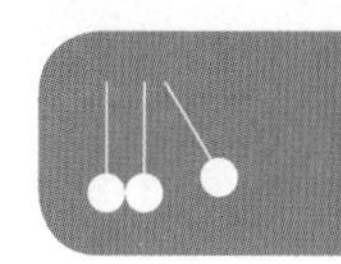

The maximum value of static friction can be calculated from the right side of the previous equation. One should not assume that objects that are stationary are experiencing a maximal static force of friction.

Consider trying to push a heavy piece of luggage. When a 25 N force is applied, the bag does not move. When a 50 N force is applied, the bag still does not move. When a 100 N force is applied, the bag slides a meter or so and slows to a rest. This setup implies that the maximal value of static friction is somewhere between 50 and 100 N; any applied force less than this threshold will not be sufficient to move the bag as there will be an equal but opposite force of static friction opposing the bag's motion.

Kinetic friction (f_k) exists between a sliding object and the surface over which the object slides. Sometimes, students misidentify the presence of kinetic friction. A wheel, for example, that is rolling along a road does not experience kinetic friction because the tire is not actually sliding against the pavement. The tire maintains an instantaneous point of static contact with the road and, therefore, experiences static friction. Only when the tire begins to slide on, say, an icy patch will kinetic friction come into play. Any time two surfaces slide against each other, kinetic friction will be present and its magnitude can be measured according to this equation:

$$f_k = \mu_k N$$

Equation 1.10

where μ_k is the coefficient of kinetic friction and N is the normal force. There are two important distinctions between this equation for kinetic friction and the previous equation for static friction. First, the kinetic friction equation has an equals sign. This means that kinetic friction will have a constant value for any given combination of a coefficient of kinetic friction and normal force. It does not matter how much surface area is in contact or even the velocity of the sliding object. Second, the two equations have a different coefficient of friction. The value of μ_s is always larger than the value of μ_k. Therefore, the maximum value for static friction will always be greater than the constant value for kinetic friction: objects will "stick" until they start moving, and then will slide more easily over one another.

As previously mentioned in the discussion of static friction, pay close attention to the conditions set in an MCAT passage or question. Does it say that friction can be assumed to be negligible, or does it provide the coefficient of friction values, which will most likely need to be used in a calculation of friction? Friction will be incorporated into our examination of translational equilibrium later in this chapter.

Mass and Weight

Mass and weight are not the same. **Mass** (m) is a measure of a body's inertia—the amount of matter in the object. Mass is a scalar quantity, and, as such, has magnitude only. The SI unit for mass is the kilogram, which is independent of gravity. One kilogram of material on Earth will have the same mass as one kilogram of material on the Moon. **Weight** ($\mathbf{F}_g$), on the other hand, is a measure of gravitational force (usually that of the Earth) on an object's mass. Because weight is a force, it is a vector quantity with units in newtons (N).

KEY CONCEPT

Contact points are the places where friction occurs between two rough surfaces sliding past each other. If the normal load—the force that squeezes the two together—rises, the total area of contact increases. That increase, more than the surface's roughness, governs the degree of friction. This is illustrated in Figure 1.7 below.

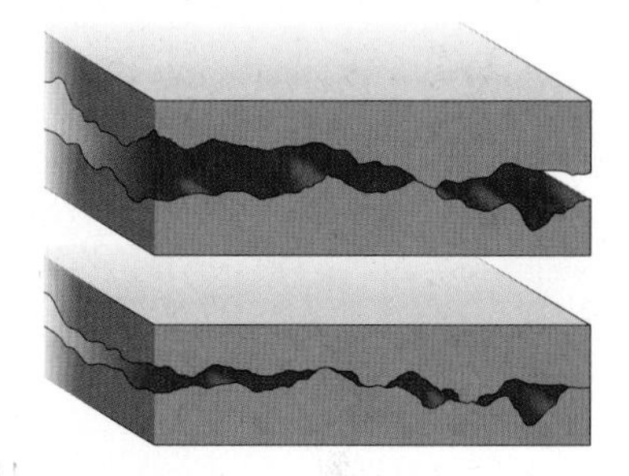

Figure 1.7 Increases in Contact Area Increase Frictional Forces

KEY CONCEPT

The coefficient of static friction will always be larger than the coefficient of kinetic friction. It always requires more force to get an object to start sliding than it takes to keep an object sliding.

While mass and weight are not synonymous, they are related by the equation:

$$\mathbf{F}_g = m\mathbf{g}$$

Equation 1.11

where $\mathbf{F}_g$ is the weight of the object, m is its mass, and $\mathbf{g}$ is acceleration due to gravity, $9.8 \frac{\text{m}}{\text{s}^2}$ (usually rounded to $10 \frac{\text{m}}{\text{s}^2}$).

The weight of an object can be thought of as being applied at a single point in that object called the **center of mass** or **gravity**. The MCAT will not directly test your ability to determine center of mass; however, such a calculation may be an important step in a problem with the larger focus of Newtonian mechanics.

To illustrate this concept and calculation, consider a tennis racquet that has been thrown into the air. Each part of the racquet moves in its own pathway, so it's not possible to represent the motion of the whole racquet as a single particle. However, one point within the racquet moves in a simple parabolic path, very similar to the flight of a ball. It is this point within the racquet that is known as the center of mass. This is clearly shown in Figure 1.8.

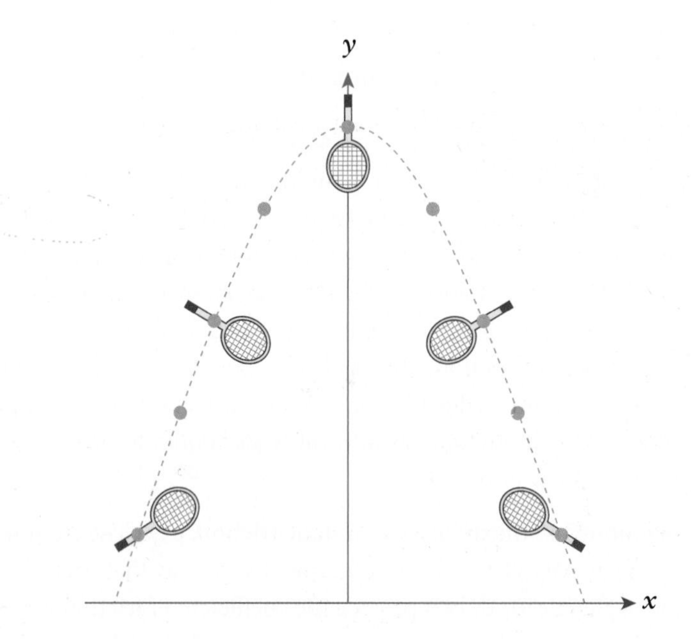

Figure 1.8 Center of Mass of a Tennis Racquet
The center of mass of a racquet thrown into the air travels along a parabolic pathway.

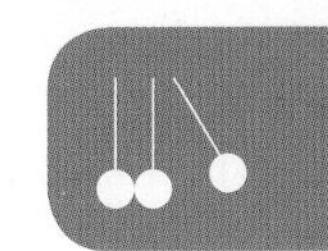

For a system in which particles are distributed in all three dimensions, the center of mass is defined by the three coordinates:

$$x = \frac{m_1x_1 + m_2x_2 + m_3x_3 + \cdots}{m_1 + m_2 + m_3 + \cdots}$$
$$y = \frac{m_1y_1 + m_2y_2 + m_3y_3 + \cdots}{m_1 + m_2 + m_3 + \cdots}$$
$$z = \frac{m_1z_1 + m_2z_2 + m_3z_3 + \cdots}{m_1 + m_2 + m_3 + \cdots}$$

Equation 1.12

where m_1, m_2, and m_3 are the three sample masses, and the x-, y- and z-values are coordinates. The center of gravity is related and corresponds to the single point at which one can conceptualize gravity acting on an object. Only for a homogeneous body (with symmetrical shape and uniform density) should one expect the center of gravity to be located at its geometric center. For example, we can approximate the center of gravity for a metal ball as the geometric center of the sphere. The same cannot be said, however, for a human body, television, or any asymmetrical, non-uniform object.

KEY CONCEPT

The center of mass of a uniform object is at the geometric center of the object.

Acceleration

Acceleration (**a**) is the rate of change of velocity that an object experiences as a result of some applied force. Acceleration, like velocity, is a vector quantity and is measured in SI units of meters per second squared. Acceleration in the direction opposite the initial velocity may be called **deceleration.** **Average acceleration** is defined as

$$\bar{\mathbf{a}} = \frac{\Delta \mathbf{v}}{\Delta t}$$

Equation 1.13

where $\bar{\mathbf{a}}$ is the average acceleration, $\Delta \mathbf{v}$ is the change in velocity, and Δt is the change in time.

Instantaneous acceleration is defined as the average acceleration as Δt approaches zero.

$$\mathbf{a} = \lim_{\Delta t \to 0} \frac{\Delta \mathbf{v}}{\Delta t}$$

Equation 1.14

On a graph of velocity *vs.* time, the tangent to the graph at any time t, which corresponds to the slope of the graph at that time, indicates the instantaneous acceleration. If the slope is positive, then the acceleration is positive and in the same direction as the velocity. If the slope is negative, then the acceleration is negative and in the opposite direction of the velocity (this is a deceleration).

MCAT CONCEPT CHECK 1.4

Before you move on, assess your understanding of the material with these questions.

1. When calculating frictional forces, how is directionality assigned?

 ____

2. When no force is being applied, the velocity must be:

 ____

3. True or False: The Earth creates a larger force on you than you create on the Earth.

4. Name two forces in addition to mechanical manipulation (pushing or pulling forces created by contact with an object):

 - ______________________________
 - ______________________________

1.5 Newton's Laws

LEARNING OBJECTIVES

After Chapter 1.5, you will be able to:

- Describe Newton's three laws of motion

Now that we have a clear understanding of force, mass, and acceleration, let's examine how they relate to each other. While it is unlikely that Newton "discovered" gravity by having an apple on his head, he did record that he was indeed inspired by watching apples fall from trees. His observations about objects in motion and at rest are the basis for the branch of physics that we now know as mechanics. Newton's laws, which are expressed as equations, concisely describe the effects forces have on objects that have mass.

First Law

$$\mathbf{F}_{\text{net}} = m\mathbf{a} = 0$$

Equation 1.15

where $\mathbf{F}_{\text{net}}$ is the net force, m is the mass, and $\mathbf{a}$ is the acceleration.

A body either at rest or in motion with constant velocity will remain that way unless a net force acts upon it. This is also known as the law of **inertia**. Newton's first law ought to be thought of as a special case of his second law, which is described next.

Second Law

$$\mathbf{F}_{\text{net}} = m\mathbf{a}$$

Equation 1.16

where $\mathbf{F}_{\text{net}}$ is the net force, m is the mass, and $\mathbf{a}$ is the acceleration.

What Newton's second law states is actually a corollary of the first: *An object of mass* m *will accelerate when the vector sum of the forces results in some nonzero resultant force vector.* No acceleration will occur when the vector sum of the forces results in a cancellation of those forces. Note that the net force and acceleration vectors necessarily point in the same direction.

Third Law

$$\mathbf{F}_{\text{AB}} = -\mathbf{F}_{\text{BA}}$$

Equation 1.17

This law is also known as the law of action and reaction: *To every action, there is always an opposed but equal reaction.* More formally, the law states that for every force exerted by object A on object B, there is an equal but opposite force exerted by object B on object A. For example, when you hit your hand against your desk, your hand exerts a force on the desk. Simultaneously, the desk exerts a force of equal magnitude in the opposite direction on your hand. Physical contact is not necessary for Newton's third law; the mutual gravitational pull between the Earth and the Moon traverses hundreds of thousands of kilometers of space.

MCAT CONCEPT CHECK 1.5

Before you move on, assess your understanding of the material with this question.

1. In your own words, provide a description of Newton's laws of motion:
    1. ____
       ____
    2. ____
       ____
    3. ____
       ____
2. During a test crash, a 500 kg car is driven at a constant velocity of 50 mph until it hits a wall without braking. Apply all three of Newton's laws to this situation.
    1. ____
    2. ____
    3. ____

1.6 Motion with Constant Acceleration

LEARNING OBJECTIVES

After Chapter 1.6, you will be able to:

- Identify which forces are active during different types of motion, including free fall and projectile motion
- Predict the angle of launch necessary to maximize horizontal or vertical displacement
- Recall the equation used to calculate centripetal acceleration

Objects can undergo only two types of motion—that which is constant (with no acceleration) or that which is changing (with acceleration). If an object's motion is changing, as indicated by a change in velocity, then the object is experiencing acceleration, and that acceleration may be constant or itself changing. A moving object that experiences constant acceleration presents a relatively simple case for analysis. The MCAT tends to restrict kinematics problems to those that involve motion with constant acceleration.

Linear Motion

In **linear motion**, the object's velocity and acceleration are along the line of motion, so the pathway of the moving object continues along a straight line. Linear motion does not need to be limited to vertical or horizontal paths; the inclined surface of a ramp will provide a path for linear motion at some angle. On the MCAT, the most common presentations of linear motion problems involve objects, such as balls, being dropped to the ground from some starting height.

Falling objects exhibit linear motion with constant acceleration. This one-dimensional motion can be fully described by the following equations:

$$\mathbf{v} = \mathbf{v}_0 + \mathbf{a}t$$

$$\mathbf{x} = \mathbf{v}_0 t + \frac{\mathbf{a}t^2}{2}$$

$$\mathbf{v}^2 = \mathbf{v}_0^2 + 2\mathbf{a}\mathbf{x}$$

$$\mathbf{x} = \overline{\mathbf{v}}t$$

Equations 1.18 to 1.21

where **x**, **v**, and **a** are the displacement, velocity, and acceleration vectors, respectively; $\mathbf{v}_0$ is the initial velocity; $\overline{\mathbf{v}}$ is the average velocity; and *t* is time. When the motion is vertical, we often use **y** instead of **x** for displacement.

MCAT EXPERTISE

When dealing with free fall problems, you can choose to make *down* either positive or negative. However, for the sake of simplicity, get in the habit of always making *up* positive and *down* negative.

To demonstrate the typical setup of a kinematics problem on the MCAT, we will consider an object falling through the air. For now, we will assume air resistance to be negligible, meaning that the only force acting on the object would be the gravitational force causing it to fall. Consequently, the object would fall with constant acceleration—the **acceleration due to gravity** $\left(\mathbf{g} = 9.8\ \frac{\text{m}}{\text{s}^2}\right)$—and would not reach **terminal velocity**. This is called **free fall**. Under these conditions of a free falling object that has not reached terminal velocity, which are typical for Test Day, we could analyze the fall, using the relevant kinematics equations.

Example: A ball is thrown vertically up into the air from a window ledge 30 meters above the ground with an initial velocity of 10 $\frac{m}{s}$.

A. Find the velocity and position of the ball after two seconds.

B. Find the distance and time at which the ball reaches its maximum height above the window ledge.

Solution:

A. Remember that velocity and acceleration are vector quantities. For this question, let's call the ball's initial position, $\mathbf{y}_0$, zero. If we consider *up* to be positive, then the initial velocity, $\mathbf{v}_0$, is $+10\ \frac{m}{s}$, and the acceleration, g, is $-9.8\ \frac{m}{s^2}$. Note that g is negative because it's oriented downward. Velocity after two seconds can be found using Equation 1.18:

$$\begin{aligned} v &= v_0 + at \\ &= \left(+10\ \frac{m}{s}\right) + \left(-9.8\ \frac{m}{s^2}\right)(2\ s) \\ &= 10\ \frac{m}{s} - 19.6\ \frac{m}{s} = -9.6\ \frac{m}{s} \end{aligned}$$

After two seconds, the position of the ball is found using Equation 1.19:

$$\begin{aligned} y &= v_0 t + \frac{at^2}{2} \\ &= \left(10\ \frac{m}{s}\right)(2\ s) + \frac{\left(-9.8\ \frac{m}{s^2}\right)(2\ s)^2}{2} \\ &= 20 - 19.6 = 0.4\ m\ (\text{above the ledge}) \end{aligned}$$

B. When the ball is at its maximum height, the velocity, which has been decreasing on the way up, is now zero. We can find the maximum height the ball reaches using Equation 1.20:

$$\begin{aligned} v^2 &= v_0^2 + 2ay \\ 0^2 &= \left(10\ \frac{m}{s}\right)^2 + 2\left(-9.8\ \frac{m}{s^2}\right)(y) \\ 19.6y &= 100 \\ y &\approx 5\ m\ (\text{actual} = 5.1\ m) \end{aligned}$$

The time at which the ball reaches its maximum height can be found from Equation 1.18:

$$\begin{aligned} v &= v_0 + at \\ 0 &= \left(+10\ \frac{m}{s}\right) + \left(-9.8\ \frac{m}{s^2}\right)(t) \\ t &\approx 1\ s(\text{actual} = 1.02\ s) \end{aligned}$$

MCAT EXPERTISE

The amount of time that an object takes to get to its maximum height is the same time it takes for the object to fall back down to the starting height (assuming air resistance is negligible); this fact makes solving these problems much easier. Because you can solve for the time to reach maximum height by setting your final velocity to zero, you can then multiply your answer by two, getting the total time in flight—as long as the object ends at the same height at which it started. Because the only force acting on the object after it is launched is gravity, the velocity it has in the x-direction will remain constant throughout its time in flight. By multiplying the time by the velocity in the x-direction, one can find the horizontal distance traveled.

KEY CONCEPT

Note that gravity is in bold, indicating it has a vector value. Gravity is unique in that it is used as both a constant and as a vector in calculations. Though gravity is not always bolded, you should recall for Test Day that gravity has a direction.

assume constant horizontal velocity!

Let's now consider what happens when air resistance is *not* negligible. **Air resistance**, like friction, opposes the motion of an object. Its value increases as the speed of the object increases. Therefore, an object in free fall will experience a growing **drag force** as the magnitude of its velocity increases. Eventually, this drag force will be equal in magnitude to the weight of the object, and the object will fall with constant velocity according to Newton's first law. This velocity is called the **terminal velocity**.

Projectile Motion

Projectile motion is motion that follows a path along two dimensions. The velocities and accelerations in the two directions (usually horizontal and vertical) are independent of each other and must, accordingly, be analyzed separately. Objects in projectile motion on Earth, such as cannonballs, baseballs, or bullets, experience the force and acceleration of gravity only in the vertical direction (along the y-axis). This means that $\mathbf{v}_y$ will change at the rate of $\mathbf{g}$ but $\mathbf{v}_x$ will remain constant. In fact, on the MCAT, you will generally be able to assume that the horizontal velocity, $\mathbf{v}_x$, will be constant because we usually assume that air resistance is negligible and, therefore, no measurable force is acting along the x-axis.

Example: A projectile is fired from ground level with an initial velocity of $50\ \frac{\text{m}}{\text{s}}$ and an initial angle of elevation of 37°, as shown below. Assuming $g = -10\ \frac{\text{m}}{\text{s}^2}$, find the following: (Note: sin 37° = 0.6; cos 37° = 0.8)

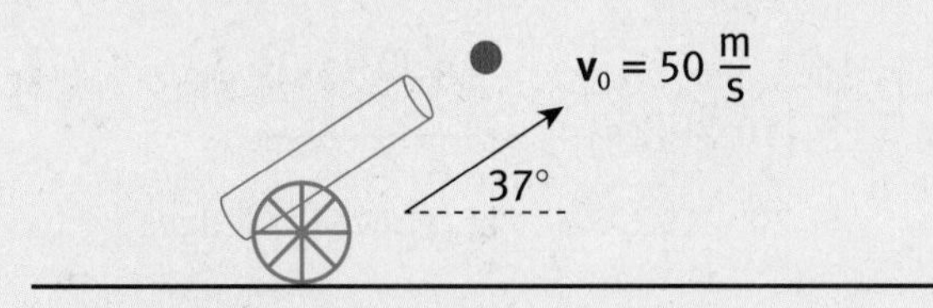

A. The projectile's total time in flight
B. The total horizontal distance traveled

Solution:

A. Let **y** equal the vertical displacement, and *up* be the positive direction. If we are going to use the vertical displacement and acceleration due to gravity (which is also in the y-direction), we must use the y-component of velocity in this part of the problem:

$$v_{0_y} = v_0 \sin 37° = \left(50\ \frac{\text{m}}{\text{s}}\right)(0.6) = 30\ \frac{\text{m}}{\text{s}}$$

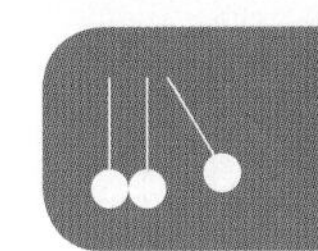

Now we can plug in:

$$y = v_{0_y} t + \frac{a_y t^2}{2}$$

$$0 = \left(30\ \frac{m}{s}\right)(t) + \frac{\left(-10\ \frac{m}{s^2}\right)(t)^2}{2}$$

$$0 = (30)(t) - 5(t)^2$$

$$5(t)^2 = 30\,t$$

$$t^2 = 6\,t$$

$$t = 0\ \text{s or } 6\ \text{s}$$

The height of the ball is zero at 0 seconds (its initial position) and 6 seconds (when it hits the ground again).

B. Now that we know the time, we can find the horizontal distance traveled. Here, we will need to consider only the *x*-component of the velocity:

$$v_{0_x} = v_0 \cos 37° = \left(50\ \frac{m}{s}\right)(0.8) = 40\ \frac{m}{s}$$

Now we can plug in:

$$x = v_x t + \frac{a_x t^2}{2} = \left(40\ \frac{m}{s}\right)(6\ s) + 0 = 240\ m$$

Note: There is only acceleration in the vertical direction due to gravity, $a_x = 0$.

Inclined Planes

Inclined planes are another example of motion in two dimensions. When working with an inclined plane question, it is often best to divide force vectors into components that are parallel and perpendicular to the plane. Most often, gravity must be split into components for these calculations. These components can be defined as:

$$\mathbf{F}_{g,||} = mg \sin \theta$$

$$\mathbf{F}_{g,\perp} = mg \cos \theta$$

Equation 1.22

where $\mathbf{F}_{g,||}$ is the component of gravity parallel to the plane (oriented down the plane), $\mathbf{F}_{g,\perp}$ is the component of gravity perpendicular to the plane (oriented into the plane), *m* is the mass, g is acceleration due to gravity, and θ is the angle of the incline. Otherwise, the same kinematics equations can be used in these problems.

Example: A 5 kg block slides down a frictionless incline at 30°. Find the normal force and acceleration of the block. (Note: $\sin 30° = 0.5$, $\sin 60° = 0.866$, $\cos 30° = 0.866$, $\cos 60° = 0.5$)

Solution: The block in this example has two forces acting on it: the normal force, which is perpendicular to the surface, and gravity, which points straight down:

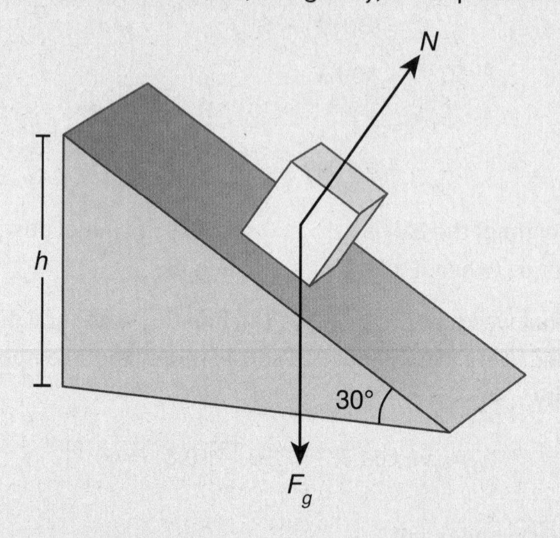

Because gravity is not in the same coordinate system as the normal force, one of the two forces must be split into components. In this case, because we are concerned with magnitude of the normal force (which is perpendicular to the plane) and the acceleration (which is parallel to the plane), we should split the force of gravity into parallel and perpendicular components:

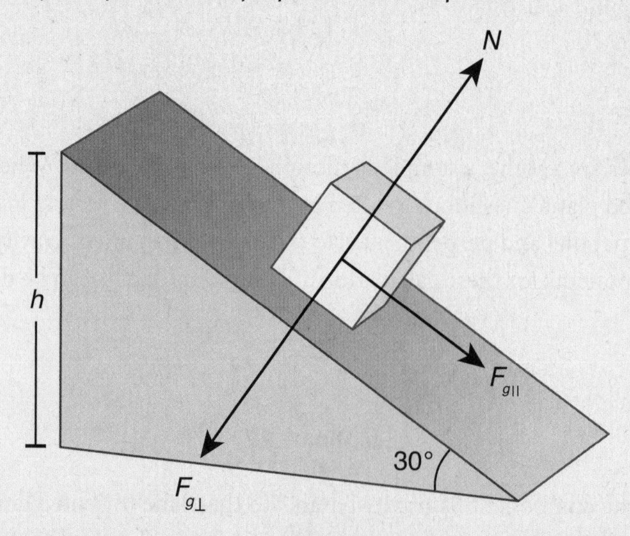

Because there is no acceleration in the perpendicular dimension, the magnitude of the normal force must be equal to that of the perpendicular component of gravity:

$$|\mathrm{N}| = |\mathrm{F}_{\mathrm{g},\perp}| = mg\cos\theta = (5\ \mathrm{kg})\left(9.8\ \frac{\mathrm{m}}{\mathrm{s}^2}\right)\cos 30° \approx 50(0.866)$$
$$= 43.3\ \mathrm{N}\ (\text{actual} = 42.4\ \mathrm{N})$$

Will need to p

The acceleration can then be determined in the parallel direction. Because the only force in this dimension is the parallel component of the force of gravity, it is the net force:

$$F_{\text{net},\parallel} = F_{g,\parallel} = ma_{\parallel}$$
$$mg\sin(\theta) = ma_{\parallel}$$
$$g\sin(\theta) = a_{\parallel} = \left(9.8\ \frac{m}{s^2}\right)\sin(30) = \left(9.8\ \frac{m}{s^2}\right)(0.5) = 4.9\ \frac{m}{s^2}$$

Circular Motion

Circular motion occurs when forces cause an object to move in a circular pathway. Upon completion of one cycle, the displacement of the object is zero. Although the MCAT focuses on uniform circular motion, in which case the speed of the object is constant, recognize that there is also nonuniform circular motion.

In **uniform circular motion**, the instantaneous velocity vector is always tangent to the circular path, as shown in Figure 1.9. What this means is that the object moving in the circular path has a tendency (inertia) to break out of its circular pathway and move in a linear direction along the tangent. It is kept from doing so by a **centripetal force**, which always points radially inward. In all circular motion, we can resolve the forces into radial and tangential components. In uniform circular motion, the tangential force is zero because there is no change in the speed of the object.

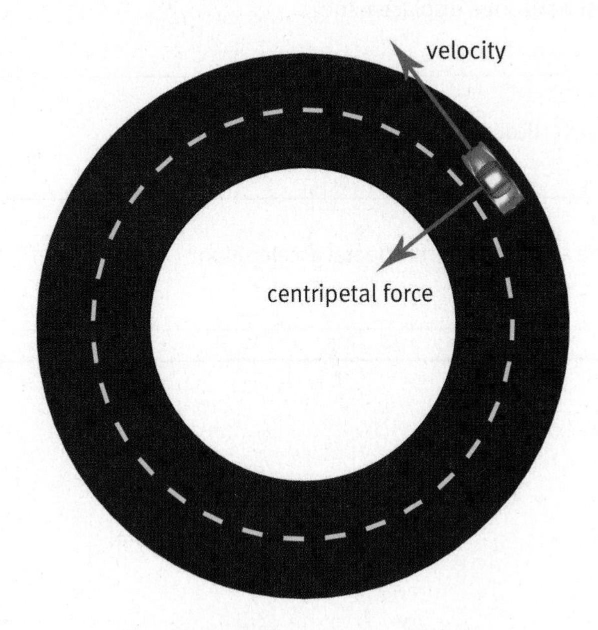

Figure 1.9 Uniform Circular Motion

As a force, the centripetal force generates **centripetal acceleration**. Remember from the discussion of Newton's laws that both force and acceleration are vectors and the acceleration is always in the same direction as the net force. Thus, it is this

acceleration generated by the centripetal force that keeps an object in its circular pathway. When the centripetal force is no longer acting on the object, it will simply exit the circular pathway and assume a path tangential to the circle at that point. The equation that describes circular motion is

$$F_c = \frac{mv^2}{r}$$

Equation 1.23

where F_c is the magnitude of the centripetal force, m is the mass, v is the speed, and r is the radius of the circular path. Note that the centripetal force can be caused by tension, gravity, electrostatic forces, or other forces.

MCAT CONCEPT CHECK 1.6

Before you move on, assess your understanding of the material with these questions.

1. How do the forces acting in free fall and projectile motion differ?

____

2. At what angle of launch is a projectile going to have the greatest horizontal displacement? What angle will result in the greatest vertical displacement, assuming a level surface?
   - Greatest horizontal displacement:

   ____

   - Greatest vertical displacement:

   ____

3. What is the equation for centripetal acceleration?

____

1.7 Mechanical Equilibrium

LEARNING OBJECTIVES

After Chapter 1.7, you will be able to:

- Identify whether an object is in equilibrium
- Calculate torque within a system:

$m_1 = 10$ kg $\quad\quad$ $m_2 = 10$ kg

So far we've been paying attention to kinematics and the special cases of linear and projectile motion. However, many times the MCAT will require you to eliminate acceleration, or otherwise maintain a system in equilibrium. To accomplish this, you must be familiar with analyzing forces, especially with free body diagrams, as well as with the special conditions for translational and rotational equilibrium. The study of forces and torques is called **dynamics**.

Free Body Diagrams

While we all have an intuitive sense of forces (and their effects) in everyday life, students often struggle to represent them diagrammatically. Drawing **free body diagrams** takes some practice but will be a valuable tool on the MCAT. On Test Day, make sure to draw a free body diagram for any problem in which you must perform calculations on forces.

MCAT EXPERTISE

When dealing with dynamics questions, always draw a quick picture of what is happening in the problem; this will keep everything in its proper relative position and help prevent you from making simple mistakes.

Example: Three people are pulling on ropes tied to a tire with forces of 100 N, 125 N, and 125 N as shown below. Find the magnitude and direction of the resultant force. (Note: $\sin 30° = 0.5$, $\cos 30° = 0.866$, $\sin 37° = 0.6$, $\cos 37° = 0.8$)

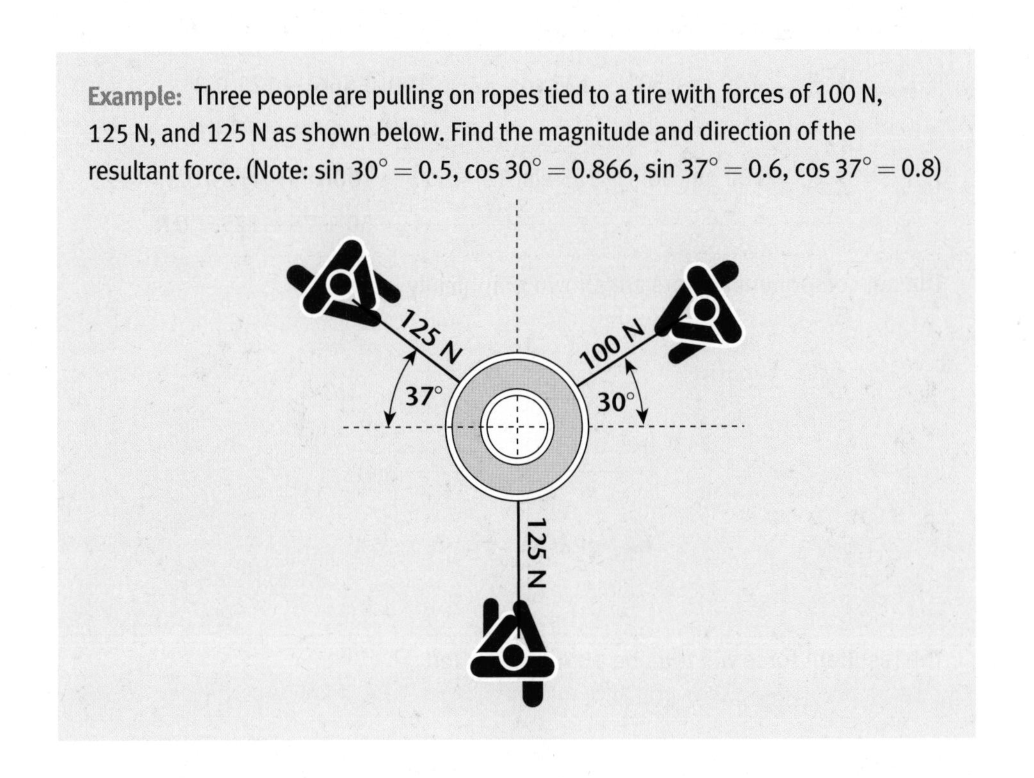

Solution: First, draw a free body diagram that shows the forces acting on the tire. Its purpose is to identify and visualize the acting forces.

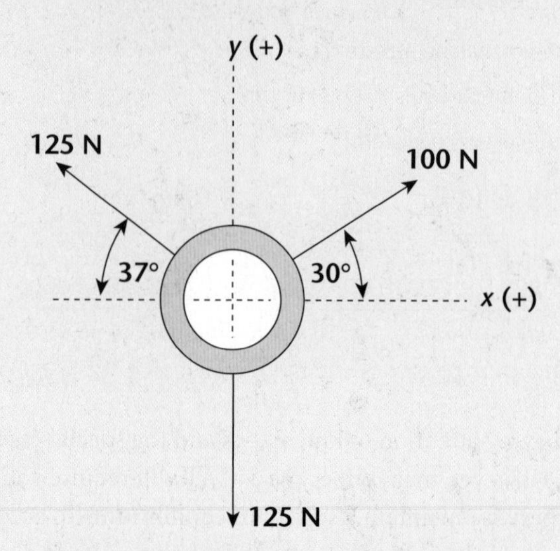

The resultant force is simply the sum of the forces. To find the resultant force vector, we need the sum of the force components, shown below.

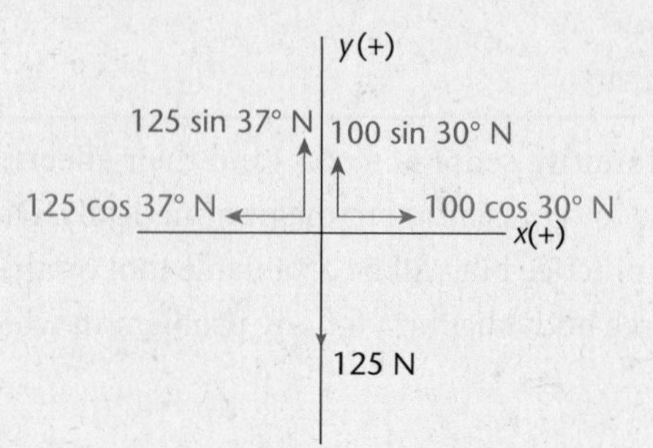

$$F_{net,x} = \sum F_x = 100\ \cos(30°) - 125\ \cos(37) = 100(0.866) - 125(0.8)$$
$$= 86.6 - 100 = -13.4\text{ N}$$

$$F_{net,y} = \sum F_y = 100\ \sin(30°) + 125\ \sin(37) - 125 = 100(0.5) + 125(0.6) - 125$$
$$= 50 + 75 - 125 = 0\text{ N}$$

The net component vectors are shown graphically below.

y(+)

$\mathbf{F}_{net,x} = -13.4$ N

x(+)

$\mathbf{F}_{net,y} = 0$ N

The resultant force will thus be 13.4 N to the left.

Translational Equilibrium

Translational motion occurs when forces cause an object to move without any rotation. The simplest pathways may be linear, such as when a child slides down a snowy hill on a sled, or parabolic, as in the case of a cannonball shot out of a cannon. Any problem regarding translational motion in the *Chemical and Physical Foundations of Biological Systems* section can be solved using free body diagrams and Newton's three laws.

Equilibrium Conditions

Translational equilibrium exists only when the vector sum of all of the forces acting on an object is zero. This is called the **first condition of equilibrium**, and it is merely a reiteration of Newton's first law. Remember that when the resultant force upon an object is zero, the object will not accelerate; that may mean that the object is stationary, but it could just as well mean that the object is moving with a constant nonzero velocity. Thus, an object experiencing translational equilibrium will have a constant velocity: both a constant speed (which could be zero or a nonzero value) and a constant direction.

KEY CONCEPT

If there is no acceleration, then there is no net force on the object. This means that any object with a constant velocity has no net force acting on it. However, just because the net force equals zero does not mean the velocity equals zero.

Example: Two blocks are in static equilibrium, as shown below:

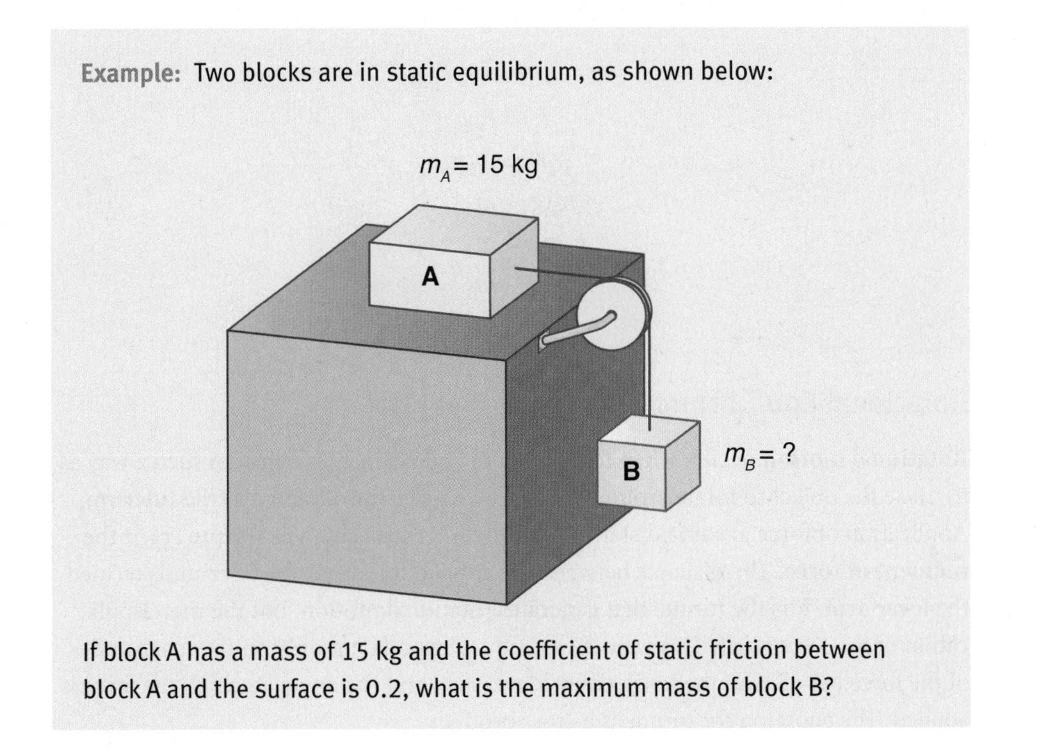

If block A has a mass of 15 kg and the coefficient of static friction between block A and the surface is 0.2, what is the maximum mass of block B?

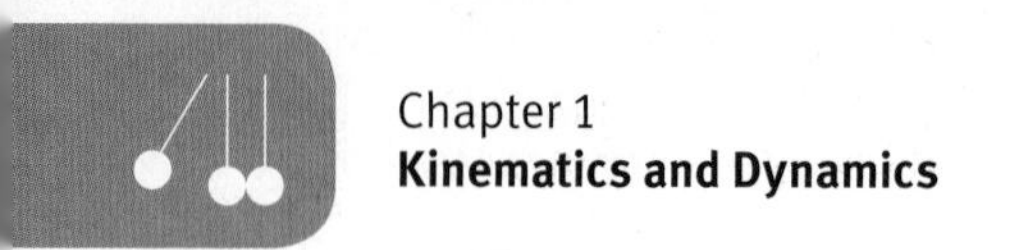

Solution: Start by making a free body diagram of each block:

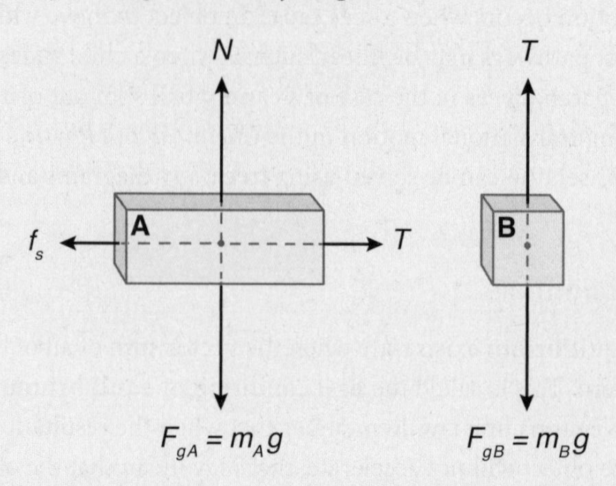

Both blocks have a net force of zero because they are in equilibrium. Therefore, the magnitude of **T** is equal to that of $\mathbf{F}_{g,B}$. Asking for the maximum mass of block B means that the force of static friction is maximized ($f_s = \mu_s N$); further, because block A is in equilibrium, $\mathbf{f}_s$ is equal in magnitude to **T** and $\mathbf{F}_{g,A}$ is equal in magnitude to **N**. Therefore:

$$f_s = T \text{ and } T = F_{g,B}$$
$$f_s = F_{g,B}$$
$$\mu_s N = m_B g$$
$$\mu_s m_A g = m_B g$$
$$\mu_s m_A = m_B$$
$$(0.2)(15 \text{ kg}) = m_B = 3.0 \text{ kg}$$

Rotational Equilibrium

Rotational motion occurs when forces are applied against an object in such a way as to cause the object to rotate around a fixed pivot point, also known as the **fulcrum**. Application of force at some distance from the fulcrum generates **torque** (τ) or the **moment of force**. The distance between the applied force and the fulcrum is termed the **lever arm**. It is the torque that generates rotational motion, not the mere application of the force itself. This is because torque depends not only on the magnitude of the force but also on the length of the lever arm and the angle at which the force is applied. The equation for torque is a cross product:

$$\boldsymbol{\tau} = \mathbf{r} \times \mathbf{F} = rF \sin \theta$$

Equation 1.24

where r is the length of the lever arm, F is the magnitude of the force, and θ is the angle between the lever arm and force vectors.

KEY CONCEPT

Remember that $\sin 90° = 1$. This means that torque is greatest when the force applied is 90 degrees (perpendicular) to the lever arm. Knowing that $\sin 0° = 0$ tells us that there is no torque when the force applied is parallel to the lever arm.

Equilibrium Conditions

Rotational equilibrium exists only when the vector sum of all the torques acting on an object is zero. This is called the **second condition of equilibrium**. Torques that generate clockwise rotation are considered negative, while torques that generate counterclockwise rotation are positive. Thus, in rotational equilibrium, it must be that all of the positive torques exactly cancel out all of the negative torques. Similar to the behavior defined by translational equilibrium, there are two possibilities of motion in the case of rotational equilibrium.

Either the object is not rotating at all (that is, it is stationary), or it is rotating with a constant angular velocity. The MCAT almost always takes rotational equilibrium to mean that the object is not rotating at all.

Example: A seesaw with a mass of 5 kg has one block of mass 10 kg two meters to the left of the fulcrum and another block 0.5 m to the right of the fulcrum, as shown below.

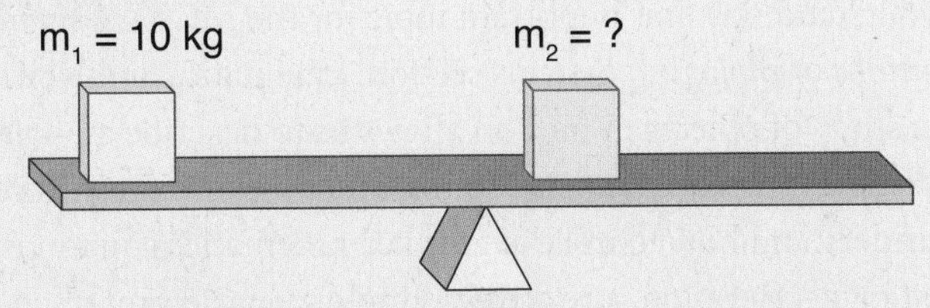

If the seesaw is in equilibrium, find the mass of block 2 and the force exerted by the fulcrum.

Solution: If the seesaw is balanced, this implies rotational equilibrium. Therefore, the positive (counterclockwise) torque exerted by block 1 is equal in magnitude to the negative (clockwise) torque exerted by block 2. Use the fulcrum as the pivot point; because the fulcrum is centered under the seesaw, both the normal force and the weight of the seesaw will be eliminated from the equation because their lever arms are 0.

$$\tau_1 = \tau_2$$
$$r_1 F_{g_1} \sin\theta_1 = r_2 F_{g_2} \sin\theta_2$$
$$r_1(m_1 g)\sin 90^\circ = r_2(m_2 g)\sin 90^\circ$$
$$r_1 m_1 = r_2 m_2 \rightarrow m_2 = \frac{r_1 m_1}{r_2} = \frac{(2\text{ m})(10\text{ kg})}{0.5\text{ m}} = 40\text{ kg}$$

To find the normal force exerted by the fulcrum, consider that the seesaw is not only in rotational equilibrium but also in translational equilibrium. Therefore, the combined weight of the seesaw and blocks (pointing down) is equal in magnitude to the normal force (pointing up):

$$N = F_{g,\text{ seesaw + blocks}}$$
$$= (m_{\text{seesaw}} + m_1 + m_2)\,g$$
$$\approx (5\text{ kg} + 10\text{ kg} + 40\text{ kg})\left(10\ \frac{\text{m}}{\text{s}^2}\right) = 550\text{ N (actual} = 539\text{ N)}$$

MCAT CONCEPT CHECK 1.7

Before you move on, assess your understanding of the material with these questions.

1. Can a moving object be in equilibrium? Why or why not?

____

2. If you have an object three times as heavy as you can lift, how could a lever be used to lift the object? Where would the fulcrum need to be placed?

____

Conclusion

In this chapter, we've equipped you with the math—the language of physics—necessary to understand our first important topic for the MCAT *Chemical and Physical Foundations of Biological Systems* section: kinematics and Newtonian mechanics. This study of objects in motion allows us to describe an object's position, displacement, distance traveled, velocity, speed, and acceleration with respect to time. We now understand how to use the four key kinematics equations when objects experience constant acceleration, a relatively simple scenario presented on Test Day.

We also learned that different kinds of forces act on objects to cause them to move in certain ways. Application of forces may cause objects to accelerate or decelerate according to Newton's second law. If the vector sum of all the forces acting on an object is equal to zero, the forces cancel out, and the object experiences no acceleration, a condition known as translational equilibrium. This is expressed in Newton's first law. Even when objects aren't touching, they can still exert forces between them, as described by Newton's third law. We considered linear motion, projectile motion, inclined planes, and circular motion. We also considered the special conditions of translational and rotational equilibrium.

We hope that you will come to appreciate the relevance that these concepts and principles have for your performance not only on the MCAT but also in medical school, residency training, and your career as a physician. Your careful consideration of the discussion topics in this chapter and your practice with the kinds of problems demonstrated here will earn you many points on Test Day.

You've reviewed the content, now test your knowledge and critical thinking skills by completing a test-like passage set in your online resources!

Concept Summary

Units

- The MCAT will test the **SI units** that are related to the metric system. The SI units include meter, kilogram, second, ampère, mole, kelvin, and candela.

Vectors and Scalars

- **Vectors** are physical quantities that have both magnitude and direction. Vector quantities include displacement, velocity, acceleration, and force, among others.
- **Scalars** are quantities without direction. Scalar quantities may be the magnitude of vectors, like speed, or may be dimensionless, like coefficients of friction.
- Vector addition may be accomplished using the tip-to-tail method or by breaking a vector into its components and using the Pythagorean theorem.
- Vector subtraction is accomplished by changing the direction of the subtracted vector and then following the procedures for vector addition.
- Multiplying a vector by a scalar changes the magnitude and may reverse the direction.
- Multiplying two vectors using the **dot product** results in a scalar quantity. The dot product is the product of the vectors' magnitudes and the cosine of the angle between them.
- Multiplying two vectors using the **cross product** results in a vector quantity. The cross product is the product of the vectors' magnitudes and the sine of the angle between them. The right-hand rule is used to determine the resultant vector's direction.

Displacement and Velocity

- **Displacement** is the vector representation of a change in position. It is path independent and is equivalent to the straight line distance between the start and end locations.
- **Distance** is a scalar quantity that reflects the path traveled.
- **Velocity** is the vector representation of the change in displacement with respect to time.
  - **Average velocity** is the total displacement divided by the total time.
  - **Average speed** is the total distance traveled divided by the total time.
  - **Instantaneous velocity** is the limit of the change in displacement over time as the change in time approaches zero.
  - **Instantaneous speed** is the magnitude of the instantaneous velocity vector.

Forces and Acceleration

- A **force** is any push or pull that has the potential to result in an acceleration.
- **Gravity** is the attractive force between two objects as a result of their masses.
- **Friction** is a force that opposes motion as a function of electrostatic interactions at the surfaces of two objects.
    - **Static friction** exists between two objects that are not in motion relative to each other.
    - **Kinetic friction** exists between two objects that are in motion relative to each other.
    - Whereas static friction can take on many values depending on the magnitude of an applied force, kinetic friction is a constant value.
    - The **coefficient of friction** depends on the two materials in contact. The coefficient of static friction is always higher than the coefficient of kinetic friction.
- Mass and weight are not synonymous.
    - **Mass** is a measure of the inertia of an object—its amount of material.
    - **Weight** is the force experienced by a given mass due to its gravitational attraction to the Earth.
- **Acceleration** is the vector representation of the change in velocity over time. Average or instantaneous acceleration may both be considered, similar to velocity.

Newton's Laws

- **Newton's first law**, or the **law of inertia**, states that an object will remain at rest or move with a constant velocity if there is no net force on the object.
- **Newton's second law** states that any acceleration is the result of the sum of the forces acting on the object and its mass.
- **Newton's third law** states that any two objects interacting with one another experience equal and opposite forces as a result of their interaction.

Motion with Constant Acceleration

- **Linear motion** includes **free fall** and motion in which the velocity and acceleration vectors are parallel or antiparallel.
- **Projectile motion** contains both an *x*- and *y*-component. Assuming negligible air resistance, the only force acting on the object is gravity.
- **Inclined planes** are another example of two-dimensional movement. It is often easiest to consider the dimensions as being parallel and perpendicular to the surface of the plane.
- **Circular motion** is best thought of as having radial and tangential dimensions. In **uniform circular motion**, the only force is the **centripetal force**, pointing radially inward. The instantaneous velocity vector always points tangentially.

Mechanical Equilibrium

- **Free body diagrams** are representations of the forces acting on an object. They are useful for equilibrium and dynamics problems.
- **Translational equilibrium** occurs in the absence of any net forces acting on an object. An object in translational equilibrium has a constant velocity, and may or may not also be in rotational equilibrium.
- **Rotational equilibrium** occurs in the absence of any net **torques** acting on an object. Rotational motion may consider any pivot point, but the center of mass is most common. An object in rotational equilibrium has a constant angular velocity; on the MCAT, the angular velocity is usually zero.

Answers to Concept Checks

1.1

1. Force will obey the same relationship with mass and acceleration, regardless of the unit system. Force is always the product of mass and acceleration, so one pound (lb) must be equal to one $\frac{\text{slug} \cdot \text{ft}}{\text{s}^2}$.
2. ångström < centimeter < inch < foot < mile

1.2

1. Vector addition, unlike vector multiplication, is a commutative function. The resultant of **A** + **B** is the same as **B** + **A**, so there would be no difference between the two resultants.
2. Vector subtraction, like vector multiplication, is not a commutative function. The resultant of **A** – **B** has the same magnitude as **B** – **A**, but is oriented in the opposite direction.
3. A scalar is calculated from two vectors by using the dot product: $\mathbf{A} \cdot \mathbf{B} = |\mathbf{A}|\,|\mathbf{B}| \cos\theta$. A vector is calculated by using the cross product: $\mathbf{A} \times \mathbf{B} = |\mathbf{A}|\,|\mathbf{B}| \sin\theta$.
4. False. This would be true of an addition problem in which both vectors have equal magnitude, but it is never true for vector multiplication. To find the direction of **C**, we must use the right-hand rule. If the thumb points in the direction of **A**, and the fingers point in the direction of **B**, then our palm, **C**, points out of the page.

1.3

1. Instantaneous speed is the magnitude of the instantaneous velocity vector. Average speed and average velocity may be unrelated because speed does not depend on displacement, but is rather the total distance traveled divided by time.
2. True. Displacement considers the most direct route between two points. Distance will always be equal to or larger in magnitude than displacement.
3. Velocity is the rate of change of the displacement of an object. Displacement is a function of velocity acting over a period of time.

1.4

1. The direction of the frictional force always opposes movement. Once the instantaneous velocity vector is known (or net force, in the case of static friction), the frictional force must be in the opposite direction.
2. If there is no net force acting on an object, then that object is not experiencing an acceleration and it has a constant velocity.
3. False. Forces are always reciprocal in nature. When the Earth exerts a force on a person, the person also exerts a force of the same magnitude on the Earth (in the opposite direction). The difference in masses gives the Earth an apparent acceleration of zero.
4. Gravity and frictional forces were discussed in this chapter. Electrostatic, magnetic, elastic, weak nuclear, and strong nuclear forces are other examples of forces.

1.5

1. Any answer which is similar to the following is acceptable:
    1. In the absence of any forces—or when the net force is zero—there will be no change in velocity.
    2. Acceleration results from the sum of the force vectors.
    3. For any two interacting objects, all forces acting on one object have an equal and opposing force acting on the other object.

2. Any answer which is similar to the following is acceptable:
    1. Prior to the collision, the vehicle is travelling at constant velocity, which (according to Newton's first law) indicates that there is no acceleration and no net force.
    2. The collision with the wall creates a sudden deceleration. Because there is acceleration, there must be a net force. The value of the net force can be calculated by multiplying the mass of the car times the acceleration.
    3. When the car collides with the wall, the car exerts a force on the wall. Simultaneously, the wall exerts a force of equal magnitude in the opposite direction on the car.

1.6

1. The only force acting in both free fall and projectile motion is gravity.
2. The product of sine and cosine is maximized when the angle is 45°. Because horizontal displacement relies on both measurements, the maximum horizontal displacement will also be achieved at this angle. Vertical displacement will always be zero as the object returns to the starting point. Objects launched vertically will experience the greatest vertical *distance*.
3. If the equation for centripetal force is $F_c = \frac{mv^2}{r}$ and force is simply mass times acceleration (from Newton's second law), then $a_c = \frac{v^2}{r}$.

1.7

1. A moving object can be in either translational or rotational equilibrium (or both). Translational equilibrium only requires the net force on an object be zero—its velocity is constant. The corresponding condition in rotational equilibrium is that net torque equals zero—its angular velocity is constant.
2. One could place the fulcrum one quarter of the way across the lever, closer to the object. The ratio of the lever arms would then be 3:1, which means that only one-third of the original force is necessary. (Alternatively, the fulcrum could be placed at the end with the object one-third of the way across the lever. This would again result in a 3:1 ratio of lever arms, meaning that only one-third of the original force is necessary.)

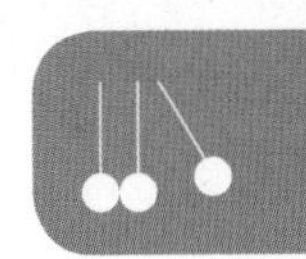

Science Mastery Assessment Explanations

1. **B**

Using the Pythagorean theorem, calculate the magnitude of the man's displacement:

$$x = \sqrt{30^2 + 40^2} = 50 \text{ m}$$

His total distance traveled is equal to $30 + 40 = 70$ m. Therefore, the difference between these two is 20 m.

2. **A**

The average force on the rocket equals its mass times the average acceleration; the average acceleration equals the change in velocity divided by the time over which the change occurs. So, the change in velocity equals the average force times the time divided by the mass:

$$\mathbf{F} = m\mathbf{a} \rightarrow \mathbf{a} = \frac{\mathbf{F}}{m}$$

$$\mathbf{a} = \frac{\Delta \mathbf{v}}{\Delta t} \rightarrow \Delta \mathbf{v} = \mathbf{a}\Delta t = \frac{\mathbf{F}\Delta t}{m} = \frac{(20 \times 10^3 \text{ N})(8 \text{ s})}{1000 \text{ kg}}$$

$$= 160 \ \frac{\text{m}}{\text{s}}$$

(B) represents the new velocity of the rocket, not its change in velocity. **(C)** and **(D)** neglect dividing by the mass of the rocket.

3. **C**

The magnitude of the average acceleration is the change in velocity divided by the time. The velocity changes by $-40 \ \frac{\text{km}}{\text{hr}}$ because the car comes to rest. The time, in hours, is $6 \text{ s} \times \left[\frac{1 \text{ hr}}{3600 \text{ s}}\right] = \frac{1}{600}$ hr. The average acceleration is then

$$\bar{\mathbf{a}} = \frac{\Delta \mathbf{v}}{\Delta t} = \frac{-40 \ \frac{\text{km}}{\text{hr}}}{\frac{1}{600} \text{ hr}} = -24{,}000 \ \frac{\text{km}}{\text{hr}^2}$$

This question asked for the magnitude of this acceleration, which is $24{,}000 \ \frac{\text{km}}{\text{hr}^2}$.

4. **A**

The forces on the elevator are the tension upward and the weight downward, so the net force on the elevator is the difference between the two. For the elevator to accelerate upwards, the tension in the cable will have to be greater than the maximum weight so that there is a net force directed upwards.

5. **C**

The unit of power is the Watt, which breaks down to $\text{kg} \cdot \text{m}^2/\text{s}^2$. Because there is no unit of temperature in this formulation, the ambient temperature is extra information, supporting **(C)** as the correct answer. Note that mass, distance, and time are all used to compute power, eliminating **(A)**, **(B)**, and **(D)**.

6. **A**

The firefighter's acceleration is always directed downward, whereas his velocity starts out horizontal and gradually rotates downwards as his downward velocity increases. Therefore, as time progresses, the angle between his velocity and acceleration decreases, which means that the maximum angle occurs at the instant he jumps.

7. **B**

The static force of friction acts parallel to the plane and is in the opposite direction from the parallel component of gravity in this setup. Because the wagon is in equilibrium, these two forces are equal in magnitude. Remember that gravity is often split into components in inclined plane problems. Rather than splitting into *x*- and *y*-components, however, it is more convenient to split the gravity vector into parallel and perpendicular components. The parallel component of gravity is given by the expression $mg \sin \theta$. Plugging in the values from the question, both the parallel component of gravity and static force of friction must be equal to $(10 \text{ kg})\left(9.8 \ \frac{\text{m}}{\text{s}^2}\right)(\sin 30°) = 49 \text{ N}$.

8. **C**

In SI units, mass is measured in kilograms (kg), velocity in meters per second $\left(\frac{\text{m}}{\text{s}}\right)$, and time in seconds (s). The newton is a derived unit, and is not considered to be a base unit of the SI system. A newton is equal to a $\frac{\text{kg} \cdot \text{m}}{\text{s}^2}$.

9. **D**

Since the equation uses the sine of the angle, this equation computes a cross product. The output of the cross product is another vector quantity. To find the direction of the resultant vector, use the right hand rule. The thumb of the right hand points in the direction of vector *A*, the fingers point in the direction of vector *B*, and the palm reveals the direction of the vector product. In this problem, the right hand rule indicates that the direction of the resulting vector is out of the page, consistent with **(D)**. Note that since the cross product always produces another vector, choices **(A)** and **(B)** can be immediately eliminated.

10. **D**

A vector is characterized by both magnitude and direction. From the given answer choices, all are vectors except for distance. Distance is a scalar because it has only a numerical value and lacks direction.

11. **B**

In order for the seesaw to be balanced, the torque due to the girl (τ_g) must be exactly counteracted by the torque due to her father (τ_f). In other words, the magnitudes of these torques must be equal ($\tau_g = \tau_f$):

$$\begin{aligned} r_g F_g \sin\theta_g &= r_f F_f \sin\theta_f \\ r_g m_g g \sin 90° &= r_f m_f g \sin 90° \\ (2\text{ m})(30\text{ kg}) &= r_f(90\text{ kg}) \\ 0.67\text{ m} &= r_f \end{aligned}$$

Because *r* represents the distance of each person from the fulcrum, the father must sit 67 cm from the fulcrum.

12. **B**

The maximum height depends only on the *y*-component of the velocity, so the horizontal distance can be ignored. The initial *y*-velocity can be calculated by: $v_f = 12 \cdot \sin(30°) =$ 6 m/s. To solve for max height, use the equation: $v_f^2 = v_i^2 + 2ax$, where, at the maximum height, the vertical velocity is equal to 0 m/s. Plugging in values yields the equation $0 = 36 - 20x$. Solving for *x* gives $x = 1.8$ m, matching **(B)**.

13. **C**

We only need to analyze the motion in the vertical dimension to answer this question. If both the rock and ball began with no vertical velocity, they would reach the ground at the same time. However, because the rock begins with an upward component of velocity, it will take time to reach a maximum height before falling back toward the ground. Functionally, the rock's free fall thus starts higher and later than the ball's. The rock will necessarily hit the ground after the ball.

14. **C**

Because the question stem indicates that centrifugal force is reactionary and acts outwardly away from the center of rotation, we can draw the conclusion that it is a reaction to the centripetal force. According to Newton's third law, these forces must have equal magnitude and opposite directions (antiparallel).

15. **B**

The presence of friction does not change the impact of Newton's laws. A net force must still be applied to cause motion. This net force is not necessarily equal to an applied force, as friction and gravity also act on the object; thus, statement I is eliminated. Static friction opposes the movement of stationary objects, and is necessarily greater than the force of kinetic friction; thus, statement II is correct. Statement III is false because the normal force is related to mass, and friction is related to the normal force.

Consult your online resources for additional practice. GO ONLINE

Equations to Remember

(1.1) **Component vectors:** $\mathbf{X} = \mathbf{V} \cos \theta$

$\mathbf{Y} = \mathbf{V} \sin \theta$

(1.2) **Pythagorean theorem:** $\mathbf{X}^2 + \mathbf{Y}^2 = \mathbf{V}^2$ or $\mathbf{V} = \sqrt{\mathbf{X}^2 + \mathbf{Y}^2}$

(1.3) **Determination of direction from component vectors:** $\theta = \tan^{-1} \frac{\mathbf{Y}}{\mathbf{X}}$

(1.4) **Dot product:** $\mathbf{A} \cdot \mathbf{B} = |\mathbf{A}|\,|\mathbf{B}| \cos \theta$

(1.5) **Cross product:** $\mathbf{A} \times \mathbf{B} = |\mathbf{A}|\,|\mathbf{B}| \sin \theta$

(1.6) **Instantaneous velocity:** $\mathbf{v} = \lim_{\Delta t \to 0} \frac{\Delta \mathbf{x}}{\Delta t}$

(1.7) **Average velocity:** $\bar{\mathbf{v}} = \frac{\Delta \mathbf{x}}{\Delta t}$

(1.8) **Universal gravitation equation:** $F_g = \frac{G m_1 m_2}{r^2}$

(1.9) **Static friction:** $0 \leq f_s \leq \mu_s N$

(1.10) **Kinetic friction:** $f_k = \mu_k N$

(1.11) **Force of gravity (weight on Earth):** $\mathbf{F}_g = mg$

(1.12) **Center of mass:**

$$x = \frac{m_1 x_1 + m_2 x_2 + m_3 x_3 + \cdots}{m_1 + m_2 + m_3 + \cdots}$$

$$y = \frac{m_1 y_1 + m_2 y_2 + m_3 y_3 + \cdots}{m_1 + m_2 + m_3 + \cdots}$$

$$z = \frac{m_1 z_1 + m_2 z_2 + m_3 z_3 + \cdots}{m_1 + m_2 + m_3 + \cdots}$$

(1.13) **Average acceleration:** $\bar{\mathbf{a}} = \frac{\Delta \mathbf{v}}{\Delta t}$

(1.14) **Instantaneous acceleration:** $\mathbf{a} = \lim_{\Delta t \to 0} \frac{\Delta \mathbf{v}}{\Delta t}$

(1.15) **Newton's first law:** $\mathbf{F}_{\text{net}} = m\mathbf{a} = 0$

(1.16) **Newton's second law:** $\mathbf{F}_{\text{net}} = m\mathbf{a}$

(1.17) **Newton's third law:** $\mathbf{F}_{\text{AB}} = -\mathbf{F}_{\text{BA}}$

(1.18) **Kinematics (no displacement):** $\mathbf{v} = \mathbf{v}_0 + \mathbf{a}t$

(1.19) **Kinematics (no final velocity):** $\mathbf{x} = \mathbf{v}_0 t + \frac{\mathbf{a}t^2}{2}$

(1.20) **Kinematics (no time):** $\mathbf{v}^2 = \mathbf{v}_0^2 + 2\mathbf{a}\mathbf{x}$

(1.21) **Kinematics (no acceleration):** $\mathbf{x} = \overline{\mathbf{v}}t$

(1.22) **Components of gravity on an inclined plane:** $\mathbf{F}_{g,\parallel} = mg \sin \theta$

$$\mathbf{F}_{g,\perp} = mg \cos \theta$$

(1.23) **Centripetal force:** $F_c = \frac{mv^2}{r}$

(1.24) **Torque:** $\boldsymbol{\tau} = \mathbf{r} \times \mathbf{F} = rF \sin \theta$

Shared Concepts

General Chemistry Chapter 1
Atomic Structure

General Chemistry Chapter 3
Bonding and Chemical Interactions

Physics and Math Chapter 2
Work and Energy

Physics and Math Chapter 4
Fluids

Physics and Math Chapter 5
Electrostatics and Magnetism

Physics and Math Chapter 10
Mathematics

CHAPTER 2

Work and Energy

SCIENCE MASTERY ASSESSMENT

Every pre-med knows this feeling: there is so much content I have to know for the MCAT! How do I know what to do first or what's important?

While the high-yield badges throughout this book will help you identify the most important topics, this Science Mastery Assessment is another tool in your MCAT prep arsenal. This quiz (which can also be taken in your online resources) and the guidance below will help ensure that you are spending the appropriate amount of time on this chapter based on your personal strengths and weaknesses. Don't worry though—skipping something now does not mean you'll never study it. Later on in your prep, as you complete full-length tests, you'll uncover specific pieces of content that you need to review and can come back to these chapters as appropriate.

How to Use This Assessment

If you answer 0–7 questions correctly:

Spend about 1 hour to read this chapter in full and take limited notes throughout. Follow up by reviewing **all** quiz questions to ensure that you now understand how to solve each one.

If you answer 8–11 questions correctly:

Spend 20–40 minutes reviewing the quiz questions. Beginning with the questions you missed, read and take notes on the corresponding subchapters. For questions you answered correctly, ensure your thinking matches that of the explanation and you understand why each choice was correct or incorrect.

If you answer 12–15 questions correctly:

Spend less than 20 minutes reviewing all questions from the quiz. If you missed any, then include a quick read-through of the corresponding subchapters, or even just the relevant content within a subchapter, as part of your question review. For questions you got correct, ensure your thinking matches that of the explanation and review the Concept Summary at the end of the chapter.

1. A weightlifter lifts a 275 kg barbell from the ground to a height of 2.4 m. How much work has he done in lifting the barbell, and how much work is required to hold the weight at that height?
   - **A.** 3234 J and 0 J, respectively
   - **B.** 3234 J and 3234 J, respectively
   - **C.** 6468 J and 0 J, respectively
   - **D.** 6468 J and 6468 J, respectively

2. A tractor pulls a log with a mass of 500 kg along the ground for 100 m. The rope (between the tractor and the log) makes an angle of 30° with the ground and is acted on by a tensile force of 5000 N. How much work does the tractor perform in this scenario? (Note: $\sin 30° = 0.5$, $\cos 30° = 0.866$, $\tan 30° = 0.577$)
   - **A.** 250 kJ
   - **B.** 289 kJ
   - **C.** 433 kJ
   - **D.** 500 kJ

3. A 2000 kg experimental car can accelerate from 0 to 30 $\frac{m}{s}$ in 6 s. What is the average power of the engine needed to achieve this acceleration?
   - **A.** 150 W
   - **B.** 150 kW
   - **C.** 900 W
   - **D.** 900 kW

4. A 40 kg block is resting at a height of 5 m off the ground. If the block is released and falls to the ground, which of the following is closest to its total mechanical energy at a height of 2 m, assuming negligible air resistance?
   - **A.** 0 J
   - **B.** 400 J
   - **C.** 800 J
   - **D.** 2000 J

5. 5 m^3 of a gas are brought from an initial pressure of 1 kPa to a pressure of 3 kPa through an isochoric process. During this process, the work performed by the gas is:
   - **A.** −10 kJ
   - **B.** −10 J
   - **C.** 0 J
   - **D.** +10 kJ

6. In the pulley system shown below, which of the following is closest to the tension force in each rope if the mass of the object is 10 kg and the object is accelerating upwards at 2 $\frac{m}{s^2}$?

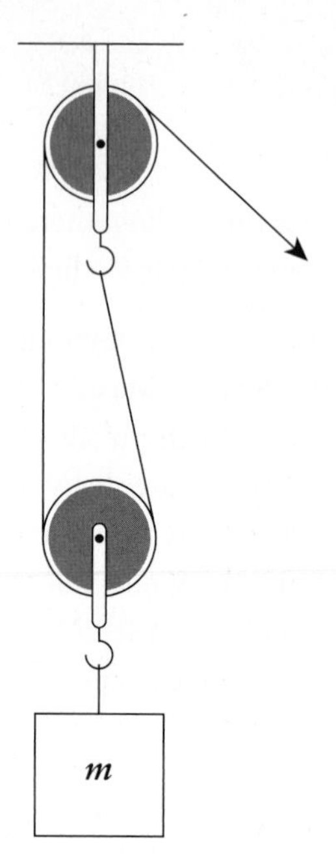

   - **A.** 50 N
   - **B.** 60 N
   - **C.** 100 N
   - **D.** 120 N

7. Which of the following is a conservative force?
   - **A.** Air resistance
   - **B.** Friction
   - **C.** Gravity
   - **D.** Convection

8. During uniform circular motion, which of the following relationships is necessarily true?
   - **A.** No work is done.
   - **B.** The centripetal force does work.
   - **C.** The velocity does work.
   - **D.** Potential energy depends on position of the object around the circle.

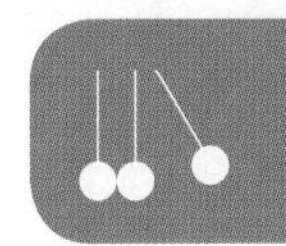

9. Which of the following best characterizes the work–energy theorem?

A. The work done by any force is proportional only to the magnitude of that force.

B. The total work done on any object is equal to the change in kinetic energy for that object.

C. The work done on an object by any force is proportional to the change in kinetic energy for that object.

D. The work done by an applied force on an object is equal to the change in kinetic energy of that object.

10. A massless spring initially compressed by a displacement of two centimeters is now compressed by four centimeters. How has the potential energy of this system changed?

A. The potential energy has not changed.

B. The potential energy has doubled.

C. The potential energy has increased by two joules.

D. The potential energy has quadrupled.

11. Josh, who has a mass of 80 kg, and Sarah, who has a mass of 50 kg, jump off a 20 m tall building and land on a fire net. The net compresses, and they bounce back up at the same time. Which of the following statements is NOT true?

A. Sarah will bounce higher than Josh.

B. For Josh, the change in speed from the start of the jump to contacting the net is 20 $\frac{m}{s}$.

C. Josh will experience a greater force upon impact than Sarah.

D. The energy in this event is converted from potential to kinetic to elastic to kinetic.

12. A parachutist jumps from a plane. Beginning at the point when she reaches terminal velocity (constant velocity during freefall), which of the following is/are true?

I. The jumper is in translational equilibrium.

II. The jumper is not being acted upon by any forces.

III. There is an equal amount of work being done by gravity and air resistance.

A. I only

B. I and III only

C. II and III only

D. I, II, and III

13. Mechanical advantage and efficiency are both ratios. Which of the following is true regarding the quantities used in these ratios?

A. Mechanical advantage compares values of work; efficiency compares values of power.

B. Mechanical advantage compares values of forces; efficiency compares values of work.

C. Mechanical advantage compares values of power; efficiency compares values of energy.

D. Mechanical advantage compares values of work; efficiency compares values of forces.

14. If the gravitational potential energy of an object has doubled in the absence of nonconservative forces, which of the following must be true, assuming the total mechanical energy of the object is constant?

A. The object has been lifted to twice its initial height.

B. The kinetic energy of the object has been halved.

C. The kinetic energy has decreased by the same quantity as the potential energy has increased.

D. The mass of the object has doubled.

15. A consumer is comparing two new cars. Car A exerts 250 horsepower, while Car B exerts 300 horsepower. The consumer is most concerned about the peak velocity that the car can reach. If nonconservative forces can be ignored, which of the following statements is true? (Note: 1 horsepower = 745.7 W)

A. Car A and Car B both have unlimited velocities.

B. Car A will reach its peak velocity more quickly than Car B.

C. Car A will dissipate less energy to the surroundings than Car B.

D. Car A will have a lower peak velocity than Car B.

Answer Key

1. **C** (Ch. 2.2)
2. **C** (Ch. 2.2)
3. **B** (Ch. 2.1)
4. **D** (Ch. 2.3)
5. **C** (Ch. 2.2)
6. **B** (Ch. 2.3)
7. **C** (Ch. 2.2)
8. **A** (Ch. 2.2)
9. **B** (Ch. 2.2)
10. **D** (Ch. 2.1)
11. **A** (Ch. 2.1)
12. **B** (Ch. 2.2)
13. **B** (Ch. 2.3)
14. **C** (Ch. 2.1)
15. **A** (Ch. 2.2)

CHAPTER 2

Work and Energy

In This Chapter

2.1 **Energy**
Kinetic Energy....................56
Potential Energy....................57
Total Mechanical Energy 59
Conservation of Mechanical Energy..............59

2.2 **Work** HY
Force and Displacement............63
Pressure and Volume..............63
Power..........................65
Work–Energy Theorem.............65

2.3 **Mechanical Advantage**
Pulleys..........................70

Concept Summary 76

CHAPTER PROFILE

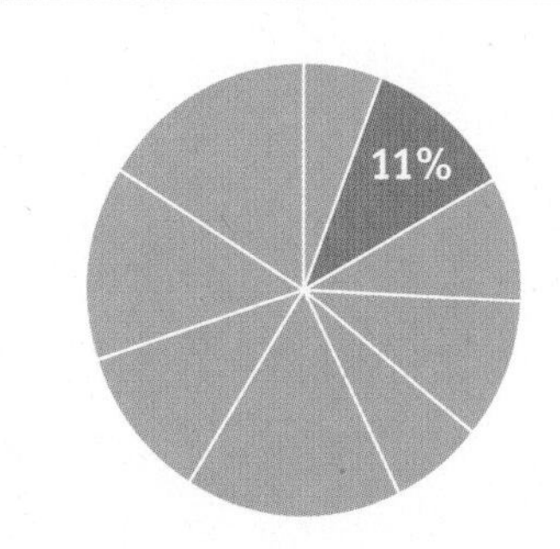

The content in this chapter should be relevant to about 11% of all questions about physics on the MCAT.

This chapter covers material from the following AAMC content category:

4A: Translational motion, forces, work, energy, and equilibrium in living systems

Introduction

The Greek myth of Sisyphus is a tale of unending, pointless work. For eternity, Sisyphus was sentenced to roll a large, heavy rock up a steep hill as penance for his crimes. Just as Sisyphus would nearly reach the top of the hill, the rock would roll back again to the bottom. The cycle continued for eternity: Sisyphus would near the top of the hill, and the boulder—enchanted by Zeus—would roll away, forcing Sisyphus to restart his task.

This is a story of work and mechanical energy transfer. Pushing that boulder up the hill, Sisyphus exerted forces that performed work on the rock, resulting in an increase in the rock's gravitational potential energy. When the rock escaped from his grasp and rolled backwards, its energy changed from gravitational potential energy into kinetic energy. While Sisyphus's punishment was futile work, it serves as a strong model of the exchange of mechanical energy between its two forms: potential and kinetic. Although a number of other forms of energy exist (thermal energy, sound, light, chemical potential energy, and electrical potential energy, to name a few), mechanical energy specifically focuses on objects in motion.

This chapter reviews the fundamental concepts of energy and work. The work–energy theorem is a powerful expression of the relationship between energy and work that is often a simpler approach to kinematics questions on Test Day. Finally, we'll discuss the topic of mechanical advantage, and we'll examine how a pulley or ramp might be helpful in raising heavy objects. We hope to convince you throughout the Kaplan MCAT program that your preparation for Test Day is in no way a Sisyphean task.

2.1 Energy

LEARNING OBJECTIVES

After Chapter 2.1, you will be able to:

- Describe kinetic energy and potential energy
- Compare and contrast conservative and nonconservative forces

Energy refers to a system's ability to do work or—more broadly—to make something happen. This broad definition helps us understand that different forms of energy have the capacity to perform different actions. For example, mechanical energy can cause objects to move or accelerate. An ice cube sitting on the kitchen counter at room temperature will absorb thermal energy through heat transfer and eventually melt into water, undergoing a phase transformation from solid to liquid. Nuclear binding energy can be released during fission reactions to run power plants. Let's turn our attention to the different forms that energy can take. After that, we will discuss the two ways in which energy can be transferred from one system to another.

Kinetic Energy

KEY CONCEPT

Kinetic energy is incredibly important on the MCAT; any time an object has a speed, think about kinetic energy and link its kinetic energy to the related concepts of work and conservation of mechanical energy.

Kinetic energy is the energy of motion. Objects that have mass and that are moving with some speed will have an associated amount of kinetic energy, calculated as follows:

$$K = \frac{1}{2}mv^2$$

Equation 2.1

where K is kinetic energy, m is the mass in kilograms, and v is speed in meters per second. The SI unit for kinetic energy, as with all forms of energy, is the **joule (J)**, which is equal to $\frac{\text{kg}\cdot\text{m}^2}{\text{s}^2}$.

KEY CONCEPT

Kinetic energy is related to speed, not velocity. An object has the same kinetic energy regardless of the direction of its velocity vector.

Recall the falling objects in Chapter 1. Such objects have kinetic energy while they fall. The faster they fall, the more kinetic energy they have. Be mindful of the fact that the MCAT is interested in testing students' comprehension of the relationship between kinetic energy and speed. From the equation, we can see that the kinetic energy is a function of the square of the speed. If the speed doubles, the kinetic energy will quadruple, assuming the mass is constant. Also note that kinetic energy is related to speed—not velocity. An object has the same kinetic energy regardless of the direction of its velocity vector.

Falling objects have kinetic energy, but so do objects that are moving in other ways. For example, the kinetic energy of a fluid flowing at some speed can be measured indirectly as the dynamic pressure, which is one of the terms in Bernoulli's equation—discussed in Chapter 4 of *MCAT Physics and Math Review*. Objects that slide down inclined planes gain kinetic energy as their speeds increase down the ramp.

Example: A 15 kg block, initially at rest, slides down a frictionless incline and comes to the bottom with a speed of 7 $\frac{m}{s}$, as shown below. What is the kinetic energy of the object at the top and bottom of the ramp?

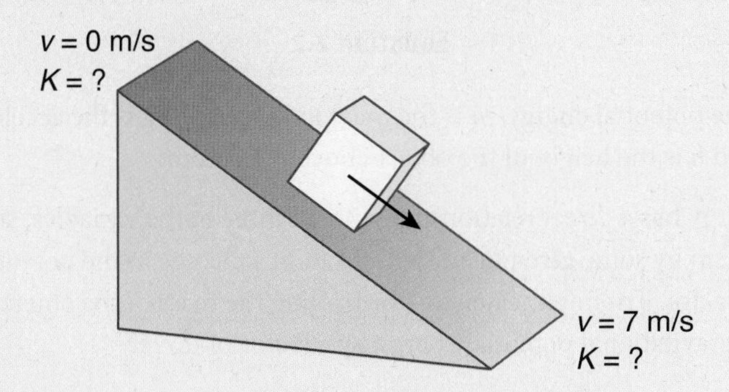

Solution: At the top, $v = 0$, so the kinetic energy is

$$K = \frac{1}{2}mv^2 = \frac{1}{2}(15\text{ kg})\left(0\ \frac{\text{m}}{\text{s}}\right)^2 = 0\text{ J}$$

At the bottom, the kinetic energy is

$$K = \frac{1}{2}mv^2 = \frac{1}{2}(15\text{ kg})\left(7\ \frac{\text{m}}{\text{s}}\right)^2 \approx 15 \times 25 = 375\text{ J}(\text{actual} = 367.5\text{ J})$$

Potential Energy

Potential energy refers to energy that is associated with a given object's position in space or other intrinsic qualities of the system. Potential energy is often said to have the *potential* to do work, and can take named forms. Energy can be stored as chemical potential energy—this is the energy we absorb from the food we eat when we digest and metabolize it. Electrical potential energy, which is discussed in Chapter 5 of *MCAT Physics and Math Review*, is based on the electrostatic attractions between charged particles. In this chapter, we'll examine the types of potential energy that are dissipated as movement: gravitational potential energy and elastic potential energy.

Gravitational Potential Energy

Gravitational potential energy depends on an object's position with respect to some level identified as the **datum** ("ground" or the zero potential energy position). This zero potential energy position is usually chosen for convenience. For example, you may find it convenient to consider the potential energy of the pencil in your hand

with respect to the floor if you are holding the pencil above the floor, or with respect to a desktop if you are holding the pencil over a desk. The equation that we use to calculate gravitational potential energy is

$$U = mgh$$

Equation 2.2

where U is the potential energy, m is the mass in kilograms, g is the acceleration due to gravity, and h is the height of the object above the datum.

Potential energy has a direct relationship with all three of the variables, so changing any one of them by some given factor will result in a change in the potential energy by the same factor. Tripling the height—or tripling the mass of the object—will increase the gravitational potential energy by a factor of three.

MCAT EXPERTISE

The height used in the potential energy equation is relative to whatever the problem states is the ground level. It will often be simply the distance to the ground, but it doesn't need to be. The zero potential energy position may be a ledge, a desktop, or a platform. Just pay attention to the question stem and use the height that is discussed.

Example: An 80 kg diver leaps from a 10 m cliff into the sea, as shown below. Find the diver's potential energy at the top of the cliff and when he is two meters underwater, using sea level as the datum.

Solution: At the top of the cliff:

$$U = mgh = (80\ \text{kg})\left(9.8\ \frac{\text{m}}{\text{s}^2}\right)(10\ \text{m}) \approx 8000\ \text{J}\ (\text{actual} = 7840\ \text{J})$$

When he is two meters underwater:

$$U = mgh = (80\ \text{kg})\left(9.8\ \frac{\text{m}}{\text{s}^2}\right)(-2\ \text{m}) \approx -1600\ \text{J}\ (\text{actual} = -1568\ \text{J})$$

Elastic Potential Energy

Springs and other elastic systems act to store energy. Every spring has a characteristic length at which it is considered relaxed, or in equilibrium. When a spring is stretched or compressed from its **equilibrium length**, the spring has **elastic potential energy**, which can be determined by

$$U = \frac{1}{2}kx^2$$

Equation 2.3

where U is the potential energy, k is the **spring constant** (a measure of the stiffness of the spring), and x is the magnitude of displacement from equilibrium. Note the similarities between this equation and the formula for kinetic energy.

Total Mechanical Energy

The sum of an object's potential and kinetic energies is its **total mechanical energy**. The equation is

$$E = U + K$$

Equation 2.4

where E is total mechanical energy, U is potential energy, and K is kinetic energy. The **first law of thermodynamics** accounts for the **conservation of mechanical energy**, which posits that energy is never created nor destroyed—it is merely transferred from one form to another. This does not mean that the total mechanical energy will necessarily remain constant, though. You'll notice that the total mechanical energy equation accounts for potential and kinetic energies but not for other forms of energy, such as thermal energy that is transferred as a result of friction (heat). If frictional forces are present, some of the mechanical energy will be transformed into thermal energy and will be "lost"—or, more accurately, dissipated from the system and not accounted for by the equation. Note that there is no violation of the first law of thermodynamics, as a full accounting of all the forms of energy (kinetic, potential, thermal, sound, light, and so on) would reveal no net gain or loss of total energy, but merely the transformation of some energy from one form to another.

BRIDGE

The transfer of energy from one form to another is a key feature of bioenergetics and metabolism, discussed in Chapters 9 through 12 of *MCAT Biochemistry Review*. When looking at carbohydrate metabolism, one can see the chemical potential energy in the bonds in glucose being converted into electrical potential energy in the high-energy electrons of NADH and $FADH_2$, which is dissipated along the electron transport chain to generate the proton-motive force (another example of electrical potential energy). This force fuels ATP synthase, trapping the energy in high-energy phosphate bonds in ATP.

Conservation of Mechanical Energy

In the absence of nonconservative forces, such as frictional forces, the sum of the kinetic and potential energies will be constant. **Conservative forces** are those that are path independent and that do not dissipate energy. Conservative forces also have potential energies associated with them. On the MCAT, the two most commonly encountered conservative forces are gravitational and electrostatic. Elastic forces can also be approximated to be conservative in many cases, although the MCAT may include spring problems in which frictional forces are *not* ignored (in actuality, springs heat up as they move back and forth due to the friction between the particles of the spring material). There are two equivalent ways to determine whether a force is conservative, as demonstrated in Figure 2.1.

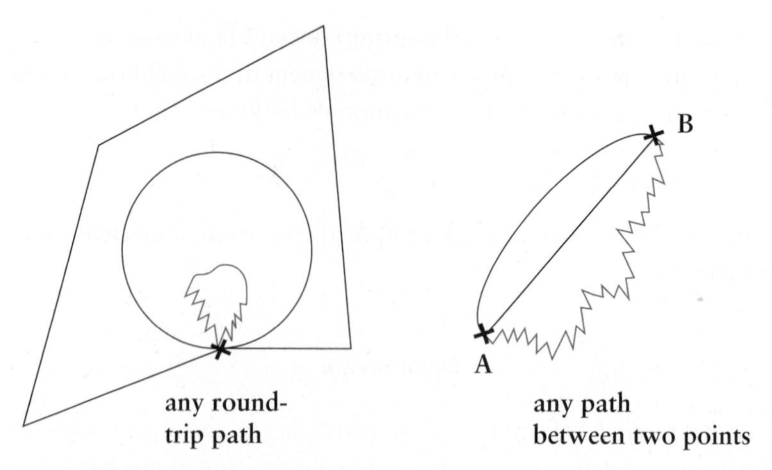

Figure 2.1 Determining if a Force Is Conservative
If the change in energy around any round-trip path is zero—or if the change in energy is equal despite taking any path between two points—then the force is conservative.

One method is to consider the change in energy of a system in which the system is brought back to its original setup. In mechanical terms, this means that an object comes back to its starting position. If the net change in energy is zero regardless of the path taken to get back to the initial position, then the forces acting on the object are conservative. Basically, this means that a system that is experiencing only conservative forces will be "given back" an amount of usable energy equal to the amount that had been "taken away" from it in the course of a closed path. For example, an object that falls through a certain displacement in a vacuum will lose some measurable amount of potential energy but will gain exactly that same amount of potential energy when it is lifted back to its original height, regardless of whether the return pathway is the same as that of the initial descent. Furthermore, at all points during the fall through the vacuum, there will be a perfect conversion of potential energy into kinetic energy, with no energy lost to nonconservative forces such as air resistance. Of course, in real life, nonconservative forces are impossible to avoid.

The other method is to consider the change in energy of a system moving from one setup to another. In mechanical terms, this means an object undergoes a particular displacement. If the energy change is equal regardless of the path taken, then the forces acting on the object are again all conservative.

When the work done by nonconservative forces is zero, or when there are no nonconservative forces acting on the system, the total mechanical energy of the system $(U + K)$ remains constant. The conservation of mechanical energy can be expressed as

$$\Delta E = \Delta U + \Delta K = 0$$

Equation 2.5

where ΔE, ΔU, and ΔK are the changes in total mechanical energy, potential energy, and kinetic energy, respectively.

KEY CONCEPT

Conservative forces (such as gravity and electrostatic forces) conserve mechanical energy. Nonconservative forces (such as friction and air resistance) dissipate mechanical energy as thermal or chemical energy.

When **nonconservative forces**, such as friction, air resistance, or viscous drag (a resistance force created by fluid viscosity) are present, total mechanical energy is not conserved. The equation is

$$W_{\text{nonconservative}} = \Delta E = \Delta U + \Delta K$$

Equation 2.6

where $W_{\text{nonconservative}}$ is the work done by the nonconservative forces only. The work done by the nonconservative forces will be exactly equal to the amount of energy "lost" from the system. In reality, this energy is simply transformed into another form of energy, such as thermal energy, that is not accounted for in the mechanical energy equation. Nonconservative forces, unlike conservative forces, are path dependent. The longer the distance traveled, the larger the amount of energy dissipated.

Example: A baseball of mass 0.25 kg is thrown in the air with an initial speed of 30 $\frac{\text{m}}{\text{s}}$, but because of air resistance, the ball returns to the ground with a speed of 27 $\frac{\text{m}}{\text{s}}$. Find the work done by air resistance.

Solution: Air resistance is a nonconservative force. To solve this problem, the energy equation for a nonconservative system is needed. The work done by air resistance is:

$$W_{\text{nonconservative}} = \Delta E = \Delta U + \Delta K$$

In this case, $\Delta U = 0$ because the initial and final heights are the same. Therefore,

$$\begin{aligned} W_{\text{nonconservative}} &= 0 + \Delta K \\ &= \frac{1}{2} m v_f^2 - \frac{1}{2} m v_i^2 \\ &= \frac{1}{2}(0.25\ \text{kg})\left[\left(27\ \frac{\text{m}}{\text{s}}\right)^2 - \left(30\ \frac{\text{m}}{\text{s}}\right)^2\right] \\ &= \frac{1}{8}(729 - 900) \approx \frac{-160}{8} \\ &= -20\ \text{J}\ (\text{actual} = -21.4\ \text{J}) \end{aligned}$$

The negative sign in the answer indicates that energy is being dissipated from the system.

MCAT CONCEPT CHECK 2.1

Before you move on, assess your understanding of the material with these questions.

1. Define kinetic energy and potential energy.
   - Kinetic energy:

   ____

   - Potential energy:

   ____

2. Compare and contrast conservative and nonconservative forces:

	Conservative Forces	Nonconservative Forces
What happens to total mechanical energy of the system?		
Does the path taken matter?		
What are some examples?		

2.2 Work

High-Yield

MCAT EXPERTISE

The "High-Yield" badge on this section indicates that the content is frequently tested on the MCAT.

LEARNING OBJECTIVES

After Chapter 2.2, you will be able to:

- Recall the units used for work
- Distinguish between work and energy
- Calculate the work done on or by a system

KEY CONCEPT

Work is not energy but a measure of energy transfer. The other form of energy transfer is heat.

Often, the term work is used erroneously to mean another form of energy. After all, the SI unit for work is the joule (J), which is the same SI unit for all forms of energy. Nevertheless, to say that work is just another form of energy is to miss something important: **work** is not actually a form of energy itself, but a process by which energy is transferred from one system to another. In fact, it is one of only two ways in which energy can be transferred. The other transfer of energy is called heat, which we will focus on quite a bit in Chapter 3 of *MCAT Physics and Math Review*.

The transfer of energy by work or heat is the only way by which anything occurs. We are familiar with both processes from everyday life. For example, as discussed in the introduction to this chapter, every time King Sisyphus pushed the rock up the hill, the rock gained kinetic and potential energy. That energy came from Sisyphus's muscles, in which the potential energy contained in the high-energy phosphate bonds of ATP molecules was converted to the mechanical energy of the contracting muscles, which exerted forces against the rock, causing it to accelerate and move up the hill.

On a chemical level, the potential energy in the ATP was harnessed by heat transfer. In fact, at the molecular level, this is no different from work because it involves the movement of molecules, atoms, and electrons, each of which exert forces that do work on other molecules and atoms. Like any transfer of energy, it's not a perfectly efficient process, and some of the energy is lost as thermal energy. Our muscles quite literally warm up when we contract them repeatedly.

Force and Displacement

Energy is transferred through the process of work when something exerts forces on or against something else. This is expressed mathematically by the equation

$$W = \mathbf{F} \cdot \mathbf{d} = Fd \cos \theta$$

Equation 2.7

where W is work, F is the magnitude of the applied force, d is the magnitude of the displacement through which the force is applied, and θ is the angle between the applied force vector and the displacement vector. You'll notice that work is a dot product; as such, it is a function of the cosine of the angle between the vectors. This also means that only forces (or components of forces) parallel or antiparallel to the displacement vector will do work (that is, transfer energy). We've already said that the SI unit for work is the joule. While this suggests that work and energy are the same thing, remember they are not: work is the process by which a quantity of energy is moved from one system to another.

Pressure and Volume

As described above, work is a process of energy transfer. In mechanics, we think of work as application of force through some distance. We will learn in our discussion of fluids in Chapter 4 of *MCAT Physics and Math Review* that pressure can be thought of as an "energy density." In systems of gases, we therefore approach work as a combination of pressure and volume changes. In Chapter 3 of *MCAT Physics and Math Review*, we'll examine how these changes also relate to heat.

For a gas system contained in a cylinder with a movable **piston**, we can analyze the relationship between pressure, volume, and work. When the gas expands, it pushes up against the piston, exerting a force that causes the piston to move up and the volume of the system to increase. When the gas is compressed, the piston pushes down on the gas, exerting a force that decreases the volume of the system. We say that work has been done when the volume of the system has changed due to an applied pressure. Gas expansion and compression processes can be represented in graphical form with volume on the x-axis and pressure on the y-axis. Such graphs, as shown in Figure 2.2, are termed **P–V graphs**.

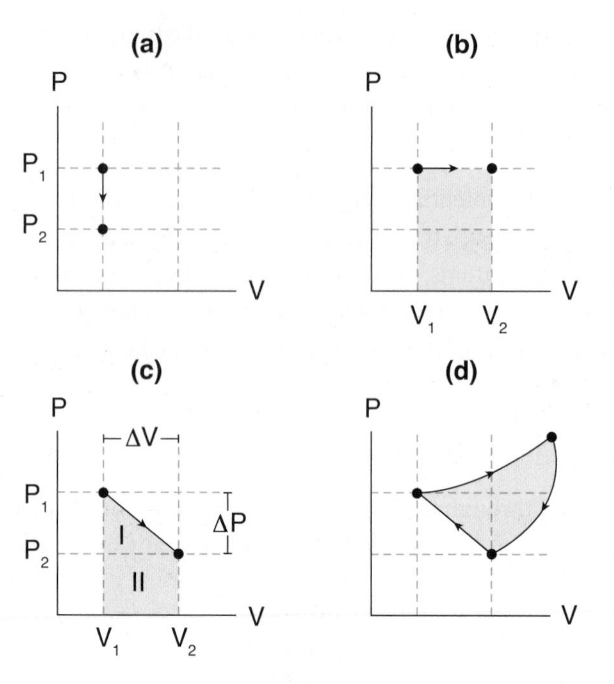

Figure 2.2 Pressure–Volume (P–V) Curves
The work done on or by a system undergoing a thermodynamic process can be determined by finding the area enclosed by the corresponding pressure–volume curve.

KEY CONCEPT

When work is done *by* a system (the gas expands), the work is said to be positive. When work is done *on* a system (the gas compresses), the work is said to be negative. The MCAT will not expect you to calculate the integral of a P–V graph using calculus, but you are expected to be able to calculate the area under a straight-line graph if necessary.

When a gas expands, we say that work was done by the gas and the work is positive; when a gas is compressed, we say that work was done on the gas and the work is negative. There are an infinite number of paths between an initial and final state. Different paths require different amounts of work. You can calculate the work done on or by a system by finding the area under the pressure–volume curve. Note that if volume stays constant as pressure changes (that is, $\Delta V = 0$), then no work is done because there is no area to calculate. This is the case in Figure 2.2a above, and is called an **isovolumetric** or **isochoric process**. On the other hand, if pressure remains constant as volume changes (that is, $\Delta P = 0$), then the area under the curve is a rectangle of length P and width ΔV as shown in Figure 2.2b. For processes in which pressure remains constant (**isobaric processes**), the work can be calculated as

$$W = P\Delta V$$

Equation 2.8

Figure 2.2c shows a process in which neither pressure nor volume is held constant. The total area under the graph (Regions I and II) gives the work done.

Region I is a triangle with base ΔV and height ΔP, so the area is

$$A_{\text{I}} = \frac{1}{2}\,\Delta V \Delta P$$

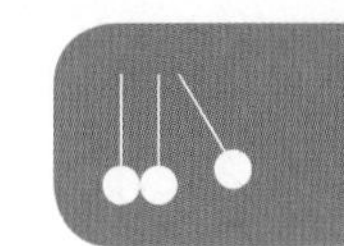

Region II is a rectangle with base ΔV and height P_2, so its area is

$$A_{II} = P_2 \Delta V$$

The work done is the sum of the areas of regions I and II:

$$W = A_I + A_{II}$$

Figure 2.2d shows a closed cycle in which, after certain interchanges of work and heat, the system returns to its initial state. Because work is positive when the gas expands and negative when the gas is compressed, the work done is the area enclosed by the curve. Calculating the work done in this situation would require calculus, but the MCAT does not test calculus-based physics.

Power

Power refers to the rate at which energy is transferred from one system to another. It is calculated with the equation

$$P = \frac{W}{t} = \frac{\Delta E}{t}$$

Equation 2.9

where P is power, W is work (which is equal to ΔE, the change in energy), and t is the time over which the work is done. The SI unit for power is the **watt (W)**, which is equal to $\frac{\text{J}}{\text{s}}$. In Chapter 6 of *MCAT Physics and Math Review*, we will identify additional ways to calculate power in electric circuits. For now, note that many of the devices we use every day—toaster ovens, light bulbs, phones, cars, and so on—are quantified by the rate at which these appliances transform electrical potential energy into other forms, such as thermal, light, sound, and kinetic energy.

BRIDGE

Power is calculated in many different situations, especially those involving circuits, resistors, and capacitors. The equation for electric power is $P = IV$, where P is power, I is current, and V is electrical potential difference (voltage). This equation is discussed in Chapter 6 of *MCAT Physics and Math Review*. Power is always a measure of the rate of energy consumption, transfer, or transformation per unit time.

Work–Energy Theorem

The **work–energy theorem** is a powerful expression of the relationship between work and energy. In its mechanical applications, it offers a direct relationship between the work done by all the forces acting on an object and the change in kinetic energy of that object. The net work done by forces acting on an object will result in an equal change in the object's kinetic energy. In other words,

$$W_{net} = \Delta K = K_f - K_i$$

Equation 2.10

This relationship is important to understand, as it allows one to calculate work without knowing the magnitude of the forces acting on an object or the displacement through which the forces act. If one calculates the change in kinetic energy experienced by an object, then—by definition—the net work done on or by an object is the same. Pressing the brake pedal in your car puts the work–energy theorem into practice. The brake pads exert frictional forces against the rotors, which are attached to the wheels. These frictional forces do work against the wheels, causing them to decelerate and bringing the car to a halt. The net work done by all these forces is equal to the change in kinetic energy of the car.

In more general iterations, the work–energy theorem can be applied to changes in other forms of energy. In fact, the first law of thermodynamics is essentially a reiteration of the work–energy theorem, in which the change in internal energy (ΔU) is equal to the heat transferred into the system (Q) minus the mechanical work done by the system (W).

Example: A lead ball of mass 0.125 kg is thrown straight up in the air with an initial velocity of 30 $\frac{\text{m}}{\text{s}}$. Assuming no air resistance, find the work done by the force of gravity by the time the ball is at its maximum height.

Solution: The answer could be calculated using kinematics and determining the maximum height of the ball ($W = Fd \cos \theta$), but it is simpler to use the work–energy theorem:

$$\begin{aligned} W_{\text{net}} &= K_{\text{f}} - K_{\text{i}} \\ &= 0 - \frac{1}{2} m v_{\text{i}}^2 \\ &= -\left(\frac{1}{2}\right)\left(\frac{1}{8}\ \text{kg}\right)\left(30\ \frac{\text{m}}{\text{s}}\right)^2 = -\frac{900}{16} \approx -\frac{900}{15} \\ &= -60\ \text{J}\ (\text{actual} = -56.25\ \text{J}) \end{aligned}$$

MCAT CONCEPT CHECK 2.2

Before you move on, assess your understanding of the material with these questions.

1. What are the units for work? How are work and energy different?

___

2. Provide three methods for calculating the work done on or by a system.

   •

   ___

   •

   ___

   •

   ___

3. While driving a vehicle at constant velocity on a flat surface, the accelerator must be slightly depressed to overcome resistive forces. How does the amount of work done by the engine (via the accelerator) compare to the amount of work done by resistance?

___

2.3 Mechanical Advantage

LEARNING OBJECTIVES

After Chapter 2.3, you will be able to:

- Explain how work can be lost in a system that is not 100% efficient
- Recall the meaning of mechanical advantage
- Recognize the six simple machines
- Predict the impact of changing effort values on effort distance in a pulley system, assuming work output remains the same:

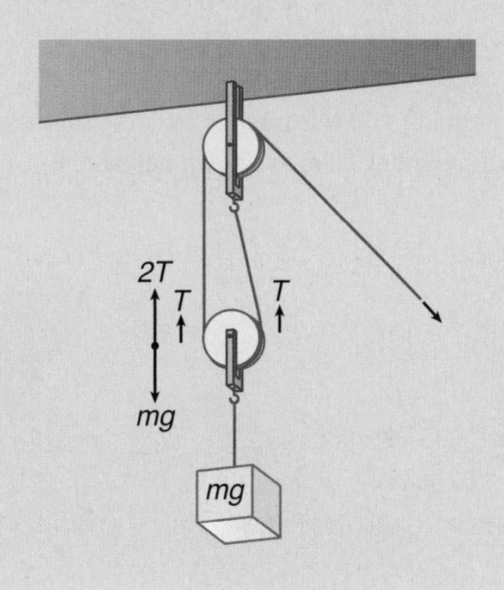

Would it make a difference whether Sisyphus lifted the rock vertically to its final position or rolled it there along an incline? The difference between these two scenarios is mechanical advantage, a measure of the increase in force accomplished by using a tool. Sloping inclines, such as hillsides and ramps, make it easier to lift objects because they distribute the required work over a larger distance, decreasing the required force. For a given quantity of work, any device that allows for work to be accomplished through a smaller applied force is thus said to provide mechanical advantage. In addition to the inclined plane, five other devices are considered the classic **simple machines** which are designed to provide mechanical advantage: wedge (two merged inclined planes), wheel and axle, lever, pulley, and screw (rotating inclined plane). Of these, the inclined plane, lever, and pulley are most frequently tested on the MCAT.

Mechanical advantage is the ratio of magnitudes of the force exerted on an object by a simple machine (F_{out}) to the force actually applied on the simple machine (F_{in}):

$$\text{Mechanical advantage} = \frac{F_{out}}{F_{in}}$$

Equation 2.11

The mechanical advantage, because it is a ratio, is dimensionless.

Reducing the force needed to accomplish a given amount of work does have a cost associated with it; however, the distance through which the smaller force must be applied in order to do the work must be increased. Inclined planes, levers, and pulleys do not magically change the amount of work necessary to move an object from one place to another. Because displacement is pathway independent, the actual distance traveled from the initial to final position does not matter, assuming all forces are conservative. Therefore, applying a lesser force over a greater distance to achieve the same change in position (displacement) accomplishes the same amount of work. We've already considered the dynamics of inclined planes and levers in Chapter 1 of *MCAT Physics and Math Review*. Here, we look at the work associated with inclined planes.

Example: A block weighing 100 N is pushed up a frictionless incline over a distance of 20 m to a height of 10 m as shown below.

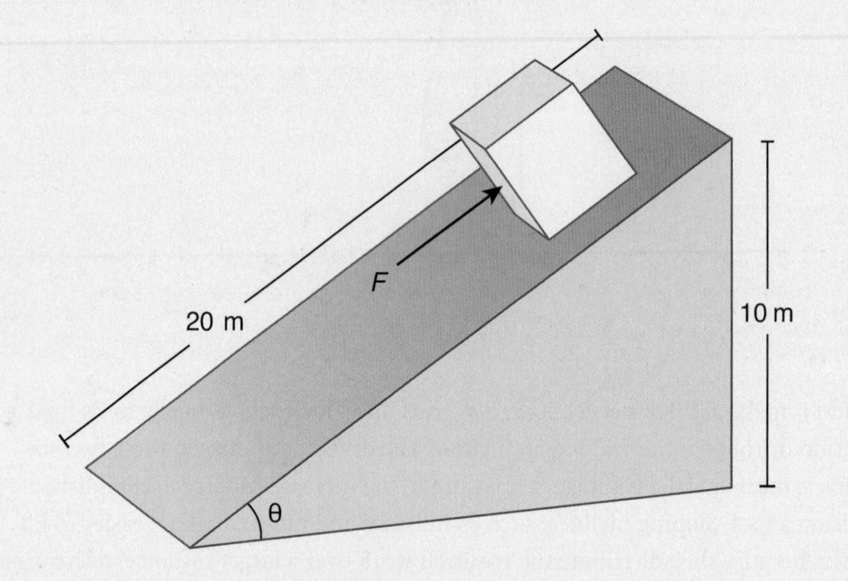

Find:

A. The minimum force required to push the block
B. The work done by the force
C. The force required and the work done by the force if the block were simply lifted vertically 10 m

Solution:

A. To find the minimum force required to push the block, we must draw a free body diagram of the situation:

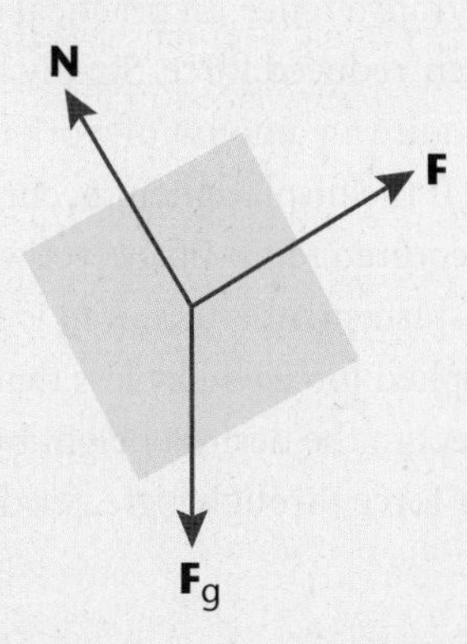

The minimum force needed is a force that will push the block with no acceleration parallel to the surface of the incline. This means the magnitude of the applied force is equal to that of the parallel component of gravity:

$$F = mg \sin \theta$$

mg represents the weight of the object, which is 100 N. Using trigonometry, $\sin \theta$ is ratio of the length of the opposite side to the hypotenuse, which is $\frac{10}{20}$. Therefore,

$$F = (100 \text{ N})\left(\frac{10}{20}\right) = 50 \text{ N}$$

B. The work done by **F** is

$$W = Fd \cos \theta$$

In this case, θ represents the angle between the force and displacement vectors, *not* the angle of the inclined plane. Because the force and displacement vectors are parallel, $\theta = 0$ and $\cos \theta = 1$. Therefore,

$$W = (50 \text{ N})(20 \text{ m})(1) = 1000 \text{ J}$$

C. To raise the block vertically, an upward force equal to the object's weight (100 N) would have to be generated. The work done by the lifting force is

$$W = Fd \cos \theta = (100 \text{ N})(10 \text{ m})(1) = 1000 \text{ J}$$

The same amount of work is required in both cases, but twice the force is needed to raise the block vertically compared with pushing it up the incline.

Pulleys

Pulleys utilize the same paradigm to provide mechanical advantage as the inclined plane: a reduction of necessary force at the cost of increased distance to achieve a given value of work or energy transference. In practical terms, pulleys allow heavy objects to be lifted using a much-reduced force. Simply lifting a heavy object of mass m to a height of h will require an amount of work equal to mgh—its change in gravitational potential energy. If the displacement occurs over a distance equal to the displacement, then the force required to lift the object will equal mg. If, however, the distance through which the displacement is achieved is greater than the displacement (an indirect path), then the applied force will be less than mg. In other words, we've been able to lift this heavy object to the desired height by using a smaller force, but we've had to apply that smaller force through a greater distance in order to lift this heavy object to its final height.

Before examining how pulleys create this mechanical advantage, let's consider first the heavy block in Figure 2.3, suspended from two ropes. Because the block is not accelerating, it is in translational equilibrium, and the force that the block exerts downward (its weight) is cancelled by the sum of the tensions in the two ropes. For a symmetrical system, the tensions in the two ropes are the same and are each equal to half the weight of the block.

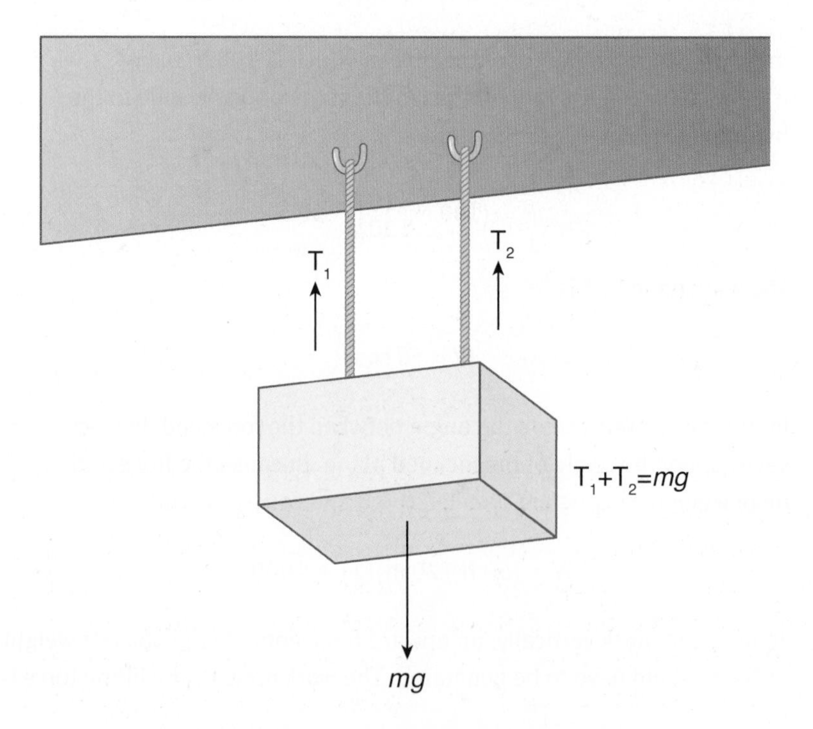

Figure 2.3 Block Suspended by Two Ropes
If the block is in translational equilibrium, the tension in each rope is equal to half the weight of the block.

Now let's imagine the heavy block in Figure 2.4 represents a heavy crate that must be lifted. Assuming that the crate is momentarily being held stationary in midair, we again have a system in translational equilibrium: the weight (the **load**) is balanced by the total tension in the ropes. The tensions in the two vertical ropes are equal to each other; if they were unequal, the pulleys would turn until the tensions were equal on both sides. Therefore, each rope supports one-half of the crate's total weight. By extension, only half the force (**effort**) is required to lift the crate. This decrease in effort is the mechanical advantage provided by the pulley, but as we've already discussed, mechanical advantage comes at the expense of distance. To lift an object to a certain height in the air (the **load distance**), one must pull through a length of rope (the **effort distance**) equal to twice that displacement. If, for example, the crate must be lifted to a shelf 3 meters above the ground, then both sides of the supporting rope must shorten by 3 meters, and the only way to accomplish this is by pulling through 6 meters of rope.

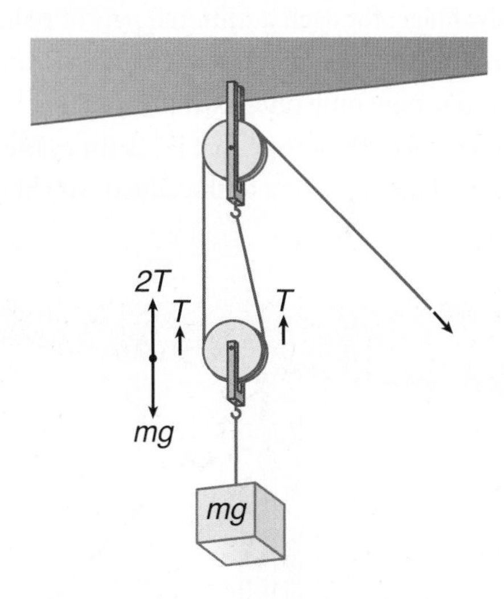

Figure 2.4 Two-Pulley System
The block is suspended from two ropes, each of which bears half of the block's weight.

All simple machines can be approximated as conservative systems if we ignore the (usually) small amount of energy that is lost due to external forces, such as friction. The idealized pulley is massless and frictionless, and under these theoretical conditions, the work put into the system (the exertion of force through a distance of rope) will exactly equal the work that comes out of the system (the displacement of the mass to some height). Real pulleys—and all real machines, for that matter—fail to conform to these idealized conditions and, therefore, do not achieve 100 percent **efficiency** in conserving energy output to input. We can define work input as the

KEY CONCEPT

When considering simple machines, load and effort are both forces. The load determines the necessary output force. From the output force and mechanical advantage, we can determine the necessary input force.

product of effort and effort distance; likewise, we can define work output as the product of load and load distance. Comparing the two as a ratio defines the efficiency of the simple machine:

$$\text{Efficiency} = \frac{W_{\text{out}}}{W_{\text{in}}} = \frac{(\text{load})(\text{load distance})}{(\text{effort})(\text{effort distance})}$$

Equation 2.12

Efficiency is often expressed as a percentage by multiplying the efficiency ratio by 100 percent. The efficiency of a machine gives a measure of the amount of useful work generated by the machine for a given amount of work put into the system. A corollary of this definition is that the percentage of the work put into the system that becomes unusable is due to nonconservative or external forces.

The pulley system in Figure 2.5 illustrates the fact that adding more pulleys further increases mechanical advantage: for each additional pair of pulleys, we can reduce the effort further still. In this case, the load has been divided among six lengths of rope, so the effort required is now only one-sixth the total load. Remember that we would need to pull through a length of rope that is six times the desired displacement, and that efficiency will decrease due to the added weight of each pulley and the additional friction forces.

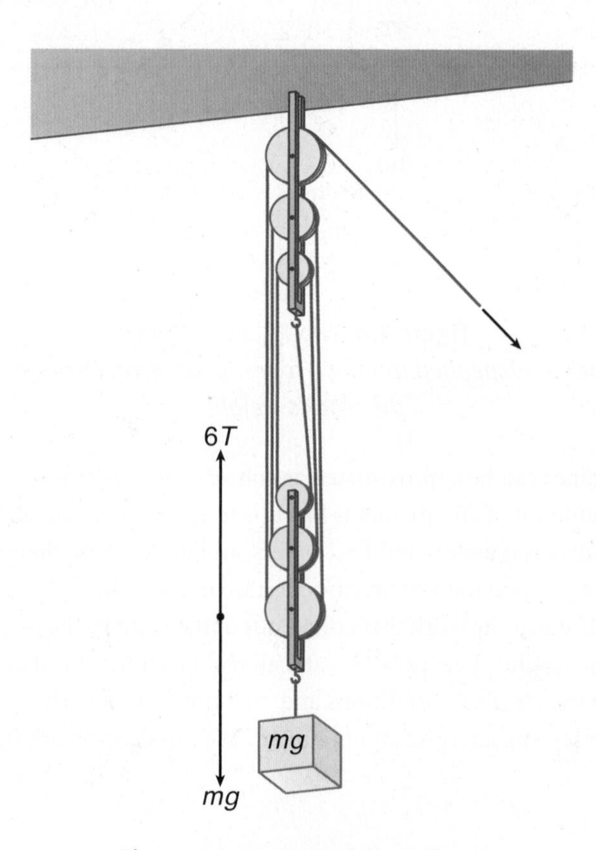

Figure 2.5 System of Six Pulleys
Increasing the number of pulleys decreases the tension in each segment of rope; this leads to an increase in the mechanical advantage of the setup.

Example: The pulley system in Figure 2.5 has an efficiency of 80 percent. A person is lifting a mass of 200 kg with the pulley.

Find:

A. The distance through which the effort must move to raise the load a distance of 4 m
B. The effort required to lift the load
C. The work done by the person lifting the load through a height of 4 m

Solution:

A. For the load to move through a vertical distance of 4 m, all six of the supporting ropes must also shorten 4 m. This may only be accomplished by pulling $6 \times 4 = 24$ m of rope through the setup. Therefore, the effort must move through a distance of 24 m.

B. To calculate the effort required, the equation for efficiency should be used. The load is the weight of the object being lifted and is equal to the mass of the object times the acceleration due to gravity g. The effort distance, calculated in part A, is 24 m.

$$\text{Efficiency} = \frac{(\text{load})(\text{load distance})}{(\text{effort})(\text{effort distance})}$$

$$0.8 = \frac{\left(200\ \text{kg} \times 9.8\ \frac{\text{m}}{\text{s}^2}\right)(4\ \text{m})}{(\text{effort})(24\ \text{m})}$$

$$\text{effort} \approx \frac{2000 \times 4}{0.8 \times 24} \approx \frac{8000}{20} = 400\ \text{N}\ (\text{actual} = 408\ \text{N})$$

C. The work done by the person is

$$W_{\text{in}} = (\text{effort})(\text{effort distance}) = 408\ \text{N} \times 24$$
$$\approx 9600\ \text{J}\ (\text{actual} = 9800\ \text{J})$$

MCAT CONCEPT CHECK 2.3

Before you move on, assess your understanding of the material with these questions.

1. As the length of an inclined plane increases, what happens to the force required to move an object the same displacement?

____

2. As the effort decreases in a pulley system, what happens to the effort distance to maintain the same work output?

____

3. What accounts for the difference between work input and work output in a system that operates at less than 100% efficiency?

____

4. What does it mean for a device to provide mechanical advantage?

____

5. Name the six simple machines:
   - ______________________________
   - ______________________________
   - ______________________________
   - ______________________________
   - ______________________________
   - ______________________________

Conclusion

The conceptualization of energy as the capacity to do something or make something happen is a very broad definition. However, such an all-encompassing definition allows us to understand everything from pushing a rock up a hill to melting an ice cube, from stopping a car at an intersection to harnessing the energy of biomolecules in metabolism, to all the forms of energy transfer. Indeed, energy on its own has little significance without considering the *transfer* of energy, either through work or heat. The work–energy theorem is a powerful expression that will guide our approach to many problems in the *Chemical and Physical Foundations of Biological Systems* section. We also covered the application of energy and work with simple machines, such as levers, inclined planes, and pulleys. These devices assist us in accomplishing work by reducing the forces necessary for displacing objects.

Preparing for the MCAT is hard (mental) work, but you are well on your way to achieving success on Test Day. This *MCAT Physics and Math Review* book (and all the materials provided in your Kaplan program) is part of a set of tools—your simple machines, if you will—that will provide you with the mechanical advantage to ease your efforts toward a higher score.

GO ONLINE

You've reviewed the content, now test your knowledge and critical thinking skills by completing a test-like passage set in your online resources!

Concept Summary

Energy

- **Energy** is the property of a system that enables it to do something or make something happen, including the capacity to do work. The SI units for all forms of energy are joules (J).
- **Kinetic energy** is energy associated with the movement of objects. It depends on mass and speed squared (not velocity).
- **Potential energy** is energy stored within a system. It exists in gravitational, elastic, electrical, and chemical forms.
  - **Gravitational potential energy** is related to the mass of an object and its height above a zero-point, called a **datum**.
  - **Elastic potential energy** is related to the **spring constant** (a measure of the stiffness of a spring) and the degree of stretch or compression of a spring squared.
  - **Electrical potential energy** exists between charged particles.
  - **Chemical potential energy** is the energy stored in the bonds of compounds.
- The total **mechanical energy** of a system is the sum of its kinetic and potential energies.
- **Conservative forces** are path independent and do not dissipate the mechanical energy of a system.
  - If only conservative forces are acting on an object, the total mechanical energy is conserved.
  - Examples of conservative forces include gravity and electrostatic forces. Elastic forces, such as those created by springs, are nearly conservative.
- **Nonconservative forces** are path dependent and cause dissipation of mechanical energy from a system.
  - While total energy is conserved, some mechanical energy is lost as thermal or chemical energy.
  - Examples of nonconservative forces include friction, air resistance, and viscous drag.

Work

- **Work** is a process by which energy is transferred from one system to another.
  - Work may be expressed as the dot product of force and displacement, or the product of force and distance traveled with the cosine of the angle between the two.
  - Work may also be expressed as the area under a **pressure–volume (P–V) curve**.
- **Power** is the rate at which work is done or energy is transferred. The SI unit for power is the watt (W).
- The **work–energy theorem** states that when net work is done on or by a system, the system's kinetic energy will change by the same amount. In more general applications, the work done on or by a system can be transferred to other forms of energy as well.

Mechanical Advantage

- **Mechanical advantage** is the factor by which a simple machine multiplies the input force to accomplish work.
- The six **simple machines** are the inclined plane, wedge, wheel and axle, lever, pulley, and screw. Simple machines provide the benefit of mechanical advantage.
- Mechanical advantage makes it easier to accomplish a given amount of work because the input force necessary to accomplish the work is reduced; the distance through which the reduced input force must be applied, however, is increased by the same factor (assuming 100% efficiency).
- The **load** is the output force of a simple machine, which acts over a given **load distance** to determine the work output of the simple machine. The **effort** is the input force of a simple machine, which acts over a given **effort distance** to determine the work input of the simple machine.
- **Efficiency** is the ratio of the machine's work output to work input when nonconservative forces are taken into account.

Answers to Concept Checks

2.1

1. Kinetic energy is the energy of motion. It is related to the mass of an object, as well as its speed squared. Potential energy is energy associated with a given position or intrinsic property of a system; it is stored in gravitational, electrical, elastic, or chemical forms. Gravitational potential energy is directly related to the mass of the object and its height above a reference point.

2.

	Conservative Forces	Nonconservative Forces
What happens to total mechanical energy of the system?	Remains constant	Decreases (energy is dissipated)
Does the path taken matter?	No	Yes; more energy is dissipated with a longer path
What are some examples?	Gravity Electrostatic forces Elastic forces (approximately conservative)	Friction Air resistance Viscous drag

2.2

1. The unit of work is the joule, which is also the unit for energy. Work and energy are related concepts. By performing work, the energy of a system is changed. Work, along with heat, is a form of energy transfer.
2. Three methods for calculating work discussed in this chapter are:
   1. $W = Fd \cos \theta$ (the dot product of the force and displacement vectors)
   2. $W = P\Delta V$ (the area under a pressure–volume curve)
   3. $W_{net} = \Delta K$ (the work–energy theorem)
3. Begin by thinking about how each form of work is affecting the vehicle. While we could try to work through what may be happening in terms of forces and displacements, this gets very tricky when considering moving engine parts. In this case, it is simpler to think about each work in terms of kinetic energy. The work done by the engine increases the kinetic energy of the car, so it's positive. Conversely, the work done by resistance decreases the kinetic energy of the car, meaning the work done on the car is negative. If the engine does more work than friction, then there is a positive change in kinetic energy. If resistance does more work, then the change is negative. If they do equal amounts of work, then there is no change in kinetic energy. We are given that the vehicle maintains a constant velocity; thus, there is no change in kinetic energy. Therefore, according to the work–energy theorem, the net work must also be zero, and we can infer that the amount of positive work done by the engine must be equal to the amount of negative work done by resistance.

2.3

1. As the length of an inclined plane increases, the amount of force necessary to perform the same amount of work (moving the object the same displacement) decreases.
2. As the effort (required force) decreases in a pulley system, the effort distance increases to generate the same amount of work.
3. The decrease in work output is due to nonconservative or external forces that generate or dissipate energy.
4. When a device provides mechanical advantage, it decreases the input force required to generate a particular output force. Generally, this is accomplished at the expense of increased distance over which the force must act.
5. The six simple machines are: inclined plane, wedge, wheel and axle, lever, pulley, and screw.

Science Mastery Assessment Explanations

1. **C**

Because the weight of the barbell (force acting downward) is $mg = 275\ \text{kg} \times 10\ \frac{\text{m}}{\text{s}^2}$, or about 2750 N, it follows that the weightlifter must exert an equal and opposite force of 2750 N on the barbell. The work done in lifting the barbell is therefore $W = Fd\cos\theta = (2750\ \text{N})(2.4\ \text{m})(\cos 0) \approx 7000\ \text{J}$. Using the same equation, it follows that the work done to hold the barbell in place is $W = Fd\cos\theta = (2750\ \text{N})(0\ \text{m})(\cos\theta) = 0\ \text{J}$. Because the barbell is held in place and there is no displacement, the work done is zero. This is closest to **(C)**.

2. **C**

The work done by the tractor can be calculated from the equation $W = Fd\cos\theta = (5000\ \text{N})(100\ \text{m})(\cos 30°) = (5000)(100)(0.866) \approx 5000 \times 90 = 450{,}000\ \text{J} = 450\ \text{kJ}$. This is closest to **(C)**. Since we estimated by rounding 0.866 up to 0.9, we can expect the actual answer to be less than the calculated answer.

3. **B**

The work done by the engine is equal to the change in kinetic energy of the car: $W = \Delta K = \frac{1}{2}m\left(v_f^2 - v_i^2\right) = \frac{1}{2}(2000\ \text{kg})\left(900\ \frac{\text{m}^2}{\text{s}^2} - 0\right) = 900{,}000\ \text{J}$. The average power therefore is $P = \frac{W}{t} = \frac{900{,}000\ \text{J}}{6\ \text{s}} = 150{,}000\ \text{W} = 150\ \text{kW}$.

4. **D**

Assuming negligible air resistance, conservation of energy states that the total mechanical energy of the block is constant as it falls. At the starting height of 5 m, the block only has potential energy equal to $U = mgh \approx 40\ \text{kg} \times 10\frac{\text{m}}{\text{s}^2} \times 5\ \text{m} = 2000\ \text{J}$. Because the kinetic energy at this point is 0 J, the total mechanical energy is 2000 J at any point during the block's descent.

5. **C**

An isochoric process, by definition, is one in which the gas system undergoes no change in volume. If the gas neither expands nor is compressed, then no work is performed. Remember that work in a thermodynamic system is the area under a P–V curve; if the change in volume is 0, then the area under the curve is also 0.

6. **B**

To calculate the tension force in each rope, first draw a free body diagram:

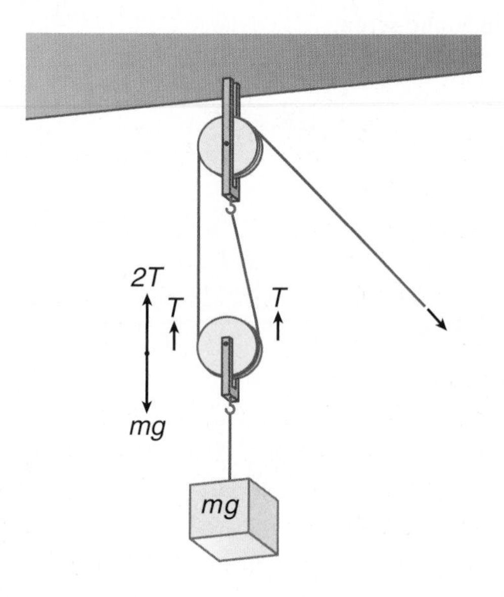

From the force diagram, notice that there are two tension forces pulling the mass up. The net force for this system ($\mathbf{F}_{\text{net}}$) is equal to $2\mathbf{T} - m\mathbf{g}$. Now we can use Newton's second law:

$$\mathbf{F}_{\text{net}} = m\mathbf{a}$$
$$2\mathbf{T} - m\mathbf{g} = m\mathbf{a}$$
$$\mathbf{T} = \frac{m(\mathbf{a} + \mathbf{g})}{2} \approx \frac{(10\ \text{kg})\left(2\ \frac{\text{m}}{\text{s}^2} + 10\ \frac{\text{m}}{\text{s}^2}\right)}{2}$$
$$= 5 \times 12 = 60\ \text{N}$$

7. **C**

Gravity is a conservative force because it is pathway independent and it does not dissipate mechanical energy. Air resistance and friction—**(A)** and **(B)**—are nonconservative forces that dissipate energy thermally. Convection is not a force, but a method of heat transfer, eliminating **(D)**.

8. **A**

In uniform circular motion, the displacement vector and force vector are always perpendicular; therefore, no work is done. Potential energy is constant for an object in uniform circular motion, whether it is the gravitational potential energy of a satellite orbiting the Earth or the electrical potential energy of an electron orbiting the nucleus of an idealized atom. In both cases, potential energy does not change and does not depend on the position of the object around the circle, eliminating **(D)**.

9. **B**

The work–energy theorem relates the total work done on an object by all forces to the change in kinetic energy experienced by the same object. While the work done by a force is indeed proportional to the magnitude of the force, it is also proportional to the displacement of the object, eliminating **(A)**. The change in kinetic energy is equal—not proportional—to the total work done on the object; further, it is the net force, not *any* force, that relates to the work done on an object, eliminating **(C)**. Finally, the change in kinetic energy of the object is equal to the work done by all of the forces acting on the object combined, not just the applied force, which eliminates **(D)**.

10. **D**

Elastic potential energy, like kinetic energy, is related to the square of another variable, as shown by the equation $U = \frac{1}{2}kx^2$. Increasing the displacement by a factor of 2 increases the potential energy by a factor of $2^2 = 4$.

11. **A**

Sarah will not bounce higher than Josh. Assuming that mechanical energy is conserved, Sarah and Josh will start with a given amount of potential energy, which is converted into kinetic energy, then elastic potential energy, then kinetic energy again with no loss of energy from the system, eliminating **(D)**. By this logic, both individuals should return to the same starting height. Josh starts with $U = mgh \approx 80 \text{ kg} \times 10 \frac{\text{m}}{\text{s}^2} \times 20 \text{ m} = 16{,}000 \text{ J}$ of potential energy. At the moment he hits the net, all of this potential energy has been converted into kinetic energy. Therefore,

$$K = \frac{1}{2}mv^2 \rightarrow v = \sqrt{\frac{2K}{m}}$$
$$= \sqrt{\frac{2 \times 16{,}000}{80}} = \sqrt{400} = 20 \frac{\text{m}}{\text{s}}$$

eliminating **(B)**. Josh will experience a greater force upon impact because the net exerts a force proportional to weight; the higher the weight, the larger the force exerted by the net, eliminating **(C)**.

12. **B**

At terminal velocity, the force of gravity and force of air resistance are equal in magnitude, leading to translational equilibrium. Thus, statement I is true. If these forces have the same magnitude and act over the same displacement, then the work performed is the same as well, making statement III true. Even though the net force is equal to zero, there are still forces acting on the parachutist, making statement II false.

13. **B**

Mechanical advantage is a ratio of the output force generated given a particular input force. Efficiency is a ratio of the useful work performed by a system compared to the work performed on the system.

14. **C**

In the absence of nonconservative forces, all changes in potential energy must be met by an equal change in kinetic energy. Note that it is the difference in potential energy that is the same as the difference in kinetic energy, not the proportionality, eliminating **(B)**. Both **(A)** and **(D)** could be true statements but do not necessarily have to be—the object's mass could have been quadrupled while its height was halved.

15. **A**

Horsepower is a unit of power, as evidenced by the name and the conversion factor given in the question stem. Power is a rate of energy expenditure over time. Given unlimited time, both cars are capable of unlimited increases in (kinetic) energy, meaning that they have unlimited maximum velocities. The fact that Car B has a higher power rating means that it will reach any given velocity faster than Car A, eliminating **(B)**. There is not enough information to make any judgments on the efficiency of the cars, eliminating **(C)**. While it may take longer for Car A to reach a given velocity, both cars have unlimited maximum velocities, eliminating **(D)**.

Consult your online resources for additional practice. GO ONLINE

Equations to Remember

(2.1) **Kinetic energy:** $K = \frac{1}{2}mv^2$

(2.2) **Gravitational potential energy:** $U = mgh$

(2.3) **Elastic potential energy:** $U = \frac{1}{2}kx^2$

(2.4) **Total mechanical energy:** $E = U + K$

(2.5) **Conservation of mechanical energy:** $\Delta E = \Delta U + \Delta K = 0$

(2.6) **Work done by nonconservative forces:** $W_{\text{nonconservative}} = \Delta E = \Delta U + \Delta K$

(2.7) **Definition of work (mechanical):** $W = \mathbf{F} \cdot \mathbf{d} = Fd\cos\theta$

(2.8) **Definition of work (isobaric gas–piston system):** $W = P\Delta V$

(2.9) **Definition of power:** $P = \frac{W}{t} = \frac{\Delta E}{t}$

(2.10) **Work–Energy theorem:** $W_{\text{net}} = \Delta K = K_f - K_i$

(2.11) **Mechanical advantage:** Mechanical advantage $= \frac{F_{\text{out}}}{F_{\text{in}}}$

(2.12) **Efficiency:** Efficiency $= \frac{W_{\text{out}}}{W_{\text{in}}} = \frac{(\text{load})(\text{load distance})}{(\text{effort})(\text{effort distance})}$

Shared Concepts

Biochemistry Chapter 6
DNA and Biotechnology

Biochemistry Chapter 9
Carbohydrate Metabolism I

Biochemistry Chapter 12
Bioenergetics and Regulation of Metabolism

General Chemistry Chapter 7
Thermochemistry

Physics and Math Chapter 1
Kinematics and Dynamics

Physics and Math Chapter 3
Thermodynamics

CHAPTER 3

Thermo-dynamics

SCIENCE MASTERY ASSESSMENT

Every pre-med knows this feeling: there is so much content I have to know for the MCAT! How do I know what to do first or what's important?

While the high-yield badges throughout this book will help you identify the most important topics, this Science Mastery Assessment is another tool in your MCAT prep arsenal. This quiz (which can also be taken in your online resources) and the guidance below will help ensure that you are spending the appropriate amount of time on this chapter based on your personal strengths and weaknesses. Don't worry though—skipping something now does not mean you'll never study it. Later on in your prep, as you complete full-length tests, you'll uncover specific pieces of content that you need to review and can come back to these chapters as appropriate.

How to Use This Assessment

If you answer 0–7 questions correctly:

Spend about 1 hour to read this chapter in full and take limited notes throughout. Follow up by reviewing **all** quiz questions to ensure that you now understand how to solve each one.

If you answer 8–11 questions correctly:

Spend 20–40 minutes reviewing the quiz questions. Beginning with the questions you missed, read and take notes on the corresponding subchapters. For questions you answered correctly, ensure your thinking matches that of the explanation and you understand why each choice was correct or incorrect.

If you answer 12–15 questions correctly:

Spend less than 20 minutes reviewing all questions from the quiz. If you missed any, then include a quick read-through of the corresponding subchapters, or even just the relevant content within a subchapter, as part of your question review. For questions you got correct, ensure your thinking matches that of the explanation and review the Concept Summary at the end of the chapter.

1. If an object with an initial temperature of 300 K increases its temperature by 1°C every minute, by how many degrees Fahrenheit will its temperature have increased in 10 minutes?
   - **A.** 6°F
   - **B.** 10°F
   - **C.** 18°F
   - **D.** 30°C

2. Which of the following choices correctly identifies the following three heat transfer processes?
   - **I.** Heat transferred from the Sun to the Earth
   - **II.** A metal spoon heating up when placed in a pot of hot soup
   - **III.** A rising plume of smoke from a fire

   - **A.** I. Radiation; II. Conduction; III. Convection
   - **B.** I. Conduction; II. Radiation; III. Convection
   - **C.** I. Radiation; II. Convection; III. Conduction
   - **D.** I. Convection; II. Conduction; III. Radiation

3. A 20 m steel rod at 10°C is dangling from the edge of a building and is 2.5 cm from the ground. If the rod is heated to 110°C, will the rod touch the ground? (Note: $\alpha = 1.1 \times 10^{-5}\ \text{K}^{-1}$)
   - **A.** Yes, because it expands by 3.2 cm.
   - **B.** Yes, because it expands by 2.6 cm.
   - **C.** No, because it expands by 2.2 cm.
   - **D.** No, because it expands by 1.8 cm.

4. What is the final temperature of a 3 kg wrought iron fireplace tool that is left in front of an electric heater, absorbing heat energy at a rate of 100 W for 10 minutes? Assume the tool is initially at 20°C and that the specific heat of wrought iron is $500\ \frac{\text{J}}{\text{kg}\cdot\text{K}}$.
   - **A.** 40°C
   - **B.** 50°C
   - **C.** 60°C
   - **D.** 70°C

5. How much heat is required to completely melt a pair of gold earrings weighing 500 g, given that their initial temperature is 25°C? (The melting point of gold is 1064°C, its heat of fusion is $6.37 \times 10^4\ \frac{\text{J}}{\text{kg}}$, and its specific heat is $126\ \frac{\text{J}}{\text{kg}\cdot\text{K}}$.)
   - **A.** 15 kJ
   - **B.** 32 kJ
   - **C.** 66 kJ
   - **D.** 97 kJ

6. Given the cycle shown, what is the total work done by the gas during the cycle?

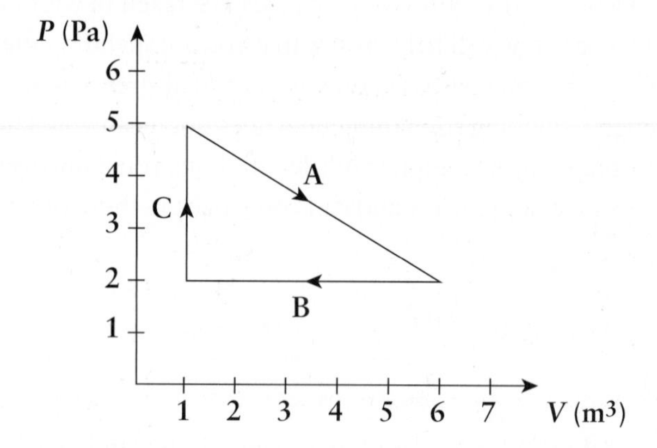

   - **A.** −10 J
   - **B.** 0 J
   - **C.** 7.5 J
   - **D.** 17.5 J

7. In an adiabatic compression process, the internal energy of the gas:
   - **A.** increases because the work done on the gas is negative.
   - **B.** increases because the work done on the gas is positive.
   - **C.** decreases because the work done on the gas is negative.
   - **D.** decreases because the work done on the gas is positive.

8. The entropy of a system can:
   - **A.** never decrease.
   - **B.** decrease when the entropy of the surroundings increases by at least as much.
   - **C.** decrease when the system is isolated and the process is irreversible.
   - **D.** decrease during an adiabatic reversible process.

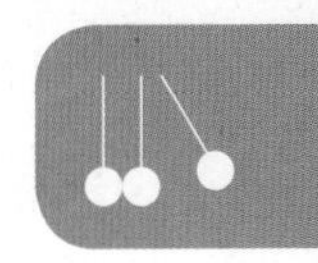

9. A student making a coffee cup calorimeter fails to use a second coffee cup and inadequately seals the lid. What was her initial goal, and what was the result of this mistake?

- **A.** She was trying to create an isolated system but created an open system instead.
- **B.** She was trying to create an isolated system but created a closed system instead.
- **C.** She was trying to create a closed system but created an open system instead.
- **D.** She was trying to create a closed system but created an isolated system instead.

10. A certain substance has a specific heat of $1\ \frac{\text{J}}{\text{mol} \cdot \text{K}}$ and a melting point of 350 K. If one mole of the substance is currently at a temperature of 349 K, how much energy must be added in order to melt it?

- **A.** More than 1 J
- **B.** Exactly 1 J
- **C.** Less than 1 J but more than 0 J
- **D.** Less than 0 J

11. Which of the following is NOT a state function?

- **A.** Internal energy
- **B.** Heat
- **C.** Temperature
- **D.** Entropy

12. The figure shown depicts a thick metal container with two compartments. Compartment A is full of a hot gas, while compartment B is full of a cold gas. What is the primary mode of heat transfer in this system?

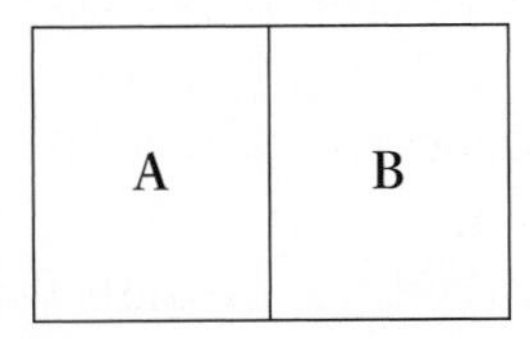

- **A.** Radiation
- **B.** Convection
- **C.** Conduction
- **D.** Enthalpy

13. Substances A and B have the same freezing and boiling points. If solid samples of both substances are heated in the exact same way, substance A boils before substance B. Which of the following would NOT explain this phenomenon?

- **A.** Substance B has a higher specific heat.
- **B.** Substance B has a higher heat of vaporization.
- **C.** Substance B has a higher heat of fusion.
- **D.** Substance B has a higher internal energy.

14. In experiment A, a student mixes ink with water and notices that the two liquids mix evenly. In experiment B, the student mixes oil with water; in this case, the liquids separate into two different layers. The entropy change is:

- **A.** positive in experiment A and negative in experiment B.
- **B.** positive in experiment A and zero in experiment B.
- **C.** negative in experiment A and positive in experiment B.
- **D.** zero in experiment A and negative in experiment B.

15. Which of the following processes is LEAST likely to be accompanied by a change in temperature?

- **A.** The kinetic energy of a gas is increased through a chemical reaction.
- **B.** Energy is transferred to a solid via electromagnetic waves.
- **C.** A boiling liquid is heated on a hot plate.
- **D.** A warm gas is mixed with a cold gas.

Answer Key

1. **C** (Ch. 3.1)
2. **A** (Ch. 3.3)
3. **C** (Ch. 3.1)
4. **C** (Ch. 3.3)
5. **D** (Ch. 3.3)
6. **C** (Ch. 3.3)
7. **B** (Ch. 3.3)
8. **B** (Ch. 3.4)
9. **A** (Ch. 3.2)
10. **A** (Ch. 3.3)
11. **B** (Ch. 3.2)
12. **C** (Ch. 3.3)
13. **D** (Ch. 3.3)
14. **B** (Ch. 3.4)
15. **C** (Ch. 3.3)

CHAPTER 3

Thermodynamics

In This Chapter

3.1 **Zeroth Law of Thermodynamics**
Temperature 92
Thermal Expansion................ 94

3.2 **Systems**
System Types 96
State Functions.................. 97

3.3 **First Law of Thermodynamics** HY
Heat............................ 99
Thermodynamic Processes 103

3.4 **Second Law of Thermodynamics and Entropy**
Energy Dispersion............... 106
Entropy 106

Concept Summary 110

CHAPTER PROFILE

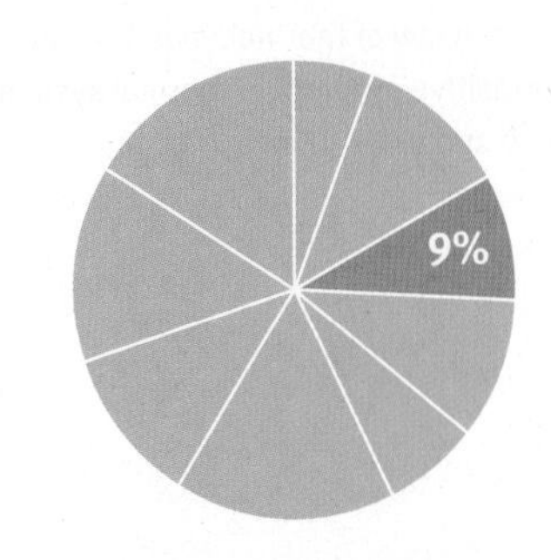

The content in this chapter should be relevant to about 9% of all questions about physics on the MCAT.

This chapter covers material from the following AAMC content categories:

4B: Importance of fluids for the circulation of blood, gas movement, and gas exchange

5E: Principles of chemical thermodynamics and kinetics

Introduction

Thermodynamics is the study of the flow of energy in the universe, as that flow relates to work, heat, entropy, and the different forms of energy. Classical thermodynamics concerns itself only with observations that can be made at the macroscopic level, such as measurements of temperature, pressure, volume, and work. Although the MCAT will test entropy from a thermodynamic rather than probabilistic understanding, we will briefly discuss the statistical model of entropy because it clarifies much of the confusion that arises from a characterization of entropy as a measure of "disorder."

This chapter reviews the laws of thermodynamics with a specific focus on the zeroth, first, and second laws. We will examine how the zeroth law leads to the formulation of temperature scales. Thermal expansion will be discussed as an example of the relationship between thermal energy and physical properties like length, volume, and conductivity. We will then take a moment to examine thermodynamic terminology and functions that are also intimately related to the concepts discussed in Chapter 7 of *MCAT General Chemistry Review*. In the context of the first law—the conservation of energy—we will discuss the relationship between internal energy, heat, and work, and characterize specific heat and heat of transformation. We will also review the various processes by which a system goes from one equilibrium state to another and connect work, discussed in Chapter 2 of *MCAT Physics and Math Review*, with heat. Finally, we will investigate the second law of thermodynamics through the concepts of entropy and its measurement. The third law of thermodynamics is not directly tested on the MCAT, so we will only briefly mention it.

3.1 Zeroth Law of Thermodynamics

LEARNING OBJECTIVES

After Chapter 3.1, you will be able to:

- Explain the zeroth law of thermodynamics
- Predict the relative expansion of an object undergoing a temperature change
- Describe the basis and the significance of the Kelvin scale

KEY CONCEPT

The zeroth law of thermodynamics states the transitive property in thermal systems: If $a = b$ and $b = c$, then $a = c$.

The **zeroth law of thermodynamics** is based on a simple observation: when one object is in thermal equilibrium with another object, say a cup of warm tea and a metal stirring stick, and the second object is in thermal equilibrium with a third object, such as your hand, then the first and third object are also in thermal equilibrium. As such, when brought into thermal contact, no net heat will flow between these objects. Note that thermal contact does not necessarily imply physical contact, as objects can be in thermal contact across space.

Temperature

KEY CONCEPT

Temperature is a physical property of matter related to the average kinetic energy of the particles. Differences in temperature determine the direction of heat transfer.

The formulation of the zeroth law—that no net heat flows between objects in thermal equilibrium, and the corollary that heat flows between two objects *not* in thermal equilibrium—actually arose from studies of temperature. At any given time, all substances have a particular **temperature**. In everyday language, we use the term temperature to describe qualitatively how hot or cold something is, but in thermodynamics, it has a more precise meaning. At the molecular level, temperature is proportional to the average kinetic energy of the particles that make up the substance. At the macroscopic level, it is the difference in temperature between two objects that determines the direction of heat flow. When possible, heat moves spontaneously from materials that have higher temperatures to materials that have lower temperatures. **Heat** itself refers to the transfer of thermal energy from a hotter object with higher temperature (energy) to a colder object with lower temperature (energy). If no net heat flows between two objects in thermal contact, then we can say that their temperatures are equal and they are in **thermal equilibrium**.

Since the 18th century, scales have been developed to quantify the temperature of matter with thermometers. Some of these systems are still in common use, including the **Fahrenheit (°F)**, **Celsius (°C)**, and **Kelvin (K)** scales. Fahrenheit and Celsius are the oldest scales still in common use and are relatively convenient because they are based on the phase changes for water, as shown in Table 3.1. In the Celsius scale, 0° and 100° define the freezing and boiling temperatures of water. In the Fahrenheit scale, these phase change temperatures are defined as 32° and 212°.

	°F	°C	K
Absolute Zero	−460	−273	0
Freezing Point of Water	32	0	273
Boiling Point of Water	212	100	373

Table 3.1 Relevant Points in the Three Major Temperature Scales

✩ Remember that 273.15K = absolute zero/ freezing point of water

The Kelvin scale is most commonly used for scientific measurements and is one of the seven SI base units. It defines as the zero reference point **absolute zero**, the theoretical temperature at which there is no thermal energy, and sets the freezing point of water as 273.15 K. The **third law of thermodynamics** states that the entropy of a perfectly organized crystal at absolute zero is zero. Note that there are no negative temperatures on the Kelvin scale because it starts from absolute zero. Although the Kelvin and Celsius scales have different zero reference points, the size of their units is the same. That is to say, a change of one degree Celsius equals a change of one unit kelvin. Because there are 180 degrees between water's phase changes on the Fahrenheit scale, rather than 100 degrees as on both the Celsius and the Kelvin scales, the size of the Fahrenheit unit is smaller. The following formulas can be used to convert from one scale to another:

$$F = \frac{9}{5}C + 32$$

$$K = C + 273$$

Equation 3.1

where *F*, *C*, and *K* are the temperatures in Fahrenheit, Celsius, and Kelvin, respectively.

MCAT EXPERTISE

The only time Fahrenheit is used routinely on the MCAT is for body temperature, which is 98.6°F or 37°C. In the rare occasion that it is used for a quantitative analysis question, conversions will be given.

Example: If a meteorologist says that the temperature will reach a high of 86°F today, what will be the high temperature in °C and in K?

Solution: To convert from Fahrenheit to Celsius, use:

$$F = \frac{9}{5}C + 32 \rightarrow C = \frac{5}{9}(F - 32)$$

$$C = \frac{5}{9}(86 - 32) = \frac{5}{9}(54) = 30^\circ\text{C}$$

Now convert from Celsius to Kelvin:

$$K = C + 273 = 30 + 273 = 303\text{ K}$$

Thermal Expansion

It has long been noted that some physical properties of matter change when the matter gets hotter or colder. Length, volume, solubility, and even the conductivity of matter change as a function of temperature. The relationship between temperature and a physical property of some matter was used to develop the temperature scales with which we are familiar today. For example, Daniel Fahrenheit developed the temperature scale that bears his name by placing a thermometer filled with mercury into a bath of ice, water, and ammonium chloride. The cold temperature caused the mercury to contract, and when the level in the glass tube stabilized at a lower level, he marked this as the zero reference on the scale. He then placed the same mercury thermometer in a mixture of ice and water (that is, at the freezing point for water). The slightly warmer temperature of this mixture caused the mercury to rise in the glass column, and when it stabilized at this higher level, Fahrenheit assigned it a value of 32°. When he stuck the thermometer under his (or someone else's) tongue, he marked the even higher mercury level as 100° (not 98.6°). The details of how and why Fahrenheit came to choose these numbers (and the history of their adjustment since Fahrenheit first developed the scale) are beyond the scope of this discussion; rather, what is important to note is that a change in some physical property of one kind of matter—in this case, the height of a column of mercury—can be correlated to certain temperature markers, such as the phase changes for water. Once the scale has been set in reference to the decided-upon temperature markers, then the thermometer can be used to take the temperature of any other matter, in accordance with the zeroth law.

REAL WORLD

It is because of thermal expansion that bridges and sidewalks have gaps between their segments; they allow for thermal expansion without damaging integrity.

Because the property of **thermal expansion** was integral to the development of thermometers, let's look a little more closely at this phenomenon. A change in the temperature of most solids results in a change in their length. Rising temperatures cause an increase in length, and falling temperatures cause a decrease in length. The amount of length change is proportional to the original length of the solid and the increase in temperature according to the equation

MNEMONIC

When the temperature of an object changes, its length changes **a lot** ($\alpha L \Delta T$).

$$\Delta L = \alpha L \Delta T$$

Equation 3.2

where ΔL is the change in length, α is the coefficient of linear expansion, L is the original length, and ΔT is the change in temperature. The **coefficient of linear expansion** is a constant that characterizes how a specific material's length changes as the temperature changes. This usually has units of K^{-1}, although it may sometimes be quoted as $°C^{-1}$. This difference is inconsequential because the unit size for the Kelvin and Celsius scales is the same.

Example: A metal rod of length 2 m has a coefficient of linear expansion of $10^{-6}\ K^{-1}$. It is cooled from 1080°C to 80°C. What is the final length of the rod?

Solution: By using the information given in the problem, we can substitute directly into the thermal expansion formula:

$$\Delta L = \alpha L \Delta T = (10^{-6}\ K)(2\ m)(80\ K - 1080\ K) = -2 \times 10^{-3}\ m$$

The negative sign represents a decrease in length. The original length was 2 m; therefore, the final length is $2 - (2 \times 10^{-3}) = 1.998$ m.

Liquids also experience thermal expansion, but the only meaningful parameter of expansion is volume expansion. The formula for volumetric thermal expansion is applicable to both liquids and solids:

$$\Delta V = \beta V \Delta T$$

Equation 3.3

where ΔV is the change in volume, β is the coefficient of volumetric expansion, V is the original volume, and ΔT is the change in temperature. The **coefficient of volumetric expansion** is a constant that characterizes how a specific material's volume changes as the temperature changes. Its value is equal to three times the coefficient of linear expansion for the same material ($\beta = 3\alpha$).

Example: Suppose that a thermometer with 1 mL of mercury is taken from a freezer with a temperature of −25°C and placed near an oven at 275°C. If the coefficient of volume expansion of mercury is $1.8 \times 10^{-4}\ K^{-1}$, by how much will the liquid expand?

Solution: Use the information given to plug into the volumetric expansion formula:

$$\Delta V = \beta V \Delta T = \left(1.8 \times 10^{-4}\ K^{-1}\right)(1\ mL)\left(275^{\circ}C - (-25^{\circ}C)\right)$$
$$= \left(1.8 \times 10^{-4}\right)(300) = 540 \times 10^{-4}\ mL = 0.054\ mL$$

MCAT CONCEPT CHECK 3.1

Before you move on, assess your understanding of the material with these questions.

1. What is the zeroth law of thermodynamics?

2. What is the maximum distance that two objects can be from one another and still adhere to the zeroth law of thermodynamics?

3. How do the initial length of an object and the amount it expands for a given temperature change relate to one another?

4. True or False: The Kelvin scale is the most accurate measurement method for temperature because it is based on absolute zero.

3.2 Systems

LEARNING OBJECTIVES

After Chapter 3.2, you will be able to:

- Distinguish between closed, isolated, and open thermodynamic systems
- Compare and contrast state and process functions
- List common state functions

Physicists and chemists tend to classify the world on the basis of observable phenomena and interactions between objects. Before moving on, we need to become familiar with some of the jargon that these fields have in common—specifically, thermodynamic systems and state functions. Note that the same jargon is discussed in Chapter 7 of *MCAT General Chemistry Review*.

System Types

A **system** is the portion of the universe that we are interested in observing or manipulating. The rest of the universe is considered the **surroundings**.

Isolated Systems

Isolated systems are not capable of exchanging energy or matter with their surroundings. As a result, the total change in internal energy must be zero. Isolated systems are very rare, although they can be approximated. A bomb calorimeter attempts to insulate a reaction from the surroundings to prevent energy transfer, and the entire universe can be considered an isolated system because there are no surroundings.

Closed Systems

Closed systems are capable of exchanging energy, but not matter, with the surroundings. The classic experiments involving gases in vessels with movable pistons are examples of closed systems. For thermodynamic purposes, most of what will be encountered on Test Day will be a closed system or will approximate a closed system.

Open Systems

Open systems can exchange both matter and energy with the environment. In an open system, not only does the matter carry energy, but more energy may be transferred in the form of heat or work. A boiling pot of water, human beings, and uncontained combustion reactions are all examples of open systems.

process functions
= work & heat

State Functions

State functions are thermodynamic properties that are a function of only the current equilibrium state of a system. In other words, state functions are defined by the fact that they are independent of the path taken to get to a particular equilibrium state. The state functions include pressure (P), density (ρ), temperature (T), volume (V), enthalpy (H), internal energy (U), Gibbs free energy (G), and entropy (S). On the other hand, **process functions**, such as work and heat, describe the path taken to get to from one state to another.

MCAT CONCEPT CHECK 3.2

Before you move on, assess your understanding of the material with these questions.

1. Which of the following thermodynamic systems transfer matter? Transfer energy?

System Type	Transfers Matter	Transfers Energy
Closed		
Isolated		
Open		

2. What is the difference between a state function and a process function?

- State function:

____

- Process function:

____

3. List at least five common state functions:

 •

 •

 •

 •

 •

3.3 First Law of Thermodynamics

High-Yield

LEARNING OBJECTIVES

After Chapter 3.3, you will be able to:

- Recall the mathematical relationships between internal energy, work, and heat
- Describe conduction, convection, and radiation
- Draw a graph of the temperature of a solid as it is heated to a gas
- Calculate work for a P–V diagram:

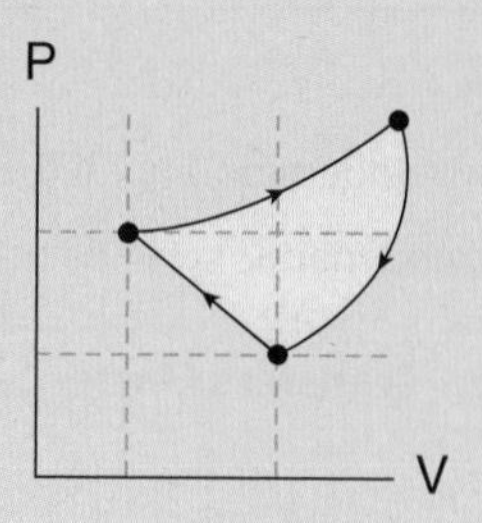

We have already encountered the first law of thermodynamics in our discussion of the conservation of mechanical energy in Chapter 2 of *MCAT Physics and Math Review*. Remember that in the absence of nonconservative forces, the sum of kinetic and potential energies is constant in a system. Now, in our present discussion of thermodynamics, we will look more closely at the relationship between internal energy, heat, and work. Essentially, the **first law of thermodynamics** states that the change in the total internal energy of a system is equal to the amount of energy transferred in the form of **heat** to the system, minus the amount of energy transferred from the

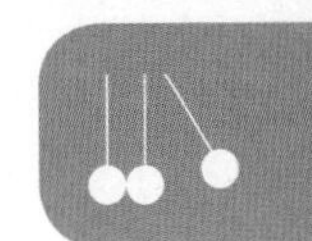

system in the form of **work**. The internal energy of a system can be increased by adding heat, doing work on the system, or some combination of both processes. The change in internal energy is calculated from the equation

$$\Delta U = Q - W$$

Equation 3.4

where ΔU is the change in the system's internal energy, Q is the energy transferred into the system as heat, and W is the work done by the system. To use this equation properly, one must carefully apply the following sign convention shown in Table 3.2.

Variable	Positive Value	Negative Value
Change in Internal Energy (ΔU)	Increasing temperature	Decreasing temperature
Heat (Q)	Heat flows into system	Heat flows out of system
Work (W)	Work is done by the system (expansion)	Work is done on the system (compression)

Table 3.2 Sign Convention for the First Law of Thermodynamics

The first law is really just a particular iteration of the more universal physical law of energy conservation: energy can be neither created nor destroyed; it can only be changed from one form to another. Because the first law accounts for all work and all heat processes impacting the system, the presence of nonconservative forces poses no problem because the energy transfer associated with friction, air resistance, or viscous drag will be accounted for in the first law equation. For example, when a car "burns rubber," all the smoke and noise coming from the back tires is a clear indication that mechanical energy is not being conserved. However, if we include the energy transfers associated with the frictional forces in our consideration of the change in internal energy of the system, then we can confidently say that no energy has been lost at all: there may be a "loss" of energy from the car as a result of the friction, but that precise amount of energy can be "found" elsewhere—as thermal energy in the atoms and molecules of the surrounding road and air.

KEY CONCEPT

The first law of thermodynamics tells us that an increase in the total internal energy of a system is caused by transferring heat into the system or performing work on the system. The total internal energy of a system will decrease when heat is lost from the system or work is performed by the system.

Heat

In Chapter 2 of *MCAT Physics and Math Review*, we defined work as the process by which energy is transferred as the result of force being applied through some distance. We noted that work and heat are the only two processes by which energy can be transferred from one object to another. As discussed earlier in this chapter, the zeroth law of thermodynamics says that objects in thermal contact are in thermal equilibrium when their temperatures are the same. The corollary of this is the **second law of thermodynamics**: objects in thermal contact and not in thermal equilibrium will exchange heat energy such that the object with a higher temperature will give off heat energy to the object with a lower temperature until both objects have the same temperature at thermal equilibrium. **Heat**, then, is defined as the process by which a quantity of energy is transferred between two objects as a result of a difference

KEY CONCEPT

Heat is the process of energy transfer between two objects at different temperatures and will continue until the two objects come into thermal equilibrium at the same temperature.

in temperature. As we will discuss further in our examination of the second law, heat can never spontaneously transfer energy from a cooler object to a warmer one without work being done on the system.

KEY CONCEPT

One calorie (little c) is the amount of heat required to raise 1 g of water one degree Celsius. One Calorie (big C) is the amount of heat required to raise 1 kg of water 1 degree Celsius, equal to 1000 calories.

The SI unit for heat is the joule (J), which should not be surprising because it is based on energy. Heat can also be measured in the units of calorie (cal), nutritional Calorie (Cal), or the British thermal unit (BTU). The nutritional Calorie ("big C") is not the same thing as the calorie ("little c"); one Calorie is equal to 1000 calories or 1 kcal.

The conversion factors between the units of heat are as follows:

$$1\ \text{Cal} \equiv 10^3\ \text{cal} = 4184\ \text{J} = 3.97\ \text{BTU}$$

Heat Transfer

For energy to be transferred between objects, they must be in thermal contact with each other. This does not necessarily mean that the objects are touching. Like force, energy can travel tremendous distances and does not require a medium to pass through. There are three means by which heat can transfer energy: conduction, convection, and radiation.

Conduction is the direct transfer of energy from molecule to molecule through molecular collisions. As this definition would suggest, there must be direct physical contact between the objects. At the atomic level, the particles of the hotter matter transfer some of their kinetic energy to the particles of the cooler matter through collisions between the particles of the two materials. Metals are described as the best heat conductors because metallic bonds contain a density of atoms embedded in a sea of electrons, which facilitate rapid energy transfer. Gases tend to be the poorest heat conductors because there is so much space between individual molecules that energy-transferring collisions occur relatively infrequently. An example of heat transfer through conduction is the heat that is rapidly, and painfully, conducted to your fingers when you touch a hot stove.

Convection is the transfer of heat by the physical motion of a fluid over a material. Because convection involves flow, only liquids and gases can transfer heat by this means. In convection, if the fluid has a higher temperature, it will transfer energy to the material. Most restaurants and some home kitchens have convection ovens, which use fans to circulate hot air inside the oven. Because the heat is being transferred to the food by both convection and radiation rather than only by radiation, convection ovens cook more rapidly than radiation-only ovens. Convection may also be used to wick heat energy away from a hot object. In laboratory experiments, for example, a running cold water bath may be used to rapidly cool a reaction.

Radiation is the transfer of energy by electromagnetic waves. Unlike conduction and convection, radiation can transfer energy through a vacuum. Radiation is the method by which the Sun is able to warm the Earth. Most home kitchens have radiant ovens, which use either electrical coils or gas flames to heat the insulated metal box that forms the body of the oven. The hot metal box then radiates the energy through the open space of the oven, where it is absorbed by whatever food is placed inside.

Specific Heat

When heat energy is added to or removed from a system, the temperature of that system will change in proportion to the amount of heat transfer, unless the system is undergoing a phase change during which the temperature is constant. This relationship between heat and temperature for a substance is called the specific heat. The **specific heat (*c*)** of a substance is defined as the amount of heat energy required to raise one gram of a substance by one degree Celsius or one unit kelvin. For example, the specific heat of liquid water is one calorie per gram per unit kelvin $\left(1\ \frac{\text{cal}}{\text{g}\cdot\text{K}}\right)$. Equivalently, this can be expressed as $4.184\ \frac{\text{J}}{\text{g}\cdot\text{K}}$. The specific heat for a substance changes according to its phase. The MCAT will generally provide specific heat values as necessary, although you are expected to know the specific heat of water in calories. The equation that relates the heat gained or lost by an object and the change in temperature of that object is

$$q = mc\Delta T$$

Equation 3.5

where m is the mass, c is the specific heat of the substance, and ΔT is the change in temperature (in Celsius or kelvins). Because the unit size for the Celsius and Kelvin scales is the same, the change in temperature will be the same for temperatures measured in Celsius or kelvins.

MCAT EXPERTISE

The specific heat of water (in calories) is a constant you are expected to know for Test Day. Its value is $1\ \frac{\text{cal}}{\text{g}\cdot\text{K}}$.

MNEMONIC

The equation for heat transfer, given a specific heat, is the same as the test you're studying for! $q = mc\Delta T$ looks a lot like "*q* equals MCAT."

Heat of Transformation

When a substance is undergoing a phase change, such as from solid to liquid or liquid to gas, the heat that is added or removed from the system does not result in a change in temperature. In other words, phase changes occur at a constant temperature, and the temperature will not begin to change until all of the substance has been converted from one phase into the other. For example, water melts at 0°C. No matter how much heat is added to a mass of ice at 0°C, the temperature of the equilibrated system will not rise until all the ice has been melted into liquid water.

We've determined that adding heat raises the temperature of a system because the particles in that system now have a greater average kinetic energy, and it's true that molecules have greater degrees of freedom of movement in the liquid state than in the solid state (and even more so in the gas state). However, phase changes are related to changes in potential energy, not kinetic energy. The molecules of water in ice, for example, aren't truly frozen in place and unable to move. The molecules rotate, vibrate, and wiggle around. The bonds within each molecule are also free to bend and stretch. Of course, the molecules are held in relatively stable positions by the hydrogen bonds that form between them, but they still have a fairly significant amount of kinetic energy. The potential energy, however, is quite low because of the stability provided by the relative closeness of one molecule to another and by the hydrogen bonds.

Now, think about what happens when one adds heat to ice that is at 0°C. The heat energy causes the water molecules to begin to move away from each other by breaking free of the hydrogen bonds between them. Because the water molecules are being held less rigidly in place, they now have greater degrees of freedom of movement and their average potential energy increases. In statistical mechanics, one would say that this increased freedom of movement permits a greater number of **microstates** for the water molecules. For example, instead of only being able to move up and down or sway side-to-side, a water molecule may now have more freedom of movement and be able to rock forward and back. In gaining additional directions and forms of motion, however, the amount of up-and-down or side-to-side motion must decrease, thus keeping the average kinetic energy of liquid water at 0°C the same as solid water at 0°C. In summary, while liquid water may have a greater number of microstates due to increased freedom of movement, its average kinetic energy is the same as solid water at the same temperature.

When heat energy is added to or removed from a system that is experiencing a phase change, the amount of heat that is added or removed cannot be calculated with the equation $q = mc\Delta T$ because there is no temperature change during a phase change. Instead, the following equation is used:

$$q = mL$$

Equation 3.6

where q is the amount of heat gained or lost from the material, m is the mass of the substance, and L is the **heat of transformation** or **latent heat** of the substance.

The phase change from liquid to solid (**freezing** or **solidification**) or solid to liquid (**melting** or **fusion**) occurs at the **melting point**. The corresponding heat of transformation is called the **heat of fusion**. The phase change from liquid to gas (**boiling**, **evaporation**, or **vaporization**) or gas to liquid (**condensation**) occurs at the **boiling point**. The corresponding heat of transformation is called the **heat of vaporization**. The relevant heats of fusion and vaporization will be provided on Test Day.

Example: Silver has a melting point of 962°C and a heat of fusion of approximately $1.05 \times 10^5 \frac{\text{J}}{\text{kg}}$. The specific heat of silver is $233 \frac{\text{J}}{\text{kg} \cdot \text{K}}$. Approximately how much heat is required to completely melt a 1 kg silver chain with an initial temperature of 20°C?

Solution: Before melting the chain, we must first heat the chain to the melting point. To figure out how much heat is required, we use this formula:

$$\begin{aligned} q &= mc\Delta T \\ &= (1\text{ kg})\left(233 \frac{\text{J}}{\text{kg} \cdot \text{K}}\right)(942\text{ K}) \approx 200 \times 975 \\ &= 1.95 \times 10^5\text{ J}\left(\text{actual} = 2.19 \times 10^5\text{ J}\right) \end{aligned}$$

BRIDGE

It is important to know the common terms used for phase changes:

- Solid to liquid: fusion or melting
- Liquid to solid: freezing or solidification
- Liquid to gas: boiling, evaporation, or vaporization
- Gas to liquid: condensation
- Solid to gas: sublimation
- Gas to solid: deposition

These phase changes are discussed in Chapter 7 of *MCAT General Chemistry Review*.

REAL WORLD

It is because of the heat of transformation that sweating is such an efficient cooling mechanism. When sweat evaporates, the heat of vaporization is absorbed from the surface of the body. This is also why a hot day seems so much more intense when it is very humid out. The sweat is less likely to evaporate due to the dampness of the environment, so less heat can be lost from the surface of the skin through sweating.

This tells us we have to add 219 kJ of heat to the chain just to get its temperature to the melting point. The chain is still in the solid phase. To melt it, we must continue to add heat in accordance with this formula:

$$q = mL$$

$$= (1\ \text{kg})\left(1.05\times 10^{5}\ \frac{\text{J}}{\text{kg}}\right) = 1.05\times 10^{5}\ \text{J}$$

The total heat needed to melt the solid silver chain is 219 kJ + 105 kJ = 324 kJ.

Thermodynamic Processes

In the last chapter, we gave significant consideration to work as a change of energy in a system, both as a function of force and displacement and as a function of volume and pressure. We will briefly review the latter and its relationship to heat transfer within a system. Keep in mind that work accomplished by a change in displacement is not likely to be motivated by heat transfer, and any heat transfer that does occur is most likely a result of friction dissipating mechanical energy from the system.

During any thermodynamic process, a system goes from some initial equilibrium state with an initial pressure, temperature, and volume to some other equilibrium state, which may be at a different final pressure, temperature, or volume. These thermodynamic processes can be represented in graphical form with volume on the *x*-axis and pressure (or temperature) on the *y*-axis.

Process	First law of thermodynamics reduces to:
Isothermal ($\Delta U = 0$)	$Q = W$
Adiabatic ($Q = 0$)	$\Delta U = -W$
Isobaric (constant pressure)	(Multiple possible forms)
Isovolumetric (isochoric) ($W = 0$)	$\Delta U = Q$

Table 3.3 Special Types of Thermodynamic Processes

The MCAT focuses on three particular thermodynamic processes as special cases of the first law (excluding isobaric processes), as shown in Table 3.3. In each of these cases, some physical property is held constant during the process. These processes are **isothermal** (constant temperature, and therefore no change in internal energy), **adiabatic** (no heat exchange), and **isovolumetric** (no change in volume, and therefore no work accomplished; also called **isochoric**). **Isobaric** processes are those that occur at a constant pressure, and are of less focus on the MCAT. Figure 3.1 shows the different types of thermodynamic behaviors for a gas.

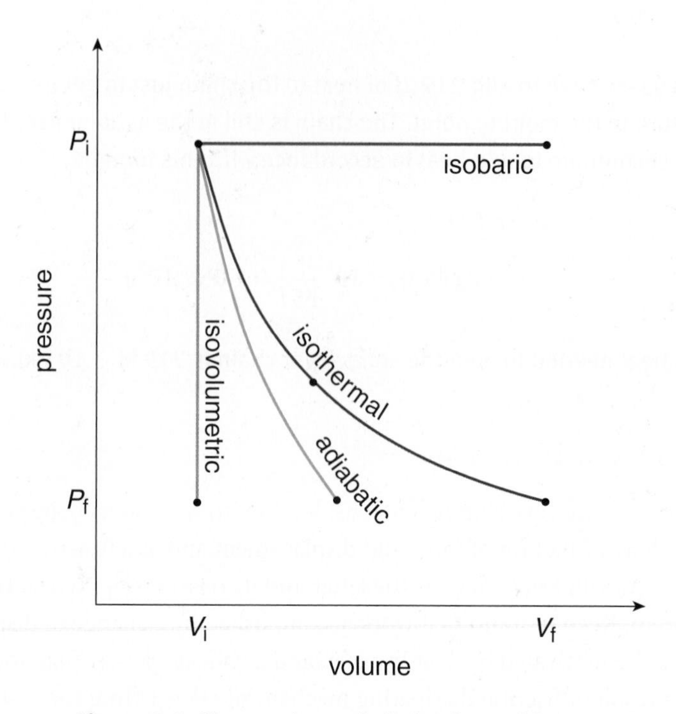

Figure 3.1 Thermodynamic Behaviors of Gases

Figure 3.2 shows a closed-loop thermodynamic process. Because the work on a P–V graph is simply the area under the curve, the work done in this closed-loop process is the area inside the loop.

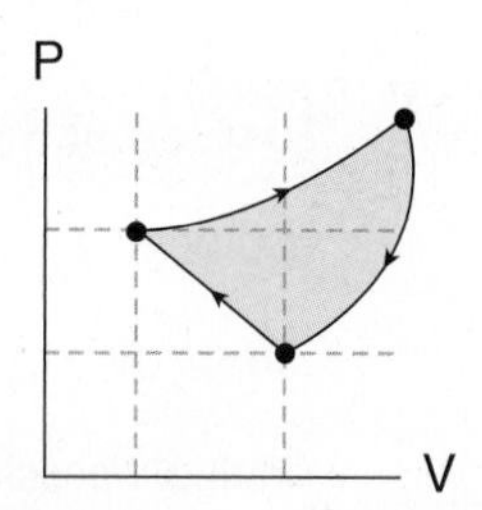

Figure 3.2 A Closed-Loop Process
The work is the area inside the loop.

Example: A gas in a cylinder is kept at a constant pressure of 3.6×10^5 Pa while 300 kJ of heat are added to it, causing the gas to expand from 1.0 m^3 to 1.5 m^3. Find the work done by the gas and the change in internal energy of the gas.

Solution: The pressure is held constant through the entire process so the work can be found using the equation:

$$W = P\Delta V$$
$$= \left(3.6 \times 10^5 \text{ Pa}\right)\left(1.5 \text{ m}^3 - 1.0 \text{ m}^3\right) = 1.8 \times 10^5 \text{ J}$$

The change in internal energy can be found from the first law of thermodynamics:

$$\Delta U = Q - W = \left(3 \times 10^5\right) - \left(1.8 \times 10^5 \text{ J}\right) = 1.2 \times 10^5 \text{ J} = 120 \text{ kJ}$$

MCAT CONCEPT CHECK 3.3

Before you move on, assess your understanding of the material with these questions.

1. Describe the relationship between internal energy, work, and heat in words.

___

2. Define the following forms of heat transfer:

- Conduction:

___

- Convection:

___

- Radiation:

___

3. Draw a representative graph of the temperature of a solid object as it is heated and goes through two phase changes to become a gas.

temperature

heat added

4. How is work calculated in P–V diagrams?

___

3.4 Second Law of Thermodynamics and Entropy

LEARNING OBJECTIVES

After Chapter 3.4, you will be able to:

- Describe entropy, on both a macroscopic level and in statistical terms
- Explain the relationship between entropy of a system and entropy of the surroundings for any thermodynamic process

The **second law of thermodynamics** states that objects in thermal contact and not in thermal equilibrium will exchange heat energy such that the object with a higher temperature will give off heat energy to the object with a lower temperature until both objects have the same temperature at thermal equilibrium. As such, energy is constantly being dispersed.

Energy Dispersion

Consider each of the following scenarios: hot tea cooling down, a frozen drink melting, iron rusting, buildings crumbling, balloons deflating, and living things dying and decaying. These scenarios share a common denominator. In each of them, energy of some form is going from being localized or concentrated to being spread out or dispersed. The thermal energy in the hot tea is spreading out to the cooler air that surrounds it. The thermal energy in the warmer air is spreading out to the cooler frozen drink. The chemical energy in the bonds of elemental iron and oxygen is released and dispersed as a result of the formation of the more stable, lower-energy bonds of iron oxide (rust). The potential energy of the building is released and dispersed in the form of light, sound, and heat as the building crumbles and falls. The energy of the pressurized air is released to the surrounding atmosphere as the balloon deflates. The chemical energy of all the molecules and atoms in living flesh is released into the environment during the process of death and decay.

Entropy

The **second law of thermodynamics** states that energy spontaneously disperses from being localized to becoming spread out if it is not hindered from doing so. Pay attention to this: *the usual way of thinking about entropy as "disorder" must not be taken too literally, a trap that many students fall into. Be very careful in thinking about entropy as disorder*. The old analogy between a messy (disordered) room and entropy is deficient and may not only hinder understanding but actually increase confusion.

Entropy is the measure of the spontaneous dispersal of energy at a specific temperature: *how much* energy is spread out, or *how widely* spread out energy becomes in a process. In the discussion of microstates earlier, we considered that when ice melts, the freedom of movement of the water molecules increases. If the water remains at the melting point, it will have the same average kinetic energy as molecules of ice; the difference between the two is the number of available microstates. That is, while both water and ice at 0°C have the same kinetic energy, the energy is dispersed over

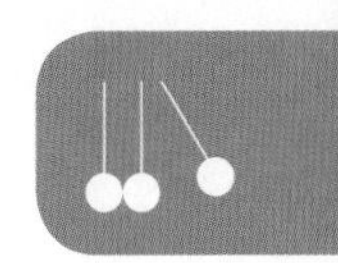

a larger number of microstates in liquid water. Liquid water therefore has higher entropy and, by extension, it is indeed less organized than ice. The equation for calculating the change in entropy is:

$$\Delta S = \frac{Q_{\text{rev}}}{T}$$

Equation 3.7

where ΔS is the change in entropy, Q_{rev} is the heat that is gained or lost in a reversible process, and T is the temperature in kelvin. The units of entropy are usually $\frac{\text{J}}{\text{mol} \cdot \text{K}}$. When energy is distributed into a system at a given temperature, its entropy increases. When energy is distributed out of a system at a given temperature, its entropy decreases.

Example: If, in a reversible process, 5.46×10^4 J of heat is used to change a 200 g block of ice to water at a temperature of 273 K, what is the change in the entropy of the system? (Note: The heat of fusion of ice = 333 $\frac{\text{J}}{\text{g}}$.)

Solution: We know that during a phase change, the temperature is constant; in this case, 273 K. From the information given,

$$\Delta S = \frac{Q_{\text{rev}}}{T} = \frac{5.46 \times 10^4 \text{ J}}{273 \text{ K}} = 200 \ \frac{\text{J}}{\text{K}}$$

The amount of heat added did not exceed the amount needed to completely melt the block of ice, as calculated by:

$$q = mL = (200 \text{ g})\left(333 \ \frac{\text{J}}{\text{g}}\right) = 6.66 \times 10^4 \text{ J}$$

Therefore, no heat was applied to change the temperature of the resulting liquid and T remained constant.

Notice that the second law states that energy will spontaneously disperse; it does not say that energy can never be localized or concentrated. However, the concentration of energy will not happen spontaneously in a closed system. Work usually must be done to concentrate energy. For example, refrigerators work against the direction of spontaneous heat flow (that is, they counteract the flow of heat from the "warm" exterior of the refrigerator to the "cool" interior), thereby "concentrating" energy outside of the system in the surroundings. As a result, refrigerators consume a lot of energy to accomplish this movement of energy against the temperature gradient.

The second law has been described as *time's arrow* because there is a unidirectional limitation on the movement of energy by which we recognize *before and after* or *new and old*. For example, you would instantly recognize whether a video recording of an explosion was running forward or backward. Another way of understanding this is

KEY CONCEPT

The universe is a closed, expanding system, so you know that the entropy of the universe is always increasing. The more space that appears with the expansion of the universe, the more space there is for the entire universe's energy to be distributed and the total entropy of the universe to increase irreversibly.

to say that energy in a closed system will spontaneously spread out and entropy will increase if it is not hindered from doing so. Remember that a system can be variably defined to include the entire universe; in fact, the second law ultimately claims that the entropy of the universe is increasing.

$$\Delta S_{\text{universe}} = \Delta S_{\text{system}} + \Delta S_{\text{surroundings}} > 0$$

Equation 3.8

When describing processes, physicists often use terms such as natural, unnatural, reversible, or irreversible. These terms confuse students but needlessly so because these terms are descriptive of observable phenomena. For example, we expect that when a hot object is brought into thermal contact with a cold object, the hot object will transfer heat energy to the cold object until both are in thermal equilibrium (that is, at the same temperature). This is a **natural process** and also one that we would describe as **irreversible**: we are not surprised that the two objects eventually reach a common temperature, but we would be shocked if all of a sudden the hot object became hotter and the cold object became colder. This would be an **unnatural process**.

To define a **reversible reaction**, let's consider a system of ice and liquid water in equilibrium at 0°C. If we place this mixture of ice and liquid water into a thermostat (device for regulating temperature) that is also at 0°C and allow infinitesimal amounts of heat to be absorbed by the ice from the thermostat so that the ice melts to liquid water at 0°C and the thermostat remains at 0°C, then the increase in the entropy of the system (the water) will be exactly equal to the entropy decrease of the surroundings (the thermostat). The net change in the entropy of the system and its surroundings is zero. Under these conditions, the process is reversible. The key to a reversible reaction is making sure that the process goes so slowly—requiring an infinite amount of time—that the system is always in equilibrium and no energy is lost or dissipated. To be frank, no real processes are reversible; we can only approximate a reversible process. Note how physicists define reversible processes: These are processes that can spontaneously reverse course. For example, while water can be put through cycles of freezing and melting innumerable times, ice melting on the warm countertop would not be expected to suddenly freeze if it remains in the warm environment. The liquid water will need to be placed in an environment that is cold enough to cause the water to freeze, and once frozen in the cold environment, the ice would not be expected to begin melting spontaneously. The freezing and melting of water in real life are therefore irreversible processes in *physics* while still being *chemically* reversible.

MCAT CONCEPT CHECK 3.4

Before you move on, assess your understanding of the material with these questions.

1. Describe entropy on a macroscopic level and in statistical terms.
   - Macroscopic:

   ____

   ____

   - Statistical:

   ____

   ____

2. What is the relationship between the entropy of a system and its surroundings for any thermodynamic process?

   ____

   ____

Conclusion

This chapter reviewed the zeroth law of thermodynamics, which reflects the observation that objects at the same temperature are in thermal equilibrium and the net heat exchanged between them is zero. We may consider the zeroth law to be *ex post facto* because it provides the thermodynamic explanation for the function of thermometers and temperature scales, which had been developed many years prior to the law's formulation. We then took some time to define basic thermodynamic terms for systems and state functions. Examination of the first law of thermodynamics revealed that the energy of a closed system (up to and including the universe) is constant, such that the total internal energy of a system (the sum of all its potential and motional energies) equals the heat gained by the system minus the work done by the system. Finally, we carefully investigated the second law of thermodynamics and the concept of entropy. We understand entropy as a measure not only of "disorder" but of the degree to which energy is spread out through a system, up to and including the universe. We now understand that the constant energy of the universe is progressively and irreversibly spreading out and will continue to spread out until there is an even distribution of energy throughout the universe. Many of these concepts will make a reappearance throughout our discussions of general chemistry, and will certainly be seen on the MCAT. In the next chapter, we'll investigate fluids, the final mechanical concept for Test Day.

GO ONLINE

You've reviewed the content, now test your knowledge and critical thinking skills by completing a test-like passage set in your online resources!

Concept Summary

Zeroth Law of Thermodynamics

- The **zeroth law of thermodynamics** states that objects are in thermal equilibrium when they are at the same temperature.
- Objects in thermal equilibrium experience no net exchange of heat energy.
- **Temperature** is a qualitative measure of how hot or cold an object is; quantitatively, it is related to the average kinetic energy of the particles that make up a substance.
- **Thermal expansion** describes how a substance changes in length or volume as a function of the change in temperature.

Systems

- A thermodynamic **system** is the portion of the universe that we are interested in observing, whereas the **surroundings** include everything that is not part of the system.
  - **Isolated systems** do not exchange matter or energy with the surroundings.
  - **Closed systems** exchange energy but not matter with their surroundings.
  - **Open systems** exchange both energy and matter with their surroundings.
- **State functions** are pathway independent and are not themselves defined by a process. Pressure, density, temperature, volume, enthalpy, internal energy, Gibbs free energy, and entropy are all state functions.
- **Process functions** describe the pathway from one equilibrium state to another. Work and heat are process functions.

First Law of Thermodynamics

- The **first law of thermodynamics** is a statement of conservation of energy: the total energy in the universe can never decrease or increase.
- For a closed system, the total internal energy is equal to the heat flow into the system minus the work done by the system.
- **Heat** is the process of energy transfer between two objects at different temperatures that occurs until the two objects come into thermal equilibrium (reach the same temperature).
  - **Specific heat** is the amount of energy necessary to raise one gram of a substance by one degree Celsius or one kelvin.
  - The specific heat of water is $1 \frac{\text{cal}}{\text{g} \cdot \text{K}}$.
  - During a phase change, heat energy causes changes in the particles' potential energy and energy distribution (entropy), but not kinetic energy. Therefore, there is no change in temperature. This is the **heat of transformation.**

- There are four special types of thermodynamic systems in which a given variable is held constant:
  - For **isothermal processes**, the temperature is constant, and the change in internal energy is therefore 0.
  - For **adiabatic processes**, no heat is exchanged.
  - For **isobaric processes**, the pressure is held constant.
  - For **isovolumetric (isochoric) processes**, the volume is held constant and the work done by or on the system is 0.

Second Law of Thermodynamics and Entropy

- The **second law of thermodynamics** states that in a closed system (up to and including the entire universe), energy will spontaneously and irreversibly go from being localized to being spread out (dispersed).
- **Entropy** is a measure of how much energy has spread out or how spread out energy has become.
- On a statistical level, as the number of available **microstates** increases, the potential energy of a molecule is distributed over that larger number of microstates, increasing entropy.
- Every **natural process** is ultimately **irreversible**; under highly controlled conditions, certain equilibrium processes such as phase changes can be treated as essentially **reversible**.

Answers to Concept Checks

3.1

1. The zeroth law of thermodynamics states that when two objects are both in thermal equilibrium with a third object, they are in thermal equilibrium with each other. By extension, no heat flows between two objects in thermal equilibrium.
2. While there may be a distance at which thermal equilibrium is impractical, there is no theoretical maximum distance. As long as two objects are in thermal contact and at the same temperature, they are in thermal equilibrium.
3. Expansion is a result of an increase in dimension at all points along an object. If an object is initially longer, it will experience a greater expansion. This is also represented in the formula for thermal expansion because there is a direct relationship between length change and the initial length of an object.
4. False. As we will discuss in Chapter 11 of *MCAT Physics and Math Review*, accuracy is related to an instrument, rather than the scale. In addition, Kelvin uses the same scale as Celsius, so there are no practical differences in terms of accuracy.

3.2

1.

System Type	Transfers Matter	Transfers Energy
Closed	No	Yes
Isolated	No	No
Open	Yes	Yes

2. State functions are variables independent from the path taken to achieve a particular equilibrium and are properties of a given system at equilibrium; they may be dependent on one another. Process functions define the path (or how the system got to its state) through variables such as Q (heat) or W (work).
3. State functions include pressure (P), density (ρ), temperature (T), volume (V), enthalpy (H), internal energy (U), Gibbs free energy (G), and entropy (S).

3.3

1. The change in the internal energy of a system is equal to heat put into a system minus the work done by the system. This is the first law of thermodynamics.
2. Conduction is heat exchange by direct molecular interactions. Convection is heat exchange by fluid movement. Radiation is heat exchange by electromagnetic waves, and does not depend on matter.
3. 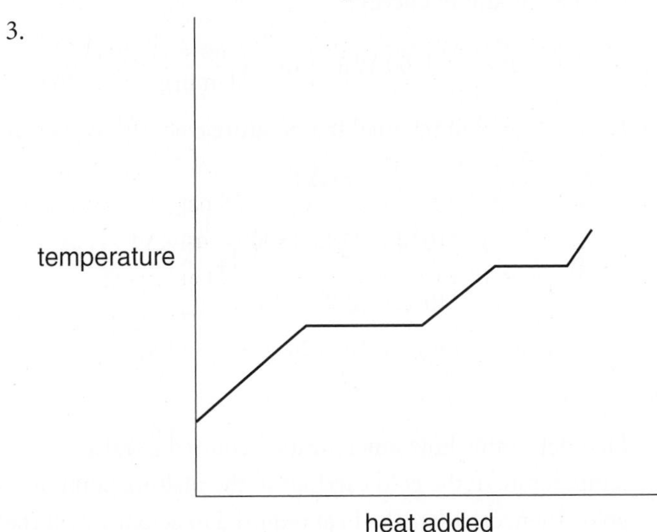

4. In a P–V graph, work is the area under the curve (or within a closed loop).

3.4

1. On a macroscopic level, entropy can be thought of as the tendency toward disorder. Statistically, entropy is the measure of the spontaneous dispersal of energy at a specific temperature, increasing the number of available microstates for a given molecule.
2. The entropy of a system and its surroundings will never decrease; entropy will always either remain constant or increase.

Science Mastery Assessment Explanations

1. **C**

The kelvin unit and Celsius degree are the same size; that is, a change of 10 K is equal to a change of 10°C. One degree Celsius is equal to 1.8 degrees Fahrenheit; therefore, 10°C = 18°F.

2. **A**

Because there is essentially only empty space between the Sun and the Earth, the only means of heat transfer is by radiation—electromagnetic waves that propagate across space. When a metal spoon is placed in a pot of hot soup, the molecules in the soup collide with those on the surface of the spoon, thereby transferring heat by conduction. Finally, fire warms the air above it, and the warmed air is less dense than the surrounding air, so it rises. A rising column of warm air means that heat is being transported in the air mass, which is simply the process of convection. The smoke particles ride along with the upward moving air current and create a plume of smoke.

3. **C**

The magnitude of one degree Celsius equals the magnitude of one unit kelvin. Therefore, a change of 100°C is equal to a change of 100 K. We can then find the change in length due to thermal expansion:

$$\Delta L = \alpha L \Delta T = \left(1.1 \times 10^{-5}\ \text{K}^{-1}\right)(20\ \text{m})(100\ \text{K})$$
$$= 0.022\ \text{m} = 2.2\ \text{cm}$$

Because the rod is originally 2.5 cm above the ground and its length increases by 2.2 cm, we can conclude that it will not touch the ground after the thermal expansion process is completed.

4. **C**

To answer this question, first remember that watts are equal to joules per second; in other words, power is energy transfer over time. In 10 minutes, the tool absorbs the following amount of energy:

$$E = P \times t = (100\ \text{W})(10\ \text{min})\left(\frac{60\ \text{s}}{1\ \text{min}}\right) = 6 \times 10^4\ \text{J}$$

Now we can find the final temperature from this equation:

$$q = mc\Delta T$$
$$6 \times 10^4\ \text{J} = (3\ \text{kg})\left(500\ \frac{\text{J}}{\text{kg}\cdot\text{K}}\right)\Delta T$$
$$40°\text{C} = \Delta T$$

The final temperature is thus 20 + 40 = 60°C.

5. **D**

First determine how much heat is required to raise the temperature of the gold earrings to the melting point of gold. Then, calculate the heat required to actually melt the earrings (the latent heat). The total heat required to melt the earrings completely will be the sum of the two heats. The heat required to raise the temperature of the earrings from 25°C to 1064°C is

$$q = mc\Delta T = (0.5\ \text{kg})\left(126\ \frac{\text{J}}{\text{kg}\cdot\text{K}}\right)(1039\ \text{K})$$
$$\approx 60 \times 1000 = 60\ \text{kJ}$$

Thus, it takes about 60 kJ of heat to bring the earrings to their melting point. The next step is to calculate how much heat is needed to melt the earrings. For this, use the heat of fusion (the latent heat) of gold:

$$q = mL = (0.5\ \text{kg})\left(6.37 \times 10^4\ \frac{\text{J}}{\text{kg}}\right) = 32\ \text{kJ}$$

So overall, it requires approximately 60 + 32 = 92 kJ of heat to melt the gold earrings. Notice that we can heavily approximate the numbers used in our calculations because the answer choices are so spread out. The closest answer is **(D)**.

6. **C**

The total work done by the cycle is the sum of the work of paths A, B, and C, or the area within the cycle. Because the area bounded by A, B, and C is a triangle with a base of 5 m^3 and a height of 3 Pa, we can calculate the area as $\frac{1}{2}\left(5\ m^3\right)(3\ Pa) = 7.5\ J$. Clockwise loops tend to do positive work on the environment, while counterclockwise loops do negative work.

7. **B**

To answer the question, make sure you understand all the terms. An adiabatic process means that there is no exchange of heat; in other words, $Q = 0$. When a gas is compressed, positive work is being done *on* the gas (rather than *by* the gas), so the value for work done *by* the gas will be negative ($W < 0$). Based on this, we can determine how the internal energy of the gas changes by using the first law of thermodynamics ($\Delta U = Q - W$). If $Q = 0$ and W is negative, then ΔU is positive.

8. **B**

The entropy of a system can decrease as long as the entropy of its surroundings increases by at least as much. On the other hand, the entropy of an isolated system increases for all real (irreversible) processes. This adheres to the second law of thermodynamics, which says that energy will be dispersed and entropy of the universe will remain constant or increase during all processes.

9. **A**

Calorimeters are our best approximations of isolated systems, where neither energy nor matter is exchanged with the environment. By failing to use an insulating layer and failing to fully contain the system, heat can be exchanged with the environment and matter may be dispersed, creating an open system.

10. **A**

To find the amount of heat needed to bring the substance to its melting point, you can use the specific heat. To heat one mole of the substance one unit kelvin, it would take 1 J of heat. After the substance reaches its melting point, additional heat is needed to actually induce the phase change. Therefore, the total amount of heat required is greater than 1 J.

11. **B**

State functions are any that are independent of the path taken to achieve a given state and which are not themselves defined as a process, such as pressure, density, temperature, volume, enthalpy, internal energy, Gibbs free energy, and entropy. Heat and work are process functions that are pathway dependent.

12. **C**

In this situation, heat will transfer from the warm gas to the metal and then to the cold gas. Convection requires flow of a fluid to cause heat transfer, invalidating **(B)** as an answer. In this case, the gas is not flowing, but rather is in contact with the metal. **(A)** is an invalid answer because heat transfer through radiation is also implausible not only because gases are unlikely to emit heat in the form of waves but also because the radiation would be unlikely to penetrate the thick metal container. Enthalpy, **(D)**, is not a form of heat transfer. Conduction, **(C)**, is the most likely option; it happens when two substances make direct contact with one another. Here, gas A makes contact with the metal container, which makes contact with gas B.

13. **D**

Saying that substance B has a higher internal energy cannot explain the phenomenon because the internal energy is irrelevant; the heat involved in the process is related only to the specific heat, the heat of fusion, and the heat of vaporization. All of the other choices could explain the phenomenon. The heat required to melt the solid is determined by the heat of fusion, **(C)**. The heat required to bring the liquid to its boiling point is determined by the specific heat, **(A)**. The heat required to boil the liquid is determined by the heat of vaporization, **(B)**.

14. **B**

When the ink randomly intersperses throughout the water, the final state is more disordered than the initial state, so the entropy change of the system is positive. When the oil separates from the water, the final state is just as ordered as the initial state (because the oil and the water are still completely separate), so the entropy change is zero. You can also answer this question by noticing the reversibility of the two experiments. Experiment A has a positive entropy change because it is irreversible, while experiment B has no entropy change because the reaction is reversible. According to the second law of thermodynamics, the overall entropy change of a system and its surroundings can never be negative in a thermodynamic process that moves from one equilibrium state to another.

15. **C**

If a substance is undergoing a phase change, any added heat will be used toward overcoming the heat of transformation of the phase change. During the phase change, the temperature will remain constant. Temperature is a measure of the kinetic energy of the molecules in a sample, so a change in kinetic energy, **(A)**, is essentially the same thing as a change in temperature. The heat transfer by radiation described in **(B)** will definitely change the temperature of the solid as long as it is not in the process of melting. **(D)** Describes heat transfer by convection, in which the warm gas will transfer heat to the cold gas until they both reach an intermediate temperature.

Consult your online resources for additional practice. GO ONLINE

Equations to Remember

(3.1) **Temperature conversions:** $F = \frac{9}{5}C + 32$

$K = C + 273$

(3.2) **Thermal expansion equation:** $\Delta L = \alpha L \Delta T$

(3.3) **Volume expansion equation:** $\Delta V = \beta V \Delta T$

(3.4) **First law of thermodynamics:** $\Delta U = Q - W$

(3.5) **Heat gained or lost (with temperature change):** $q = mc\Delta T$

(3.6) **Heat gained or lost (phase change):** $q = mL$ → Latent heat

(3.7) **Entropy and heat:** $\Delta S = \frac{Q_{\text{rev}}}{T}$

(3.8) **Second law of thermodynamics:** $\Delta S_{\text{universe}} = \Delta S_{\text{system}} + \Delta S_{\text{surroundings}} > 0$

Shared Concepts

Biochemistry Chapter 12
Bioenergetics and Regulation of Metabolism

General Chemistry Chapter 6
Equilibrium

General Chemistry Chapter 7
Thermochemistry

General Chemistry Chapter 8
The Gas Phase

General Chemistry Chapter 12
Electrochemistry

Physics and Math Chapter 2
Work and Energy

Q → Latent heat

CHAPTER 4

Fluids

SCIENCE MASTERY ASSESSMENT

Every pre-med knows this feeling: there is so much content I have to know for the MCAT! How do I know what to do first or what's important?

While the high-yield badges throughout this book will help you identify the most important topics, this Science Mastery Assessment is another tool in your MCAT prep arsenal. This quiz (which can also be taken in your online resources) and the guidance below will help ensure that you are spending the appropriate amount of time on this chapter based on your personal strengths and weaknesses. Don't worry though—skipping something now does not mean you'll never study it. Later on in your prep, as you complete full-length tests, you'll uncover specific pieces of content that you need to review and can come back to these chapters as appropriate.

How to Use This Assessment

If you answer 0–7 questions correctly:

Spend about 1 hour to read this chapter in full and take limited notes throughout. Follow up by reviewing **all** quiz questions to ensure that you now understand how to solve each one.

If you answer 8–11 questions correctly:

Spend 20–40 minutes reviewing the quiz questions. Beginning with the questions you missed, read and take notes on the corresponding subchapters. For questions you answered correctly, ensure your thinking matches that of the explanation and you understand why each choice was correct or incorrect.

If you answer 12–15 questions correctly:

Spend less than 20 minutes reviewing all questions from the quiz. If you missed any, then include a quick read-through of the corresponding subchapters, or even just the relevant content within a subchapter, as part of your question review. For questions you got correct, ensure your thinking matches that of the explanation and review the Concept Summary at the end of the chapter.

1. Objects A and B are submerged at a depth of 1 m in a liquid with a specific gravity of 0.877. Given that the density of object B is one-third that of object A and that the gauge pressure of object A is 3 atm, what is the gauge pressure of object B? (Note: Assume atmospheric pressure is 1 atm and $g = 9.8\ \frac{\text{m}}{\text{s}^2}$.)
   - **A.** 1 atm
   - **B.** 2 atm
   - **C.** 3 atm
   - **D.** 9 atm

2. An anchor made of iron weighs 833 N on the deck of a ship. If the anchor is now suspended in seawater by a massless chain, what is the tension in the chain? (Note: The density of iron is $7800\ \frac{\text{kg}}{\text{m}^3}$ and the density of seawater is $1025\ \frac{\text{kg}}{\text{m}^3}$.)
   - **A.** 100 N
   - **B.** 724 N
   - **C.** 833 N
   - **D.** 957 N

3. Two wooden balls of equal volume but different density are held beneath the surface of a container of water. Ball A has a density of $0.5\ \frac{\text{g}}{\text{m}^3}$, and ball B has a density of $0.7\ \frac{\text{g}}{\text{cm}^3}$. When the balls are released, they will accelerate upward to the surface. What is the relationship between the acceleration of ball A and that of ball B?
   - **A.** Ball A has the greater acceleration.
   - **B.** Ball B has the greater acceleration.
   - **C.** Balls A and B have the same acceleration.
   - **D.** It cannot be determined from information given.

4. Water flows from a pipe of diameter 0.15 m into one of diameter 0.2 m. If the speed in the 0.15 m pipe is $8\ \frac{\text{m}}{\text{s}}$, what is the speed in the 0.2 m pipe?
   - **A.** $3\ \frac{\text{m}}{\text{s}}$
   - **B.** $3.7\ \frac{\text{m}}{\text{s}}$
   - **C.** $4.5\ \frac{\text{m}}{\text{s}}$
   - **D.** $6\ \frac{\text{m}}{\text{s}}$

5. A hydraulic lever is used to lift a heavy hospital bed, requiring an amount of work *W*. When the same bed with a patient is lifted, the work required is doubled. How can the cross-sectional area of the platform on which the bed is lifted be changed so that the pressure on the hydraulic lever remains constant?
   - **A.** The cross-sectional area must be doubled.
   - **B.** The cross-sectional area must be halved.
   - **C.** The cross-sectional area must be divided by four.
   - **D.** The cross-sectional area must remain constant.

6. The figure shown represents a section through a horizontal pipe of varying diameters into which four open vertical pipes connect. If water is allowed to flow through the pipe in the direction indicated, in which of the vertical pipes will the water level be lowest?

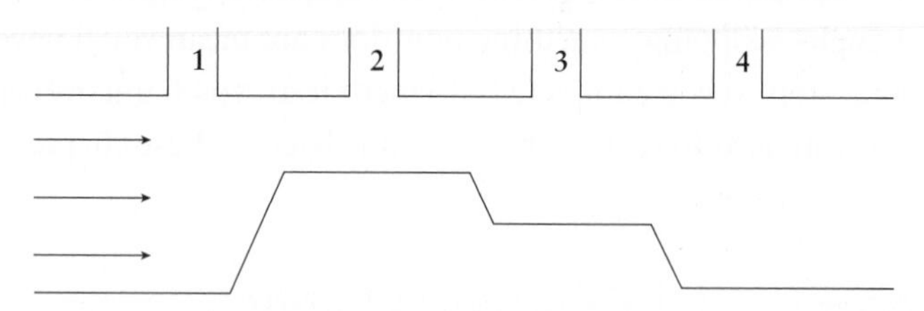

   - **A.** Pipe 1
   - **B.** Pipe 2
   - **C.** Pipe 3
   - **D.** Pipe 4

7. The speed of blood in the aorta is much higher than the speed of blood through a capillary bed. How can this fact be explained using the continuity equation, assuming that we are interested in average flow and that there is no net fluid loss?
   - **A.** The aorta is located higher than the capillary bed.
   - **B.** The pressure in the aorta is the same as the pressure in the capillary bed.
   - **C.** The cross-sectional area of all the capillaries added together is much greater than the cross-sectional area of the aorta.
   - **D.** The cross-sectional area of a capillary is much smaller than the cross-sectional area of the aorta.

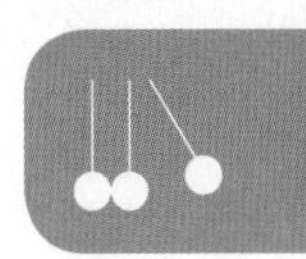

8. Which of the following data sets is sufficient to determine the linear speed through an area of a rigid pipe?
   - **A.** The cross-sectional area in another segment of pipe and the cross-sectional area in the region of interest
   - **B.** The Reynolds number, viscosity of the fluid, density, and diameter of the pipe
   - **C.** The radius of the pipe, pressure gradient, viscosity, and length of the pipe
   - **D.** The absolute pressure and density

9. An object with a density of 2 g/cm^3 is submerged to a depth of 25 cm in a container of dichloromethane. If the specific gravity of dichloromethane is 1.33, what is the total pressure exerted on the submerged object?
   - **A.** 3.3 kPa
   - **B.** 104 kPa
   - **C.** 332 kPa
   - **D.** 433 kPa

10. A hydraulic system is designed to allow water levels to change depending on a force applied at the top of the tank as shown. If a force, F_1, of 4 N is applied to a square, flexible cover where $A_1 = 16$, and the area $A_2 = 64$, what force must be applied to A_2 to keep the water levels from changing?

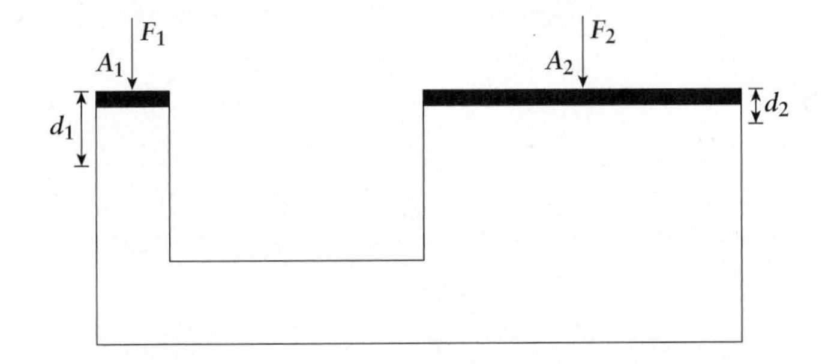

   - **A.** 4 N
   - **B.** 16 N
   - **C.** 32 N
   - **D.** No force needs to be applied.

11. Balls A and B of equal mass (shown below) are fully submerged in a swimming pool. Which ball will produce the greater buoyant force?

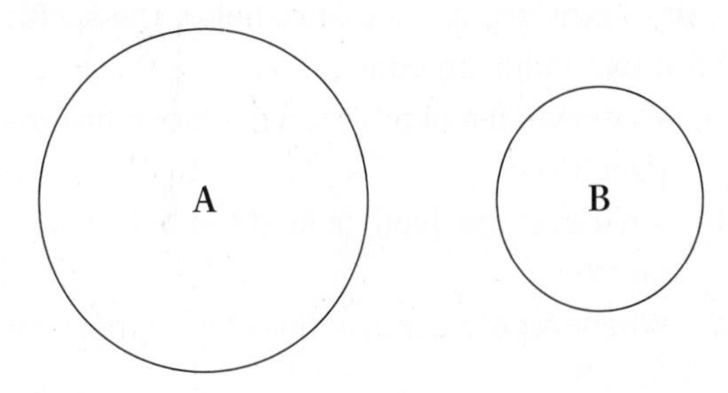

   - **A.** Ball A
   - **B.** Ball B
   - **C.** The forces will be equal.
   - **D.** It is impossible to know without knowing the exact volume of each ball.

12. Which of the following correctly describes blood flow through the circulatory system?
   - **A.** The flow rate is constant.
   - **B.** Pressure created by the heart moves blood through venous circulation.
   - **C.** The volume of blood entering and exiting the heart in a single cycle is equal.
   - **D.** The resistance of an artery is greater than the resistance of an arteriole.

13. A low-pressure weather system can decrease the atmospheric pressure from 1 atm to 0.99 atm. By what percent will this decrease the force on a rectangular window from the outside? (Note: Assume the window is 6 m by 3 m and the glass is 3 cm thick.)
   - **A.** 1%
   - **B.** 10%
   - **C.** $\frac{1}{3}$%
   - **D.** 30%

14. Two fluids, A and B, have densities of x and $2x$, respectively. They are tested independently to assess absolute pressure at varying depths. At what depths will the pressure below the surface of these two fluids be equal?
   A. Whenever the depth of fluid A is one-half that of fluid B
   B. Whenever the depth of fluid A equals that of fluid B
   C. Whenever the depth of fluid A is 2 times that of fluid B
   D. Whenever the depth of fluid A is 4 times that of fluid B

15. A water tower operator is interested in increasing the pressure of a column of water that is applied to a piston. She hopes that increasing the pressure will increase the force being applied to the piston. The only way to increase the pressure is to alter the speed of the water as it flows through the pipe to the piston. How should the speed of the water be changed to increase the pressure and force?
   A. Increase the speed
   B. Decrease the speed
   C. Release water intermittently against the pipe
   D. The speed of water will not change pressure at the piston.

Answer Key follows on next page.

Answer Key

1. **C** (Ch. 4.1)
2. **B** (Ch. 4.2)
3. **A** (Ch. 4.2)
4. **C** (Ch. 4.3)
5. **A** (Ch. 4.3)
6. **B** (Ch. 4.3)
7. **C** (Ch. 4.4)
8. **C** (Ch. 4.3)
9. **B** (Ch. 4.1)
10. **B** (Ch. 4.3)
11. **A** (Ch. 4.2)
12. **C** (Ch. 4.4)
13. **A** (Ch. 4.3)
14. **C** (Ch. 4.2)
15. **B** (Ch. 4.3)

CHAPTER 4

Fluids

In This Chapter

4.1 **Characteristics of Fluids and Solids**

Density 128

Pressure 129

4.2 **Hydrostatics**

Pascal's Principle 133

Archimedes' Principle 136

Molecular Forces in Liquids 137

4.3 **Fluid Dynamics** HY

Viscosity 140

Laminar and Turbulent Flow 140

Streamlines 142

Bernoulli's Equation 143

4.4 **Fluids in Physiology**

Circulatory System 148

Respiratory System 149

Concept Summary 151

CHAPTER PROFILE

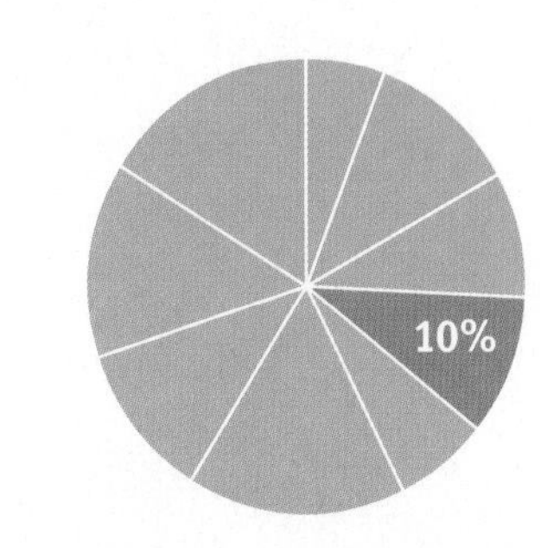

The content in this chapter should be relevant to about 10% of all questions about physics on the MCAT.

This chapter covers material from the following AAMC content category:

4B: Importance of fluids for the circulation of blood, gas movement, and gas exchange

4E: Atoms, nuclear decay, electronic structure, and atomic chemical behavior

Introduction

Hidden beneath the waves of the Mediterranean Sea, at depths of more than 4,000 meters, lie three lakes. The water in these "seas under the sea" is so salty—five to ten times saltier than the seawater that sits above it—that its extreme density prevents it from mixing with the ocean water above, forming a layer of separation not unlike that between the oil and water in a bottle of salad dressing. These underwater lakes behave eerily like their more common cousins found at sea level. They have tides, shore lines, beach ridges, and swash zones. When deep sea exploratory vessels set down on their surfaces, the vessels bob up and down, causing ripples to emanate outward like a stone dropped in a pond.

Suboceanic lakes and rivers present a particularly fascinating opportunity to illustrate the physics of fluids and solids. This chapter covers the important concepts and principles of fluid mechanics as they are tested on the MCAT. We will begin with a review of some important terms and measurements, including density and pressure. Our next topic will be hydrostatics, the branch of fluid mechanics that characterizes the behavior of fluids at rest. We'll then turn our attention to fluid dynamics, including Bernoulli's equation and the aerodynamics of flight. Finally, the chapter concludes with a discussion of fluid dynamics in physiology, examining the properties that motivate the movement of blood and air within the body.

4.1 Characteristics of Fluids and Solids

LEARNING OBJECTIVES

After Chapter 4.1, you will be able to:

- Predict when gauge pressure will be equal to fluid pressure for a column of fluid
- Relate weight and density for an object
- Recall the common units for pressure, as well as the equations for gauge pressure and absolute pressure

Fluids are characterized by their ability to flow and conform to the shapes of their containers. **Solids**, on the other hand, do not flow and are rigid enough to retain a shape independent of their containers. Both liquids and gases are fluids. The natural gas (methane) that many of us use to cook flows through pipes to the burners of our stove and ovens, and the air that we breathe flows in and out of our lungs, filling the spaces of our respiratory tract and the alveoli.

Fluids and solids share certain characteristics. Both can exert forces perpendicular to their surface, although only solids can withstand **shear (tangential) forces**. Fluids can impose large perpendicular forces; falling into water from a significant height can be just as painful as falling onto a solid surface.

Density

All fluids and solids are characterized by the ratio of their mass to their volume. This is known as **density**, which is a scalar quantity and therefore has no direction. The equation for density is

$$\rho = \frac{m}{V}$$

Equation 4.1

where ρ (rho) represents density, m is mass, and V is volume. The SI units for density are $\frac{\text{kg}}{\text{m}^3}$, but you may find it convenient to use $\frac{\text{g}}{\text{mL}}$ or $\frac{\text{g}}{\text{cm}^3}$, both of which may be seen on the MCAT. Remember that a milliliter and a cubic centimeter are the same volume. A word of caution: students sometimes assume that if the mL and the cm^3 are equivalent, then so must be the liter and the m^3. This is absolutely not the case; in fact, there are 1000 liters in a cubic meter. For the MCAT, it is important to know the density of water, which is $1\ \frac{\text{g}}{\text{cm}^3} \equiv 1000\ \frac{\text{kg}}{\text{m}^3}$.

The weight of any volume of a given substance with a known density can be calculated by multiplying the substance's density by its volume and the acceleration due to gravity. This is a calculation that appears frequently when working through buoyancy problems on Test Day:

$$\mathbf{F}_g = \rho V g$$

Equation 4.2

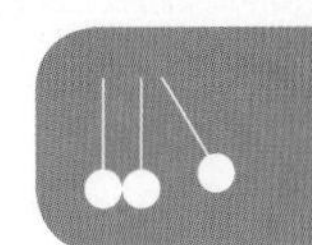

The density of a fluid is often compared to that of pure water at 1 atm and 4°C, a variable called **specific gravity**. It is at this combination of pressure and temperature that water has a density of exactly $1\ \frac{\text{g}}{\text{cm}^3}$. The specific gravity is given by

$$SG = \frac{\rho}{1\ \frac{\text{g}}{\text{cm}^3}}$$

specific gravity is dimensionless

Equation 4.3

This is a unitless number that is usually expressed as a decimal. The specific gravity can be used as a tool for determining if an object will sink or float in water, as described later in this chapter.

MCAT EXPERTISE

If an object's density is given in $\frac{\text{g}}{\text{cm}^3}$, its specific gravity is simply its density as a dimensionless number. This is because the density of water in $\frac{\text{g}}{\text{cm}^3}$ is 1.

Example: Find the specific gravity of benzene, given that its density is $877\ \frac{\text{kg}}{\text{m}^3}$.

Solution: The ratio of the density of benzene to the density of water is the specific gravity. Either the numerator must be converted to $\frac{\text{g}}{\text{cm}^3}$ or the denominator (the density of water) must be given in $\frac{\text{kg}}{\text{m}^3}$:

$$SG = \frac{\rho}{1000\ \frac{\text{kg}}{\text{m}^3}} = \frac{877}{1000} = 0.877$$

Pressure

Pressure is a ratio of the force per unit area. The equation for pressure is

$$P = \frac{F}{A}$$

Equation 4.4

where P is pressure, F is the magnitude of the normal force vector, and A is the area. The SI unit of pressure is the **pascal (Pa)**, which is equivalent to the newton per square meter $\left(1\ \text{Pa} = 1\ \frac{\text{N}}{\text{m}^2}\right)$. Other commonly used units of pressure are millimeters of mercury (mmHg), torr, and the atmosphere (atm). Millimeters of mercury and torr are identical units. The unit of atmosphere is based on the average atmospheric pressure at sea level. The conversions between Pa, mmHg, torr, and atm are as follows:

$$1.013 \times 10^5\ \text{Pa} = 760\ \text{mmHg} \equiv 760\ \text{torr} = 1\ \text{atm}$$

MCAT EXPERTISE

If you ever forget the units of a variable, you can derive them from equations. You know that pressure equals force over area. Because you know the units of force (N) and area (m^2), you can solve for the base units of pascal by plugging these units into the equation: $\frac{\text{N}}{\text{m}^2}$.

Pressure is a scalar quantity, and therefore has a magnitude but no direction. It is easy to assume that pressure has a direction because it is related to a force, which is a vector. However, note that it is the magnitude of the normal force that is used. No matter where one positions a given surface, the pressure exerted on that surface within a closed container will be the same, neglecting gravity. For example, if we placed a surface inside a closed container filled with gas, the individual molecules, which are moving randomly within the space, will exert pressure that is the same at all points within the container. Because the pressure is the same at all points along the walls of the container and within the space of the container itself, pressure applies in all directions at any point and, therefore, is a scalar rather than a vector. Of course, because pressure is a ratio of force to area, when unequal pressures are exerted against objects, the forces acting on the object will add in vectors, possibly resulting in acceleration. It's this difference in pressure that causes air to rush into and out of the lungs during respiration, windows to burst outward during a tornado, and the plastic covering a broken car window to bubble outward when the car is moving. Note that when gravity is present, this also results in a pressure differential, which we will explore with hydrostatics later in this chapter.

Example: The window of a skyscraper measures 2.0 m by 3.5 m. If a storm passes by and lowers the pressure outside the window to 0.997 atm while the pressure inside the building remains at 1 atm, what is the net force pushing on the window?

Solution: Because the pressures are different on the two sides of this window, there will be a net force pushing on it in the direction of the lower pressure (outside the window). The difference in pressure itself can be used to determine the net force:

$$\begin{aligned} F_{\text{net}} &= P_{\text{net}}A \\ &= (P_{\text{inside}} - P_{\text{outside}})A \\ &= (1 - 0.997\text{ atm})\left(\frac{1.013 \times 10^5\text{ Pa}}{1\text{ atm}}\right)(2.0\text{ m} \times 3.5\text{ m}) \\ &\approx (0.003)\left(10^5\right)(7.0) = \left(3 \times 10^{-3}\right)\left(10^5\right)(7.0) = 21 \times 10^2 \\ &= 2100\text{ N}\,(\text{actual} = 2128\text{ N}) \end{aligned}$$

Absolute Pressure

At this very moment, countless trillions of air molecules are exerting tremendous pressure on our bodies, with a total force of about 2×10^5 N! Of course, we don't actually feel all this pressure because our internal organs exert a pressure that perfectly balances it.

Atmospheric pressure changes with altitude. Residents of Denver (5280 feet above sea level) experience atmospheric pressure equal to 632 mmHg (0.83 atm), whereas travelers making their way through Death Valley (282 feet below sea level) experience atmospheric pressure equal to 767 mm Hg (1.01 atm). Atmospheric pressure impacts a number of processes, including hemoglobin's affinity for oxygen and the boiling of liquids.

Absolute (hydrostatic) pressure is the total pressure that is exerted on an object that is submerged in a fluid. Remember that fluids include both liquids and gases. The equation for absolute pressure is

$$P = P_0 + \rho gz$$

Equation 4.5

where P is the absolute pressure, P_0 is the **incident** or **ambient pressure** (the pressure at the surface), ρ is the density of the fluid, g is acceleration due to gravity, and z is the depth of the object. Do not make the mistake of assuming that P_0 always stands for atmospheric pressure. In open air and most day-to-day situations P_0 is equal to 1 atm, but in other fluid systems, the surface pressure may be higher or lower than atmospheric pressure. In a closed container, such as a pressure cooker, the pressure at the surface may be much higher than atmospheric pressure. This is, in fact, exactly the point of a pressure cooker, which allows food to cook at higher temperatures. This is because the increased pressure raises the boiling point of water in the food, thus reducing the cooking time and preventing loss of moisture.

REAL WORLD

A useful way to remember the two parts of the absolute pressure equation is to think of diving into a swimming pool. At the surface of the water, the absolute pressure is usually equal to the atmospheric pressure (P_0). But if you dive into the pool, the water exerts an extra pressure on you (ρgz), in addition to the surface pressure. You feel this extra pressure on your eardrums.

Gauge Pressure

When you check the pressure in your car or bike tires using a device known as a gauge, you are measuring the **gauge pressure,** which is the difference between the absolute pressure inside the tire and the atmospheric pressure outside the tire. In other words, gauge pressure is the amount of pressure in a closed space above and beyond atmospheric pressure. This is a more common pressure measurement than absolute pressure, and the equation is:

$$P_{gauge} = P - P_{atm} = (P_0 + \rho gz) - P_{atm}$$

Equation 4.6

Note that when $P_0 = P_{atm}$, then $P_{gauge} = P - P_0 = \rho gz$ at a depth z.

Example: A diver in the ocean is 20 m below the surface. What is the gauge pressure at her depth? What is the absolute pressure she experiences? (Note: The density of sea water is 1025 $\frac{\text{kg}}{\text{m}^3}$.)

Solution: Since the pressure at the surface (P_0) is equal to atmospheric pressure (P_{atm}), we can first solve for gauge pressure using the equation:

$$P_{gauge} = \rho g z = \left(1025\ \frac{\text{kg}}{\text{m}^3}\right)\left(9.8\ \frac{\text{m}}{\text{s}^2}\right)(20\ \text{m})$$
$$\approx (1000)(10)(20)$$
$$= 2 \times 10^5\ \text{Pa}\left(\text{actual} = 2.01 \times 10^5\ \text{Pa}\right)$$

Then, we can solve for absolute pressure using the absolute pressure equation:

$$P = P_{atm} + P_{gauge} = 1.013 \times 10^5\ \text{Pa} + 2.01 \times 10^5\ \text{Pa} = 3.02 \times 10^5\ \text{Pa}$$

MCAT CONCEPT CHECK 4.1

Before you move on, assess your understanding of the material with these questions.

1. How does gauge pressure relate to the pressure exerted by a column of fluid?

____

____

2. What is the relationship between weight and density?

____

____

3. What is the SI unit for pressure? What are other common units of pressure?
   - SI unit:

   ____

   - Other units:

   ____

4. True or False: Density is a scalar quantity.

4.2 Hydrostatics

LEARNING OBJECTIVES

After Chapter 4.2, you will be able to:

- Distinguish between cohesion and adhesion
- Predict the appearance of the meniscus of a fluid given knowledge of its cohesive and adhesive properties
- Calculate the buoyant force acting on an object
- Apply the concept of specific gravity
- Solve hydraulic lift problems using Pascal's principle:

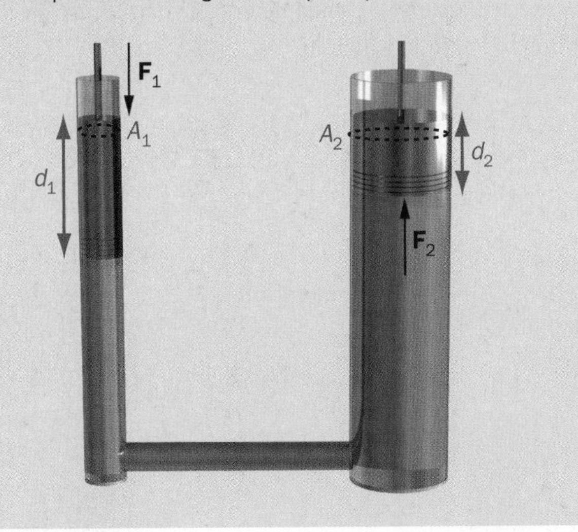

Hydrostatics is the study of fluids at rest and the forces and pressures associated with standing fluids. A proper understanding of hydrostatics is important for the MCAT because the testmakers frequently include passages and questions on hydraulics and buoyancy.

Pascal's Principle

For fluids that are incompressible—that is, fluids with volumes that cannot be reduced by any significant degree through application of pressure—a change in pressure will be transmitted undiminished to every portion of the fluid and to the walls of the containing vessel. This is **Pascal's principle**. For example, an unopened carton of milk could be considered an incompressible fluid in a closed container. If one were to squeeze the container, exerting an increased pressure on the sides of the milk carton, the applied pressure would be transmitted through the entire volume of milk. If the cap were to suddenly pop off, the resulting geyser of milk would be evidence of this increased pressure.

One application of Pascal's principle can be seen in **hydraulic systems**. These systems take advantage of the near-incompressibility of liquids to generate mechanical advantage, which, as we've seen in our discussion of inclined planes and pulleys in Chapter 2 of *MCAT Physics and Math Review*, allows us to accomplish a certain amount of work more easily by applying reduced forces. Many heavy machines use hydraulics, including car brakes, bulldozers, cranes, and lifts.

REAL WORLD

When the air pressure changes above a large body of water, the water level rises or falls to re-establish pressure equilibrium between the air and the water. The surface of a water body directly below a high-pressure air pocket forms a very small but measurable valley of water. A low-pressure air system has the opposite effect, creating a hill of water.

Figure 4.1 shows a simple diagram of a hydraulic lift. Let's determine how such a lift could allow an auto mechanic to raise a heavy car with far less force than the weight of the car. We have a closed container that is filled with an incompressible liquid. On the left side of the lift, there is a piston of cross-sectional area A_1. When this piston is pushed down the column, it exerts a force with a magnitude equal to F_1 and generates a pressure equal to P_1. The piston displaces a volume of liquid equal to A_1d_1 (the cross-sectional area times the distance gives a volume). Because the liquid inside is incompressible, the same volume of fluid must be displaced on the right side of the hydraulic lift, where we find a second piston with a much larger surface area, A_2. The pressure generated by piston 1 is transmitted undiminished to all points within the system, including to A_2. As A_2 is larger than A_1 by some factor, the magnitude of the force, F_2, exerted against A_2 must be greater than F_1 by the same factor so that $P_1 = P_2$, according to Pascal's principle.

$$P = \frac{F_1}{A_1} = \frac{F_2}{A_2}$$

$$F_2 = F_1\left(\frac{A_2}{A_1}\right)$$

Equation 4.7

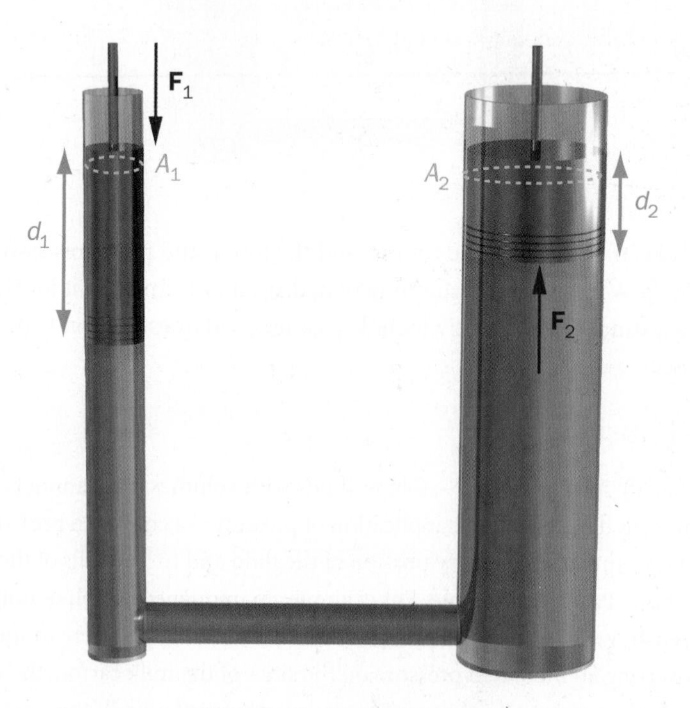

Figure 4.1 Hydraulic Lift

What this series of equations shows us is that hydraulic machines generate output force by magnifying an input force by a factor equal to the ratio of the cross-sectional area of the larger piston to that of the smaller piston. This does not violate the law of energy conservation; an analysis of the input and output work in a frictionless system reveals that there is indeed conservation of energy. As mentioned above, the volume of fluid displaced by piston 1 is equal to the volume of fluid displaced at piston 2.

$$V = A_1 d_1 = A_2 d_2$$

$$d_2 = d_1\left(\frac{A_1}{A_2}\right)$$

Combining the equations for pressure and volume, we can generate an equation for work as the product of constant pressure and volume change, as this is an isobaric process.

$$W = P\Delta V = \frac{F_1}{A_1}(A_1 d_1) = \frac{F_2}{A_2}(A_2 d_2)$$

$$= F_1 d_1 = F_2 d_2$$

This shows us the familiar form of work as the product of the magnitude of force and displacement (times the cosine of the angle between them, which is 0° in this case). Because the factor by which d_1 is larger than d_2 is equal to the factor by which F_2 is larger than F_1, we see that no additional work has been done or unaccounted for; the greater force F_2 is moving through a smaller distance d_2. Therefore, an auto mechanic needs only to exert a small force over a small area through a large distance to generate a much larger force over a larger area through a smaller distance.

KEY CONCEPT

Remember when applying Pascal's principle that the larger the area, the larger the force, although this force will be exerted through a smaller distance.

Example: A hydraulic press has a piston of radius 5 cm, which pushes down on an enclosed fluid. A 50 kg weight rests on this piston. Another piston in contact with this system has a radius of 20 cm. Taking $g = 10\ \frac{m}{s^2}$, what force is needed on the larger piston to keep the press in equilibrium?

Solution: Use Pascal's principle:

$$P = \frac{F_1}{A_1} = \frac{F_2}{A_2}$$

$$F_2 = F_1\left(\frac{A_2}{A_1}\right) = m_1 g\left(\frac{\pi r_2^2}{\pi r_1^2}\right) = m_1 g\left(\frac{r_2}{r_1}\right)^2$$

$$F_2 = (50\text{ kg})\left(10\ \frac{m}{s^2}\right)\left(\frac{20\text{ cm}}{5\text{ cm}}\right)^2 = (500)(4)^2 = (500)(16) = 8000\text{ N}$$

Archimedes' Principle

You've probably heard some version of this story before: Archimedes, a physicist in ancient Greece, was tasked by his king to determine the metallic composition of a certain crown given to the king as a gift. Archimedes knew that he could do this by finding the crown's volume and mass, which would allow him to find its density and compare that density to those of known metals. Weighing the crown would be easy enough, but he was having trouble finding a way to measure its volume without melting it down and ruining its workmanship. Then one day, while getting into his bath, the water that overflowed from the tub gave him the idea to submerge the crown in water and measure the volume of the displaced liquid—*Eureka!*

The principle that derives from the story is one of Archimedes' lasting contributions to the field of physics. **Archimedes' principle** deals with the **buoyancy** of objects when placed in a fluid. It helps us understand how ships stay afloat and why we feel lighter when we're swimming. The principle states that a body wholly or partially immersed in a fluid will be buoyed upwards by a force equal to the weight of the fluid that it displaces.

Just as Archimedes' body and his crown caused the water level to rise in the tub, any object placed in a fluid will cause a volume of fluid to be displaced equal to the volume of the object that is submerged. Because all fluids have density, the volume of fluid displaced will correspond to a certain mass of that fluid. The mass of the fluid displaced exerts a force equal to its weight against the submerged object. This force, which is always directed upward, is called the buoyant force, and its magnitude is given by:

$$F_{\text{buoy}} = \rho_{\text{fluid}} V_{\text{fluid displaced}} g = \rho_{\text{fluid}} V_{\text{submerged}} g$$

Equation 4.8

MCAT EXPERTISE

The most common mistake students make using the buoyancy equation is to use the density of the object rather than the density of the fluid. Remember always to use the density of the fluid itself.

When an object is placed in a fluid, it will sink into the fluid only to the point at which the volume of displaced fluid exerts a force that is equal to the weight of the object. If the object becomes completely submerged and the volume of displaced fluid still does not exert a buoyant force equal to the weight of the object, the object will accelerate downward and sink to the bottom. This will be the case if an object is more dense than the fluid it's in—a gold crown will sink to the bottom of the bathtub because it is denser than water. On the other hand, an object that is less dense than water, such as a block of wood or an ice cube, will stop sinking (and start floating) because it is less dense than water. These objects will submerge enough of their volume to displace a volume of water equal to the object's weight.

KEY CONCEPT

An object will float if its average density is less than the average density of the fluid it is immersed in. It will sink if its average density is greater than that of the fluid.

One way to conceptualize the buoyant force is that it is the force of the liquid trying to return to the space from which it was displaced, thus trying to push the object up and out of the water. This is an important concept because the buoyant force is due to the liquid itself, not the object. If two objects placed in a fluid displace the same volume of fluid, they will experience the same magnitude of buoyant force even if the objects themselves have different masses.

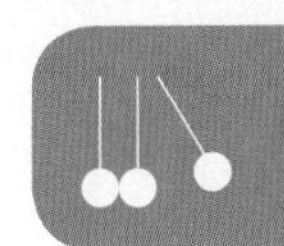

How can one determine how much of a floating object lies below the surface? To do this, one can make comparisons of density or specific gravity. Remember that an object will float, no matter what it is made of and no matter how much mass it has, if its average density is less than or equal to the density of the fluid into which it is placed. If we express the object's specific gravity as a percent, this directly indicates the percent of the object's volume that is submerged (when the fluid is pure water). For instance, the density of ice is 0.92 $\frac{\text{g}}{\text{cm}^3}$, so its specific gravity is 0.92. An ice cube floating in a glass of water has 92 percent of its volume submerged in the water—only 8 percent is sitting above the surface. Therefore, any object with a specific gravity less than or equal to 1 will float in water and any object with a specific gravity greater than 1 will sink in water. A specific gravity of exactly 1 indicates that 100 percent of the object will be submerged but it will not sink.

Example: A wooden block floats in the ocean with half its volume submerged. Find the density of the wood ρ_b. (Note: The density of sea water is 1025 $\frac{\text{kg}}{\text{m}^3}$.)

Solution: The magnitude of the weight of the block of total volume V_b is

$$F_{g,b} = m_b g = \rho_b V_b g$$

The weight of the displaced seawater is the buoyant force and is given by

$$F_{buoy} = m_{water} g = \rho_{water} V_{water} g$$

V_{water} is the volume of displaced water, which is also the volume of the part of the block that is submerged $\left(\frac{V_b}{2}\right)$. Because the block is floating, the buoyant force equals the block's weight:

$$F_{g,b} = F_{buoy}$$

$$\rho_b V_b g = \rho_{water}\left(\frac{V_b}{2}\right)g$$

$$\rho_b = \frac{\rho_{water}}{2} = \frac{1025\ \frac{\text{kg}}{\text{m}^3}}{2} = 512.5\ \frac{\text{kg}}{\text{m}^3}$$

Molecular Forces in Liquids

Water striders are insects that have the ability to walk on water. Water striders are able to glide across the water's surface without sinking, even though they are denser than water, because of a special physical property of liquids at the interface between a liquid and a gas. **Surface tension** causes the liquid to form a thin but strong layer like a "skin" at the liquid's surface. Surface tension results from **cohesion**, which is

REAL WORLD

At first it may seem strange that cruise ships, constructed of dense metals and weighing thousands of kilograms, can float on water. But remember that any object will float if its average density is less than that of water. The steel hull of the ship would sink by itself, but all the air submerged beneath the water level, between the ship's lower decks, lowers the ship's average density to be less than that of water.

REAL WORLD

The Dead Sea is the deepest hypersaline lake in the world. Having a salt content of about 35 percent, it is almost nine times saltier than the ocean. All of this dissolved salt makes for some of the densest water on the surface of the Earth, with a specific gravity of 1.24. Humans have a specific gravity around 1.1; thus, in most bodies of water, we have a tendency to sink—but in the Dead Sea, one is unable to do anything but float.

REAL WORLD

Remember that cohesion occurs between molecules with the same properties. In a container of both water and oil, the water molecules will be cohesive with other water molecules, and the oil will be cohesive with other oil molecules.

the attractive force that a molecule of liquid feels toward other molecules of the same liquid. Consider the intermolecular forces between the separate molecules of liquid water. For those molecules below the surface, there are attractive intermolecular forces coming from all sides; these forces balance out. However, on the surface, the molecules only have these strong attractive forces from the molecules below them, which pulls the surface of the liquid toward the center. This establishes tension in the plane of the surface of the water; when there is an indentation on the surface (say, caused by a water strider's foot), then the cohesion can lead to a net upward force.

Another force that liquid molecules experience is **adhesion**, which is the attractive force that a molecule of the liquid feels toward the molecules of some other substance. For example, adhesive forces cause water molecules to form droplets on the windshield of a car even though gravity is pulling them downward. When liquids are placed in containers, a **meniscus**, or curved surface in which the liquid "crawls" up the side of the container a small amount, will form when the adhesive forces are greater than the cohesive forces. A **backwards (convex) meniscus** (with the liquid level higher in the middle than at the edges) occurs when the cohesive forces are greater than the adhesive forces. Mercury, the only metal that is liquid at room temperature, forms a backward meniscus when placed in a container. Both types of menisci are shown in Figure 4.2.

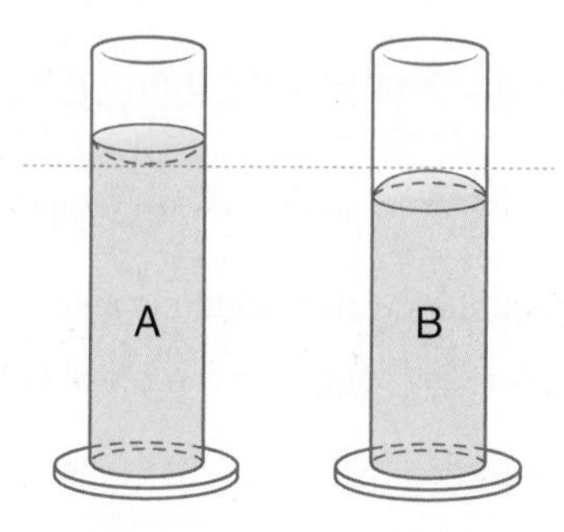

Figure 4.2 Types of Menisci
(A) A concave meniscus (more common); (B) A convex (backwards) meniscus. The dotted line indicates where measurements of depth or volume should be taken with each type of meniscus.

MCAT CONCEPT CHECK 4.2

Before you move on, assess your understanding of the material with these questions.

1. Contrast cohesion and adhesion.
   - Cohesion:

   - Adhesion:

   ---

2. What would the meniscus of a liquid that experiences equal cohesive and adhesive forces look like?

---

3. A block is fully submerged three inches below the surface of a fluid, but is not experiencing any acceleration. What can be said about the displaced volume of fluid and the buoyant force?

---

4. True or False: To determine the volume of an object by fluid displacement it must have a specific gravity greater than 1.
5. To which side of a hydraulic lift would the operator usually apply a force—the side with the larger cross-sectional area, or the side with the smaller cross-sectional area? Why?

4.3 Fluid Dynamics

LEARNING OBJECTIVES

After Chapter 4.3, you will be able to:

- Describe laminar flow, turbulent flow, dynamic and static pressure, pitot tubes, and viscosity
- Predict the behavior of fluids using the continuity equation, Bernoulli's equation, and the Venturi effect
- Recall the variables involved in flow rate

As the term suggests, **fluid dynamics** is the study of fluids in motion. This is perhaps one of the most fascinating areas of physics because its applications to real life are everywhere. Many aspects of our world, from water delivery to our homes to blood flow through our arteries and veins, can be analyzed and explained (at least in part) by the principles of fluid dynamics. The MCAT presents a relatively simplified version of the topic, making important assumptions such as rigid-walled containers and uniform density of fluids.

Viscosity

KEY CONCEPT

Low-viscosity fluids have low internal resistance to flow and behave like ideal fluids. Assume conservation of energy in low-viscosity fluids with laminar flow.

Some fluids flow very easily, while others barely flow at all. The resistance of a fluid to flow is called **viscosity** (η). Increased viscosity of a fluid increases its **viscous drag**, which is a nonconservative force that is analogous to air resistance. Thin fluids, like gases, water, and dilute aqueous solutions, have low viscosity and so they flow easily. Objects can move through these fluids with low viscous drag. Whole blood, vegetable oil, honey, cream, and molasses are thick fluids and flow more slowly. Objects can move through these fluids, but with significantly more viscous drag.

All fluids (except superfluids, which are not tested on the MCAT) are viscous to one degree or another; those with lower viscosities are said to behave more like ideal fluids, which have no viscosity and are described as **inviscid**. Because viscosity is a measure of a fluid's internal resistance to flow, more viscous fluids will "lose" more energy while flowing. Unless otherwise indicated, viscosity should be assumed to be negligible on Test Day, thus allowing Bernoulli's equation (explained later in this chapter) to be an expression of energy conservation for flowing fluids.

The SI unit of viscosity is the pascal–second $\left(\text{Pa}\cdot\text{s} = \frac{\text{N}\cdot\text{s}}{\text{m}^2}\right)$.

Laminar and Turbulent Flow

When a fluid is moving, its flow can be laminar or turbulent. **Laminar flow** is smooth and orderly, and is often modeled as layers of fluid that flow parallel to each other, as shown in Figure 4.3.

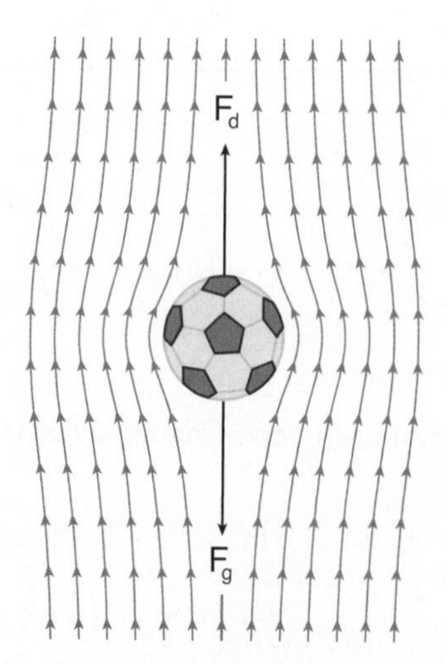

Figure 4.3 Laminar Flow around an Object in a Fluid
When the gravitational force is larger than the buoyant force, an object will sink. Laminar flow is characterized by smooth flow lines around the object.

The layers will not necessarily have the same linear speed. For example, the layer closest to the wall of a pipe flows more slowly than the more interior layers of fluid.

Poiseuille's Law

With laminar flow through a pipe or confined space, it is possible to calculate the rate of flow using **Poiseuille's law:**

$$Q = \frac{\pi r^4 \Delta P}{8 \eta L}$$

Equation 4.9

where Q is the flow rate (volume flowing per time), r is the radius of the tube, ΔP is the pressure gradient, η (eta) is the viscosity of the fluid, and L is the length of the pipe. This equation is rarely tested in full; most often, MCAT passages and questions focus on the relationship between the radius and pressure gradient. Note that the relationship between the radius and pressure gradient is inverse exponential to the fourth power—even a very slight change in the radius of the tube has a significant effect on the pressure gradient, assuming a constant flow rate.

Turbulence and Speed

Turbulent flow is rough and disorderly. Turbulence causes the formation of **eddies,** which are swirls of fluid of varying sizes occurring typically on the downstream side of an obstacle, as shown in Figure 4.4.

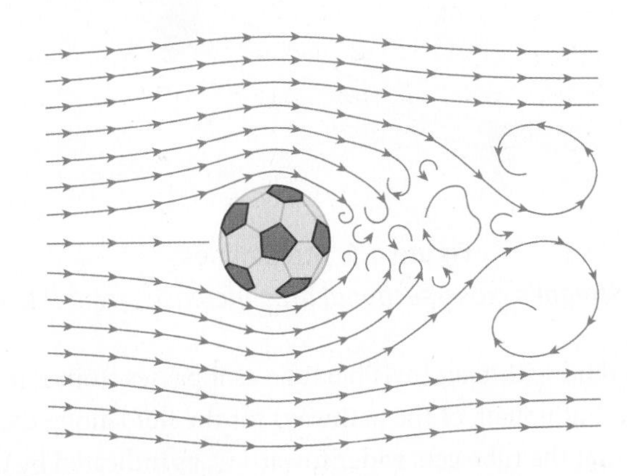

Figure 4.4 Turbulent Flow around an Object in a Fluid
Eddy formation downstream of an object obstructing laminar flow.

In unobstructed fluid flow, turbulence can arise when the speed of the fluid exceeds a certain **critical speed.** This critical speed depends on the physical properties of the fluid, such as its viscosity and the diameter of the tube. When the critical speed for a fluid is exceeded, the fluid demonstrates complex flow patterns, and laminar flow occurs only in the thin layer of fluid adjacent to the wall, called the **boundary layer.** The flow speed immediately at the wall is zero and increases uniformly throughout the layer. Beyond the boundary layer, however, the motion is highly irregular and turbulent. A significant amount of energy is dissipated from the system as a result of the increased frictional forces. Calculations of energy conservation, such as Bernoulli's equation, cannot be applied to turbulent flow systems. Luckily, the MCAT always assumes laminar (nonturbulent) flow for such questions.

For a fluid flowing through a tube of diameter D, the critical speed, v_c, can be calculated as

$$v_c = \frac{N_R \eta}{\rho D}$$

Equation 4.10

where v_c is the critical speed, N_R is a dimensionless constant called the Reynolds number, η is the viscosity of the fluid, and ρ is the density of the fluid. The **Reynolds number** depends on factors such as the size, shape, and surface roughness of any objects within the fluid.

Streamlines

Because the movement of individual molecules of a fluid is impossible to track with the unaided eye, it is often helpful to use representations of the molecular movement called **streamlines**. Streamlines indicate the pathways followed by tiny fluid elements (sometimes called fluid particles) as they move. The velocity vector of a fluid particle will always be tangential to the streamline at any point. Streamlines never cross each other.

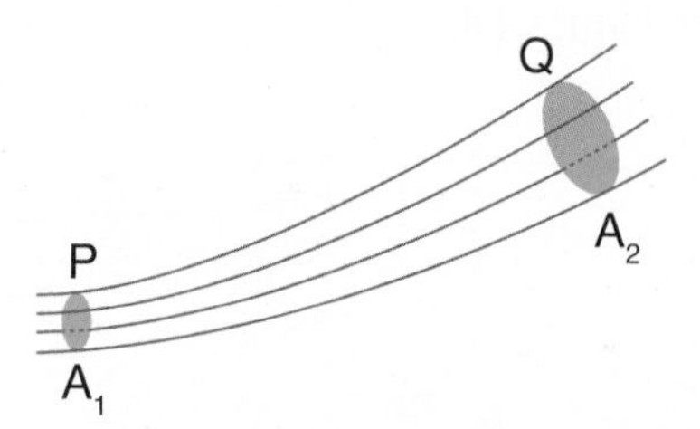

Figure 4.5 Streamlines
The stream's cross-sectional area increases from P to Q.

Figure 4.5 shows a fluid within an invisible tube as it passes from P to Q. The streamlines indicate some, but not all, of the pathways for the fluid along the walls of the tube. You'll notice that the tube gets wider toward Q, as indicated by the streamlines that are spreading out over the increased cross-sectional area. This leads us to consider the relationship between flow rate and the cross-sectional area of the container through which the fluid is moving. Once again, we can assume that the fluid is incompressible (which means that we are not considering a flowing gas). Because the fluid is incompressible, the rate at which a given volume (or mass) of fluid passes by one point must be the same for all other points in the closed system. This is essentially an expression of conservation of matter: if x liters of fluid pass a point in a given amount of time, then x liters of fluid must pass all other points in the system in the same amount of time. Thus, we can very clearly state, without any exceptions, the **flow rate** (that is, the volume per unit time) is constant for a closed system and is independent of changes in cross-sectional area.

While the flow rate is constant, the linear speed of the fluid does change relative to cross-sectional area. **Linear speed** is a measure of the linear displacement of fluid particles in a given amount of time. Notably, the product of linear speed and cross-sectional area is equal to the flow rate. We've already said that the volumetric rate of flow for a fluid must be constant throughout a closed system. Therefore,

$$Q = v_1 A_1 = v_2 A_2$$

Equation 4.11

where Q is the flow rate, v_1 and v_2 are the linear speeds of the fluid at points 1 and 2, respectively, and A_1 and A_2 are the cross-sectional areas at these points. This equation is known as the **continuity equation**, and it tells us that fluids will flow more quickly through narrow passages and more slowly through wider ones. Therefore, in Figure 4.5 earlier, while the flow rate at points P and Q are the same, the linear speed is faster at point P than point Q.

KEY CONCEPT

While flow rate is constant in a tube regardless of cross-sectional area, linear speed of a fluid will increase with decreasing cross-sectional area.

Bernoulli's Equation

Before we cover Bernoulli's equation itself, let's approach a flowing fluid from two perspectives that we've already discussed. First, the continuity equation arises from the conservation of mass of fluids. Liquids are essentially incompressible, so the flow rate within a closed space must be constant at all points. The continuity equation shows us that for a constant flow rate, there is an inverse relationship between the linear speed of the fluid and the cross-sectional area of the tube: fluids have higher speeds through narrower tubes.

Second, fluids that have low viscosity and demonstrate laminar flow can also be approximated to be conservative systems. The total mechanical energy of the system is constant if we discount the small viscous drag forces that occur in all real liquids.

Combining these principles of conservation, we arrive at **Bernoulli's equation**:

$$P_1 + \frac{1}{2}\rho v_1^2 + \rho g h_1 = P_2 + \frac{1}{2}\rho v_2^2 + \rho g h_2$$

Equation 4.12

where P is the absolute pressure of the fluid, ρ is the density of the fluid, v is the linear speed, g is acceleration due to gravity, and h is the height of the fluid above some datum. Some of the terms of Bernoulli's equation should look vaguely familiar. The term $\frac{1}{2}\rho v^2$ is sometimes called the **dynamic pressure**, and is the pressure associated with the movement of a fluid. This term is essentially the kinetic energy of the fluid divided by volume $\left(\rho = \frac{m}{V}\right)$. The term ρgh looks like the expression for gravitational potential energy, and is essentially the pressure associated with the mass of fluid sitting above some position. Finally, let's consider how the absolute pressure fits into this conservation equation. If one multiplies the unit of pressure $\left(\frac{\text{N}}{\text{m}^2}\right)$ by meters over meters, we obtain $\frac{\text{N}\cdot\text{m}}{\text{m}^3} = \frac{\text{J}}{\text{m}^3}$. Pressure can therefore be thought of as a ratio of

energy per cubic meter, or **energy density**. Systems at higher pressure have a higher energy density than systems at lower pressure. Finally, the combination of $P + \rho gh$ gives us the **static pressure**, and is the same equation as that for absolute pressure (although h is used here to imply height above a certain point, whereas z was used earlier to imply depth below a certain point). Bernoulli's equation states, then, that the sum of the static pressure and dynamic pressure will be constant within a closed container for an incompressible fluid not experiencing viscous drag. In the end, Bernoulli's equation is nothing other than a statement of energy conservation: more energy dedicated toward fluid movement means less energy dedicated toward static fluid pressure. The inverse of this is also true—less movement means more static pressure. One example of this principle that you may have previously encountered is how the shape of an airplane's wing helps generate lift, as shown in Figure 4.6.

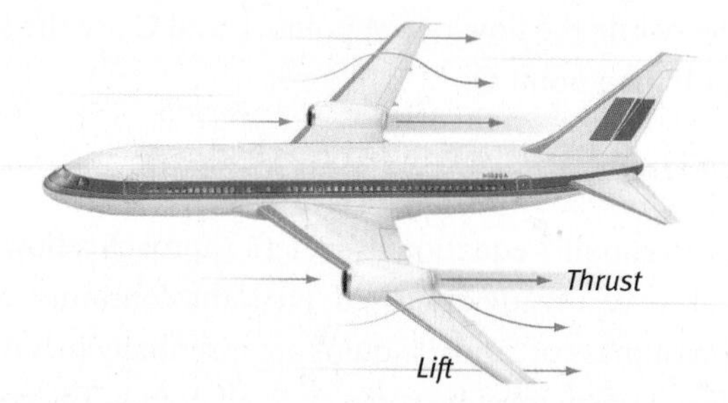

Figure 4.6 Aerodynamics of an Airplane

Propeller and jet engines generate thrust by pushing air backward. In both cases, because the wing top is curved, air streaming over it must travel farther and thus faster than air passing underneath the flat bottom. According to Bernoulli's equation, the slower air below exerts more force on the wing than the faster air above, thereby lifting the plane. Another example of Bernoulli's equation in action is the use of **pitot tubes**. These are specialized measurement devices that determine the speed of fluid flow by determining the difference between the static and dynamic pressure of the fluid at given points along a tube.

A common application of Bernoulli's equation on the MCAT is the **Venturi flow meter**, as shown in Figure 4.7.

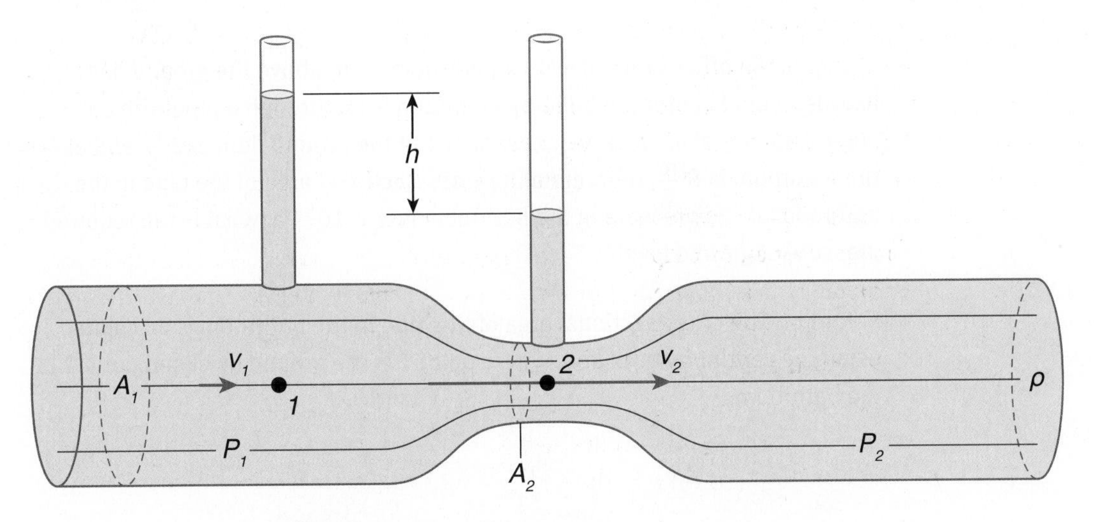

Figure 4.7 Venturi Flow Meter
As the tube narrows, the linear speed increases at point 2. Thus, the pressure exerted on the walls decreases, causing the column above the tube to have a lower height at point 2.

When considering Bernoulli's equation in this example, start by noting that the average height of the tube itself remains constant. Therefore, the ρgh term remains constant at points 1 and 2. Note that the h shown in Figure 4.7 is the difference in height between the two columns at points 1 and 2, not h from Bernoulli's equation, which corresponds to the average height of the tube above a datum. As the cross-sectional area decreases from point 1 to point 2, the linear speed must increase according to the continuity equation. Then, as the dynamic pressure increases, the absolute pressure must decrease at point 2. With a lower absolute pressure, the column of fluid sticking up from the Venturi tube will be lower at point 2. This phenomenon is often called the **Venturi effect**.

Example: An office building with a bathroom 40 m above the ground has its water supply enter the building at ground level through a pipe with an inner diameter of 4 cm. If the linear speed at the ground floor is 2 $\frac{m}{s}$ and at the bathroom is 8 $\frac{m}{s}$, determine the cross-sectional area of the pipe in the bathroom. If the pressure in the bathroom is 3×10^5 Pa, what is the required pressure at ground level?

Solution: The cross-sectional area of the pipe in the bathroom is calculated using the continuity equation, where point 1 is the ground level and point 2 is the bathroom:

$$A_1v_1 = A_2v_2 \rightarrow A_2 = \frac{A_1v_1}{v_2}$$

$$A_2 = \frac{\left(\pi r_1^2\right)v_1}{v_2} = \frac{(\pi)\left(2 \times 10^{-2}\ \text{m}\right)^2\left(2\ \frac{\text{m}}{\text{s}}\right)}{\left(8\ \frac{\text{m}}{\text{s}}\right)} = \frac{(\pi)\left(4 \times 10^{-4}\right)(2)}{8}$$
$$= 3.14 \times 10^{-4}\ \text{m}^2$$

The pressure can be found from Bernoulli's equation:

$$P_1 + \frac{1}{2}\rho v_1^2 + \rho g h_1 = P_2 + \frac{1}{2}\rho v_2^2 + \rho g h_2$$
$$P_1 = P_2 + \rho\left[\frac{v_2^2 - v_1^2}{2} + g(h_2 - h_1)\right]$$
$$= 3 \times 10^5\ \text{Pa} + \left(1000\ \frac{\text{kg}}{\text{m}^3}\right)\left[\frac{\left(8\ \frac{\text{m}}{\text{s}}\right)^2 - \left(2\ \frac{\text{m}}{\text{s}}\right)^2}{2} + \left(9.8\ \frac{\text{m}}{\text{s}^2}\right)(40\ \text{m})\right]$$
$$\approx 3 \times 10^5 + (1000)\left[\frac{64 - 4}{2} + 400\right] = 3 \times 10^5 + (1000)(430)$$
$$= 3 \times 10^5 + 4.3 \times 10^5 = 7.3 \times 10^5\ \text{Pa}\ \left(\text{actual} = 7.22 \times 10^5\ \text{Pa}\right)$$

MCAT CONCEPT CHECK 4.3

Before you move on, assess your understanding of the material with these questions.

1. Define the following terms:
   - Dynamic pressure:

   - Static pressure:

   - Pitot tube:

   - Viscosity:

   - Laminar flow:

   - Turbulence:

   ---

2. How do the following concepts relate to one another: Venturi effect, Bernoulli's equation, and continuity equation? What relationship does each describe?

   ---

3. What effect would increasing each of the following have on flow rate: the radius of the tube, pressure gradient, viscosity, and length of the tube?

4.4 Fluids in Physiology

LEARNING OBJECTIVES

After Chapter 4.4, you will be able to:

- Recall the conditions in which the continuity equation applies to human circulation
- Describe how total resistance of the airways changes during exhalation
- Compare flow volume and flow rate in different areas of the circulatory system

As a future student of medicine, you may feel that the abstract application of physics and math can often seem unimportant or tedious. However, these disciplines are exceptionally important in physiology. The movement of blood, lymph, and air throughout the body and lungs follow basic principles of fluid dynamics and pressure, with some minor alterations. We will focus primarily on the circulatory system, but also briefly discuss pressure and flow as they relate to gas exchange.

Circulatory System

The circulatory system is a **closed loop** that has a nonconstant flow rate. This nonconstant flow is a result of valves, gravity, the physical properties of our vessels (elasticity, in particular), and the mechanics of the heart. In particular, the nonconstant flow can be felt and measured as a **pulse**. In addition to these features, there is a loss of volume from the circulation as a result of a difference between osmotic (oncotic) and hydrostatic pressures. This fluid is eventually returned to the circulation as a result of lymphatic flow, but it is problematic for applications of the continuity equation. An important point to note is that despite these differences, blood volume entering the heart is always equal to blood volume leaving the heart during a single cycle.

As blood flows away from the heart, each vessel has a progressively higher resistance until the capillaries; however, the total resistance of the system decreases because the increased number of vessels are in parallel with each other. Like parallel resistors in circuits, the equivalent resistance is therefore lower for the capillaries in parallel than in the aorta. Return flow to the heart is facilitated by mechanical squeezing of the skeletal muscles, which increases pressure in the limbs and pushes blood to the heart, and the expansion of the heart, which decreases pressure in the heart and pulls blood in. Finally, the pressure gradients created in the thorax by inhalation and exhalation also motivate blood flow. Venous circulation holds approximately three times as much blood as arterial circulation. Heart murmurs, which result from structural defects of the heart, are heard because of turbulent blood flow.

Respiratory System

The respiratory system is also mediated by changes in pressure, and follows the same resistance relationship as the circulatory system. During inspiration, there is a negative pressure gradient that moves air into the lungs. During expiration, this gradient reverses. An additional point to note is that when air reaches the alveoli, it has essentially no speed.

MCAT CONCEPT CHECK 4.4

Before you move on, assess your understanding of the material with these questions.

1. Under what conditions could the continuity equation be applied to human circulation?

____

____

2. During exhalation, how does the total resistance of the encountered airways change as air leaves the alveoli to escape the nose and mouth?

____

____

____

3. How does flow in the venae cavae relate to flow in the main pulmonary artery?

____

____

Conclusion

The behavior of fluids impacts every moment of our lives. Even if we are nowhere near an ocean or a lake, we are quite literally submerged in a vast expanse of fluid, a mix of gases known as the atmosphere, which exerts forces on all of the surfaces of our bodies. Whenever we take a bath or submerge an object in water, we experience the effect of buoyant forces exerted by the displaced fluid. When we water our gardens, take a shower, or ride in a car with open windows, we experience the speeds, forces, and pressures of a fluid on the move. In the world of medicine, one must consider fluids, flowing and at rest, when evaluating the function of the respiratory and circulatory systems: conditions as varied as asthma and heart murmurs are related to the way in which the body causes fluids to flow. The balance of hydrostatic and oncotic pressures is important for maintaining the proper balance of fluid in the peripheral tissues of the body.

Now that you have the basic concepts of hydrostatics and fluid dynamics, learn to apply them to MCAT passages and questions through your Kaplan practice materials. Don't be intimidated by the seeming complexity of buoyant force problems and applications of Bernoulli's equation. Remember that all fluids, whether liquid or gas, exert buoyant forces against objects that are placed in them as a function of the weight of the fluid displaced. Remember that incompressible fluids demonstrate an inverse relationship between their dynamic pressure (as a function of speed) and their static pressure. This chapter concludes the section of this book focusing on mechanics; in the next two chapters, we'll turn our attention to electrostatics and electricity.

You've reviewed the content, now test your knowledge and critical thinking skills by completing a test-like passage set in your online resources!

Concept Summary

Characteristics of Fluids and Solids

- **Fluids** are substances that have the ability to flow and conform to the shape of their containers.
  - Fluids can exert perpendicular forces, but cannot exert shear forces.
  - Liquids and gases are the two phases of matter that are fluids.
- **Solids** do not flow and they retain their shape regardless of their containers.
- **Density** is the mass per unit volume of a substance (fluid or solid).
- **Pressure** is defined as a measure of force per unit area; it is exerted by a fluid on the walls of its container and on objects placed in the fluid.
  - It is a scalar quantity; its value has magnitude only, and no direction.
  - The pressure exerted by a gas against the walls of its container will always be perpendicular (normal) to the container walls.
- **Absolute pressure** is the sum of all pressures at a certain point within a fluid; it is equal to the pressure at the surface of the fluid (usually atmospheric pressure) plus the pressure due to the fluid itself.
- **Gauge pressure** is the name for the difference between absolute pressure and atmospheric pressure. In liquids, gauge pressure is caused by the weight of the liquid above the point of measurement.

Hydrostatics

- **Pascal's principle** states that a pressure applied to an incompressible fluid will be distributed undiminished throughout the entire volume of the fluid
- **Hydraulic machines** operate based on the application of Pascal's principle to generate mechanical advantage.
- **Archimedes' principle** governs the buoyant force. When an object is placed in a fluid, the fluid generates a **buoyant force** against the object that is equal to the weight of the fluid displaced by the object.
  - The direction of the buoyant force is always opposite to the direction of gravity.
  - If the maximum buoyant force is larger than the force of gravity on the object, the object will float. This will be true if the object is less dense than the fluid it is in.
  - If the maximum buoyant force is smaller than the force of gravity on the object, the object will sink. This will be true if the object is more dense than the fluid it is in.
- Fluids experience **cohesive** forces with other molecules of the same fluid and **adhesive** forces with other materials; cohesive forces give rise to **surface tension**.

Fluid Dynamics

- Fluid dynamics is a set of principles regarding actively flowing fluids.
- **Viscosity** is a measurement of a fluid's internal friction. **Viscous drag** is a nonconservative force generated by viscosity.
- Fluids can move with either **laminar flow** or **turbulent flow**.
  - The rate of laminar flow is determined by the relationships in **Poiseuille's law**.
  - On the MCAT, incompressible fluids are assumed to have laminar flow and very low viscosity while flowing, allowing us to assume conservation of energy.
- The **continuity equation** is a statement of the conservation of mass as applied to fluid dynamics.
- **Bernoulli's equation** is an expression of conservation of energy for a flowing fluid. This equation states that the sum of the **static pressure** and the **dynamic pressure** will be constant between any two points in a closed system.
- For a horizontal flow, there is an inverse relationship between pressure and speed, and in a closed system, there is a direct relationship between cross-sectional area and pressure exerted on the walls of the tube known as the **Venturi effect**.

Fluids in Physiology

- The circulatory system behaves as a closed system with nonconstant flow.
- Resistance decreases as the total cross-sectional area increases.
  - Arterial circulation is primarily motivated by the heart.
  - Venous circulation has three times the volume of arterial circulation and is motivated by the skeletal musculature and expansion of the heart.
- Inspiration and expiration create a pressure gradient not only for the respiratory system, but for the circulatory system as well.
- Air at the alveoli has essentially zero speed.

Answers to Concept Checks

4.1

1. Gauge pressure is equal to the pressure exerted by a column of fluid plus the ambient pressure above the fluid, minus atmospheric pressure. When atmospheric pressure is the only pressure above the fluid column, then gauge pressure equals the fluid pressure.
2. Weight is density times volume and acceleration due to gravity.
3. The SI unit of pressure is the pascal. Other common units include mmHg, torr, and atm.
4. True. Density is directionless, and is thus a scalar quantity.

4.2

1. Cohesion is the attractive force experienced by molecules of a fluid for one another. Adhesion is the attractive force experienced by molecules of a fluid for a different material (usually a solid).
2. If adhesive and cohesive forces are equal, then no meniscus would form and the liquid surface would be flat.
3. The displaced volume is equal to the volume of the block. The buoyant force is equal to the weight of the block, and is equal to the weight of the displaced fluid. By extension, the block and the fluid in which it is immersed must have the same density.
4. False. A fluid with a low specific gravity can be used instead of water to determine volumes of objects that would otherwise float in water.
5. The operator usually applies a force to the side with the smaller cross-sectional area. Because pressure is the same on both sides of the lift, a smaller force can be applied on the smaller surface area to generate the desired pressure.

4.3

1. Dynamic pressure is the pressure associated with flow, and is represented by $\frac{1}{2}\rho v^2$. Static pressure is the pressure associated with position; static pressure is sacrificed for dynamic pressure during flow. A pitot tube is a device that measures static pressure during flow to calculate speed. Viscosity is a measure of the resistance of a liquid to flow. Laminar flow is flow in which there are no eddies and in which streamlines roughly parallel each other. Turbulence is the presence of backflow or current eddies.
2. The continuity equation describes the relationship of flow and cross-sectional area in a tube, while Bernoulli's equation describes the relationship between height, pressure, and flow. The Venturi effect is the direct relationship between cross-sectional area and pressure, and results from the combined relationships of the Bernoulli and continuity equations.

3. Flow rate would increase when increasing either the radius of the tube or the pressure gradient, but would decrease with increasing viscosity or length of the tube.

4.4

1. The continuity equation cannot be applied to human circulation. The presence of pulses, the elasticity of the vessels, and the nature of the pressure gradient preclude this type of analysis. Poiseuille's law should instead be used for isolated segments.
2. Total resistance increases as the air exits the body despite the increase in the diameter of the airways. This is because there are fewer airways in parallel with each other.
3. In theory, there should be equal flow in the venae cavae and the main pulmonary trunk. In reality, the flow in the venae cavae is actually slightly less than in the pulmonary trunk because some of the blood entering the right side of the heart is actually from cardiac (coronary) circulation, not systemic circulation.

Science Mastery Assessment Explanations

1. **C**

The absolute and gauge pressures depend only on the density of the fluid, not on that of the object. When the pressure at the surface is equal to atmospheric pressure, the gauge pressure is given by $P_{gauge} = \rho g z$, where ρ represents the density of the fluid, not the object. These objects are also at the same depth, so they must have the same gauge pressure.

2. **B**

The tension in the chain is the difference between the anchor's weight and the buoyant force because the object is in translational equilibrium: $\mathbf{T} = \mathbf{F}_g - \mathbf{F}_{buoy}$. The object's weight is 833 N, and the buoyant force can be found using Archimedes' principle. The magnitude of the buoyant force is equal to the weight of the seawater that the anchor displaces:

$$F_{buoy} = \rho_w V_w g$$

Because the anchor is submerged entirely, the volume of the water displaced is equal to the volume of the anchor, which is equal to its mass divided by its density $\left(V_A = \frac{m_A}{\rho_A}\right)$. We are not given the anchor's mass, but its value must be the magnitude of the weight of the anchor divided by g. Putting all of this together, we can obtain the buoyant force:

$$F_{buoy} = \rho_w V_w g = \rho_w V_A g = \rho_w \left(\frac{m_A}{\rho_A}\right) g = \rho_w \left(\frac{F_g}{\rho_A g}\right) g = F_g \left(\frac{\rho_w}{\rho_A}\right)$$

$$= 833\text{ N}\left(\frac{1025\ \frac{\text{kg}}{\text{m}^3}}{7800\ \frac{\text{kg}}{\text{m}^3}}\right) \approx 100\text{ N}\,(\text{actual} = 109\text{ N})$$

Lastly, we can obtain the tension from the initial equation $\mathbf{T} = \mathbf{F}_g - \mathbf{F}_{buoy}$:

$$\mathbf{T} = 833\text{ N} - 109\text{ N} = 724\text{ N}$$

The key to quickly solving this problem on Test Day is recognizing that the answer choices contain an outlier **(A)**, a value slightly less than the weight of the anchor **(B)**, the weight of the anchor **(C)**, and a value slightly higher than the weight of the anchor **(D)**. Since buoyant force is in the same direction as tension and their sum must equal the weight of the anchor, **(B)** is the most likely answer.

3. **A**

Using Newton's second law, $F_{net} = ma$, we obtain the following equation:

$$F_{buoy} - mg = ma$$

Thus,

$$a = \frac{F_{buoy} - mg}{m} = \frac{F_{buoy}}{m} - g$$

Both balls experience the same buoyant force because they are in the same liquid and have the same volume ($F_{buoy} = \rho V g$). Thus, the ball with the smaller mass experiences the greater acceleration. Because both balls have the same volume, the ball with the smaller density has the smaller mass ($m = \rho V$), which is ball A.

4. **C**

It is known that water flows faster through a narrower pipe. The speed is inversely proportional to the cross-sectional area of the pipe because the same volume of water must pass by each point at each time interval. Let A be the 0.15 m pipe and B the 0.20 m pipe, and use the continuity equation:

$$v_A A_A = v_B A_B$$

where v is the speed and A is the cross-sectional area of the pipe. Because v is inversely proportional to the cross-sectional area, and the area is proportional to the square of the diameter $A = \pi r^2 = \frac{\pi d^2}{4}$, we obtain the following:

$$v_B = \frac{v_A\left(\frac{\pi d_A^2}{4}\right)}{\left(\frac{\pi d_B^2}{4}\right)} = v_A\left(\frac{d_A}{d_B}\right)^2$$

$$v_B = (8)\left(\frac{0.15}{0.20}\right)^2 = (8)\left(\frac{3}{4}\right)^2$$

$$v_B = (8)\left(\frac{9}{16}\right) = \frac{9}{2} = 4.5\ \frac{\text{m}}{\text{s}}$$

5. **A**

This question tests our understanding of Pascal's principle, which states that a change in pressure applied to an enclosed fluid is transmitted undiminished to every portion of the fluid and to the walls of the containing vessel. We are told that the work required to lift the bed with the patient is double the work needed to lift just the bed. In other words, the force required doubles when both the bed and the patient have to be lifted. To maintain the same pressure, we must double the cross-sectional area of the platform of the hydraulic lever on which the patient and the bed are lifted.

6. **B**

It is not necessary to do any calculations to answer this question. The open vertical pipes are exposed to the same atmospheric pressure; therefore, differences in the heights of the columns of water in the vertical pipes are dependent only on the differences in hydrostatic pressures in the horizontal pipe. Because the horizontal pipe has variable cross-sectional area, water will flow the fastest and the hydrostatic pressure will have its lowest value where the horizontal pipe is narrowest; this is called the Venturi effect. As a result, pipe 2 will have the lowest water level.

7. **C**

The continuity equation states that the flow rate of a fluid must remain constant from one cross-section to another. In other words, when an ideal fluid flows from a pipe with a large cross-sectional area to one that is narrower, its speed increases. This can be illustrated through the equation $A_1 v_1 = A_2 v_2$. If blood flows much more slowly through the capillaries, we can infer that the cross-sectional area is larger. This might seem surprising at first glance, but given that each blood vessel divides into thousands of little capillaries, it is not hard to imagine that adding the cross-sectional areas of each capillary from an entire capillary bed results in an area that is larger than the cross-sectional area of the aorta.

8. **C**

The data given in **(C)** are sufficient to determine the flow rate through Poiseuille's law, which can then be used to determine the linear speed by dividing by the cross-sectional area (which could be determined from the radius, as well). **(A)** would be sufficient if we also knew the flow rate in the other segment of pipe; one could use the continuity equation to determine the linear speed. The data in **(B)** could be used to determine the critical speed at which turbulent flow begins, but there is no indication that there is turbulent flow. The data in **(D)** could be used to determine the depth of an object in a fluid.

9. **B**

To calculate the total pressure, use the hydrostatic pressure formula: $P_{tot} = P_{atm} + \rho gz$. The unit of depth in this equation is meters, so first convert 25 cm to 0.25 m. The specific gravity is equal to the density of a medium divided by the density of water, so, with a specific gravity of 1.33, the density of dichloromethane is 1,330 kg/m^3. Finally, plug in these values and solve for total pressure: $P_{tot} = 101 \text{ kPa} + (1{,}330 \text{ kg/m}^3)(0.25 \text{ m})(10 \text{ m/s}^2) = 104 \text{ kPa}$, matching **(B)**.

10. **B**

This is a basic restatement of Pascal's principle that a force applied to an area will be transmitted through a fluid. This will result in changing fluid levels through the system. The relationship is stated as $F_2 = \frac{F_1 A_2}{A_1}$. Plugging in the numbers gives an answer of 16 N.

11. **A**

The buoyant force (F_{buoy}) is equal to the weight of water displaced, which is quantitatively expressed as

$$F_{buoy} = m_{\text{fluid displaced}}g = \rho_{\text{fluid}}V_{\text{fluid displaced}}g$$

The volume of displaced fluid is equal to the volume of the ball. The density of the fluid remains constant. Therefore, because ball A has a larger volume, it will displace more water and experience a larger buoyant force.

12. **C**

The volume of blood entering and exiting the heart is always equal in a single cardiac cycle, supporting **(C)** as the correct answer. By contrast, the flow rate of blood through the circulatory system is not constant, as some fluids exit the circulation in tissue and is later returned via the lymphatic system, eliminating **(A)**. Furthermore, the contraction of the heart pushing blood through arterial circulation creates an uneven flow rate which can be observed as the pulse, eliminating **(B)**. Finally, as the diameter of the vessels decreases, resistance increases, and therefore arteries have significantly less resistance to blood flow compared to arterioles, eliminating **(D)**.

13. **A**

This question is a simple application of the definition of pressure, which is force per area. If pressure decreases 1 percent and area does not change, the force will be decreased by 1 percent. Note that the other measurements given do not play a role in our calculations.

14. **C**

The equation for absolute (hydrostatic) pressure is $P = P_0 + \rho gz$, where P_0 is the pressure at the surface, ρ is the density of the fluid, g is acceleration due to gravity, and z is the depth in the fluid. If the density of fluid B is twice that of fluid A, then the depth in fluid A will have to be twice that in fluid B to obtain the same absolute pressure:

$$\begin{aligned} P_0 + \rho_A g z_A &= P_0 + \rho_B g z_B \\ (x) g z_A &= (2x) g z_B \\ z_A &= 2 z_B \end{aligned}$$

15. **B**

This is a basic interpretation of Bernoulli's equation that states, at equal heights, speed and pressure of a fluid are inversely related (the Venturi effect). Decreasing the speed of the water will therefore increase its pressure. An increase in pressure over a given area will result in increased force being transmitted to the piston.

Consult your online resources for additional practice. GO ONLINE

Equations to Remember

(4.1) **Density:** $\rho = \frac{m}{V}$

(4.2) **Weight of a volume of fluid:** $F_g = \rho V g$

(4.3) **Specific gravity:** $SG = \frac{\rho}{1 \frac{g}{cm^3}}$

(4.4) **Pressure:** $P = \frac{F}{A}$

(4.5) **Absolute pressure:** $P = P_0 + \rho g z$ → depth

(4.6) **Gauge pressure:** $P_{gauge} = P - P_{atm} = (P_0 + \rho g z) - P_{atm}$

(4.7) **Pascal's principle:** $P = \frac{F_1}{A_1} = \frac{F_2}{A_2}$

$$F_2 = F_1\left(\frac{A_2}{A_1}\right)$$

(4.8) **Buoyant force:** $F_{buoy} = \rho_{fluid} V_{fluiddisplaced} g = \rho_{fluid} V_{submerged} g$

(4.9) **Poiseuille's law:** $Q = \frac{\pi r^4 \Delta P}{8 \eta L}$

(4.10) **Critical speed:** $v_c = \frac{N_R \eta}{\rho D}$

(4.11) **Continuity equation:** $Q = v_1 A_1 = v_2 A_2$

(4.12) **Bernoulli's equation:** $P_1 + \frac{1}{2} \rho v_1^2 + \rho g h_1 = P_2 + \frac{1}{2} \rho v_2^2 + \rho g h_2$

Shared Concepts

Biology Chapter 6
The Respiratory System

Biology Chapter 7
The Cardiovascular System

Biology Chapter 8
The Immune System

General Chemistry Chapter 8
The Gas Phase

Physics and Math Chapter 2
Work and Energy

Physics and Math Chapter 3
Thermodynamics

CHAPTER 5

Electrostatics and Magnetism

SCIENCE MASTERY ASSESSMENT

Every pre-med knows this feeling: there is so much content I have to know for the MCAT! How do I know what to do first or what's important?

While the high-yield badges throughout this book will help you identify the most important topics, this Science Mastery Assessment is another tool in your MCAT prep arsenal. This quiz (which can also be taken in your online resources) and the guidance below will help ensure that you are spending the appropriate amount of time on this chapter based on your personal strengths and weaknesses. Don't worry though—skipping something now does not mean you'll never study it. Later on in your prep, as you complete full-length tests, you'll uncover specific pieces of content that you need to review and can come back to these chapters as appropriate.

How to Use This Assessment

If you answer 0–7 questions correctly:

Spend about 1 hour to read this chapter in full and take limited notes throughout. Follow up by reviewing **all** quiz questions to ensure that you now understand how to solve each one.

If you answer 8–11 questions correctly:

Spend 20–40 minutes reviewing the quiz questions. Beginning with the questions you missed, read and take notes on the corresponding subchapters. For questions you answered correctly, ensure your thinking matches that of the explanation and you understand why each choice was correct or incorrect.

If you answer 12–15 questions correctly:

Spend less than 20 minutes reviewing all questions from the quiz. If you missed any, then include a quick read-through of the corresponding subchapters, or even just the relevant content within a subchapter, as part of your question review. For questions you got correct, ensure your thinking matches that of the explanation and review the Concept Summary at the end of the chapter.

Questions 1–3 refer to the figure below, in which F represents the electrostatic force exerted on charged particle S by charged particle R.

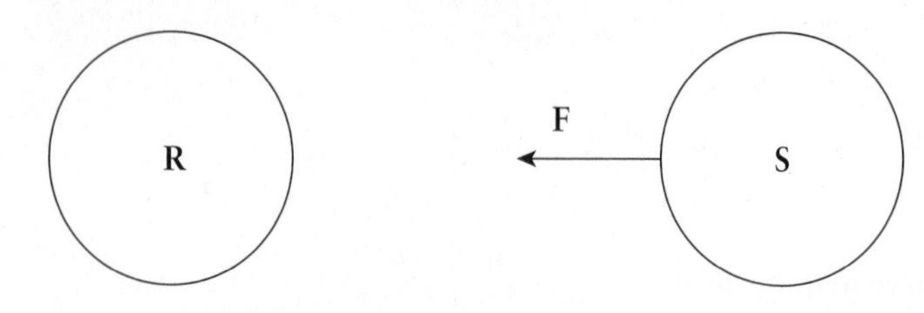

1. In the figure, the magnitude of the electric force on **R** due to **S** is:
   A. $\frac{F}{2}$.
   B. F.
   C. 2F.
   D. 4F.

2. If the distance between the centers of the spheres is halved, the magnitude of the force on **S** due to **R** will be:
   A. $\frac{F}{2}$.
   B. $\frac{F}{4}$.
   C. 2F.
   D. 4F.

3. Assume the direction of **F** is the same direction as the electric field between **R** and **S**. If an electron were placed midway between **R** and **S**, the resultant electric force on the electron would be:
   A. toward **R**.
   B. toward **S**.
   C. upward in the plane of the page.
   D. downward in the plane of the page.

4. If the electric field at a distance r away from charge Q is 36 $\frac{N}{C}$, what is the ratio of the electric fields at r, $2r$, and $3r$?
   A. 9:3:1
   B. 36:9:4
   C. 36:18:9
   D. 36:18:12

5. A positive charge of $+Q$ is fixed at point R a distance d away from another positive charge of $+2Q$ fixed at point S. Point A is located midway between the charges, and point B is a distance $\frac{d}{2}$ from $+2Q$, as shown below. In which direction will a positive charge move if placed at point A and point B, respectively?

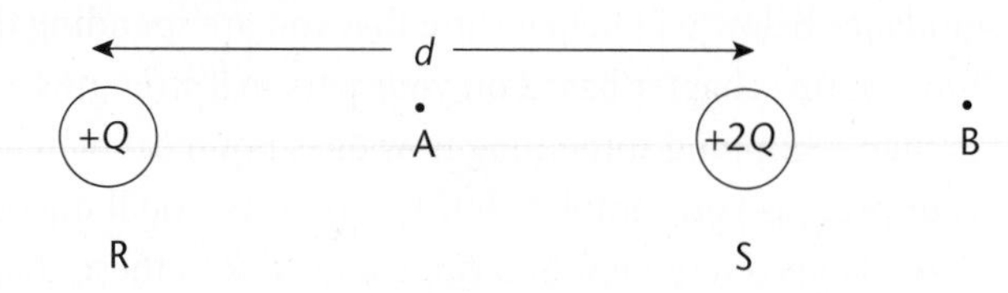

   A. Toward the $+Q$ charge for both
   B. Toward the $+2Q$ charge for both
   C. Toward the $+Q$ charge at point A, and toward the right at point B
   D. Toward the $+2Q$ charge at point A, and toward the right at point B

6. Two parallel conducting plates are separated by a distance d. One plate carries a charge $+Q$ and the other carries a charge $-Q$. The voltage between the plates is 12 V. If a $+2\ \mu C$ charge is released from rest at the positive plate, how much kinetic energy does it have when it reaches the negative plate?
   A. 2.4×10^{-6} J
   B. 4.8×10^{-6} J
   C. 2.4×10^{-5} J
   D. 4.8×10^{-5} J

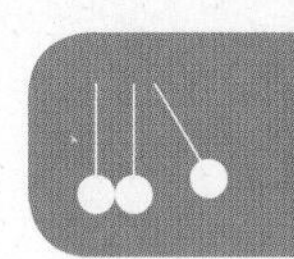

7. The negative charge in the figure below (−1 μC) moves from $y = -5$ to $y = +5$ and is made to follow the dashed line. What is the work required to move the negative charge along this dashed line?

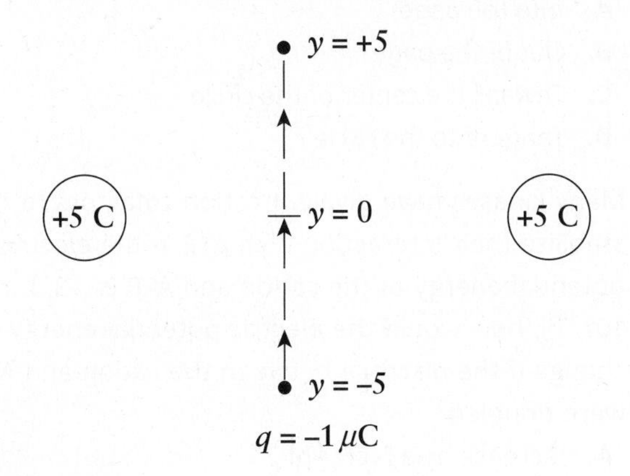

A. −10 J
B. −5 J
C. 0 J
D. 10 J

8. If the magnetic field a distance *r* away from a current-carrying wire is 10 T, what will be the net magnetic field at *r* if another wire is placed a distance 2*r* from the original wire (with *r* in the middle) and has a current twice as strong flowing in the opposite direction?
A. 0 T
B. 15 T
C. 20 T
D. 30 T

9. Given an electric dipole, the electric potential is zero:
A. only at the midpoint of the dipole axis.
B. anywhere on the perpendicular bisector of the dipole axis and at infinity.
C. anywhere on the dipole axis.
D. only for points at infinity.

10. An electron is accelerated over a distance *d* by an electric potential *V*. The electric potential applied to this electron is then increased by a factor of 4 and the electron is accelerated over the same distance *d*. The speed of the electron at the end of the second trial will be larger than at the end of the first trial by a factor of:
A. 16.
B. 8.
C. 4.
D. 2.

11. Which of the following accurately depicts the field lines created by a proton that is moving toward the right on this page?

A.

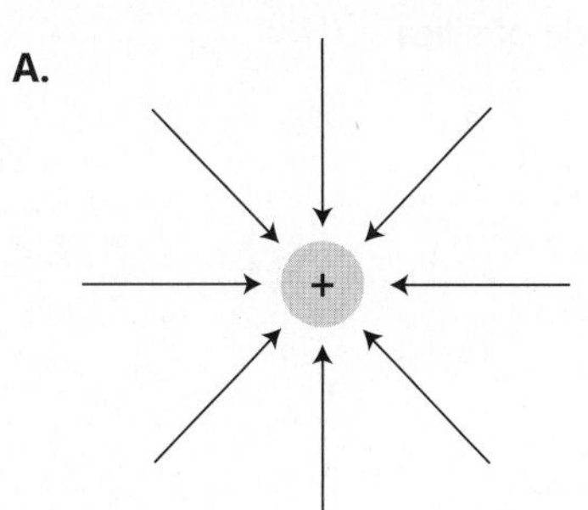

B.

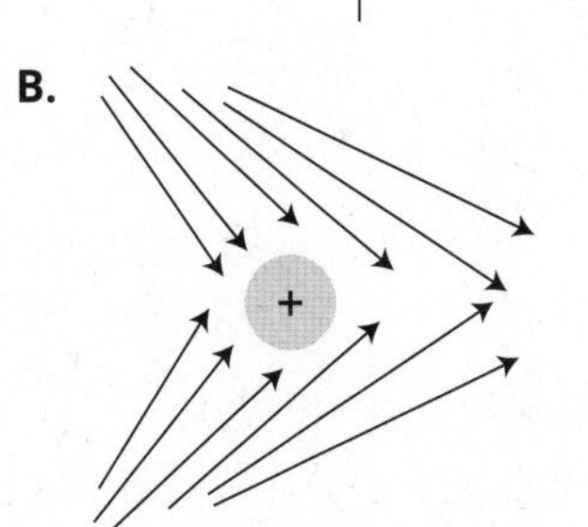

C.

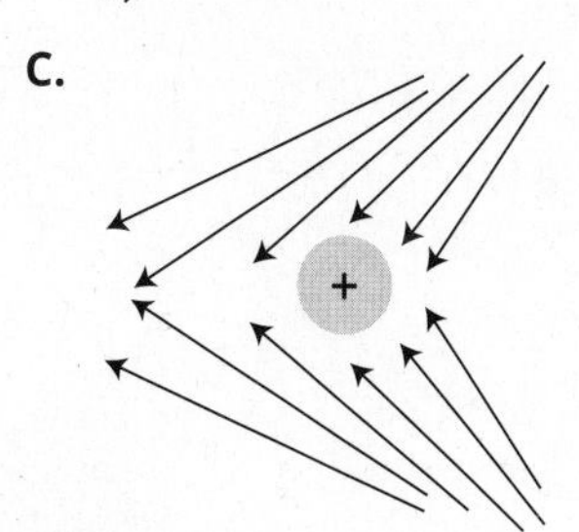

D.

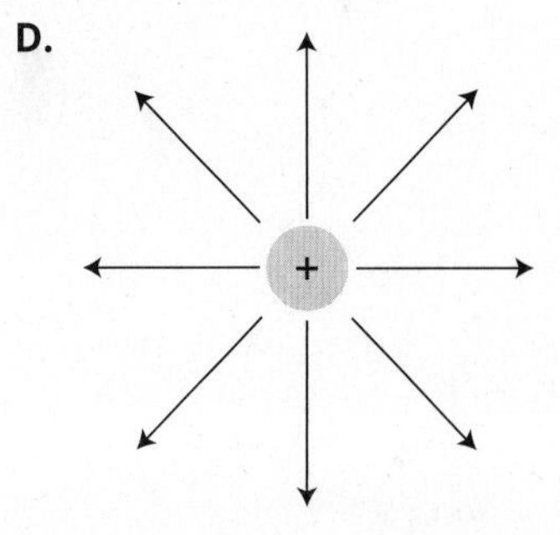

12. A certain 9 V battery is used as a power source to move a 2 C charge. How much work is done by the battery?
    - **A.** 4.5 J
    - **B.** 9 J
    - **C.** 18 J
    - **D.** 36 J

13. To increase the capacitance of a given capacitor, a dielectric is inserted between the two plates. Dielectrics are materials that block the flow of charge. Which of the following materials would be the best dielectric?
    - **A.** Salt water
    - **B.** Steel, an alloy primarily composed of iron
    - **C.** Glass, an oxide of silica
    - **D.** Copper

14. A moving negative charge placed in an external magnetic field circulates counterclockwise in the plane of the paper. In which direction is the magnetic field pointing?
    - **A.** Into the page
    - **B.** Out of the page
    - **C.** Toward the center of the circle
    - **D.** Tangent to the circle

15. Many kinases have divalent cation cofactors to stabilize their interaction with ATP. If the electric potential energy of the cation and ATP is -3.3×10^{-19} J, how would the electric potential energy change if the distance between the cation and ATP were doubled?
    - **A.** Increase by a factor of 2
    - **B.** Increase by a factor of 4
    - **C.** Decrease by a factor of 2
    - **D.** Decrease by a factor of 4

Answer Key follows on next page.

Answer Key

1. **B** (Ch. 5.2)
2. **D** (Ch. 5.2)
3. **B** (Ch. 5.2)
4. **B** (Ch. 5.2)
5. **C** (Ch. 5.1)
6. **C** (Ch. 5.3)
7. **C** (Ch. 5.4)
8. **D** (Ch. 5.6)
9. **B** (Ch. 5.5)
10. **D** (Ch. 5.4)
11. **D** (Ch. 5.2)
12. **C** (Ch. 5.4)
13. **C** (Ch. 5.1)
14. **B** (Ch. 5.6)
15. **A** (Ch. 5.3)

CHAPTER 5

Electrostatics and Magnetism

In This Chapter

5.1 **Charges**
Insulators and Conductors.........168

5.2 **Coulomb's Law**
Electric Field....................172

5.3 **Electric Potential Energy**......174

5.4 **Electric Potential**..............176

5.5 **Special Cases in Electrostatics**
Equipotential Lines 179
Electric Dipoles..................180

5.6 **Magnetism**
Magnetic Fields 185
Magnetic Forces 187

Concept Summary 192

CHAPTER PROFILE

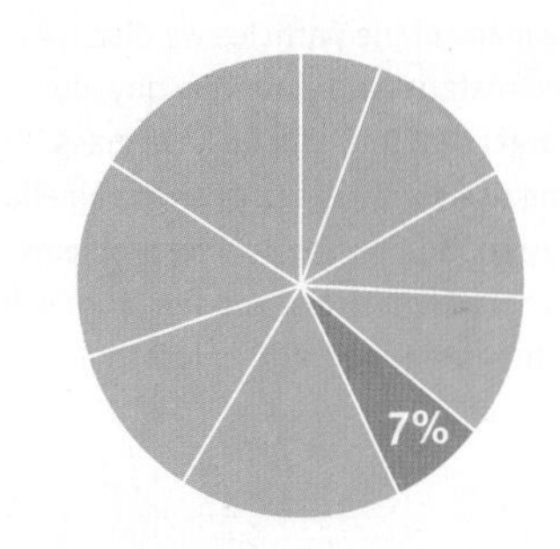

The content in this chapter should be relevant to about 7% of all questions about physics on the MCAT.

This chapter covers material from the following AAMC content category:

4C: Electrochemistry and electrical circuits and their elements

Introduction

Electrostatics is the study of stationary charges and the forces that are created by and which act upon these charges. Without electrical charge, we would not be able to do many of the activities that we enjoy or consider essential to basic living. But living with electrical charge can also be dangerous and even deadly: magnify the small shock you receive from a doorknob after walking across a carpet, and you have the equivalent of a lightning bolt strong enough to stop a heart. This same concept can be used in life-saving therapy as well: cardioversion and defibrillation create a strong electrical current through the heart's conduction system that attempts to resynchronize a pulse.

In this chapter, we will review the basic concepts essential to understanding charges and electrostatic forces including conductors and insulators. We will review Coulomb's law, which describes the attractions and repulsions between charged objects. Next, we will describe the electric fields that all charges create, which allow them to exert forces on other charges. After we've discussed how charges set up these fields, we'll observe the behavior of charges that are placed into these fields. In particular, we will note the motional behavior of these test charges inside a field in relation to the electrical potential difference, or voltage, between two points in space. We can then determine the change in electric potential energy as the charge moves from a position of some electric potential to another. Next, we will describe the electric dipole and solve a problem involving one of the molecular dipoles most important to life on this planet: the water molecule. Finally, we will explore the topic of magnetic fields and forces.

5.1 Charges

LEARNING OBJECTIVES

After Chapter 5.1, you will be able to:

- Contrast the behavior of protons and electrons in charged environments
- Categorize materials, including glass and copper, as conductors or insulators
- Recognize the charge of an electron in coulombs

MCAT EXPERTISE

While many of the particles we discuss in electrostatics are very, very tiny, do not forget that they still do have mass. We can use equations such as the kinetic energy equation when solving problems with charged particles, and the MCAT will sometimes require us to do just that.

Charged subatomic particles come in two varieties. One, the **proton**, has a positive charge; the other, the **electron**, has a negative charge. While opposite charges exert **attractive** forces, like charges—those that have the same sign—exert **repulsive** forces. Unlike the force of gravity, which is always an attractive force, the electrostatic force may be repulsive or attractive depending on the signs of the charges that are interacting.

Most matter is electrically neutral, as a balance of positive and negative charges ensures a relative degree of stability. When charges are out of balance, the system can become electrically unstable. Even materials that are normally electrically neutral can acquire a net charge as result of friction. When you shuffle your feet across the carpet, negatively charged particles are transferred from the carpet to your feet, and these charges spread out over the total surface of your body. The shock that occurs when your hand gets close enough to a metal doorknob allows that excess charge to jump from your fingers to the knob, which acts as a **ground**—a means of returning charge to the earth. **Static charge buildup** or **static electricity** is more significant in drier air because lower humidity makes it easier for charge to become and remain separated.

KEY CONCEPT

The fundamental unit of charge is $e = 1.60 \times 10^{-19}$ C. A proton and an electron each have this amount of charge; the proton is positively charged ($q = +e$), while the electron is negatively charged ($q = -e$).

The SI unit of charge is the **coulomb**, and the fundamental unit of charge is

$$e = 1.60 \times 10^{-19}\ \text{C}$$

A proton and an electron each have this amount of charge, although the proton is positively charged ($q = +e$), while the electron is negatively charged ($q = -e$). Even though the proton and the electron share the same magnitude of charge, they do not share the same mass; the proton has a much greater mass than the electron.

Like mass and energy, electric charge is governed by a law of conservation of charge. This law states that charge can neither be created nor destroyed.

Insulators and Conductors

Insulators and conductors vary in their ability to both hold and transfer charges. An **insulator** will not easily distribute a charge over its surface and will not transfer that charge to another neutral object very well—especially not to another insulator. On a molecular level, the electrons of insulators tend to be closely linked with their respective nuclei. By extension, most nonmetals are insulators. Experimentally, insulators serve as dielectric materials in capacitors, as well as in isolating electrostatic experiments from the environment to prevent grounding.

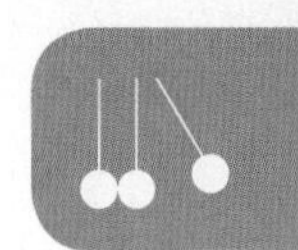

In contrast, when a **conductor** is given a charge, the charges will distribute approximately evenly upon the surface of the conductor. Conductors are able to transfer and transport charges and are often used in circuits or electrochemical cells. Conductors are often conceptualized as nuclei surrounded by a sea of free electrons that are able to move rapidly throughout the material and are only loosely associated with the positive charges. Conductors are generally metals, although ionic (electrolyte) solutions are also effective conductors. Figure 5.1 demonstrates the behaviors of an insulator and a conductor when a negative charge is placed on them.

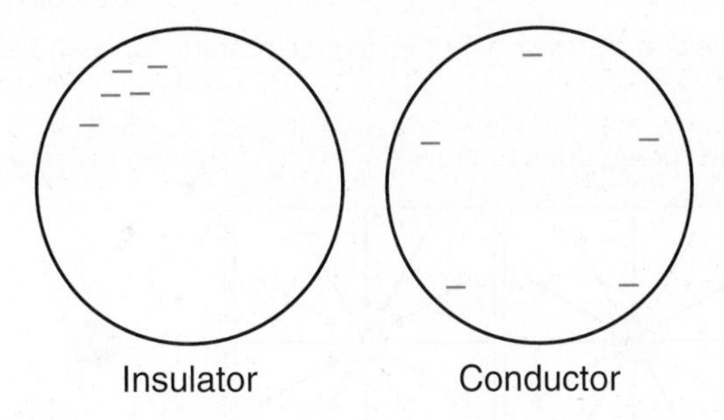

Figure 5.1 A Negatively Charged Insulator and Conductor
Insulators will not distribute charge over their surface; conductors will.

MCAT CONCEPT CHECK 5.1

Before you move on, assess your understanding of the material with these questions.

1. When placed one meter apart from each other, which will experience a greater acceleration: one coulomb of electrons or one coulomb of protons?

   ____

2. Categorize the following materials as either conductors or insulators: blood, hair, copper, glass, iron, sulfuric acid, and distilled water

   - Conductors:

   ____

   - Insulators:

   ____

3. What is the net charge of an object with one coulomb of electrons and 3 moles of neutrons?

   ____

5.2 Coulomb's Law

LEARNING OBJECTIVES

After Chapter 5.2, you will be able to:

- Calculate the electric field that a charge generates and the electric force between it and another charge
- Recall the direction in which a negative or a positive electrostatic force will move two charges relative to one another
- Relate distance and charge quantities to electrostatic force and electrostatic field magnitudes
- Apply direction conventions to draw electric fields generated by a charged object:

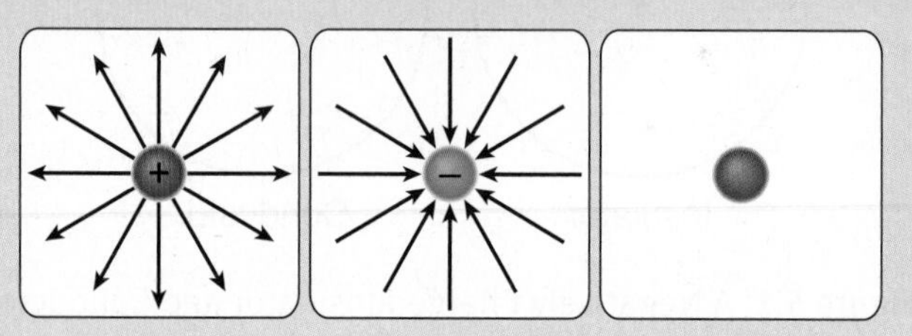

Coulomb's law quantifies the magnitude of the electrostatic force F_e between two charges:

$$F_e = \frac{kq_1q_2}{r^2}$$

Equation 5.1

where F_e is the magnitude of the electrostatic force, k is Coulomb's constant, q_1 and q_2 are the magnitudes of the two charges, and r is the distance between the charges. **Coulomb's constant** (also called the **electrostatic constant**) is a number that depends on the units used in the equation. In SI units, $k = \frac{1}{4\pi\varepsilon_0} = 8.99 \times 10^9 \frac{\text{N}\cdot\text{m}^2}{\text{C}^2}$, where ε_0 represents the **permittivity of free space**, $8.85 \times 10^{-12} \frac{\text{C}^2}{\text{N}\cdot\text{m}^2}$. The direction of the force may be obtained by remembering that unlike charges attract and like charges repel. The force always points along the line connecting the centers of the two charges.

Example: A positive charge is attracted to a negative charge a certain distance away. The charges are then moved so that they are separated by twice the distance. How has the force of attraction changed between them?

Solution: Coulomb's law states that the force between two charges varies as the inverse of the square of the distance between them. Therefore, if the distance is doubled, the square of the distance is quadrupled, and the force is reduced to one-fourth of what it was originally. Note that it was not necessary to know the distance or the units being used.

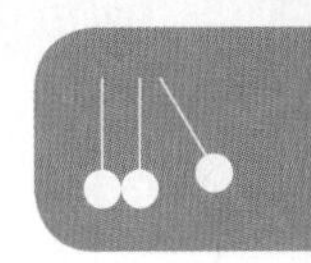

A close examination of Coulomb's law reveals that it is remarkably similar in form to the equation for gravitational force. In the electrostatic force equation, the force magnitude is proportional to the charge magnitudes, and this is similar to the proportional relationship between gravitational force and mass. In both equations, the force magnitude is inversely proportional to the square of the distance of separation. These similarities ought to help you remember both equations on Test Day.

BRIDGE

Notice how Coulomb's law looks very similar to the gravitational force equation, but with a different constant and using charge rather than mass. It is this fact that should remind us that this equation is dealing with electrostatic force between two charges, just as the gravitation equation is dealing with the gravitational force between two bodies of mass. The gravitation equation is discussed in Chapter 1 of *MCAT Physics and Math Review*.

Example: Negatively charged electrons are electrostatically attracted to positively charged protons. Because electrons and protons have mass, they will be gravitationally attracted to each other as well. What is the ratio of the electrostatic force to the gravitational force between an electron and proton? (Note: $m_p = 1.67 \times 10^{-27}$ kg, $m_e = 9.11 \times 10^{-31}$ kg, $e = 1.60 \times 10^{-19}$ C, $k = 8.99 \times 10^9 \frac{\text{N} \cdot \text{m}^2}{\text{C}^2}$, and $\text{G} = 6.67 \times 10^{-11} \frac{\text{N} \cdot \text{m}^2}{\text{kg}^2}$)

No calculator, would I need to know these?

Solution: Both Coulomb's law and the universal law of gravitation state that the attractive forces between the electron and proton vary as the inverse of the square of the distance between them. The ratio between these forces can be calculated by dividing their magnitudes:

$$\frac{F_e}{F_g} = \frac{\left[\frac{kq_1q_2}{r^2}\right]}{\left[\frac{Gm_1m_2}{r^2}\right]} = \frac{kq_1q_2}{Gm_1m_2}$$

Now the values can be plugged in:

$$\frac{F_e}{F_g} = \frac{\left(8.99 \times 10^9 \frac{\text{N} \cdot \text{m}^2}{\text{C}^2}\right)\left(1.60 \times 10^{-19}\ \text{C}\right)\left(1.60 \times 10^{-19}\ \text{C}\right)}{\left(6.67 \times 10^{-11} \frac{\text{N} \cdot \text{m}^2}{\text{kg}^2}\right)\left(1.67 \times 10^{-27}\ \text{kg}\right)\left(9.11 \times 10^{-31}\ \text{kg}\right)}$$

$$\approx \frac{(9)(1.6)(1.6)\left(10^{-29}\right)}{(6.67)(1.6)(9)\left(10^{-69}\right)}$$

$$= \frac{1.6}{6.67} \times 10^{40} \approx \frac{1.6}{6.4} \times 10^{40} = 0.25 \times 10^{40}$$

$$= 2.5 \times 10^{39} \left(\text{actual} = 2.27 \times 10^{39}\right)$$

Note that the electrostatic attraction between the electron and proton is stronger than the gravitational attraction by a factor of almost 10^{40}. Also, note that setting up all of the variables before working out the math simplifies the process because a number of the variables cancel out during the division.

Electric Field

KEY CONCEPT

Electric fields are produced by source charges (Q). When a test charge (q) is placed in an electric field (**E**), it will experience an electrostatic force ($\mathbf{F_e}$) equal to $q\mathbf{E}$.

Every electric charge sets up a surrounding electric field, just like every mass creates a gravitational field. **Electric fields** make their presence known by exerting forces on other charges that move into the space of the field. Whether the force exerted through the electric field is attractive or repulsive depends on whether the stationary **test charge** q (the charge placed in the electric field) and the stationary **source charge** Q (which actually creates the electric field) are opposite charges (attractive) or like charges (repulsive).

The magnitude of an electric field can be calculated in one of two ways, both of which can be seen in the definitional equation for the electric field:

$$E = \frac{F_e}{q} = \frac{kQ}{r^2}$$

Equation 5.2

where E is the electric field magnitude in newtons per coulomb, F_e is the magnitude of the force felt by the test charge q, k is the electrostatic constant, Q is the source charge magnitude, and r is the distance between the charges. The electric field is a vector quantity, and we will discuss the process of determining the direction of the electric field vector in a moment. Look closely: you can see that this equation for the electric field magnitude is derived simply by dividing both sides of Coulomb's law by the test charge q. In doing so, we arrive at two different methods for calculating the magnitude of the electric field at a particular point in space. The first method is to place a test charge q at some point within the electric field, measure the force exerted on that test charge, and define the electric field at that point in space as the ratio of the force magnitude to test charge magnitude $\left(\frac{F_e}{q}\right)$. One of the disadvantages of this method of calculation is that a test charge must actually be present in order for a force to be generated and measured. Sometimes, however, no test charge is actually within the electric field, so we need another way to measure the magnitude of that field.

KEY CONCEPT

By dividing Coulomb's law by the magnitude of the test charge, we arrive at two ways of determining the magnitude of the electric field at a point in space around the source charge.

The second method of calculating the electric field magnitude at a point in space does not require the presence of a test charge. We only need to know the magnitude of the source charge and the distance between the source charge and point in space at which we want to measure the electric field $\left(\frac{kQ}{r^2}\right)$. In this method, we need to know the value of the source charge to be able to calculate the electric field.

By convention, the direction of the electric field vector is given as the direction that a positive test charge would move in the presence of the source charge. If the source charge is positive, then the test charge would experience a repulsive force and would accelerate away from the positive source charge. On the other hand, if the source charge is negative, then the test charge would experience an attractive force and would accelerate toward the negative source charge. Therefore, positive charges have electric field vectors that radiate outward (that is, point away) from the charge, whereas negative charges have electric field vectors that radiate inward (point toward) the charge. These electric field vectors may be represented using field lines, as shown in Figure 5.2.

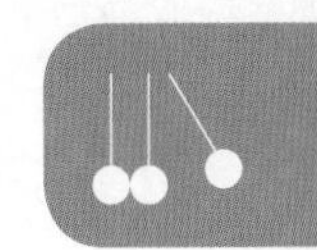

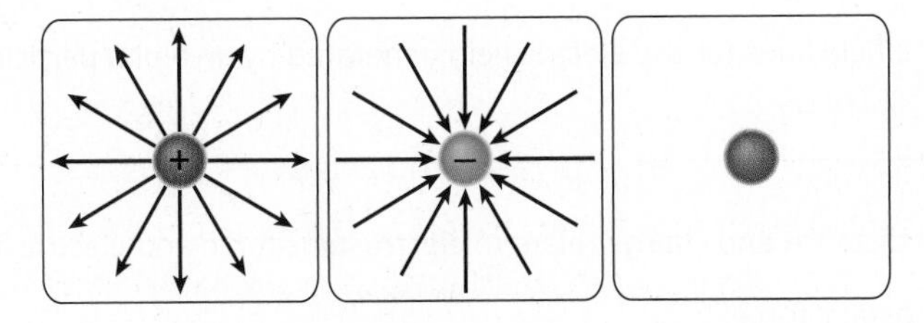

Figure 5.2 Field Lines around a Positive, a Negative, and a Neutral Source Charge

Field lines are imaginary lines that represent how a positive test charge would move in the presence of the source charge. The field lines are drawn in the direction of the actual electric field vectors and also indicate the relative strength of the electric field at a given point in the space of the field. When drawn on a sheet of paper, field lines look like the metal spokes of a bicycle wheel: the lines are closer together near the source charge and spread out at distances farther from the charge. Where the field lines are closer together, the field is stronger; where the lines are farther apart, the field is weaker. Because every charge exerts its own electric field, a collection of charges will exert a net electric field at a point in space that is equal to the vector sum of all the electric fields.

Because electric field and electrostatic force are both vector quantities, it is important to remember the conventions for their direction. If the test charge within a field is positive, then the force will be in the same direction as the electric field vector of the source charge; if the test charge is negative, then the force will be in the direction opposite to the field vector of the source charge.

KEY CONCEPT

Field lines are used to represent the electric field vectors for a charge. They point away from a positive charge and point toward a negative charge. The denser the field lines, the stronger the electric field. Note that field lines of a single charge never cross each other.

MCAT CONCEPT CHECK 5.2

Before you move on, assess your understanding of the material with these questions.

1. What is the electric field midway between two negative charges in isolation?

___

2. What direction does a negative electrostatic force point? What direction does a positive electrostatic force point?
   - Negative electrostatic force:

   ___

   - Positive electrostatic force:

   ___

3. Draw the field lines for the electric field generated by an alpha particle $\left({}^{4}_{2}\text{He}^{2+}\right)$.

4. How do distance and charge relate to electrostatic force and electric field?
   - Electrostatic force:
   - Electric field:

5.3 Electric Potential Energy

LEARNING OBJECTIVES

After Chapter 5.3, you will be able to:

- Describe how a change in electric potential energy will affect the stability of a system
- Calculate electric potential energy
- Compare and contrast electric potential energy with electrostatic force, conceptually and mathematically
- Predict change in electric potential energy given a change in distance

We have already defined potential energy as stored energy that can be used to do something or make something happen. There are different types of potential energy; gravitational, elastic, and chemical are three forms that you will need to know for Test Day. A fourth form is **electric potential energy**. Similar to gravitational potential energy, this is a form of potential energy that is dependent on the relative position of one charge with respect to another charge or to a collection of charges. Electric potential energy is given by the equation

$$U = \frac{kQq}{r}$$

Equation 5.3

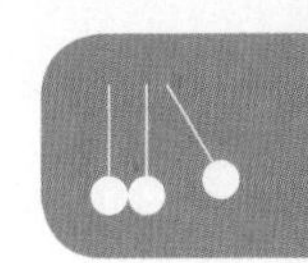

If the charges are like charges (both positive or both negative), then the potential energy will be positive. If the charges are unlike (one positive and the other negative), then the potential energy will be negative. Remember that work and energy have the same unit (the joule), so we can define electric potential energy for a charge at a point in space in an electric field as the amount of work necessary to bring the charge from infinitely far away to that point. Because $F_e = \frac{kQq}{r^2}$ and $W = Fd\cos\theta$, if we define d as the distance r that separates two charges and assume the force and displacement vectors to be parallel, then:

$$\Delta U = W = Fd\cos\theta = Fr \times 1 = \left(\frac{kQq}{r^2}\right)r = \frac{kQq}{r}$$

KEY CONCEPT

Electric potential energy is the work necessary to move a test charge from infinity to a point in space in an electric field surrounding a source charge.

Consider two charges: a stationary negative source charge and a positive test charge that can be moved. Because these two charges are unlike, they will exert attractive forces between them. Therefore, the closer they are to each other, the more stable they will be. Opposite charges will have negative potential energy, and this energy will become increasingly negative as the charges are brought closer and closer together. Increasingly negative numbers are actually decreasing values because they are moving farther to the left of 0 on the number line. This decrease in energy represents an increase in stability.

Now let's consider two positive charges. As like charges, these will exert repulsive forces, and the potential energy of the system will be positive. Because like charges repel each other, the closer they are to each other, the less stable they will be. Remember that unlike gravitational systems, the forces of electrostatics can be either attractive or repulsive. In this case, the like charges will become more stable the farther apart they move because the magnitude of the electric potential energy becomes a smaller and smaller positive number.

KEY CONCEPT

The electric potential energy of a system will increase when two like charges move toward each other or when two opposite charges move apart. Conversely, the electric potential energy of a system will decrease when two like charges move apart or when two opposite charges move toward each other.

Example: If a charge of $+2e$ and a charge of $-3e$ are separated by a distance of 3 nm, what is the potential energy of the system? (Note: e is the fundamental unit of charge equal to 1.6×10^{-19} C, and k is the electrostatic constant equal to $8.99 \times 10^9 \frac{\text{N}\cdot\text{m}^2}{\text{C}^2}$.)

Solution: The equation for potential energy is $U = \frac{kQq}{r}$. From the question stem, we know that the charges are $+2e$ and $-3e$, and $r = 3\text{ nm} = 3 \times 10^{-9}$ m. Plugging into the equation, we get:

$$U = \frac{\left(8.99 \times 10^9 \frac{\text{N}\cdot\text{m}^2}{\text{C}^2}\right)\left(2 \times 1.6 \times 10^{-19}\text{ C}\right)\left(-3 \times 1.6 \times 10^{-19}\text{ C}\right)}{3 \times 10^{-9}\text{ m}}$$

$$\approx -45 \times 10^{-20}\text{ J} = -4.5 \times 10^{-19}\text{ J}\left(\text{actual} = -4.6 \times 10^{-19}\text{ J}\right)$$

MCAT CONCEPT CHECK 5.3

Before you move on, assess your understanding of the material with these questions.

1. How does a change in electric potential energy from −4 J to −7 J reflect on the stability of a system?

 ____

 ____

2. Compare the relationship between electric potential energy and Coulomb's law to the relationship between gravitational potential energy and the universal law of gravitation.

 ____

 ____

 ____

3. How does electric potential energy change between two particles as the distance between them increases?

 ____

 ____

 ____

4. By what factor would electric potential energy change if the magnitude of both charges were doubled and the distance between them was halved?

 ____

5.4 Electric Potential

LEARNING OBJECTIVES

After Chapter 5.4, you will be able to:

- Calculate electric potential
- Distinguish between electric potential and voltage
- Predict the movement of a charge relative to a source charge, given the electric potential of the test charge's location
- Relate electric potential to electric potential energy, electric field, and Coulomb's law

Electric potential, discussed here, and electric potential energy, discussed previously, sound like the same (or nearly the same) thing. They are not, although they are very closely related. In fact, **electric potential** is defined as the ratio of the magnitude of a charge's electric potential energy to the magnitude of the charge itself. This can be expressed as:

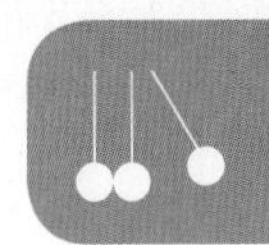

$$V = \frac{U}{q}$$

Equation 5.4

where V is the electric potential measured in **volts (V)** and $1\text{ V} = 1\frac{\text{J}}{\text{C}}$. Even if there is no test charge q, we can still calculate the electric potential of a point in space in an electric field as long as we know the magnitude of the source charge and the distance from the source charge to the point in space in the field. By dividing $U = \frac{kQq}{r}$ by q, we get:

$$V = \frac{kQ}{r}$$

Equation 5.5

Electric potential is a scalar quantity, and its sign is determined by the sign of the source charge Q. For a positive source charge, V is positive, but for a negative source charge, V is negative. For a collection of charges, the total electric potential at a point in space is the scalar sum of the electric potential due to each charge.

Because electric potential is inversely proportional to the distance from the source charge, a potential difference will exist between two points that are at different distances from the source charge. If V_a and V_b are the electric potentials at points a and b, respectively, then the **potential difference** between them, known as **voltage**, is $V_b - V_a$. From the equation for electric potential above, we can further define potential difference as:

$$\Delta V = V_b - V_a = \frac{W_{ab}}{q}$$

Equation 5.6

where W_{ab} is the work needed to move a test charge q through an electric field from point a to point b. The work depends only on the potentials at the two points a and b and is independent of the actual pathway taken between a and b. Like gravitational force, the electrostatic force is a conservative force.

We've already seen that charges, if allowed, will move spontaneously in whatever direction results in a decrease in electric potential energy. For a positive test charge, this means moving from a position of higher electric potential to a position of lower electric potential. The voltage, $\Delta V = V_b - V_a$, is negative in this case; because q is positive (for a positive test charge), thus, W_{ab} must be negative, which represents a decrease in electric potential energy.

Now let's consider a negative test charge. A negative test charge will spontaneously move from a position of lower electric potential to a position of higher electric potential. The voltage, $\Delta V = V_b - V_a$, is positive in this case; because q is negative (for a negative test charge), W_{ab} must also be negative, which again represents a decrease in electric potential energy. The takeaway: positive charges will spontaneously move in the direction that *decreases* their electric potential (negative voltage), whereas negative charges will spontaneously move in the direction that *increases* their electric potential (positive voltage)—yet, in both cases, the electric potential *energy* is decreasing.

MCAT EXPERTISE

$F_e = \frac{kQq}{r^2}$	$U = \frac{kQq}{r}$
$E = \frac{kQ}{r^2}$	$V = \frac{kQ}{r}$

These are essential equations for Test Day. Know how they relate to each other and when to use them. By memorizing Coulomb's law, you should be able to recreate the table through mathematical manipulation. From left to right, multiply by r; from top to bottom, divide by q.

MNEMONIC

The "plus" end of a battery is the high-potential end, and the "minus" end of a battery is the low-potential end. Positive charge moves from + to − (the definition of current) while negative charge moves from − to +.

KEY CONCEPT

Electric potential is the ratio of the work done to move a test charge from infinity to a point in an electric field surrounding a source charge divided by the magnitude of the test charge.

BRIDGE

Create analogies between mechanics and electrostatics to familiarize yourself with these concepts. Electric field is like a gravitational field, and it exerts forces on charges much like a gravitational field exerts forces on masses. A test charge has a particular electric potential energy at a given electric potential, depending on the magnitude of its charge, much like a mass has a particular gravitational potential energy, depending on the magnitude of its mass. These mechanics concepts are discussed in Chapters 1 and 2 of *MCAT Physics and Math Review*.

MCAT CONCEPT CHECK 5.4

Before you move on, assess your understanding of the material with these questions.

1. What is the difference between electric potential and voltage?

____

____

____

2. How will a charge that is placed at a point of zero electric potential move relative to a source charge?

____

____

____

3. True or False: The units of electric potential energy and electric potential are different.

5.5 Special Cases in Electrostatics

LEARNING OBJECTIVES

After Chapter 5.5, you will be able to:

- Describe equipotential lines and electric dipoles
- Recall the electrical potential at points along the bisector of dipole
- Predict the voltage at two distinct points on an equipotential line
- Predict the behavior of a dipole when exposed to an external field:

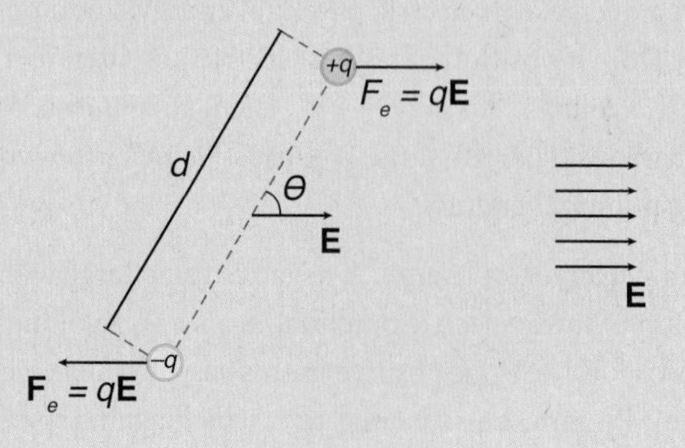

In this section, we will explore some of the unique setups in electrostatics that are common on the MCAT.

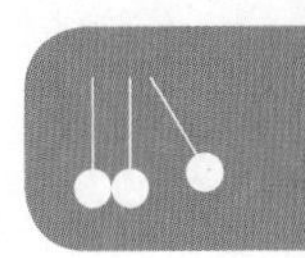

Equipotential Lines

An **equipotential line** is a line on which the potential at every point is the same. That is, the potential difference between any two points on an equipotential line is zero. Drawn on paper, equipotential lines may look like concentric circles surrounding a source charge. In three-dimensional space, these equipotential lines would actually be spheres surrounding the source charge. From the equation for electrical potential, we can see that no work is done when moving a test charge q from one point on an equipotential line to another. Work will be done in moving a test charge q from one line to another, but the work depends only on the potential difference of the two lines and not on the pathway taken between them. This is entirely analogous to the displacement of an object horizontally on a level surface. Because the object's height above the ground has not changed, its gravitational potential energy is unchanged. Furthermore, a change in the object's gravitational potential energy will not depend on the pathway taken from one height to another but only on the actual vertical displacement.

BRIDGE

Because the work to move a charge from one equipotential line to another does not depend on the path between them, we know that we are dealing only with conservative forces when moving the charge. Conservative and nonconservative forces are discussed in Chapter 2 of *MCAT Physics and Math Review*.

Example: In the diagram below, an electron goes from point a to point b in the vicinity of a very large positive charge. The electron could be made to follow any of the paths shown. Which path requires the least work to get the electron charge from a to b?

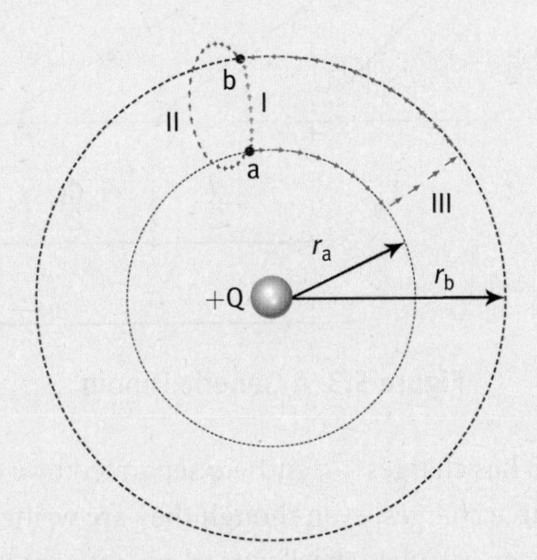

Solution: As stated, the work depends only on the potential difference and not on the path, so any of the paths shown would require the same amount of work in moving the electron from a to b. Note that because the source charge is positive, point b is at a *lower* electrical potential than point a. However, because the test charge is negative, the electrical potential energy is *higher* at point b than point a. This should make sense: the electron will have to gain energy to be moved farther away from positive source charge.

Work is not path-dependent!

Electric Dipoles

Much of the reactivity of organic compounds is based on separation of charge. The **electric dipole**, which results from two equal and opposite charges being separated a small distance d from each other, can be transient (as in the case of the moment-to-moment changes in electron distribution that create London dispersion forces) or permanent (as in the case of the molecular dipole of water or the carbonyl functional group).

The electric dipole can be visualized as a barbell: the equal weights on either end of the bar represent the equal and opposite charges separated by a small distance, represented by the length of the bar. We'll analyze the generic dipole in Figure 5.3 and then work through the specific example of one of the most important electric dipoles, the water molecule.

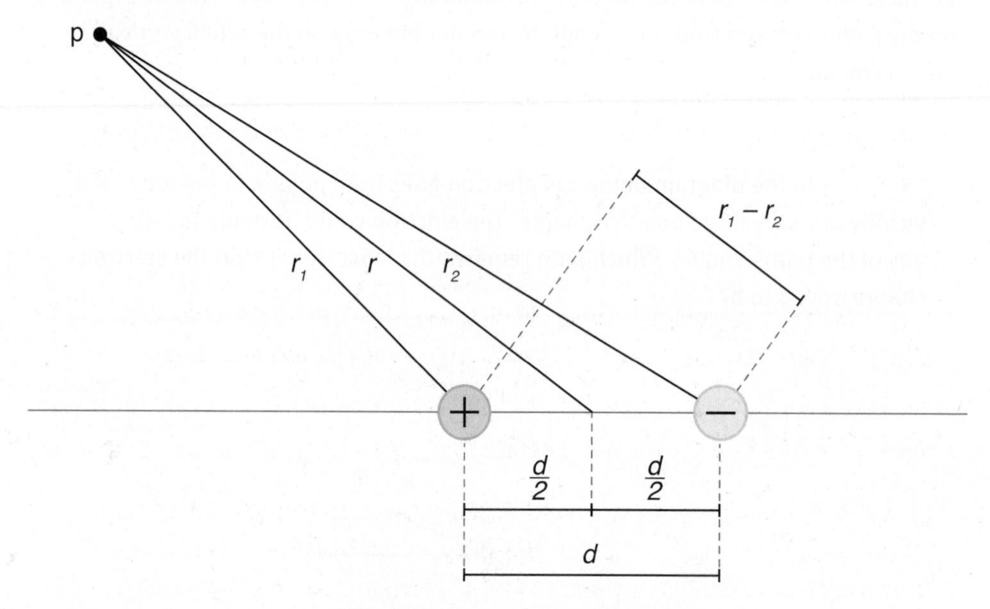

Figure 5.3 A Generic Dipole

The dipole in Figure 5.3 has charges $+q$ and $-q$ separated by a distance d. Notice that $+q$ and $-q$ are source charges, even though they are written in lowercase. Given the dipole, we may want to calculate the electrical potential at some point P near the dipole. The distance between the point in space and $+q$ is r_1; the distance between the point in space and $-q$ is r_2; the distance between the point in space and the midpoint of the dipole is r. We know that for a collection of charges, the electrical potential P is the scalar sum of the potentials due to each charge at that point. In other words,

$$V = \frac{kq}{r_1} - \frac{kq}{r_2}$$
$$= \frac{kq(r_2 - r_1)}{r_1 r_2}$$

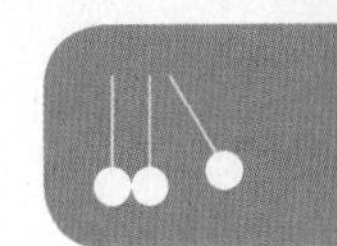

For points in space relatively distant from the dipole (compared to d), the product of r_1 and r_2 is approximately equal to the square of r, and $r_1 - r_2$ is approximately equal to $d \cos \theta$. When we plug these approximations into the equation above, we get

$$V = \frac{kqd}{r^2} \cos \theta$$

Equation 5.7

The product of charge and separation distance is defined as the **dipole moment** (**p**) with SI units of C · m:

$$\mathbf{p} = q\mathbf{d}$$

Equation 5.8

The dipole moment is a vector, but its direction is defined differently by physicists and chemists. Physicists define the vector along the line connecting the charges (the dipole axis), with the vector pointing from the negative charge toward the positive charge. Chemists usually reverse this convention, having **p** point from the positive charge toward the negative charge. Sometimes, chemists draw a crosshatch at the tail end of the dipole vector to indicate that the tail end is the positive charge.

Example: The H_2O molecule has a dipole moment of 1.85 D. Calculate the electrical potential due to a water molecule at a point 89 nm away along the axis of the dipole. (Note: $k = 8.99 \times 10^9 \frac{\text{N} \cdot \text{m}^2}{\text{C}^2}$ and 1 D (debye) = 3.34×10^{-30} C · m)

Solution: Because the question asks for the potential along the axis of the dipole, the angle θ is 0°. Substitute the values into the equation for the dipole potential and multiply 1.85 D by 3.34×10^{-30} to convert it to C · m:

$$V = \frac{kqd}{r^2} \cos \theta = \frac{kp}{r^2} \cos \theta$$

$$= \frac{\left(8.99 \times 10^9 \frac{\text{N} \cdot \text{m}^2}{\text{C}^2}\right)(1.85\ \text{D})\left(\frac{3.34 \times 10^{-30}\ \text{C} \cdot \text{m}}{1\ \text{D}}\right)}{\left(89 \times 10^{-9}\ \text{m}\right)^2}$$

$$\approx \frac{(9)(2)(3)\left(10^{-21}\right)}{\left(9 \times 10^{-8}\right)^2} = \frac{(9)(2)(3)\left(10^{-21}\right)}{(9)(9)\left(10^{-16}\right)}$$

$$= \frac{\left(2 \times 10^{-21}\right)}{\left(3 \times 10^{-16}\right)} = 0.67 \times 10^{-5}\ \text{V} = 6.7 \times 10^{-6}\ \text{V} \left(\text{actual} = 7.01 \times 10^{-6}\ \text{V}\right)$$

One very important equipotential line to be aware of is the plane that lies halfway between $+q$ and $-q$, called the **perpendicular bisector of the dipole**. Because the angle between this plane and the dipole axis is 90° (and $\cos 90° = 0$), the electrical potential at any point along this plane is 0. The magnitude of the electric field on the perpendicular bisector of the dipole can be approximated as

$$E = \frac{1}{4\pi\varepsilon_0} \times \frac{p}{r^3}$$

Equation 5.9

The electric field vectors at the points along the perpendicular bisector will point in the direction opposite to **p** (as defined directionally by physicists).

The dipole is a classic example of a setup upon which torques can act. In the absence of an electric field, the dipole axis can assume any random orientation. However, when the electric dipole is placed in a uniform external electric field, each of the equal and opposite charges of the dipole will experience a force exerted on it by the field. Because the charges are equal and opposite, the forces acting on the charges will also be equal in magnitude and opposite in direction, resulting in a situation of translational equilibrium. There will be, however, a net torque about the center of the dipole axis:

$$\begin{aligned}\tau &= \left(\frac{d}{2}\right)F_e \sin\theta + \left(\frac{d}{2}\right)F_e \sin\theta \\ &= dF_e \sin\theta \\ &= d(qE)\sin\theta \\ &= pE\sin\theta\end{aligned}$$

MCAT EXPERTISE

The dipole is a great example of how the MCAT can test kinematics and dynamics in an electrostatics setting. For a dipole at some angle in an external electric field, there will be translational equilibrium, but not rotational equilibrium. This is because the forces are in opposite directions (left and right in Figure 5.4), but the torques are in the same direction (clockwise for both).

Thus, the net torque on a dipole can be calculated from the equation

$$\tau = p\mathbf{E}\sin\theta$$

Equation 5.10

where p is the magnitude of the dipole moment ($p = qd$), E is the magnitude of the uniform external electric field, and θ is the angle the dipole moment makes with the electric field. This torque will cause the dipole to reorient itself so that its dipole moment, **p**, aligns with the electric field **E**, as shown in Figure 5.4.

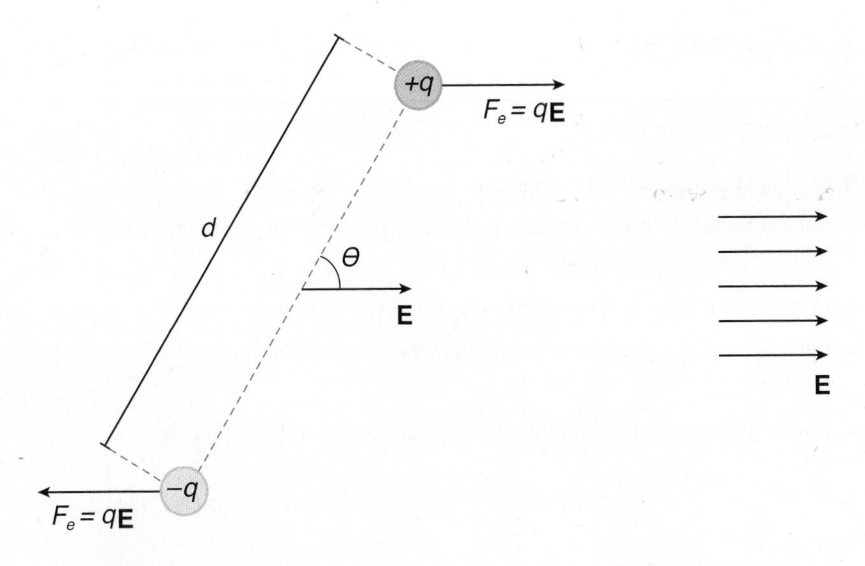

Figure 5.4 Torque on a Dipole from an External Electric Field

MCAT EXPERTISE

The electric dipole is most likely to be tested qualitatively or in the context of a passage or reaction on Test Day. It is unlikely that these mathematical relations will be presented or tested without background.

MCAT CONCEPT CHECK 5.5

Before you move on, assess your understanding of the material with these questions.

1. Define the following terms:
   - Equipotential lines:

   ___

   - Electric dipole:

   ___

2. What is the voltage between two points on an equipotential line? Will this voltage cause a charge to move along the line?

   ___

3. Why is the electrical potential at points along the perpendicular bisector of a dipole zero?

   ___

4. What is the behavior of an electric dipole when exposed to an external electric field?

   ___

5.6 Magnetism

LEARNING OBJECTIVES

After Chapter 5.6, you will be able to:

- Recall the requirements to have a nonzero electric field, a nonzero magnetic field, or a nonzero magnetic force
- Predict the impact of a magnetic field on a nearby object
- Calculate the magnitudes of a magnetic field and the magnetic force exerted by the field
- Predict the direction of a magnetic force using the right-hand rule:

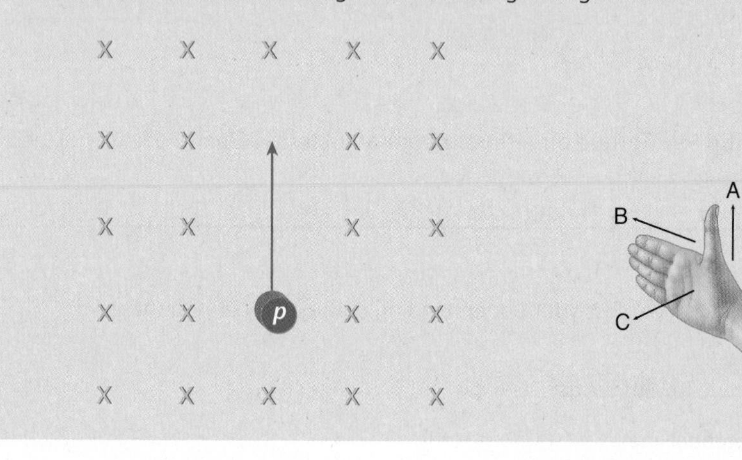

Any moving charge creates a **magnetic field**. Magnetic fields may be set up by the movement of individual charges, such as an electron moving through space; by the mass movement of charge in the form of a current though a conductive material, such as a copper wire; or by permanent magnets. The SI unit for magnetic field strength is the **tesla (T)**, where $1\,\text{T} = 1\,\frac{\text{N} \cdot \text{s}}{\text{m} \cdot \text{C}}$. The size of the tesla unit is quite large, so small magnetic fields are sometimes measured in **gauss**, where $1\ \text{T} = 10^4$ gauss.

KEY CONCEPT

Any moving charge, whether a single electron traveling through space or a current through a conductive material, creates a **magnetic field**. The SI unit for magnetic field strength is the tesla (T).

All materials can be classified as diamagnetic, paramagnetic, or ferromagnetic. **Diamagnetic materials** are made of atoms with no unpaired electrons and that have no net magnetic field. These materials are slightly repelled by a magnet and so can be called weakly antimagnetic. Diamagnetic materials include common materials that you wouldn't expect to get stuck to a magnet: wood, plastics, water, glass, and skin, just to name a few.

The atoms of both paramagnetic and ferromagnetic materials have unpaired electrons, so these atoms do have a net magnetic dipole moment, but the atoms in these materials are usually randomly oriented so that the material itself creates no net magnetic field. **Paramagnetic materials** will become weakly magnetized in the presence of an external magnetic field, aligning the magnetic dipoles of the material with the external field. Upon removal of the external field, the thermal energy of the individual atoms will cause the individual magnetic dipoles to reorient randomly. Some paramagnetic materials include aluminum, copper, and gold.

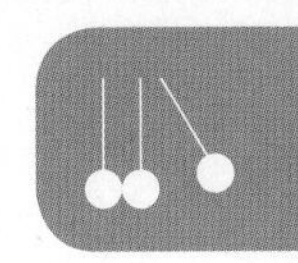

Ferromagnetic materials, like paramagnetic materials, have unpaired electrons and permanent atomic magnetic dipoles that are normally oriented randomly so that the material has no net magnetic dipole. However, unlike paramagnetic materials, ferromagnetic materials will become strongly magnetized when exposed to a magnetic field or under certain temperatures. Common ferromagnetic materials include iron, nickel, and cobalt. Bar magnets are ferromagnetic materials with a north and south pole. Field lines exit the north pole and enter the south pole. Because magnetic field lines are circular, it is impossible to have a monopole magnet. If two bar magnets are allowed to interact, opposite poles will attract each other, while like poles will repel each other.

ferromagnetic materials are more easily magnetized than paramagnetic materials!

Magnetic Fields

Because any moving charge creates a magnetic field, we would certainly expect that a collection of moving charges, in the form of a current through a conductive wire, would produce a magnetic field in its vicinity. The configuration of the magnetic field lines surrounding a current-carrying wire will depend on the shape of the wire. Two special cases that are commonly tested on the MCAT include a long, straight wire and a circular loop of wire (with particular attention paid to the magnetic field at the center of that loop).

For an infinitely long and straight current-carrying wire, we can calculate the magnitude of the magnetic field produced by the current I in the wire at a perpendicular distance, r, from the wire as:

$$B = \frac{\mu_0 I}{2\pi r}$$

Equation 5.11

where B is the magnetic field at a distance r from the wire, μ_0 is the **permeability of free space** $\left(4\pi \times 10^{-7}\ \frac{\text{T}\cdot\text{m}}{\text{A}}\right)$, and I is the current. The equation demonstrates an inverse relationship between the magnitude of the magnetic field and the distance from the current. Straight wires create magnetic fields in the shape of concentric rings. To determine the direction of the field vectors, use a **right-hand rule**. (This is one of two right-hand rules used in magnetism.) Point your thumb in the direction of the current and wrap your fingers around the current-carrying wire. Your fingers then mimic the circular field lines, curling around the wire.

For a circular loop of current-carrying wire of radius r, the magnitude of the magnetic field at the center of the circular loop is given as:

$$B = \frac{\mu_0 I}{2r}$$

Equation 5.12

You'll notice that the two equations are quite similar—the obvious difference being that the equation for the magnetic field at the center of the circular loop of wire does not include the constant π. The less obvious difference is that the first expression gives the magnitude of the magnetic field at any perpendicular distance, r, from the current-carrying wire, while the second expression gives the magnitude of the magnetic field only at the center of the circular loop of current-carrying wire with radius r.

Example: Suppose a wire is formed into a loop that carries a current of 0.25 A in a clockwise direction, as shown here:

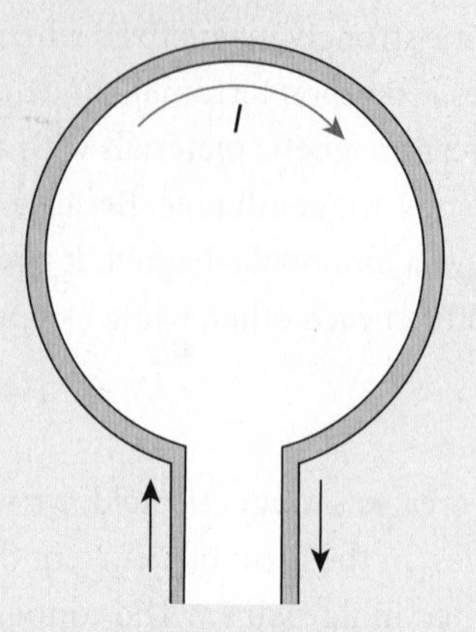

Determine the direction of the magnetic field produced by this loop within the loop and outside the loop. If the loop has a diameter of 1 m, what is the magnitude of the magnetic field at the center of the loop?

Solution: Use the right-hand rule to determine the direction of the magnetic field within and outside the loop, as shown here:

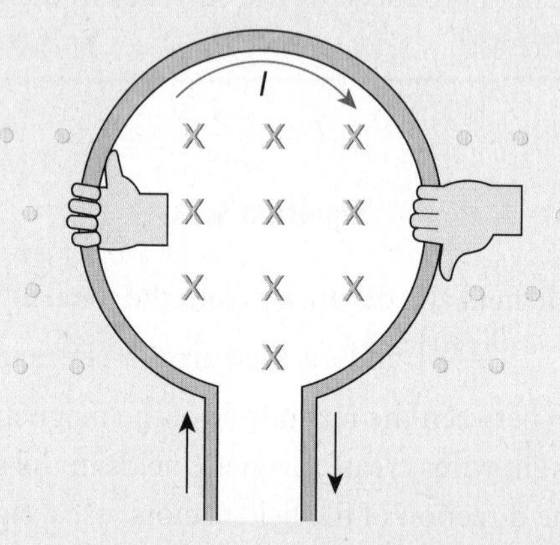

Align your right thumb with the current at any point in the loop. When you encircle the wire with the remaining fingers of your right hand, your fingers should point into the page within the loop and out of the page outside of the loop.

To determine the magnitude of the magnetic field at the center, use the equation for a loop of wire:

$$B = \frac{\mu_0 I}{2r}$$

$$= \frac{\left(4\pi \times 10^{-7}\ \frac{\text{T}\cdot\text{m}}{\text{A}}\right)(0.25\ \text{A})}{2 \times 0.5\ \text{m}} \approx 3.14 \times 10^{-7}\ \text{T} = 3.14 \times 10^{-3}\ \text{gauss}$$

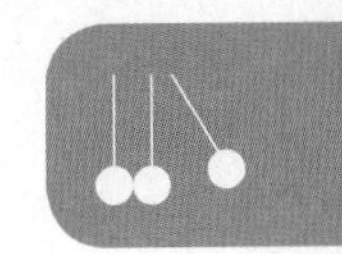

Magnetic Forces

Now that we've reviewed the ways in which magnetic fields can be created, let's examine the forces that are exerted by magnetic fields on moving charges. Magnetic fields exert forces only on other moving charges. That is, charges do not "sense" their own fields; they only sense the field established by some external charge or collection of charges. Therefore, in our discussion of the magnetic force on moving charges and on current-carrying wires, we will assume the presence of a fixed and uniform external magnetic field. Note that charges often have both electrostatic and magnetic forces acting on them at the same time; the sum of these electrostatic and magnetic forces is known as the **Lorentz force**.

Force on a Moving Charge

When a charge moves in a magnetic field, a **magnetic force** may be exerted on it, the magnitude of which can be calculated as follows:

$$F_B = qvB \sin \theta$$

Equation 5.13

where q is the charge, v is the magnitude of its velocity, B is the magnitude of the magnetic field, and θ is the smallest angle between the velocity vector **v** and the magnetic field vector **B**. Notice that the magnetic force is a function of the sine of the angle, which means that the charge must have a perpendicular component of velocity in order to experience a magnetic force. If the charge is moving parallel or antiparallel to the magnetic field vector, it will experience no magnetic force.

Here we will introduce the second **right-hand rule** that you should practice in anticipation of Test Day. To determine the direction of the magnetic force on a moving charge, first position your right thumb in the direction of the velocity vector. Then, put your fingers in the direction of the magnetic field lines. Your palm will point in the direction of the force vector for a positive charge, whereas the back of your hand will point in the direction of the force vector for a negative charge.

KEY CONCEPT

Remember that sin 0° and sin 180° equal zero. This means that any charge moving parallel or antiparallel to the direction of the magnetic field will experience no force from the magnetic field.

MNEMONIC

Parts of the right-hand rule for magnetic force:

- Thumb—velocity (indicates direction of movement, like a hitchhiker's thumb)
- Fingers—field lines (fingers are parallel like the uniform magnetic field lines)
- Palm—force on a positive charge (you might give a "high five" to a positive person)
- Back of hand—force on a negative charge (you might give a backhand to a negative person)

Example: Suppose a proton is moving with a velocity of 15 $\frac{\text{m}}{\text{s}}$ toward the top of the page through a uniform magnetic field of 3.0 T directed into the page, as shown here:

What is the magnitude and direction of the magnetic force on the proton? Describe the motion that will result from this setup. (Note: The charge of a proton is 1.60×10^{-19} C, and its mass is 1.67×10^{-27} kg.)

Solution: Start by determining the magnitude of the force:

$$\begin{aligned} F_B &= qvB \sin \theta \\ &= \left(1.60 \times 10^{-19}\ \text{C}\right)\left(15\ \frac{\text{m}}{\text{s}}\right)(3.0\ \text{T}) \sin 90° \\ &= 7.2 \times 10^{-18}\ \text{N} \end{aligned}$$

To determine the direction, use the right-hand rule. Your thumb should point up the page in the direction of **v**. Your fingers should point into the page in the direction of **B**. Protons are positively charged; thus the force, $\boldsymbol{F}_B$, is in the direction of your palm, which is to the left. Note that $\boldsymbol{v}$ and $\boldsymbol{F}_B$ will always be perpendicular to each other; this implies that uniform circular motion will occur in this field, with $\boldsymbol{F}_B$ pointing radially inward toward the center of the circle.

If the centripetal force is the magnetic force, then we can set these two equations equal to each other:

$$\begin{aligned} F_c &= F_B \\ \frac{mv^2}{r} &= qvB \sin \theta \\ mv &= qBr \sin 90° \\ r &= \frac{mv}{qB} = \frac{\left(1.67 \times 10^{-27}\ \text{kg}\right)\left(15\ \frac{\text{m}}{\text{s}}\right)}{\left(1.60 \times 10^{-19}\ \text{C}\right)(3.0\ \text{T})} \approx 5 \times 10^{-8}\ \text{m} \left(\text{actual} = 5.2 \times 10^{-8}\ \text{m}\right) \end{aligned}$$

Thus, the proton will move in a circle with a radius of 52 nm.

Force on a Current-Carrying Wire

We've just examined the force that can be created by a magnetic field when a point charge moves through the field, so it should not come as a surprise that a current-carrying wire placed in a magnetic field may also experience a magnetic force. For a straight wire, the magnitude of the force created by an external magnetic field, F_B, is:

$$F_B = ILB \sin \theta$$

Equation 5.14

where I is the current, L is the length of the wire in the field, B is the magnitude of the magnetic field, and θ is the angle between **L** and **B**. The same right-hand rule can be used for a current-carrying wire in a field as for a moving point charge; just remember that current is considered the flow of positive charge.

Example: Suppose a wire of length 2.0 m is conducting a current of 5.0 A toward the top of the page and through a 30 gauss uniform magnetic field directed into the page. What is the magnitude and direction of the magnetic force on the wire?

Solution: Because 1 T $= 10^4$ gauss, 1 gauss $= 10^{-4}$ T, and 30 gauss $= 30 \times 10^{-4}$ T $= 3 \times 10^{-3}$ T. The wire is conducting current toward the top of the page, and the magnetic field points into the page; therefore, the current is perpendicular to the magnetic field. The angle between them is $\theta = 90°$. Now, plug into the equation:

$$\begin{aligned} F_B &= ILB \sin \theta \\ &= (5.0\text{ A})(2.0\text{ m})\left(3 \times 10^{-3}\text{ T}\right)\left(\sin 90°\right) \\ &= 30 \times 10^{-3} = 3 \times 10^{-2}\text{ N} \end{aligned}$$

To determine the direction, use the right-hand rule. Your thumb should point up the page in the direction of **L**. Your fingers should point into the page in the direction of **B**. Current is a flow of positive charge; thus, the force, $\mathbf{F_B}$, is in the direction of your palm, which is to the left.

MCAT CONCEPT CHECK 5.6

Before you move on, assess your understanding of the material with these questions.

1. What are the requirements to have a nonzero electric field? A nonzero magnetic field? A nonzero magnetic force?
    - Nonzero electric field:

    - Nonzero magnetic field:

    - Nonzero magnetic force:

    ---

2. Which would experience a larger magnetic field: an object placed five meters to the left of a current carrying wire, or an object placed at the center of a circle with a radius of five meters? (Note: Assume the current is constant; $\mu_0 = 4\pi \times 10^{-7} \frac{\text{T} \cdot \text{m}}{\text{A}}$)

    ---

3. For each of the following combinations of velocity and magnetic field directions, determine the direction of the magnetic force on the given particle:

v	B	Particle	F
Up the page	Left	Electron	
Into the page	Out of the page	Proton	
Right	Into the page	Proton	
Out of the page	Left	Electron	
Down the page	Right	Neutron	

Conclusion

In this chapter, we reviewed the very notion of charge, reminding ourselves that charge comes in two varieties: positive and negative. We also explored the fact that charges travel differently within insulators and conductors. We learned that charges establish electric fields through which they can exert forces on other charges. We relied on similarities between electrical and gravitational systems to better understand Coulomb's law and the nature of the forces that exist between charged particles. Don't forget that electrical forces can be repulsive as well as attractive, which is one of the differences between electrical and gravitational systems. Charges contain electrical potential energy, which we defined as their energy of position with respect to other charges. Charges move within an electric field from one position of electrical potential to another; they will move spontaneously through an electrical potential difference, or voltage, in whichever direction results in a decrease in the charge's electrical potential energy. Then, we considered the geometry of the electric dipole and derived the equation for calculating the electrical potential at any point in space around the dipole. Finally, we considered magnetic fields and forces. In the next chapter, we'll examine moving charges as they interact with circuit elements and complete our understanding of electricity.

GO ONLINE

You've reviewed the content, now test your knowledge and critical thinking skills by completing a test-like passage set in your online resources!

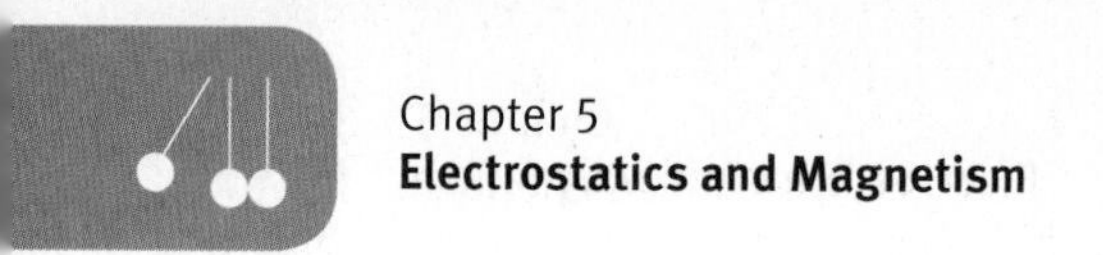

Concept Summary

Charges

- The SI unit of charge is the **coulomb**.
- **Protons** have a positive charge and **electrons** have a negative charge.
  - Both protons and electrons possess the fundamental unit of charge ($e = 1.60 \times 10^{-19}$ C).
  - Protons and electrons have different masses.
- Opposite charges exert **attractive** forces, and like charges exert **repulsive** forces.
- **Conductors** allow the free and uniform passage of electrons when charged.
- **Insulators** resist the movement of charge and will have localized areas of charge that do not distribute over the surface of the material.

Coulomb's Law

- **Coulomb's law** gives the magnitude of the electrostatic force vector between two charges. The force vector always points along the line connecting the centers of the two charges.
- Every charge generates an **electric field**, which can exert forces on other charges.
- The electric field is the ratio of the force that is exerted on a test charge to the magnitude of that charge.
  - Electric field vectors can be represented as **field lines** that radiate outward from positive source charges and radiate inward to negative source charges.
  - Positive test charges will move in the direction of the field lines; negative test charges will move in the direction opposite of the field lines.

Electric Potential Energy

- **Electric potential energy** is the amount of work required to bring the test charge from infinitely far away to a given position in the vicinity of a source charge.
- The electric potential energy of a system will increase when two like charges move toward each other or when two opposite charges move farther apart.
- The electric potential energy of a system will decrease when two opposite charges move toward each other or when two like charges move farther apart.

Electric Potential

- **Electric potential** is the electric potential energy per unit charge.
- Different points in the space of an electric field surrounding a source charge will have different electric potential values.
- Potential difference (**voltage**) is the change in electric potential that accompanies the movement of a test charge from one position to another.
  - Potential difference is path independent and depends only on the initial and final positions of the test charge.
  - The units for both electric potential and voltage are volts.
- Test charges will move spontaneously in whichever direction results in a decrease in their electric potential energy.
  - Positive test charges will move spontaneously from high potential to low potential.
  - Negative test charges will move spontaneously from low potential to high potential.

Special Cases in Electrostatics

- **Equipotential lines** designate the set of points around a source charge or multiple source charges that have the same electric potential.
  - Equipotential lines are always perpendicular to electric field lines.
  - Work will be done when a charge is moved from one equipotential line to another; the work is independent of the pathway taken between the lines.
  - No work is done when a charge moves from a point on an equipotential line to another point on the same equipotential line.
- Two charges of opposite sign separated by a fixed distance, d, generate an **electric dipole**.
  - In an external electric field, an electric dipole will experience a net torque until it is aligned with the electric field vector.
  - An electric field will not induce any translational motion in the dipole regardless of its orientation with respect to the electric field vector.

Magnetism

- **Magnetic fields** are created by magnets and moving charges.
- The SI unit for the magnetic field is the **tesla** (T; 1 T = 10,000 gauss).
- **Diamagnetic materials** possess no unpaired electrons and are slightly repelled by a magnet.
- **Paramagnetic materials** possess some unpaired electrons and become weakly magnetic in an external magnetic field.
- **Ferromagnetic materials** possess some unpaired electrons and become strongly magnetic in an external magnetic field.
- Magnets have a north and a south pole; field lines point from the north to the south pole.
- Current-carrying wires create magnetic fields that are concentric circles surrounding the wire.
- External magnetic fields exert forces on charges moving in any direction except parallel or antiparallel to the field.
- Point charges may undergo uniform circular motion in a uniform magnetic field wherein the centripetal force is the magnetic force acting on the point charge.
- The direction of the magnetic force on a moving charge or current-carrying wire is determined using the right-hand rule.
- The **Lorentz force** is the sum of the electrostatic and magnetic forces acting on a body.

Answers to Concept Checks

5.1

1. The electrons will experience the greater acceleration because they are subject to the same force as the protons but have a significantly smaller mass.
2. Conductors: blood, copper, iron, sulfuric acid; insulators: hair, glass, distilled water
3. The net charge will be -1 C; neutrons do not contribute charge.

5.2

1. The electric field would be 0 because the two charges are the same. In this case, the fields exerted by each charge at the midpoint will cancel out and there will be no electric field.
2. For a pair of charges, a negative electrostatic force points from one charge to the other (attractive), while a positive electrostatic force points from one charge away from the other (repulsive).
3. 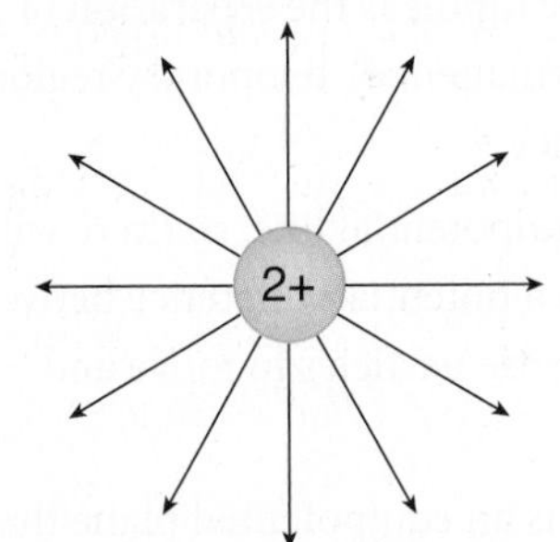

4. Electrostatic force is directly related to each charge and related to the distance by an inverse square relationship. Electric field is unrelated to test charge but is still related to distance by an inverse square relationship. Note that it is the source charge that creates the electric field—not the test charge—so we cannot use the equation $E = \frac{F_e}{q}$ to determine a relationship.

5.3

1. A decrease in potential energy indicates that the system has become more stable. Keep in mind that negative numbers with larger absolute values are more negative, and represent a decrease in value from negative numbers with smaller absolute values (that is, $-4 > -7$ even though $|-4| < |-7|$).
2. Electrical potential energy is Coulomb's law multiplied by distance, whereas gravitational potential energy is the universal law of gravitation multiplied by distance.
3. If both particles have the same charge, the electrical potential energy decreases as distance increases. If the two particles have opposite charges, then the electrical potential energy increases as distance increases.
4. As given by the equation $U = \frac{kQq}{r}$, the electric potential energy would increase by a factor of eight if both charges are doubled and the radius is halved.

5.4

1. Electrical potential is the ratio of a charge's electrical potential energy to the magnitude of the charge itself. Voltage, or potential difference, is a measure of the change in electrical potential between two points, which provides an indication of the tendency toward movement in one direction or the other.
2. A charge will move in such a way to minimize its potential energy. Placing a charge at a point of zero electrical potential does not indicate that there is zero potential difference, so the charge may or may not move—and if it moves, it may move toward or away from the source charge depending on the sign of the source charge and test charge.
3. True. Electrical potential energy is measured in joules (J), while electrical potential and potential difference (voltage) are measured in volts (V).

5.5

1. Equipotential lines are the sets of points within space at which the potential difference between any two points is zero. This is best visualized as concentric spheres surrounding a source charge. An electric dipole is the separation of charge within a molecule such that there is a permanent or temporary region of equal and opposite charges at a particular distance.
2. There is no voltage between two points on an equipotential line, so there will be no acceleration along the line. However, there is a potential difference between different sets of equipotential lines, which can cause particles to move and accelerate.
3. The perpendicular bisector of an electric dipole is an equipotential plane that is perpendicular to the axis of the dipole. As such, the equation $V = \frac{kqd}{r^2} \cos \theta$ is necessarily equal to 0 because $\cos 90° = 0$.
4. A dipole will rotate within an external electric field such that its dipole moment aligns with the field.

5.6

1. To create an electric field, one needs a charge. To create a magnetic field, one needs a charge that must also be moving. To create a magnetic force, one needs an external electric field acting on a charge moving any direction except parallel or antiparallel to the external field.
2. We need not determine the actual values of the magnetic fields in these two cases and can compare the two equations instead. The magnetic field created by the current-carrying wire is given by $B = \frac{\mu_0 I}{2\pi r}$; the magnetic field created by the loop of wire is given by $B = \frac{\mu_0 I}{2r}$ and μ_0, I, and r are the same in both equations. Therefore, the magnetic field at the center of the loop must be larger because the denominator in that equation does *not* include π.
3.

v	B	Particle	F
Up the page	Left	Electron	Into the page
Into the page	Out of the page	Proton	None ($\sin 180° = 0$)
Right	Into the page	Proton	Up the page
Out of the page	Left	Electron	Up the page
Down the page	Right	Neutron	None ($q = 0$)

Science Mastery Assessment Explanations

1. **B**

According to Newton's third law, if R exerts a force on S, then S exerts a force with equal magnitude but opposite direction back on R. Therefore, the magnitude of the force on R due to S is F.

2. **D**

The force is inversely proportional to r^2. Cutting the distance in half will therefore multiply the force by 2^2, making it four times its original value:

$$F_{\text{old}} \propto \frac{1}{r^2}$$

$$F_{\text{new}} \propto \frac{1}{\left(\frac{r}{2}\right)^2} = \frac{4}{r^2} = 4 \times F_{\text{old}}$$

3. **B**

An electric field's direction at a given point is defined as the direction of the force that would be exerted on a positive test charge in that position. Because electrons are negatively charged particles, they will therefore feel a force in the opposite direction of the electric field's vector. In this case, because the force points to the left (toward R), an electron will feel a force pointing to the right (toward S) if **E** is in the same direction as **F**.

4. **B**

The first step in answering this question is to remember that the magnitude of the electric field is inversely proportional to the square of the distance:

$$E = \frac{kQ}{r^2} \rightarrow E \propto \frac{1}{r^2}$$

Therefore, if the electric field at radius r, E_r, is 36 $\frac{\text{N}}{\text{C}}$, then the electric field at radius $2r$ will be

$$E_{2r} \propto \frac{1}{(2r)^2} = \frac{1}{4r^2} = \frac{E_r}{4}$$

$$= \frac{\left(36\ \frac{\text{N}}{\text{C}}\right)}{4} = 9\ \frac{\text{N}}{\text{C}}$$

Similarly, the electric field at radius $3r$ is equal to

$$E_{3r} \propto \frac{1}{(3r)^2} = \frac{1}{9r^2} = \frac{E_r}{9}$$

$$= \frac{\left(36\ \frac{\text{N}}{\text{C}}\right)}{9} = 4\ \frac{\text{N}}{\text{C}}$$

Therefore, the ratio of E_r:E_{2r}:E_{3r} is 36:9:4.

5. **C**

A positive charge placed at A will experience two forces: a force to the left due to $+2Q$ and a force to the right due to $+Q$. Because point A is the same distance from $+Q$ and $+2Q$, the force due to $+2Q$ will be larger than that due to $+Q$, and there will be a net force to the left (toward $+Q$). At point B, the forces from both $+Q$ and $+2Q$ will point to the right, so there will be a net force to the right.

6. **C**

Recall that the change in potential energy, ΔU, and the change in potential, ΔV, are related by $W = \Delta U = q\Delta V$. Therefore, $\Delta U = (2 \times 10^{-6}\ \text{C}) \times (-12\ \text{V}) = -2.4 \times 10^{-5}\ \text{J}$. The positive charge is moving from the positive to the negative plate, and is therefore decreasing in potential energy; this is reflected by the fact that the voltage is -12 V rather than $+12$ V. The potential energy that is lost is converted into kinetic energy, so the charge must gain 2.4×10^{-5} J of kinetic energy.

7. **C**

There will be work done in moving the negative charge from its initial position to $y = 0$. However, in moving the negative charge from $y = 0$ to the final position, the same amount of work is done but with the opposite sign. This is because the force changes direction as the electron crosses $y = 0$. Therefore, the two quantities of work cancel each other out. This argument depends crucially on the symmetry of the initial and final positions.

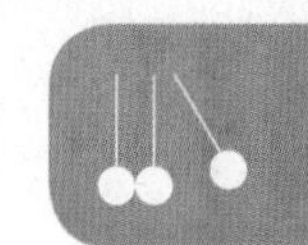

8. **D**

The safest way to answer this question is to quickly draw a diagram:

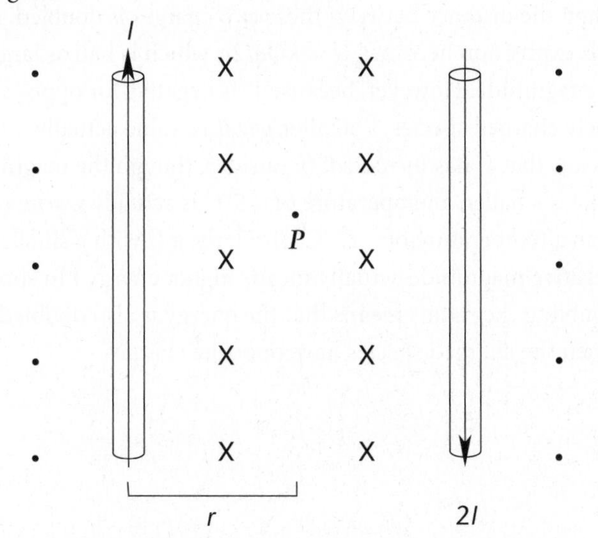

Notice right away that between the two wires, the direction of the magnetic field is the same: into the page. Therefore, because the vector direction is the same, we can just focus on the magnitudes of the two magnetic fields. We know that $B_1 =$ 10 T at a distance r. Consider the relationships in the equation $B = \frac{\mu_0 I}{2\pi r}$. Magnetic field and current are directly proportional, whereas magnetic field and distance are inversely proportional. Therefore, doubling the current will result in double the magnetic field of the first wire, or 20 T. The overall magnitude of the magnetic field is $10\text{ T} + 20\text{ T} = 30\text{ T}$ into the page.

9. **B**

Potential is a scalar quantity. The total potential is the sum of the potentials of the positive and negative charges:

$$V_T = V_+ + V_- = \frac{+kq}{r_+} + \left(\frac{-kq}{r_-}\right)$$

where r_+ and r_- represent the distances from the positive and negative charge, respectively. The sum of these terms will be zero at any point where $r_+ = r_-$. This will be at any point along the perpendicular bisector of the dipole axis, as well as at infinity.

10. **D**

The electric potential (V) is equal to the amount of work done (W) divided by the test charge (q). This means that the potential is directly proportional to the amount of work done, which is equal to the amount of energy gained by the particle; therefore, the overall amount of energy increases by a factor of 4. Because energy is directly proportional to the square of the speed (according to $\frac{1}{2}mv^2$), the speed must increase by a factor of 2.

11. **D**

You should know that the field lines for a positively charged particle will always point away from the particle in a radial pattern, regardless of the direction in which the particle is moving. This is because field lines point in the direction a positive test charge would move in that field (that is, the direction that a force would be exerted on a positive test charge in that field).

12. **C**

Voltage (ΔV) is equal to the quotient of the amount of work done (W) divided by the charge of the particle on which the work is done (q), according to the equation $V = \frac{\Delta U}{q} = \frac{W}{q}$. Because the voltage equals 9 V and the charge equals 2 C, the work done must equal $9\text{ V} \times 2\text{ C} = 18\text{ J}$.

13. **C**

To prevent the flow of charge between the two plates, the dielectric should be an insulative material. Insulative materials contain atoms that tightly bind to their electrons, often nonmetals. Glass is composed of two nonmetals, silicon and oxygen, and would be the best insulator, supporting choice **(C)**. On the other hand, while pure water itself can be an insulator, the presence of ions greatly increases the conductivity of salt water, eliminating **(A)**.

14. **B**

This problem is an application of the right-hand rule. The velocity vector **v** is always tangent to the circle. The magnetic force must always point radially toward the center of the circle. Consider when the negative charge is at the "12 o'clock" position in its circle and apply the right-hand rule. Your thumb points to the left, tangent to the circle at this point. The back of your hand, which represents the force on a negative charge, points down the page, radially toward the center of the circle. Your fingers must point out of the page to get your hand into this position. Therefore, the direction of the magnetic field must be out of the page.

15. **A**

Often, with potential energy, it can be easier to think in terms of common sense, rather than in terms of rigorous mathematical formulas: opposite charges attract. When oppositely charged species like ATP and a cation are moved apart, their potential energy will increase, reflecting the fact that these oppositely charged species would like to move back to a closer distance. Furthermore, the formula for electric potential energy is $U = kQq/r$. From this formula, observe that the relationship between U and r is given by $U \propto 1/r$, meaning that if r is changed by a factor of 2, then U will also change by a factor of 2, and not by a factor of 4. These observations together are enough to justify **(A)** as the correct answer.

For a more mathematically rigorous explanation, consider that the opposite charges of ATP and the cation will cause the output of the formula $U = kQq/r$ to be negative. Then, when the distance between these two charges is doubled, this expression becomes $U = kQq/2r$, which is half as large in magnitude. However, because U is negative for oppositely charged species, a smaller *negative* value actually means that U has *increased*! (Consider: though the magnitude is smaller, a temperature of $-5°C$ is actually *warmer* than a temperature of $-25°C$. Similarly, a U with a smaller negative magnitude actually means higher energy.) In short, doubling the radius means that the energy is also doubled when the charged species have opposite charges.

Consult your online resources for additional practice. GO ONLINE

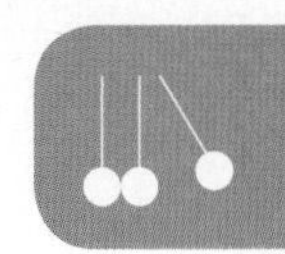

Equations to Remember

(5.1) **Coulomb's law:** $F_e = \frac{kq_1q_2}{r^2}$

(5.2) **Electric field:** $E = \frac{F_e}{q} = \frac{kQ}{r^2}$

(5.3) **Electric potential energy:** $U = \frac{kQq}{r}$

(5.4) **Electric potential (from electric potential energy):** $V = \frac{U}{q}$

(5.5) **Electric potential (from source charge):** $V = \frac{kQ}{r}$

(5.6) **Voltage:** $\Delta V = V_b - V_a = \frac{W_{ab}}{q}$

(5.7) **Electric potential near a dipole:** $V = \frac{kqd}{r^2}\cos\theta$

(5.8) **Dipole moment:** $\mathbf{p} = q\mathbf{d}$

(5.9) **Electric field on the perpendicular bisector of a dipole:** $E = \frac{1}{4\pi\varepsilon_0} \times \frac{p}{r^3}$

(5.10) **Torque on a dipole in an electric field:** $\tau = pE \sin\theta$

(5.11) **Magnetic field from a straight wire:** $B = \frac{\mu_0 I}{2\pi r}$

(5.12) **Magnetic field from a loop of wire:** $B = \frac{\mu_0 I}{2r}$

(5.13) **Magnetic force on a moving point charge:** $F_B = qvB \sin\theta$

(5.14) **Magnetic force on a current-carrying wire:** $F_B = ILB \sin\theta$

Shared Concepts

General Chemistry Chapter 1
Atomic Structure

General Chemistry Chapter 3
Bonding and Chemical Interactions

General Chemistry Chapter 12
Electrochemistry

Physics and Math Chapter 1
Kinematics and Dynamics

Physics and Math Chapter 2
Work and Energy

Physics and Math Chapter 6
Circuits

CHAPTER 6

Circuits

SCIENCE MASTERY ASSESSMENT

Every pre-med knows this feeling: there is so much content I have to know for the MCAT! How do I know what to do first or what's important?

While the high-yield badges throughout this book will help you identify the most important topics, this Science Mastery Assessment is another tool in your MCAT prep arsenal. This quiz (which can also be taken in your online resources) and the guidance below will help ensure that you are spending the appropriate amount of time on this chapter based on your personal strengths and weaknesses. Don't worry though—skipping something now does not mean you'll never study it. Later on in your prep, as you complete full-length tests, you'll uncover specific pieces of content that you need to review and can come back to these chapters as appropriate.

How to Use This Assessment

If you answer 0–7 questions correctly:

Spend about 1 hour to read this chapter in full and take limited notes throughout. Follow up by reviewing **all** quiz questions to ensure that you now understand how to solve each one.

If you answer 8–11 questions correctly:

Spend 20–40 minutes reviewing the quiz questions. Beginning with the questions you missed, read and take notes on the corresponding subchapters. For questions you answered correctly, ensure your thinking matches that of the explanation and you understand why each choice was correct or incorrect.

If you answer 12–15 questions correctly:

Spend less than 20 minutes reviewing all questions from the quiz. If you missed any, then include a quick read-through of the corresponding subchapters, or even just the relevant content within a subchapter, as part of your question review. For questions you got correct, ensure your thinking matches that of the explanation and review the Concept Summary at the end of the chapter.

1. If a defibrillator passes 15 A of current through a patient's body for 0.1 seconds, how much charge goes through the patient's skin?

- **A.** 0.15 C
- **B.** 1.5 C
- **C.** 15 C
- **D.** 150 C

2. A student places an ammeter with negligible resistance in parallel with a resistor to determine the amount of current that passes through the resistor. Does the student obtain an accurate reading?

- **A.** Yes, because he used an ammeter with negligible resistance.
- **B.** Yes, because the current going through parallel paths is equal.
- **C.** No, because the ammeter should have infinite resistance.
- **D.** No, because an ammeter in parallel changes the current through the resistor.

3. The resistance of two conductors of equal cross-sectional area and equal lengths are compared, and are found to be in the ratio 1:2. The resistivities of the materials from which they are constructed must therefore be in what ratio?

- **A.** 1:1
- **B.** 1:2
- **C.** 2:1
- **D.** 4:1

4. A voltaic cell provides a current of 0.5 A when in a circuit with a 3 Ω resistor. If the internal resistance of the cell is 0.1 Ω, what is the voltage across the terminals of the battery when there is no current flowing?

- **A.** 0.05 V
- **B.** 1.5 V
- **C.** 1.505 V
- **D.** 1.55 V

5. A transformer is a device that takes an input voltage and produces an output voltage that can be either larger or smaller than the input voltage, depending on the transformer design. Although the voltage is changed by the transformer, energy is not, so the input power equals the output power. A particular transformer produces an output voltage that is 300 percent of the input voltage. What is the ratio of the output current to the input current?

- **A.** 1:3
- **B.** 3:1
- **C.** 1:300
- **D.** 300:1

6. Given that $R_1 = 20\ \Omega$, $R_2 = 4\ \Omega$, $R_3 = R_4 = 32\ \Omega$, $R_5 = 15\ \Omega$, and $R_6 = 5\ \Omega$, what is the total resistance in the setup shown below?

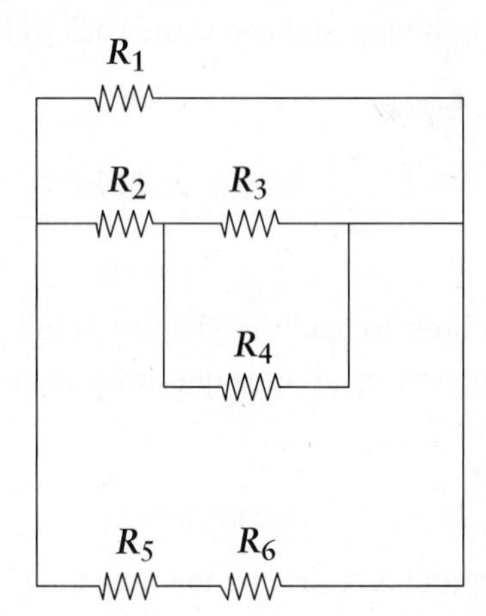

- **A.** 0.15 Ω
- **B.** 6.67 Ω
- **C.** 16.7 Ω
- **D.** 60 Ω

7. How many moles of electrons pass through a circuit containing a 100 V battery and a 2 Ω resistor over a period of 10 seconds? (Note: $F = 9.65 \times 10^4 \frac{C}{mol\ e^-}$.)

- **A.** 5.18×10^{-3} moles
- **B.** 500 moles
- **C.** 5.18×10^{3} moles
- **D.** 5.2×10^{6} moles

8. In the circuit below, what is the voltage drop across the $\frac{2}{3}\ \Omega$ resistor?

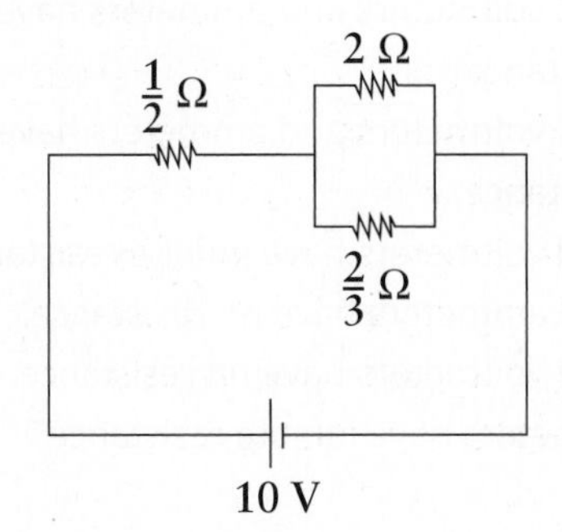

A. $\frac{1}{2}$ V

B. $\frac{2}{3}$ V

C. 5 V

D. 7.5 V

9. If the area of a capacitor's plates is doubled while the distance between them is halved, how will the final capacitance (C_f) compare to the original capacitance (C_i)?

A. $C_f = C_i$

B. $C_f = \frac{1}{2} C_i$

C. $C_f = 2C_i$

D. $C_f = 4C_i$

10. The energy stored in a fully charged capacitor is given by $U = \frac{1}{2} CV^2$. In a typical cardiac defibrillator, a capacitor charged to 7500 V has a stored energy of 400 J. Based on this information, what is the charge on the capacitor in the cardiac defibrillator?

A. 1.1×10^{-5} C

B. 5×10^{-2} C

C. 1.1×10^{-1} C

D. 3.1×10^{6} C

11. A 10 Ω resistor carries a current that varies as a function of time as shown. How much energy has been dissipated by the resistor after 5 s?

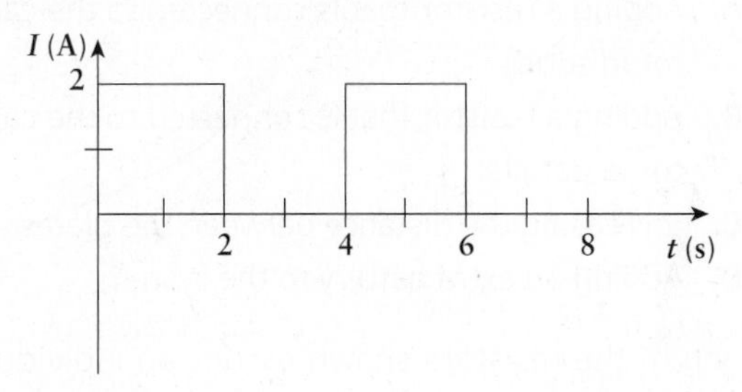

A. 40 J

B. 50 J

C. 120 J

D. 160 J

12. In the figure below, six currents meet at point *P*. What is the magnitude and direction of the current between points *P* and *x*?

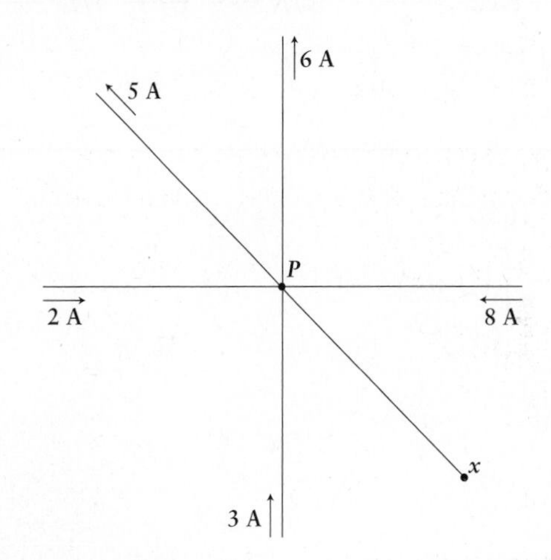

A. 2 A, toward *x*

B. 2 A, toward *P*

C. 10 A, toward *x*

D. 10 A, toward *P*

13. Which of the following will most likely increase the electric field between the plates of a parallel plate capacitor?

A. Adding a resistor that is connected to the capacitor in series

B. Adding a resistor that is connected to the capacitor in parallel

C. Increasing the distance between the plates

D. Adding an extra battery to the system

14. Each of the resistors shown carries an individual resistance of 4 Ω. Assuming negligible resistance in the wire, what is the overall resistance of the circuit?

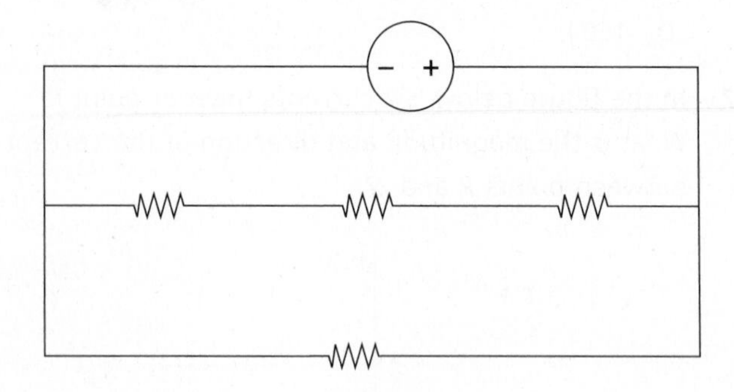

A. 16 Ω

B. 8 Ω

C. 4 Ω

D. 3 Ω

15. Which of the following best characterizes ideal voltmeters and ammeters?

A. Ideal voltmeters and ammeters have infinite resistance.

B. Ideal voltmeters and ammeters have no resistance.

C. Ideal voltmeters have infinite resistance, and ideal ammeters have no resistance.

D. Ideal voltmeters have no resistance, and ideal ammeters have infinite resistance.

Answer Key follows on next page.

Answer Key

1. **B** (Ch. 6.1)
2. **D** (Ch. 6.4)
3. **B** (Ch. 6.2)
4. **D** (Ch. 6.2)
5. **A** (Ch. 6.1)
6. **B** (Ch. 6.2)
7. **A** (Ch. 6.2)
8. **C** (Ch. 6.2)
9. **D** (Ch. 6.3)
10. **C** (Ch. 6.3)
11. **C** (Ch. 6.2)
12. **A** (Ch. 6.1)
13. **D** (Ch. 6.3)
14. **D** (Ch. 6.2)
15. **C** (Ch. 6.4)

CHAPTER 6

Circuits

In This Chapter

6.1 **Current**
- Conductivity 212
- Current 213
- Circuit Laws.................... 213

6.2 **Resistance** HY
- Properties of Resistors............ 216
- Ohm's Law and Power 217
- Resistors in Series and Parallel.................. 219

6.3 **Capacitance and Capacitors**
- Properties of Capacitors........... 225
- Dielectric Materials 226
- Capacitors in Series and Parallel.... 228

6.4 **Meters**
- Ammeters 231
- Voltmeters....................... 231
- Ohmmeters...................... 231

Concept Summary 233

CHAPTER PROFILE

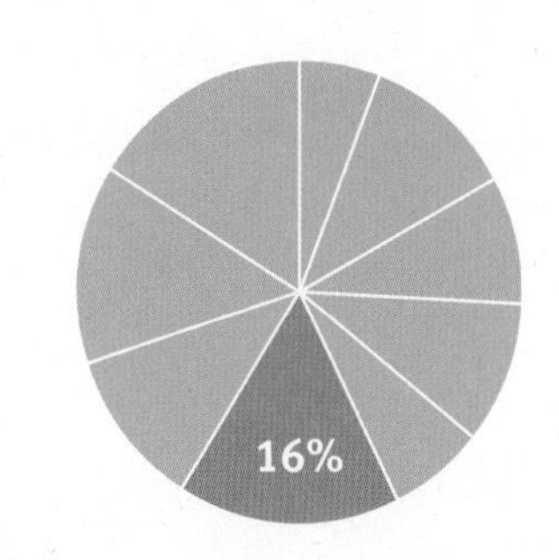

The content in this chapter should be relevant to about 16% of all questions about physics on the MCAT.

This chapter covers material from the following AAMC content category:

4C: Electrochemistry and electrical circuits and their elements

Introduction

Batteries, electric circuits, and electrical equipment pervade our everyday world. Think of any piece of equipment, tool, or toy that has a battery or a power cord, and you've identified an object that depends on the movement of electrons and the delivery of electric potential energy to carry out its function. Turn on a light, watch TV, or toast bread, and you can literally watch electrons at work as they emit light. Electricity is not restricted to the inorganic, material world—even in our bodies, we find electricity serving a key role in a number of physiological functions. Not only do the neurons in our brain and conduction system in our heart rely on electricity, but so does every cell that utilizes mitochondria to carry out oxidative phosphorylation.

This chapter reviews the essentials of circuits. From this broad knowledge base, we will draw on specific topics within circuit theory: conductivity, electromotive force (emf), resistance, power, Kirchhoff's laws, resistors, capacitors, meters, and series and parallel arrangements of circuit components. As you encounter the concepts of this chapter and the equations associated with them, remember this: the MCAT approaches the topic of circuits with a greater emphasis on the concepts than on the math. You will be expected to calculate, say, the equivalent resistance for resistors in series or parallel, but the circuits you encounter on Test Day will, on the whole, be simpler than what you may have seen in your college physics class.

6.1 Current

LEARNING OBJECTIVES

After Chapter 6.1, you will be able to:

- Recall the definitions of current, voltage, electromotive force (emf), and conductivity, and the SI units for each
- Compare the conductivity of two solutions, given their component ions
- Describe the relationship between voltage sources and voltage drop
- Use Kirchhoff's laws to describe the flow of electrons through a circuit

In the last chapter we examined the behaviors of still charges, but in most cases we are interested in the movement of charge, or current. Because of historical conventions, **current** is considered the flow of positive charge—even though only negative charges are actually moving. Any conductive substance may act as a medium through which current can pass.

Conductivity

Conductivity can be divided into two categories: **metallic conductivity**, as seen in solid metals and the molten forms of some salts, or **electrolytic conductivity**, as seen in solutions. **Conductance** is the reciprocal of resistance, a property we will examine in detail later. The SI unit for conductance is the **siemens** (S), sometimes given as siemens per meter $\left(\frac{S}{m}\right)$ for conductivity.

BRIDGE

Remember that the metals are found on the left side of the periodic table. These are the atoms with the lowest ionization energies; thus, it is easiest for these atoms to lose electrons. Due to this weak hold, electrons are free to move around in the metal, conducting electrical charges. Periodic trends are discussed in Chapter 2 of *MCAT General Chemistry Review*.

Metallic Conductivity

Some materials allow free flow of electric charge within them; these materials are called electrical conductors. Metal atoms can easily lose one or more of their outer electrons, which are then free to move around in the larger collection of metal atoms. This makes most metals good electrical and thermal conductors. The **metallic bond** has often been visualized as a sea of electrons flowing over and past a rigid lattice of metal cations. While this model is generally appropriate for the MCAT, metallic bonding is more accurately described as an equal distribution of the charge density of free electrons across all of the neutral atoms within the metallic mass.

Electrolytic Conductivity

While not substantially different from metallic conductivity, it is important to note that electrolytic conductivity depends on the strength of a solution. Distilled deionized water has such a low ion concentration that it may be considered an insulator, while sea water and orange juice are excellent conductors. Conductivity in an electrolyte solution is measured by placing the solution as a resistor in a circuit and measuring changes in voltage across the solution. Because concentration and conductivity are directly related, this method is often used to determine ionic concentrations in solutions, such as blood. One caveat is that conductivity in nonionic solutions is always lower than in ionic solutions. While the concentration of total dissolved solids does relate to conductivity, the contribution of nonionic solids is much, much less important than ion concentration.

Current

Chapter 5 of *MCAT Physics and Math Review* introduced the concept of electrical current: the flow of charge between two points at different electrical potentials connected by a conductor, such as a copper wire. The magnitude of the **current** I is the amount of charge Q passing through the conductor per unit time Δt, and it can be calculated as:

$$I = \frac{Q}{\Delta t}$$

Equation 6.1

current = charge / time

The SI unit of current is the **ampère** $\left(1\ \text{A} = 1\ \frac{\text{C}}{\text{s}}\right)$. Charge is transmitted by a flow of electrons in a conductor, and because electrons are negatively charged, they move from a point of lower electrical potential to a point of higher electrical potential (and, in doing so, reduce their electrical potential energy). By convention, however, the direction of current is the direction in which positive charge would flow (from higher potential to lower potential). Thus, the direction of current is opposite to the direction of actual electron flow. The two patterns of current flow are **direct current (DC)**, in which the charge flows in one direction only, and **alternating current (AC)**, in which the flow changes direction periodically. Direct current is produced by household batteries, while the current supplied over long distances to homes and other buildings is alternating current. Our discussion of circuits will assume direct current, which is tested on the MCAT to the exclusion of alternating current.

A **potential difference (voltage)** can be produced by an electrical generator, a galvanic (voltaic) cell, a group of cells wired into a battery, or—as seen in classic science fair projects—even a potato. When no charge is moving between the two terminals of a cell that are at different potential values, the voltage is called the **electromotive force (emf** or ε**)**. Do not be misled by this term, as emf is not actually a force; it is a potential difference (voltage) and, as such, has units of joules per coulomb $\left(1\ \text{V} = 1\ \frac{\text{J}}{\text{C}}\right)$—not newtons. It may be helpful to think of emf as a "pressure to move" that results in current, in much the same way that a pressure difference between two points in a fluid-filled tube causes the fluid to flow.

BRIDGE

The standard batteries in flashlights and remote controls are examples of galvanic (voltaic) cells. These house spontaneous oxidation–reduction reactions that generate emf as a result of differences in the reduction potentials of two electrodes. Electrochemistry is discussed in Chapter 12 of *MCAT General Chemistry Review*.

Circuit Laws

Currents (and circuits in general) are governed by the laws of conservation. Charge and energy must be fully accounted for at all times and can be neither created nor destroyed. An electric circuit is a conducting path that usually has one or more voltage sources (such as a battery) connected to one or more passive circuit elements (such as resistors). **Kirchhoff's laws** are two rules that deal with the conservation of charge and energy within a circuit.

Kirchhoff's Junction Rule

At any point or junction in a circuit, the sum of currents directed into that point equals the sum of currents directed away from that point. This is an expression of conservation of electrical charge and can be expressed as

$$I_{\text{into junction}} = I_{\text{leaving junction}}$$

Equation 6.2

KEY CONCEPT

Kirchhoff's junction rule is just like a fork in a river. There are a certain number of water molecules in a river, and at any junction, that number has to go in one of the diverging directions; no water molecules spontaneously appear or disappear. The same holds true for the amount of current at any junction.

Example: Three wires (a, b, and c) meet at a junction point P, as shown below. A current of 5 A flows into P along wire a, and a current of 3 A flows away from P along wire b. What is the magnitude and direction of the current along wire c?

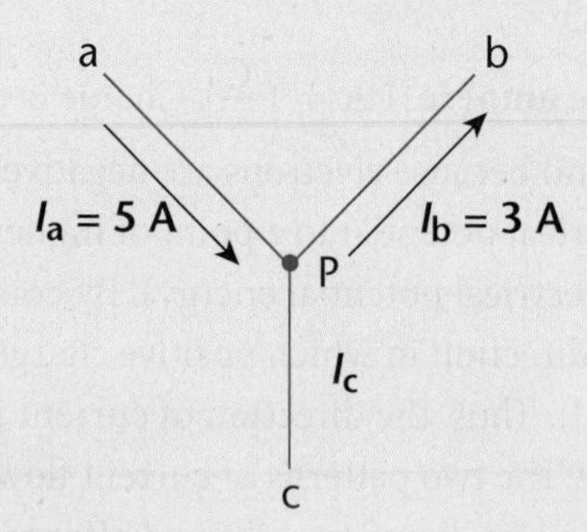

Solution: The sum of currents entering P must equal the sum of the currents leaving P. Assume for now that I_c flows out of P. If we find that it is negative, then we know the current must be going the other direction (into P).

$$I_a = I_b + I_c$$
$$5\text{ A} = 3\text{ A} + I_c$$
$$I_c = 2\text{ A}$$

Thus, a current of 2 A flows out of P along wire c.

Kirchhoff's Loop Rule

KEY CONCEPT

If all of the voltage wasn't "used up" in each loop of the circuit, then the voltage would build after each trip around the circuit, which is impossible.

Around any closed circuit loop, the sum of voltage sources will always be equal to the sum of voltage (potential) drops. This is a consequence of the conservation of energy. All the electrical energy supplied by a source gets fully used up by the other elements within that loop. No excess energy appears, and no energy disappears that cannot be accounted for. Of course, energy can be changed from one form to another, so the kinetic energy of the electrons can be converted to thermal energy, light, or sound by the particular apparatus that is connected to the circuit. Remember that although

Kirchhoff's loop rule is a consequence of the law of conservation of energy, this law is in terms of voltage (joules per coulomb), not just energy (joules). This can be expressed mathematically as

$$V_{\text{source}} = V_{\text{drop}}$$

Equation 6.3

MCAT CONCEPT CHECK 6.1

Before you move on, assess your understanding of the material with these questions.

1. Define the following terms and provide their SI units.
   - Current:

   ____

   - Voltage:

   ____

   - Electromotive force (emf):

   ____

   - Conductivity:

   ____

2. Which likely has a higher conductivity: 1 *M* glucose or 0.25 *M* NaCl? Why?

   ____

3. True or False: In a circuit, the number of electrons entering a point and leaving that point are the same.

4. True or False: The sum of the voltage sources in a circuit is equal to the sum of the voltage drops in that circuit.

6.2 Resistance

High-Yield

LEARNING OBJECTIVES

After Chapter 6.2, you will be able to:

- Recall how the physical properties of a resistor determine its resistance
- Apply the formulas that connect power to current, voltage, and resistance
- Describe how internal resistance of a battery impacts the circuit system
- Contrast the effects of a resistor on a circuit in series as compared to one in parallel
- Calculate the total resistance of a given circuit:

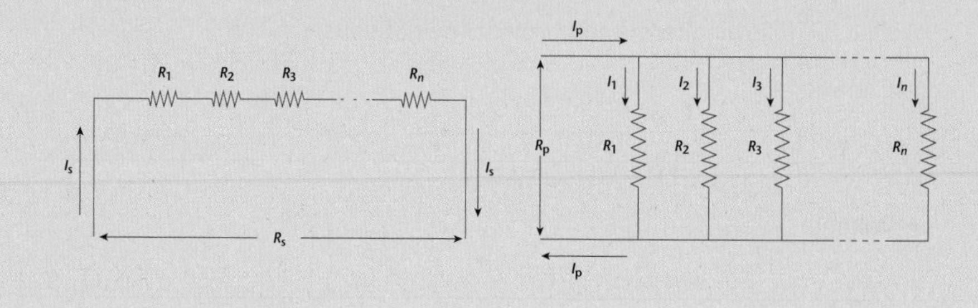

Resistance is the opposition within any material to the movement and flow of charge. Electrical resistance can be thought of like friction, air resistance, or viscous drag: in all of these cases, motion is being opposed. Materials that offer almost no resistance are called conductors, and those materials that offer very high resistance are called insulators. Conductive materials that offer amounts of resistance between these two extremes are called **resistors**.

Properties of Resistors

The resistance of a resistor is dependent upon certain characteristics of the resistor, including resistivity, length, cross-sectional area, and temperature. Three of these are summarized by the equation for resistance:

$$R = \frac{\rho L}{A}$$

Equation 6.4

where ρ is the resistivity, L is the length of the resistor, and A is its cross-sectional area. We will explore the effects of each of these variables and temperature in this section.

MCAT EXPERTISE

On the MCAT, the most common resistors you will see outside of generic, unlabeled resistors are light bulbs, although all appliances function as resistors. You may also see resistance applied atypically, as in resistance to air flow in the lungs or to blood moving in the circulatory system. The same mathematical relationships will be useful in both circumstances.

MCAT EXPERTISE

It is always good practice to be able to derive the units of a variable from equations you know. If you were to solve for resistivity (ρ) in the resistance equation, you'd end up with $\frac{AR}{L}$. Because A is in square meters, R is in ohms, and L is in meters, it simplifies to meters times ohms, which are the units for resistivity.

Resistivity

Some materials are intrinsically better conductors of electricity than others. For example, copper conducts electricity better than plastic, which is why electrical wires have a copper core surrounded by a layer of plastic rather than the other way around. The number that characterizes the intrinsic resistance to current flow in a material is called the **resistivity** (ρ), for which the SI unit is the **ohm–meter** ($\Omega \cdot \text{m}$).

Length

According to the resistance equation, the resistance of a resistor is directly proportional to its length. A longer resistor means that electrons will have to travel a greater distance through a resistant material. This factor scales linearly: if a resistor doubles its length, it will also double its resistance.

Cross-Sectional Area

The equation for resistance also demonstrates an inverse proportionality between resistance and the cross-sectional area of the resistor: if a resistor's cross-sectional area is doubled, its resistance will be cut in half. This is because the increase in cross-sectional area increases the number of pathways through the resistor, called **conduction pathways**. The wider the resistor, the more current that can flow. This is analogous to a river, where the wider the river, the less resistance there is to water flow. Note, however, that electrical current does not follow the continuity equation that applies to incompressible fluids ($A_1v_1 = A_2v_2$); it instead obeys Kirchhoff's laws.

Temperature

Although not evident from the resistance equation, most conductors have greater resistance at higher temperatures. This is due to increased thermal oscillation of the atoms in the conductive material, which produces a greater resistance to electron flow. Because temperature is an intrinsic quality of all matter, we can think of the resistivity as a function of temperature. A few materials do not follow this general rule, including glass, pure silicon, and most semiconductors.

Ohm's Law and Power

Now that we've covered voltage, current, and resistance, we can begin to bring these variables together to solve circuits.

Ohm's Law

Electrical resistance results in an energy loss, which reflects a drop in electrical potential. The voltage drop between any two points in a circuit can be calculated according to Ohm's law:

$$V = IR$$

Equation 6.5

where V is the voltage drop, I is the current, and R is the magnitude of the resistance, measured in **ohms** (Ω). **Ohm's law** is the basic law of electricity because it states that for a given magnitude of resistance, the voltage drop across the resistor will be proportional to the magnitude of the current. Likewise, for a given resistance, the magnitude of the current will be proportional to the magnitude of the emf (voltage) impressed upon the circuit. The equation applies to a single resistor within a circuit, to any part of a circuit, or to an entire circuit (provided one can calculate the equivalent resistance from all of the resistors in the circuit). As current moves through a set of resistors in a circuit, the voltage drops some amount in each resistor; the current (or sum of currents for a divided circuit) is constant. No charge is gained or lost through a resistor; thus, if resistors are connected in series, all of the current must pass through each resistor.

MCAT EXPERTISE

Most batteries on the MCAT will be considered "perfect batteries," and you will not have to account for their internal resistances.

Conductive materials, such as copper wires, act as weak resistors themselves, offering some magnitude of resistance to current and causing a drop in electrical potential (voltage). Even the very sources of emf, such as batteries, have some small but measurable amount of **internal resistance, r_{int}**. As a result of this internal resistance, the voltage supplied to a circuit is reduced from its theoretical emf value by some small amount. The actual voltage supplied by a cell to a circuit can be calculated from

$$V = E_{cell} - ir_{int}$$

Equation 6.6

where V is the voltage provided by the cell, E_{cell} is the emf of the cell, i is the current through the cell, and r_{int} is its internal resistance.

REAL WORLD

Superconductors, a special class of materials, are the one major exception to the rules of internal resistance. When these elements and compounds are cooled to very low temperatures (usually well below 100 K, but the exact threshold varies by material), the resistivity of the material (ρ) completely dissipates, dropping to zero. Because these materials break a generally applicable law of physics, superconductors have many interesting and unusual applications.

If the cell is not actually driving any current (such as when a switch is in the open position), then the internal resistance is zero, and the voltage of the cell is equal to its emf. For cases when the current is not zero and the internal resistance is not negligible, then voltage will be less than emf.

When a cell is discharging, it supplies current, and the current flows from the positive, higher potential end of the cell around the circuit to the negative, lower potential end. Certain types of cells (called **secondary batteries**) can be recharged. When these batteries are being recharged, an external voltage is applied in such a way to drive current toward the positive end of the secondary battery. In electrochemical terms, the cell acts as a galvanic (voltaic) cell when it discharges and as an electrolytic cell when it recharges.

Measuring Power

In Chapter 2 of *MCAT Physics and Math Review*, we briefly mentioned that power is the rate at which energy is transferred or transformed. Power is measured as the ratio of work (energy expenditure) to time and can be expressed as follows:

$$P = \frac{W}{t} = \frac{\Delta E}{t}$$

Equation 6.7

In electric circuits, energy is supplied by the cell that houses a spontaneous oxidation–reduction reaction, which when allowed to proceed (by the closing of a switch, for example), generates a flow of electrons. These electrons, which have electrical potential energy, convert that energy into kinetic energy as they move around the circuit, driven by the emf of the cell. As mentioned above, emf is not a force, but is better thought of as a pressure to move, exerted by the cell on the electrons. Current delivers energy to the various resistors, which convert this energy to some other form, depending on the particular configuration of the resistor.

One particularly recognizable example of resistors at work is the coils inside a toaster. The coils turn red-hot when the toaster is powered on and dissipate thermal energy, which is a direct consequence of the resistance that the coils pose to the current running through them.

The rate at which energy is dissipated by a resistor is the power of the resistor and can be calculated from

$$P = IV = I^2R = \frac{V^2}{R}$$

Equation 6.8

where I is the current through the resistor, V is the voltage drop across the resistor, and R is the resistance of the resistor. Note that these different versions of the power equation can be interconverted by substitution using Ohm's law ($V = IR$).

MCAT EXPERTISE

These equations for calculating the power of a resistor or collection of resistors are extremely helpful for the MCAT. Commit them to memory—and, more importantly, understand them—and your efforts will be rewarded as points on Test Day.

REAL WORLD

Because power equals voltage times current, power companies can manipulate these two values while keeping power constant. One option is to increase current, which results in a decrease in voltage. The other option would be to increase voltage, thus decreasing the current. Power lines are high-voltage lines, which allows them to carry a smaller current—thus decreasing the amount of energy lost from the system.

Resistors in Series and Parallel

Resistors can be connected into a circuit in one of two ways: either in **series**, in which all current must pass sequentially through each resistor connected in a linear arrangement, or in **parallel**, in which the current will divide to pass through resistors separately.

Resistors in Series

For resistors connected in series, the current has no choice but to travel through each resistor in order to return to the cell, as shown in Figure 6.1.

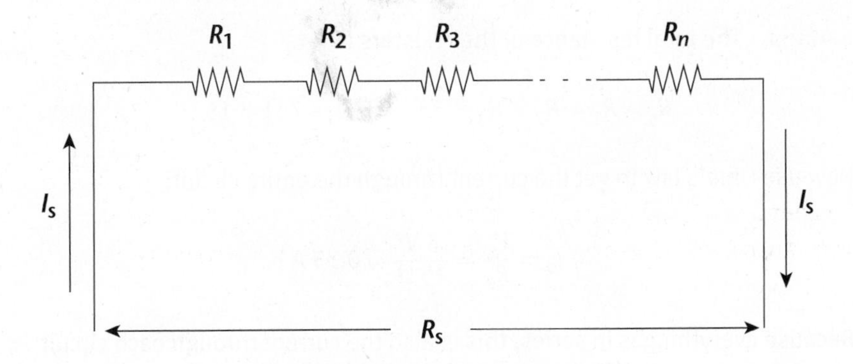

Figure 6.1 Resistors in Series
R_s increases as more resistors are added.

As the electrons flow through each resistor, energy is dissipated, and there is a voltage drop associated with each resistor. The voltage drops are additive; that is, for a series of resistors, R_1, R_2, R_3, $\cdots R_n$, the total voltage drop will be

$$V_s = V_1 + V_2 + V_3 + \cdots + V_n$$

Equation 6.9

KEY CONCEPT

When there is only one path for the current to take, the current will be the same at every point in the line, including through every resistor. Once you know the current of the whole circuit, you can use $V = IR$ to solve for the voltage drop across each resistor (assuming you know the resistances of the resistors).

Because $V = IR$, we can also see that the resistances of resistors in series are also additive, such that

$$R_s = R_1 + R_2 + R_3 + \cdots + R_n$$

Equation 6.10

The set of resistors wired in series can be treated as a single resistor with a resistance equal to the sum of the individual resistances, termed the **equivalent** or **resultant resistance**. Note that R_s will always increase as more resistors are added.

Example: A circuit is wired with one cell supplying 5 V in series with three resistors of 3 Ω, 5 Ω, and 7 Ω, also wired in series as shown below. What is the resulting voltage across and current through each resistor of this circuit, as well as the entire circuit?

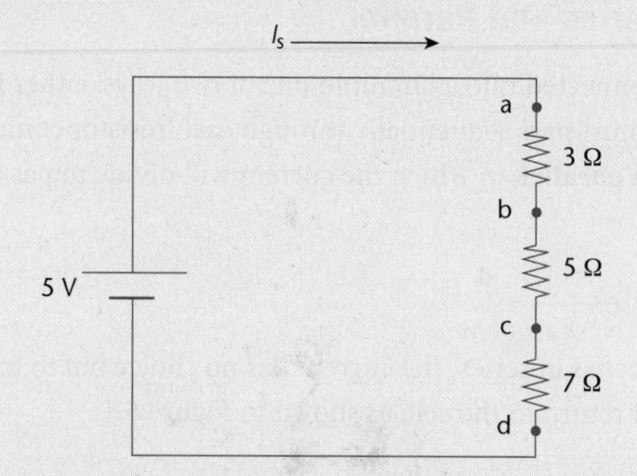

Solution: The total resistance of the resistors is

$$R_s = R_1 + R_2 + R_3 = 3\ \Omega + 5\ \Omega + 7\ \Omega = 15\ \Omega$$

Now use Ohm's law to get the current through the entire circuit:

$$I_s = \frac{V_s}{R_s} = \frac{5\text{ V}}{15\ \Omega} = 0.33\text{ A}$$

Because everything is in series, this is also the current through each circuit element. Now, use Ohm's law for each of the resistors in turn. From a to b, the voltage drop across R_1 is

$$IR_1 = (0.33\text{ A})(3\ \Omega) = 1.0\text{ V}$$

From b to c, the voltage drop across R_2 is

$$IR_2 = (0.33\text{ A})(5\ \Omega) = 1.67\text{ V}$$

From c to d, the voltage drop across R_3 is

$$IR_3 = (0.33\text{ A})(7\ \Omega) = 2.33\text{ V}$$

Resistors in Parallel

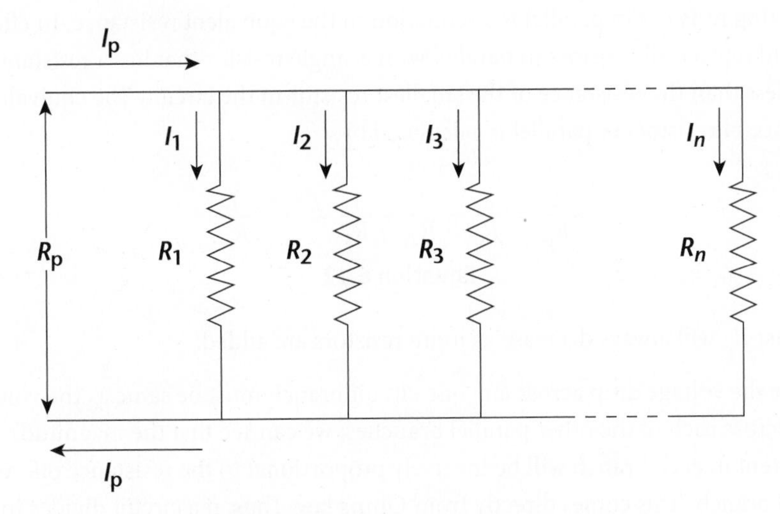

Figure 6.2 Resistors in Parallel
R_p decreases as more resistors are added.

When resistors are connected in parallel, they are wired with a common high-potential terminal and a common low-potential terminal, as shown in Figure 6.2. This configuration allows charge to follow different parallel paths between the high-potential terminal and the low-potential terminal. In this arrangement, electrons have a "choice" regarding which path they will take: some will choose one pathway, while others will choose a different pathway. No matter which path is taken, however, the voltage drop experienced by each division of current is the same because all pathways originate from a common point and end at a common point within the circuit. This is analogous to a river that splits into multiple streams before plunging over different waterfalls, which then come back together to re-form the river at a lower height. If all the water starts at some common height and ends at a lower common height, then it doesn't matter how many "steps" the water fell over to get to the bottom of the falls: the change in height is the same for each stream. In circuits with parallel arrangements of resistors, this is expressed mathematically as:

$$V_p = V_1 = V_2 = V_3 = \cdots = V_n$$

Equation 6.11

KEY CONCEPT

Remember Kirchhoff's loop rule: if every resistor is in parallel, then the voltage drop across each pathway alone must be equal to the voltage of the source.

While the voltage is the same for all parallel pathways, the resistance of each pathway may differ. In this case, electrons prefer the path of least resistance; in other words, the current will be largest through the pathways with the lowest resistance. In fact, there is an inverse relationship between the portion of the current that travels through a particular pathway and the resistance offered by that pathway.

The resistance equation previously discussed shows us that there is an inverse relationship between the cross-sectional area of a resistor and the resistance of that resistor. Like opening up rush-hour lanes to reduce traffic congestion or performing cardiac bypass to perfuse hypoxic heart tissue, the configuration of resistors in parallel allows for a greater total number of conduction paths, and the effect of connecting resistors in parallel is a reduction in the equivalent resistance. In effect, we could replace all resistors in parallel with a single resistor that has a resistance that is less than the resistance of the smallest resistor in the circuit. The equivalent resistance of resistors in parallel is calculated by

$$\frac{1}{R_p} = \frac{1}{R_1} + \frac{1}{R_2} + \frac{1}{R_3} + \cdots + \frac{1}{R_n}$$

Equation 6.12

Note that R_p will always decrease as more resistors are added.

Because the voltage drop across any one circuit branch must be same as the voltage drops across each of the other parallel branches, we can see that the magnitude of the current in each branch will be inversely proportional to the resistance offered by each branch. This comes directly from Ohm's law. Thus, if a circuit divides into two branches and one branch has twice the resistance of the other, the one with twice the resistance will have half the magnitude of current compared to the other. Remember that the sum of the currents going into each division, according to Kirchhoff's junction rule, must equal the total current going into the point at which the current divides.

Example: Consider two equal resistors wired in parallel. What is the equivalent resistance of the setup?

Solution: The equation for summing resistors in parallel is

$$\frac{1}{R_p} = \frac{1}{R_1} + \frac{1}{R_2}$$

Next, find the common denominator of the right side:

$$\frac{1}{R_p} = \frac{R_2}{R_1R_2} + \frac{R_1}{R_1R_2} = \frac{R_1 + R_2}{R_1R_2}$$

Then, take the inverse:

$$R_p = \frac{R_1R_2}{R_1 + R_2}$$

This is a special case where $R_1 = R_2$. Substituting in, we get:

$$R_p = \frac{R^2}{2R} = \frac{R}{2}$$

In the example above, we can see that the total resistance is halved by wiring two identical resistors in parallel. More generally, when n identical resistors are wired in parallel, the total resistance is given by $\frac{R}{n}$. Note that the voltage across each of the parallel resistors is equal and that, for equal resistances, the current flowing through each of the resistors is also equal (that is, a current of $\frac{I_{total}}{n}$ runs through each).

Example: Consider two resistors wired in parallel with $R_1 = 5\ \Omega$ and $R_2 = 10\ \Omega$. If the voltage across them is 10 V, what is the current through each of the two resistors?

Solution: First, the current flowing through the whole circuit must be found. To do this, the equivalent resistance must be calculated:

$$\frac{1}{R_p} = \frac{1}{R_1} + \frac{1}{R_2} = \frac{1}{5\ \Omega} + \frac{1}{10\ \Omega} = \frac{3}{10\ \Omega}$$

$$R_p = \frac{10}{3}\ \Omega$$

Using Ohm's law to calculate the current flowing through the circuit gives

$$I_p = \frac{V_p}{R_p} = \frac{10\ \text{V}}{\left(\frac{10}{3}\ \Omega\right)} = 3\ \text{A}$$

Three amps flow through the combination of R_1 and R_2. Because the resistors are in parallel, $V_p = V_1 = V_2 = 10$ V. Apply Ohm's law to each resistor individually:

$$I_1 = \frac{V_p}{R_1} = \frac{10\ \text{V}}{5\ \Omega} = 2\ \text{A}$$

$$I_2 = \frac{V_p}{R_2} = \frac{10\ \text{V}}{10\ \Omega} = 1\ \text{A}$$

As a check, note that $I_p = 3\ \text{A} = I_1 + I_2$. More current flows through the smaller resistor. In particular, note that R_1, with half the resistance of R_2, has twice the current. Once I_p was found to be 3 A, the problem could have been solved by noting the ratio of the resistances of the two branches.

MCAT EXPERTISE

When approaching circuit problems, the first things you need to find are the total (circuit) values: the total voltage (almost always given as the voltage of the battery), the total (equivalent) resistance, and the total current. To find the total current, first find the total resistance of the circuit.

MCAT CONCEPT CHECK 6.2

Before you move on, assess your understanding of the material with these questions.

1. How does adding or removing a resistor change the total resistance of a circuit with resistors in series? In parallel?
   - Series:
   - Parallel:
2. What four physical quantities determine the resistance of a resistor?
   -
   -
   -
   -
3. How does power relate to current, voltage, and resistance?
4. True or False: The internal resistance of a battery will lower the amount of current it can provide.
5. A circuit is set up with three resistors. The circuit has one branch through R_1, then splits with R_2 and R_3 set up parallel to each other. If $R_1 = 3\ \Omega$, $R_2 = 2\ \Omega$, and $R_3 = 6\ \Omega$, then what proportion of the total current will travel through each resistor? What will be the total resistance of the circuit?

6.3 Capacitance and Capacitors

LEARNING OBJECTIVES

After Chapter 6.3, you will be able to:

- Predict the behavior of a capacitor when charging and discharging
- Describe the impact of a dielectric on capacitance, voltage, and charge
- Recognize the physical properties that impact capacitance of a capacitor
- Contrast the effects of a capacitor on a circuit in series as compared to a circuit in parallel:

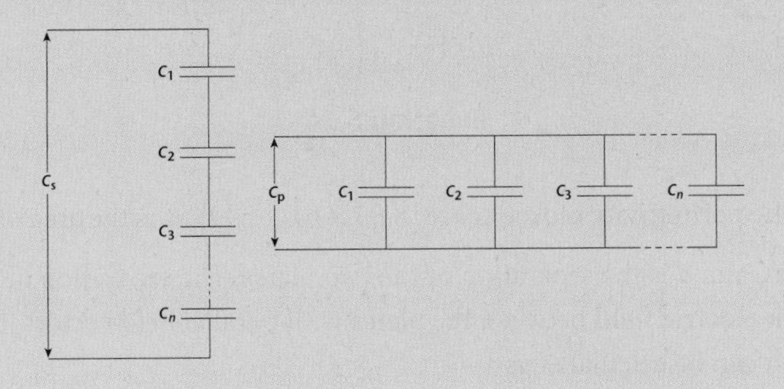

Aside from batteries and resistors, the other major circuit element tested on the MCAT is the capacitor. **Capacitors** are characterized by their ability to hold charge at a particular voltage. There are excellent real-world examples of capacitors. Perhaps the most important capacitor you'll encounter in the clinics is the defibrillator. While a defibrillator is charging, a high-pitched electronic tone sounds as electrons build up on the capacitor. When the defibrillator is fully charged, that charge can be released in one surge of power (after the operator yells *Clear!*). The clouds and the ground during a lightning storm also act as a capacitor, with the charge building up between them eventually **discharging** as a bolt of lightning. The MCAT focuses on a particular type of capacitor called a parallel plate capacitor, and all of our discussion will center on capacitors of this type.

Properties of Capacitors

When two electrically neutral metal plates are connected to a voltage source, positive charge builds up on the plate connected to the positive (higher potential) terminal, and negative charge builds up on the plate connected to the negative (lower potential) terminal. The two-plate system is a capacitor because it can store a particular amount of charge at a particular voltage. The **capacitance** of a capacitor is defined as the ratio of the magnitude of the charge stored on one plate to the potential difference (voltage) across the capacitor. Therefore, if a voltage V is applied across the plates of a capacitor and a charge Q collects on it (with $+Q$ on the positive plate and $-Q$ on the negative plate), then the capacitance is given by

$$C = \frac{Q}{V}$$

Equation 6.13

The SI unit for capacitance is the **farad** $\left(1\text{ F} = 1\,\frac{\text{C}}{\text{V}}\right)$. Because one coulomb is such a large quantity of charge, one farad is a very large capacitance. Capacitances are usually given in microfarads ($1\ \mu\text{F} = 1 \times 10^{-6}$ F) or picofarads ($1\text{ pF} = 1 \times 10^{-12}$ F). Be careful not to confuse the farad with the Faraday constant from electrochemistry, F, which is the amount of charge in one mole of electrons $\left(96{,}485\ \frac{\text{C}}{\text{mol } e^-}\right)$.

The capacitance of a parallel plate capacitor is dependent upon the geometry of the two conduction surfaces. For the simple case of the parallel plate capacitor, the capacitance is given by

$$C = \varepsilon_0 \left(\frac{A}{d}\right)$$

Equation 6.14

where ε_0 is the permittivity of free space $\left(8.85 \times 10^{-12}\ \frac{\text{F}}{\text{m}}\right)$. A is the area of overlap of the two plates, and d is the separation of the two plates. The separation of charges sets up a **uniform electric field** between the plates with parallel field vectors, the magnitude of which can be calculated as

$$E = \frac{V}{d}$$

Equation 6.15

KEY CONCEPT

If you look back to the equations discussed in Chapter 5 of *MCAT Physics and Math Review*, you can see that the equation for E here can be derived from the other fundamental electrostatics equations. If $V = \frac{kQ}{r}$ and $E = \frac{kQ}{r^2}$, then $V = E \times r$. r in this setup is the distance between the plates, d, so we can rewrite this as $V = Ed$.

The direction of the electric field at any point between the plates is from the positive plate toward the negative plate. If we imagine placing a positively charged particle between the oppositely charged plates, we would expect the particle to accelerate in that same direction. This should not be surprising, as electric field lines always point in the direction that indicates the direction of a force exerted on a positive charge.

Regardless of the particular geometry of a capacitor (parallel plate or otherwise), the function of a capacitor is to store an amount of energy in the form of charge separation at a particular voltage. This is akin to the function of a dam, the purpose of which is to store gravitational potential energy by holding back a mass of water at a given height. The potential energy stored in a capacitor is

$$U = \frac{1}{2}CV^2$$

Equation 6.16

Dielectric Materials

The term **dielectric material** is just another way of saying insulation. When a dielectric material, such as air, glass, plastic, ceramic, or certain metal oxides, is introduced between the plates of a capacitor, it increases the capacitance by a factor called the **dielectric constant** (κ). The dielectric constant of a material is a measure of its insulating ability, and a vacuum has a dielectric constant of 1, by definition. For reference, the dielectric constant of air is just slightly above 1, glass is 4.7, and rubber is 7. These numbers need not be memorized; any relevant dielectric constants will be given on Test Day.

The capacitance due to a dielectric material is

$$C' = \kappa C$$

Equation 6.17

where C' is the new capacitance with the dielectric present and C is the original capacitance.

MNEMONIC

Incorporating the dielectric constant into Equation 6.14 reveals that capacitors are **CA$\kappa\varepsilon$d** with charge ($C = A\kappa\varepsilon_0/d$).

KEY CONCEPT

A dielectric material can never decrease the capacitance; thus, κ can never be less than 1.

Dielectrics in Isolated Capacitors

When a dielectric material is placed in an isolated, charged capacitor—that is, a charged capacitor disconnected from any circuit—the voltage across the capacitor decreases. This is the result of the dielectric material shielding the opposite charges from each other. By lowering the voltage across a charged capacitor, the dielectric has increased the capacitance of the capacitor by a factor of the dielectric constant. Thus, when a dielectric material is introduced into an isolated capacitor, the increase in capacitance arises from a decrease in voltage.

Dielectrics in Circuit Capacitors

When a dielectric material is placed in a charged capacitor within a circuit—that is, still connected to a voltage source—the charge on the capacitor increases. The voltage must remain constant because it must be equal to that of the voltage source. By increasing the amount of charge stored on the capacitor, the dielectric has increased the capacitance of the capacitor by a factor of the dielectric constant. Thus, when a dielectric material is introduced into a circuit capacitor, the increase in capacitance arises from an increase in stored charge.

The stored energy in a capacitor is only useful if it is allowed to discharge. The charge can be released from the plates either by discharging across the plates or through some conductive material with which the plates are in contact. For example, capacitors can discharge into wires, causing a current to pass through the wires in much the same way that batteries cause current to move through a circuit. The paddles of the defibrillator machine, once charged, are placed on either side of a patient's heart that has gone into a life-threatening arrhythmia (such as *ventricular fibrillation*). The reason the doctor yells *Clear!* before discharging the paddles is because the current needs to travel through the patient's heart—not through any other people who might be touching the patient and creating a parallel pathway. On a much larger scale, lightning occurs when a very, very large amount of charge exceeds the capacitance of the Earth's surface and the underside of the cloud (the two serving, approximately, as a parallel plate capacitor). The large rapid discharge across the plates of a capacitor is termed a failure of the capacitor, while creating a current through the attached wires is the normal function of a capacitor.

Example: The voltage across the terminals of an isolated 3 μF capacitor is 4 V. If a piece of ceramic having dielectric constant $\kappa = 2$ is placed between the plates, find the new charge, capacitance, and voltage of the capacitor.

Solution: The introduction of a dielectric by itself has no effect on the charge stored on the isolated capacitor. There is no new charge, so the charge is the same as before. The charge stored is therefore given by

$$Q' = Q = CV = (3\ \mu\text{F})(4\ \text{V}) = 12\ \mu\text{C}$$

By introducing a dielectric with a dielectric constant of 2, the capacitance of the capacitor is multiplied by 2 ($C' = \kappa C$). Hence, the new capacitance is 6 μF.

Now, the new voltage across the capacitor can be determined:

$$V' = \frac{Q'}{C'} = \frac{12\ \mu\text{C}}{6\ \mu\text{F}} = 2\ \text{V}$$

Example: A 3 μF capacitor is connected to a 4 V battery. If a piece of ceramic having dielectric constant $\kappa = 2$ is placed between the plates, find the new charge, capacitance, and voltage of the capacitor.

Solution: This question is very similar to the previous one, but the voltage is held constant here by a battery. Thus, the new voltage is still 4 V.

By introducing a dielectric with a dielectric constant of 2, the capacitance of the capacitor is multiplied by 2 ($C' = \kappa C$). Hence, the new capacitance is 6 μF.

Now, the new charge on the capacitor can be determined:

$$Q' = C'V' = (6\ \mu\text{F})(4\ \text{V}) = 24\ \mu\text{C}$$

Capacitors in Series and Parallel

Just like resistors, capacitors can be arranged within a circuit either in parallel or in series. They can also be arranged with resistors, although this is beyond the scope of the MCAT in most cases.

Capacitors in Series

When capacitors are connected in series, the total capacitance decreases in similar fashion to the decreases in resistance seen in parallel resistors, as shown in Figure 6.3.

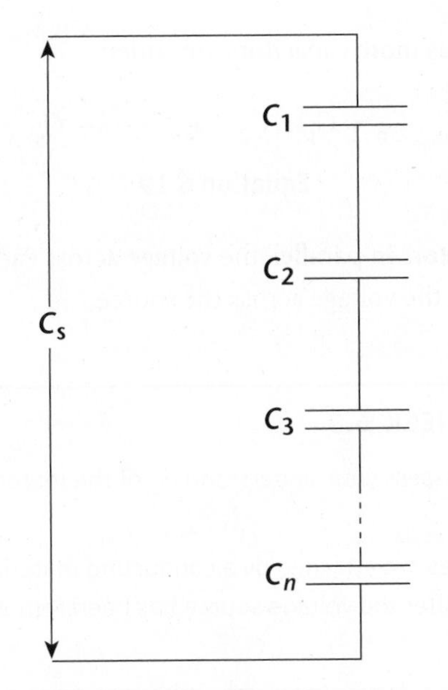

Figure 6.3 Capacitors in Series
C_s decreases as more capacitors are added.

★ Remember that capacitors & resistors have opposite rules!

This is because the capacitors must share the voltage drop in the loop and therefore cannot store as much charge. Functionally, a group of capacitors in series acts like one equivalent capacitor with a much larger distance between its plates (in fact, with a distance equal to those of each of the series capacitors added together). This increase in distance, as seen earlier, means a smaller capacitance.

Rather than memorizing the following equations independently, understand the conceptual basis for the mathematics of resistors in series and in parallel, and then simply reverse that mathematical approach for capacitors. The equation for calculating the equivalent capacitance for capacitors in series is

$$\frac{1}{C_s} = \frac{1}{C_1} + \frac{1}{C_2} + \frac{1}{C_3} + \cdots + \frac{1}{C_n}$$

Equation 6.18

which shows that C_s decreases as more capacitors are added. Note that for capacitors in series, the total voltage is the sum of the individual voltages, just like resistors in series.

Capacitors in Parallel

Capacitors wired in parallel, shown in Figure 6.4, produce a resultant capacitance that is equal to the sum of the individual capacitances.

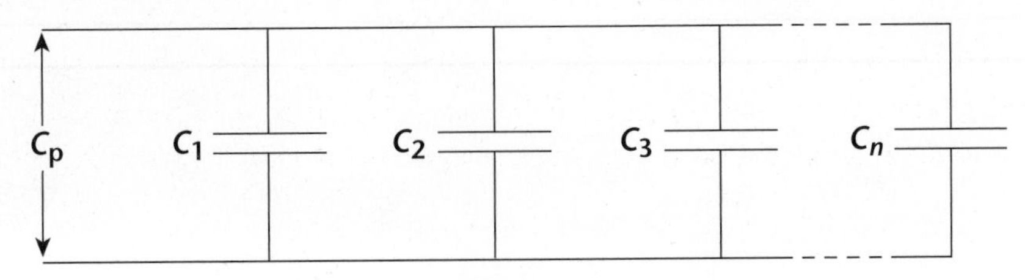

Figure 6.4 Capacitors in Parallel
C_p increases as more capacitors are added.

Therefore, C_p increases as more capacitors are added:

$$C_p = C_1 + C_2 + C_3 + \cdots + C_n$$

Equation 6.19

Just as we saw with resistors in parallel, the voltage across each parallel capacitor is the same and is equal to the voltage across the source.

MCAT CONCEPT CHECK 6.3

Before you move on, assess your understanding of the material with these questions.

1. Assuming the plates are attached by a conducting material, how does a capacitor behave after the voltage source has been removed from a circuit?

____

2. How does a dielectric material impact capacitance? Voltage? Charge?

   - Capacitance:

   ____

   - Voltage:

   ____

   - Charge:

   ____

3. How does adding or removing a capacitor change the total capacitance of a circuit with capacitors in series? In parallel?

   - Series:

   ____

   - Parallel:

   ____

4. What physical qualities contribute to the capacitance of a capacitor?

____

6.4 Meters

LEARNING OBJECTIVES

After Chapter 6.4, you will be able to:

- Recall key details about ammeters, voltmeters, and ohmmeters, including what they measure, where they should be placed in a circuit, and their ideal resistances
- Determine whether multiple meters should be placed together in a circuit

Although we've been focused on calculations of hypothetical circuits so far, it's important for us to spend some time considering real ones. While we will not analyze any complex circuits here, it is important to be familiar with **meters**, the devices that are used to measure circuit quantities in the real world.

Ammeters

Ammeters are used to measure the current at some point within a circuit. Using an ammeter requires the circuit to be on, or the current will be 0 A. Ammeters are inserted in series where the current is being measured and use the magnetic properties of a current-carrying wire to cause a visible needle movement or a calibrated display of the current. If there is a particularly high current, this will overwhelm the ammeter, and a special low resistance shunt is used in parallel with the ammeter to allow a reading. Ideally, an ammeter will not change circuit mathematics when it is inserted into the circuit. To do so, it must have an extremely low resistance. Ideal ammeters have zero resistance and no voltage drop across themselves.

Voltmeters

A **voltmeter**, like an ammeter, requires a circuit to be active. Voltmeters also use magnetic properties of current-carrying wires. However, voltmeters are used to measure the voltage drop across two points in a circuit. They are wired in parallel to these two points. Because the goal with any meter is to minimize its impact on the rest of the circuit, and voltmeters are wired in parallel, an ideal voltmeter has infinite resistance.

Ohmmeters

Unlike voltmeters and ammeters, an **ohmmeter** does not require a circuit to be active (in fact, some ohmmeters will give false readings or can be damaged by an active circuit). Ohmmeters will often have their own battery of known voltage and then function as ammeters through another point in the circuit. Because only one circuit element is being analyzed, Ohm's law can be used to calculate resistance by knowing the ohmmeter's voltage and the current created through another point in the circuit.

MCAT CONCEPT CHECK 6.4

Before you move on, assess your understanding of the material with these questions.

1. What do each of the following types of meters measure? Where are they placed in circuits? What are their ideal resistances?

Meter Type	Measures . . .	Placement	Ideal Resistance
Ammeter			
Voltmeter			
Ohmmeter			

2. True or False: A voltmeter and an ammeter should not be placed in the same circuit.

Conclusion

This chapter covered a lot of material. We began with a review of current, taking special note of the conventional definition of current as the movement of positive charge (when, in fact, negatively charged electrons are actually moving). We considered the basic laws of electricity and circuits: Kirchhoff's laws, which are expressions of conservation of charge and energy, and Ohm's law, which relates voltage, current, and resistance. We defined resistance and analyzed the relationships between resistance and resistivity (directly proportional), resistance and length (directly proportional), and resistance and cross-sectional area (inversely proportional). We also defined capacitance as the ability to store charge at some voltage, thereby storing energy. Throughout, we stressed the importance of the both the conceptual and mathematical treatment of resistors and capacitors in series and in parallel as a major testing topic on the MCAT. Finally, we covered the different meters that can be used to measure circuit quantities.

Electricity is often a challenging concept for MCAT students. Unlike kinematics, thermodynamics, and fluids, which are often more tangible, electricity is often best understood through schematics and models. Take time to review these last two chapters, as they will assuredly pay off as points on Test Day. In the next chapter, we turn our attention to a completely different topic that is no more tangible—but is far more audible: sound.

You've reviewed the content, now test your knowledge and critical thinking skills by completing a test-like passage set in your online resources!

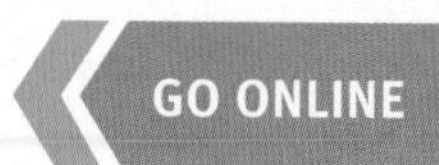

Concept Summary

Current

- **Current** is the movement of charge that occurs between two points that have different electrical potentials.
  - By convention, current is defined as the movement of positive charge from the high-potential end of a voltage source to the low-potential end.
  - In reality, it is negatively charged particles (electrons) that move in a circuit, from low potential to high potential.
- Current flows only in **conductive materials**.
  - **Metallic conduction** relies on uniform movement of free electrons in metallic bonds.
  - **Electrolytic conduction** relies on the ion concentration of a solution.
  - **Insulators** are materials that do not conduct a current.
- **Kirchhoff's laws** express conservation of charge and energy.
  - **Kirchhoff's junction rule** states that the sum of currents directed into a point within a circuit equals the sum of the currents directed away from that point.
  - **Kirchhoff's loop rule** states that in a closed loop, the sum of voltage sources is always equal to the sum of voltage drops.

Resistance

- **Resistance** is opposition to the movement of electrons through a material.
- **Resistors** are conductive materials with a moderate amount of resistance that slow down electrons without stopping them.
- Resistance is calculated using the resistivity, length, and cross-sectional area of the material in question.
- **Ohm's law** states that for a given resistance, the magnitude of the current through a resistor is proportional to the voltage drop across the resistor.
- Resistors in circuits can be combined to calculate the equivalent resistance of a full or partial circuit.
  - Resistors in **series** are additive and sum together to create the total resistance of a circuit.
  - Resistors in **parallel** cause a decrease in equivalent resistance of a circuit.
- Across each resistor in a circuit, a certain amount of power is dissipated, which is dependent on the current through the resistor and the voltage drop across the resistor.

Capacitance and Capacitors

- **Capacitors** have the ability to store and discharge electrical potential energy.
- **Capacitance** in parallel plate capacitors is determined by the area of the plates and the distance between the plates.
- Capacitors in series cause a decrease in the equivalent capacitance of a circuit.
- Capacitors in parallel sum together to create a larger equivalent capacitance.
- **Dielectric materials** are insulators placed between the plates of a capacitor that increase capacitance by a factor equal to the material's **dielectric constant**, κ.

Meters

- **Ammeters** are inserted in series in a circuit to measure current; they have negligible resistance.
- **Voltmeters** are inserted in parallel in a circuit to measure a voltage drop; they have very large resistances.
- **Ohmmeters** are inserted around a resistive element to measure resistance; they are self-powered and have negligible resistance.

Answers to Concept Checks

6.1

1. Current is the movement of positive charge through a conductive material over time and is given in ampères $\left(\frac{C}{s}\right)$. Voltage is a potential difference between two points and is given in volts $\left(\frac{J}{C}\right)$. Electromotive force (emf) refers to the potential difference of the voltage source for a circuit, usually a battery, and is given in volts. Conductivity is the reciprocal of resistance and is a measure of permissiveness to current flow; it is measured in siemens (S).
2. The sodium chloride solution likely has a higher conductivity because it is a salt and will increase the ion content of water. Glucose does not dissociate, and therefore it has a near-zero impact on conductivity.
3. True. This is a restatement of Kirchhoff's junction rule.
4. False. While the voltage sources and voltage drops are equal in any closed loop, this is not necessarily true for the entire circuit. For example, a 9 V battery that powers 10 light bulbs in parallel has a 9 V voltage source and a 9 V drop across each light bulb—a total of 90 V of drop across all of the light bulbs combined.

6.2

1. Adding a resistor in series increases the total resistance of a circuit; removing one in series decreases the total resistance in the circuit. These relationships are reversed in parallel: adding a resistor decreases resistance while removing one increases it.
2. Resistivity, length, cross-sectional area, and temperature all contribute to the resistance of a resistor.
3. Power is related to current, voltage, and resistance through the equations $P = IV = I^2R = \frac{V^2}{R}$.
4. True. The internal resistance will lower the available voltage for the circuit. Lowering the available voltage will also lower current for any given resistance.
5. All current must travel through the first resistor, regardless of its resistance. Since the ratio of resistance for $R_2{:}R_3$ is 1:3, the ratio of current passing through them will be 3:1. In other words, $\frac{3}{4}$ of the current will pass through R_2 while $\frac{1}{4}$ of the current will pass through R_3. To calculate the total resistance, first calculate the resistance of the resistors in parallel: $\frac{1}{R_{2+3}} = \frac{1}{2} + \frac{1}{6} = \frac{2}{3} \rightarrow R_{2+3} = 1.5\ \Omega$. Add this to the resistance of R_1 to get the total resistance: 4.5 Ω.

6.3

1. The capacitor discharges, providing a current in the opposite direction of the initial current.
2. A dielectric material will always increase capacitance. If the capacitor is isolated, its voltage will decrease when a dielectric material is introduced; if it is in a circuit, its voltage is constant because it is dictated by the voltage source. If a capacitor is isolated, the stored charge will remain constant because there is no additional source of charge; if it is in a circuit, the stored charge will increase.
3. Adding a capacitor in series decreases the total capacitance of a circuit; removing one in series increases the total capacitance in the circuit. These relationships are reversed in parallel: adding a capacitor increases capacitance while removing one decreases it.
4. Surface area, distance, and dielectric constant all contribute to the capacitance of a capacitor.

6.4

1.

Meter Type	Measures...	Placement	Ideal Resistance
Ammeter	Current	In series with point of interest	0
Voltmeter	Potential difference (voltage)	Parallel with circuit element of interest	∞
Ohmmeter	Resistance	Two points in series with circuit element of interest	0

2. False. Voltmeters and ammeters are designed to have minimum impact on a circuit; thus, they can be used together.

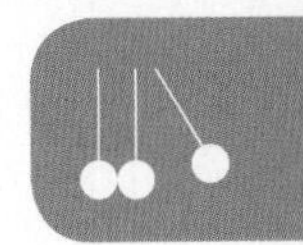

Science Mastery Assessment Explanations

1. **B**

Electrical current is defined as charge flow, or in mathematical terms, charge transferred per time: $I = \frac{Q}{\Delta t}$. A 15 A current that acts for 0.1 s will transfer 15 A × 0.1 s = 1.5 C of charge.

2. **D**

To measure the current at any point in a circuit, an ammeter should be placed in series. If placed in parallel, a new path for current would be created, so the measured value would not be reflective of actual current in the circuit. This observation matches **(D)**. Note that an ideal ammeter should have zero resistance so that it has no effect on the current of the circuit.

3. **B**

The resistance of a resistor is given by the formula $R = \frac{\rho L}{A}$. Thus, there is a direct proportionality between resistance and resistivity. Because the other variables are equal between the two resistors, we can determine that if $R_1{:}R_2$ is a 1:2 ratio, then $\rho_1{:}\rho_2$ is also a 1:2 ratio.

4. **D**

This question tests our understanding of batteries in a circuit. The voltage across the terminals of the battery when there is no current flowing is referred to as the electromotive force (emf or ε of the battery). However, when a current is flowing through the circuit, the voltage across the terminals of the battery is decreased by an amount equal to the current multiplied by the internal resistance of the battery. Mathematically, this is given by the equation

$$V = \varepsilon - ir_{\text{int}}$$

To determine the emf of the battery, first calculate the voltage across the battery when the current is flowing. For this, we can use Ohm's law:

$$\begin{aligned} V &= IR \\ &= (0.5\ \text{A})(3\ \Omega) = 1.5\ \text{V} \end{aligned}$$

Because we know the internal resistance of the battery, the current, and the voltage, we can calculate the emf:

$$\begin{aligned} \varepsilon &= V + ir_{\text{int}} \\ &= 1.5\ \text{V} + (0.5\ \text{A})(0.1\ \Omega) \\ &= 1.5 + 0.05 = 1.55\ \text{V} \end{aligned}$$

The answer makes sense in the context of a real battery because its internal resistance is supposed to be very small so that the voltage provided to the circuit is as close as possible to the emf of the cell when there is no current running.

5. **A**

We are told that transformers conserve energy so that the output power equals the input power. Thus, $P_{\text{out}} = P_{\text{in}}$, or $I_{\text{out}}V_{\text{out}} = I_{\text{in}}V_{\text{in}}$. There is therefore an inverse proportionality between current and voltage. If the output voltage is 300% of the input voltage (3 times its amount), then the output current must be $\frac{1}{3}$ of the input voltage. This can be represented as a 1:3 ratio.

6. **B**

The fastest way to tackle these kinds of questions is to simplify the circuit bit by bit. For example, notice that R_3 and R_4 are in parallel with each other and are in series with R_2; similarly, R_5 and R_6 are in series. If we determine the total resistance in each branch, we will be left with three branches in parallel. To start, find the total resistance in the middle branch:

$$\frac{1}{R_{3+4}} = \frac{1}{R_3} + \frac{1}{R_4} = \frac{1}{32\ \Omega} + \frac{1}{32\ \Omega} \rightarrow R_{3+4} = 16\ \Omega$$

$$R_{2+3+4} = R_2 + R_{3+4} = 4\ \Omega + 16\ \Omega = 20\ \Omega$$

Next, take a look at the total resistance in the bottom branch:

$$R_{5+6} = R_5 + R_6 = 15\ \Omega + 5\ \Omega = 20\ \Omega$$

The circuit can now be viewed as three resistors in parallel, each providing a resistance of 20 Ω. The total resistance in the circuit is thus

$$\frac{1}{R_{tot}} = \frac{1}{R_1} + \frac{1}{R_{2+3+4}} + \frac{1}{R_{5+6}} =$$

$$\frac{1}{20\ \Omega} + \frac{1}{20\ \Omega} + \frac{1}{20\ \Omega} \rightarrow \frac{20}{3}\ \Omega = 6.67\ \Omega$$

7. **A**

To determine the moles of charge that pass through the circuit over a period of 10 s, we will have to calculate the amount of charge running through the circuit. Charge is simply current multiplied by time, and the current can be calculated using Ohm's law:

$$V = IR \text{ and } I = \frac{Q}{\Delta t} \rightarrow Q = \frac{V\Delta t}{R}$$

$$Q = \frac{(100\ \text{V})(10\ \text{s})}{2\ \Omega} = 500\ \text{C}$$

Then, calculate the number of moles of charge that this represents by using the Faraday constant and approximating F as $10^5\ \frac{\text{C}}{\text{mol}\ e^-}$:

$$(500\ \text{C})\left[\frac{\text{mol}\ e^-}{10^5\ \text{C}}\right] = 5 \times 10^{-3}\ \text{mol}\ e^-$$

This is closest to **(A)**.

8. **C**

To determine the voltage drop across the $\frac{2}{3}$ Ω resistor, start by calculating the total resistance in the circuit. For the resistors in parallel, the equivalent resistance is

$$\frac{1}{R_p} = \frac{1}{2\ \Omega} + \frac{3}{2\ \Omega} \rightarrow R_p = \frac{1}{2}\ \Omega$$

The total resistance in the circuit is the sum of the remaining resistor and the equivalent resistance of the other two:

$$R_s = \frac{1}{2}\ \Omega + \frac{1}{2}\ \Omega = 1\ \Omega$$

Now that we know the equivalent resistance, we can calculate the total current using Ohm's law:

$$I = \frac{V}{R} = \frac{10\ \text{V}}{1\ \Omega} = 10\ \text{A}$$

Finally, we can determine the voltage drop across the parallel resistors. The voltage drop across the $\frac{1}{2}$ Ω resistor must be $(10\ \text{A})\left(\frac{1}{2}\ \Omega\right) = 5$ V. Therefore, there must be a 5 V drop across both the $\frac{2}{3}$ Ω resistor and 2 Ω resistor, according to Kirchhoff's loop rule. Each of these resistors forms a complete loop in combination with the $\frac{1}{2}$ Ω resistor and 10 V voltage source, and the net potential difference around any closed loop must be 0 V.

9. **D**

This question should bring to mind the equation $C = \varepsilon_0\left(\frac{A}{d}\right)$, where ε_0 is the permittivity of free space, A is the area of the plates, and d is the distance between the plates. From this equation, we can infer that doubling the area will double the capacitance, and halving the distance will also double the capacitance. Therefore, the new capacitance is four times larger than the initial capacitance.

10. **C**

Because the question is asking us to calculate the charge on the capacitor, use the formula $Q = CV$. We are given $V = 7500$ V and can calculate C from the formula for energy, $U = \frac{1}{2}CV^2$:

$$U = \frac{1}{2}CV^2 = \frac{1}{2}\left(\frac{Q}{V}\right)V^2 = \frac{QV}{2}$$

$$Q = \frac{2U}{V} = \frac{2(400\ \text{J})}{7500\ \text{V}} \approx \frac{800}{8000} = 0.1\ \text{C}$$

Thus, the charge is close to 0.1 C, which is closest to **(C)**.

11. **C**

Power is energy dissipated per unit time; therefore, the energy dissipated is $E = P\Delta t$. In the five-second interval during which the resistor is active, it has a 2 A current for three of those seconds. The power dissipated by a resistor R carrying a current I is $P = I^2R$. Therefore, the energy dissipated is

$$E = I^2R\Delta t = (2\ \text{A})^2(10\ \Omega)(3\ \text{s}) = 4 \times 10 \times 3 = 120\ \text{J}$$

12. **A**

Kirchhoff's junction rule states that the sum of all currents directed into a point is always equal to the sum of all currents directed out of the point. The currents directed into point P are 8 A, 2 A, and 3 A, so the sum is 13 A. The currents directed out of point P are 5 A and 6 A, so the total is 11 A. Because the two numbers must always be equal, an additional current of 2 A must be directed away from point P toward point x.

13. **D**

The electric field between two plates of a parallel plate capacitor is related to the potential difference between the plates of the capacitor and the distance between the plates, as shown in the formula $E = \frac{V}{d}$. The addition of another battery will increase the total voltage applied to the circuit, which, consequently, will increase the electric field. The addition of a resistor in series will increase the resistance and decrease the voltage applied to the capacitor, eliminating **(A)**. Adding a resistor in parallel will not change the voltage drop across the capacitor and should not change the electric field, eliminating **(B)**. Increasing the distance between the plates, **(C)**, would decrease the electric field, not increase it.

14. **D**

The resistance of the three resistors wired in series is equal to the sum of the individual resistances (12 Ω). This means that the circuit functionally contains a 12 Ω resistor and a 4 Ω resistor in parallel. To determine the overall resistance of this system, use the formula

$$\frac{1}{R_\text{p}} = \frac{1}{R_1} + \frac{1}{R_2} = \frac{1}{12\ \Omega} + \frac{1}{4\ \Omega} = \frac{1+3}{12\ \Omega} \rightarrow R_\text{p} = 3\ \Omega$$

15. **C**

While this is primarily a recall question, it should also be intuitive. Voltmeters are attempting to determine a change in potential from one point to another. To do this, they should not provide an alternate route for charge flow and should therefore have infinite resistance. Ammeters attempt to determine the flow of charge at a single point and should not contribute to the resistance of a series circuit; therefore, they should have no resistance.

Consult your online resources for additional practice. GO ONLINE

Equations to Remember

(6.1) **Current:** $I = \frac{Q}{\Delta t}$

(6.2) **Kirchhoff's junction rule:** $I_{\text{into junction}} = I_{\text{leaving junction}}$

(6.3) **Kirchhoff's loop rule:** $V_{\text{source}} = V_{\text{drop}}$

(6.4) **Definition of resistance:** $R = \frac{\rho L}{A}$

(6.5) **Ohm's law:** $V = IR$

(6.6) **Voltage and cell emf:** $V = E_{\text{cell}} - ir_{\text{int}}$

(6.7) **Definition of power:** $P = \frac{W}{t} = \frac{\Delta E}{t}$

(6.8) **Electric power:** $P = IV = I^2R = \frac{V^2}{R}$

(6.9) **Voltage drop across circuit elements (series):** $V_s = V_1 + V_2 + V_3 + \cdots + V_n$

(6.10) **Equivalent resistance (series):** $R_s = R_1 + R_2 + R_3 + \cdots + R_n$

(6.11) **Voltage drop across circuit elements (parallel):** $V_p = V_1 = V_2 = V_3 = \cdots = V_n$

(6.12) **Equivalent resistance (parallel):** $\frac{1}{R_p} = \frac{1}{R_1} + \frac{1}{R_2} + \frac{1}{R_3} + \cdots + \frac{1}{R_n}$

(6.13) **Definition of capacitance:** $C = \frac{Q}{V}$

(6.14) **Capacitance based on parallel plate geometry:** $C = \varepsilon_0\left(\frac{A}{d}\right)$

(6.15) **Electric field in a capacitor:** $E = \frac{V}{d}$

(6.16) **Potential energy of a capacitor:** $U = \frac{1}{2}CV^2$

(6.17) **Capacitance with a dielectric material:** $C' = \kappa C$

(6.18) **Equivalent capacitance (series):** $\frac{1}{C_s} = \frac{1}{C_1} + \frac{1}{C_2} + \frac{1}{C_3} + \cdots + \frac{1}{C_n}$

(6.19) **Equivalent capacitance (parallel):** $C_p = C_1 + C_2 + C_3 + \cdots + C_n$

Shared Concepts

Biology Chapter 6
The Respiratory System

Biology Chapter 7
The Cardiovascular System

General Chemistry Chapter 12
Electrochemistry

Physics and Math Chapter 2
Work and Energy

Physics and Math Chapter 4
Fluids

Physics and Math Chapter 5
Electrostatics and Magnetism

CHAPTER 7

Waves and Sound

SCIENCE MASTERY ASSESSMENT

Every pre-med knows this feeling: there is so much content I have to know for the MCAT! How do I know what to do first or what's important?

While the high-yield badges throughout this book will help you identify the most important topics, this Science Mastery Assessment is another tool in your MCAT prep arsenal. This quiz (which can also be taken in your online resources) and the guidance below will help ensure that you are spending the appropriate amount of time on this chapter based on your personal strengths and weaknesses. Don't worry though—skipping something now does not mean you'll never study it. Later on in your prep, as you complete full-length tests, you'll uncover specific pieces of content that you need to review and can come back to these chapters as appropriate.

How to Use This Assessment

If you answer 0–7 questions correctly:

Spend about 1 hour to read this chapter in full and take limited notes throughout. Follow up by reviewing **all** quiz questions to ensure that you now understand how to solve each one.

If you answer 8–11 questions correctly:

Spend 20–40 minutes reviewing the quiz questions. Beginning with the questions you missed, read and take notes on the corresponding subchapters. For questions you answered correctly, ensure your thinking matches that of the explanation and you understand why each choice was correct or incorrect.

If you answer 12–15 questions correctly:

Spend less than 20 minutes reviewing all questions from the quiz. If you missed any, then include a quick read-through of the corresponding subchapters, or even just the relevant content within a subchapter, as part of your question review. For questions you got correct, ensure your thinking matches that of the explanation and review the Concept Summary at the end of the chapter.

1. An opera singer has two precisely identical glasses. The singer produces as pure a tone as possible and shatters the first glass at a frequency of 808 Hz. She then sings a frequency of 838 Hz in the presence of the second glass. The second glass will likely:
   - **A.** shatter after a longer amount of time because the applied frequency is higher.
   - **B.** shatter after a shorter amount of time because the applied frequency is higher.
   - **C.** not shatter because the applied frequency is not equal to the natural frequency of the glass.
   - **D.** not shatter because higher-frequency sounds are associated with more attenuation.

2. A child is practicing the first overtone on his flute. If his brother covers one end of the flute for a brief second, how will the sound change, assuming that the new pitch represents the first overtone in the new setup?
   - **A.** The pitch of the sound will go up.
   - **B.** The pitch of the sound will go down.
   - **C.** The pitch of the sound will not change.
   - **D.** The change in the pitch depends on the starting pitch.

3. Which of the following is necessarily true regarding frequency, angular frequency, and period of a given wave?
   - **A.** The magnitude of the angular frequency is larger than the magnitude of the period.
   - **B.** The product of the frequency and period is equal to the angular frequency.
   - **C.** The magnitude of the angular frequency is larger than the magnitude of the frequency.
   - **D.** The product of the angular frequency and period is 1.

4. Ultrasound machines calculate distance based upon:
   - **A.** intensity of the reflected sound.
   - **B.** travel time of the reflected sound.
   - **C.** angle of incidence of the sound.
   - **D.** the detected frequency of the sound.

5. The period for a certain wave is 34 ms. If there is a Doppler shift that doubles the perceived frequency, which of the following must be true?
   - **I.** The detector is moving toward the source at a velocity equal to the speed of sound.
   - **II.** The source is moving toward the detector at a velocity equal to half the speed of sound.
   - **III.** The perceived period is 17 ms.
   - **IV.** The perceived period is 68 ms.

   - **A.** III only
   - **B.** I and IV only
   - **C.** II and III only
   - **D.** I, II, and IV only

6. If the speed of a wave is 3 $\frac{\text{m}}{\text{s}}$ and its wavelength is 10 cm, what is its period?
   - **A.** 0.01 s
   - **B.** 0.03 s
   - **C.** 0.1 s
   - **D.** 0.3 s

7. What is the angular frequency of the third harmonic in a pipe of length 1.5 m with one closed end? (Note: The speed of the sound is approximately 340 $\frac{\text{m}}{\text{s}}$.)
   - **A.** 170 radians per second
   - **B.** 170π radians per second
   - **C.** 340 radians per second
   - **D.** 340π radians per second

8. A certain sound level is increased by 20 dB. By what factor does its intensity increase?

A. 2
B. 20
C. 100
D. log 2

9. In some forms of otosclerosis, the stapedial foot plate, which transmits vibrations from the bones of the middle ear to the fluid within the cochlea, can become fixed in position. This limits the displacement of the stapedial foot plate during vibration. Based on this mechanism, which of the following symptoms would most likely be seen in an individual with otosclerosis?

A. An increase in the perceived volume of sounds
B. A decrease in the perceived volume of sounds
C. An increase in the perceived pitch of sounds
D. A decrease in the perceived pitch of sounds

10. If two waves with the same frequency are 180° out of phase, what is the amplitude of the resultant wave if the amplitudes of the original waves are 5 cm and 3 cm?

A. 2 cm
B. 3 cm
C. 5 cm
D. 8 cm

11. A student is measuring sound frequencies from the side of a road while walking east. For which of the following situations could the student determine that the difference between the perceived frequency and the actual emitted frequency is zero?

A. A plane flying directly above him from east to west
B. A police car passing the student with its siren on
C. A person playing piano in a house on the street
D. A dog barking in a car that moves east

12. In which of the following media does sound travel the fastest?

A. Vacuum
B. Air
C. Water
D. Glass

13. Shock waves have the greatest impact when the source is traveling:

A. just below the speed of sound.
B. exactly at the speed of sound.
C. just above the speed of sound.
D. well above the speed of sound.

14. As an officer approaches a student who is studying with his radio playing loudly beside him, he experiences the Doppler effect. Which of the following statements remains true while the officer moves closer to the student?

I. The apparent frequency of the music is increased.
II. The same apparent frequency would be produced if the officer were stationary and the student approached him at the same speed.
III. The apparent velocity of the wave is decreased.

A. I only
B. II only
C. I and III only
D. I, II, and III

15. Ignoring attenuation, how does the intensity of a sound change as the distance from the source doubles?

A. It is four times as intense.
B. It is twice as intense.
C. It is half as intense.
D. It is one-quarter as intense.

Answer Key

1. **C** (Ch. 7.1)
2. **B** (Ch. 7.2)
3. **C** (Ch. 7.1)
4. **B** (Ch. 7.2)
5. **A** (Ch. 7.2)
6. **B** (Ch. 7.1)
7. **D** (Ch. 7.2)
8. **C** (Ch. 7.2)
9. **B** (Ch. 7.2)
10. **A** (Ch. 7.1)
11. **D** (Ch. 7.2)
12. **D** (Ch. 7.2)
13. **B** (Ch. 7.2)
14. **A** (Ch. 7.2)
15. **D** (Ch. 7.2)

CHAPTER 7

Waves and Sound

In This Chapter

7.1 **General Wave Characteristics**
Transverse and Longitudinal Waves 250
Describing Waves 251
Phase 252
Principle of Superposition 253
Traveling and Standing Waves...... 254
Resonance 254

7.2 **Sound** HY
Production of Sound 257
Frequency and Pitch............. 258
Intensity and Loudness of Sound 261
Standing Waves 264
Ultrasound 268

Concept Summary 271

CHAPTER PROFILE

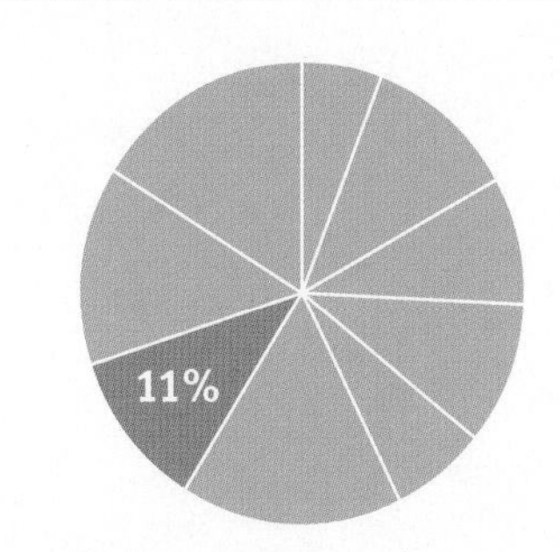

The content in this chapter should be relevant to about 11% of all questions about physics on the MCAT.

This chapter covers material from the following AAMC content category:

4A: Translational motion, forces, work, energy, and equilibrium in living systems

4D: How light and sound interact with matter

Introduction

As a species, our interactions with sound are amazingly complex. The human ear developed as a means of detecting longitudinal waves carried in the air—this likely served an evolutionary purpose. A rustle in the leaves could indicate not only a potential meal, but also a potential predator. Our brains are highly attuned to analyze the sounds around us, as discussed in Chapter 2 of *MCAT Behavioral Sciences Review*. This includes not only the normal auditory pathway from the pinna through the tympanic membrane, ossicles, cochlea, and vestibulocochlear nerve to the temporal lobe, but also secondary structures such as the superior olive, which helps localize sound, and the inferior colliculus, which is involved in the startle reflex.

Language is also inextricably linked to sound. Through changes in pitch and timbre, we can imply or evoke dozens of complex feelings. Through music, our relationship with sound becomes even more profound. As E.T.A. Hoffman, a musicologist and pedagogue, wrote in his vivid description of Beethoven's opening motif for *Symphony No. 5 in C minor*, op. 67:

Radiant beams shoot through this region's deep night, and we become aware of gigantic shadows which, rocking back and forth, close in on us and destroy everything within us except the pain of endless longing—a longing in which every pleasure that rose up in jubilant tones sinks and succumbs, and only through this pain, which, while consuming but not destroying love, hope, and joy, tries to burst our breasts with full-voiced harmonies of all the passions, we live on and are captivated beholders of the spirits.

Indeed, sound can create entire worlds that we can explore. This chapter, however, aims only to lay the foundation for understanding wave phenomena. The general properties of waves will be introduced, including a discussion of wavelength, frequency, wave speed, amplitude, and resonance. We will also review the interactions of waves meeting at a point in space through constructive and destructive interference and examine the mathematics of standing waves—the means by which musical instruments produce their characteristic sounds. The subject of sound is reviewed as a specific example of the longitudinal waveform with a focus on wave phenomena such as the Doppler effect. Finally, we provide a brief discussion of the use of ultrasound and shock waves in medicine.

7.1 General Wave Characteristics

LEARNING OBJECTIVES

After Chapter 7.1, you will be able to:

- Define key terms applying to waves and sound, such as frequency, pitch, and amplitude
- Distinguish between common examples of transverse and longitudinal waves
- Predict the impact of applying a force at the natural frequency of a given system
- Predict the relative amplitude of a resultant wave created by two interfering waves:

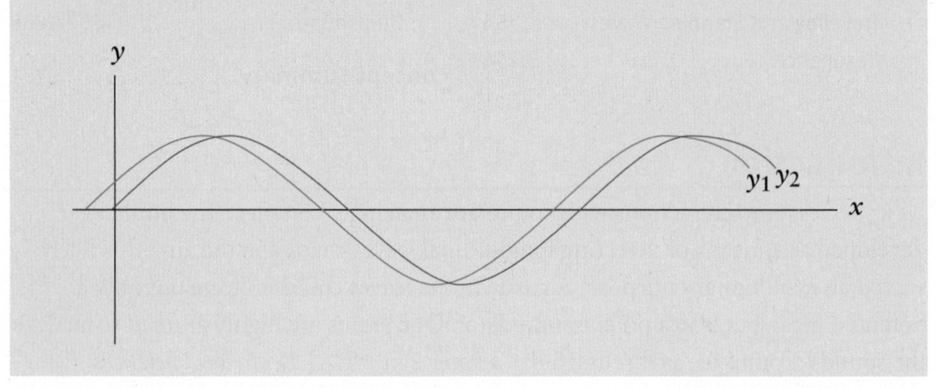

It is important to use a common language when describing waves. We'll establish the terminology associated with wave phenomena, and then spend the rest of this chapter looking at the application of wave principles to sound. In the next chapter, we'll shift our focus to electromagnetic waves.

Transverse and Longitudinal Waves

The MCAT is primarily concerned with **sinusoidal waves**. In these waves, which may be transverse or longitudinal, the individual particles oscillate back and forth with a displacement that follows a sinusoidal pattern. **Transverse waves** are those in which the direction of particle oscillation is perpendicular to the **propagation** (movement) of the wave. To visualize this, consider "The Wave" in a stadium. While "The Wave" moves around the stadium, individuals in the stands do not run around the stadium themselves. Rather, they move perpendicular to the direction of "The Wave"—by standing up and sitting down. More common examples on the MCAT include electromagnetic waves, such as visible light, microwaves, and X-rays. You could also form a transverse wave by attaching a string to a fixed point, and then moving your hand up and down, as is demonstrated in Figure 7.1a. In any waveform, energy is delivered in the direction of wave travel, so we can say that for a transverse wave, the particles are oscillating perpendicular to the direction of energy transfer.

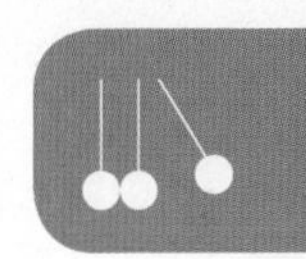

Longitudinal waves are ones in which the particles of the wave oscillate parallel to the direction of propagation; that is, the wave particles are oscillating in the direction of energy transfer. Sound waves are the classic example of longitudinal waves, but because we can't see sound, this waveform is a little more difficult to picture. Figure 7.1b helps us visualize what a longitudinal waveform traveling through air would look like. In this case, the longitudinal wave created by the person moving the piston back and forth causes air molecules to oscillate through cycles of **compression** and **rarefaction** (**decompression**) along the direction of motion of the wave. You could also form a longitudinal wave by laying a Slinky flat on a table top and tapping it on the end.

KEY CONCEPT

Transverse waves have particle oscillation perpendicular to the direction of propagation and energy transfer. Longitudinal waves have particle oscillation parallel to the direction of propagation and energy transfer.

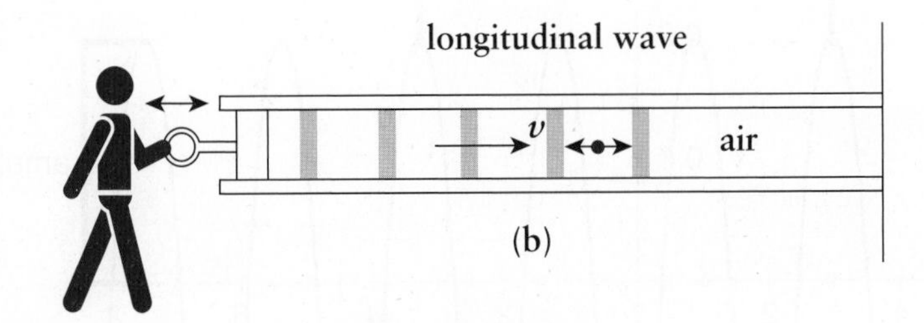

Figure 7.1 Wave Types

(a) Transverse: particles oscillate perpendicular to the direction of propagation; (b) Longitudinal: particles oscillate parallel to the direction of propagation.

Describing Waves

Waves can be described mathematically or graphically. To do so, we must first assign meaning to the physical quantities that waves represent. The distance from one maximum (**crest**) of the wave to the next is called the **wavelength** (λ). The **frequency** (f) is the number of wavelengths passing a fixed point per second, and is measured in **hertz (Hz)** or cycles per second (cps). From these two values, one can calculate the **propagation speed** (v) of a wave:

$$v = f\lambda$$

Equation 7.1

If frequency defines the number of cycles per second, then its inverse—**period** (T)—is the number of seconds per cycle:

$$T = \frac{1}{f}$$

Equation 7.2

Frequency is also related to **angular frequency** (ω), which is measured in radians per second, and is often used in consideration of simple harmonic motion in springs and pendula:

$$\omega = 2\pi f = \frac{2\pi}{T}$$

Equation 7.3

MCAT EXPERTISE

Even if simple harmonic motion in springs and strings (pendula) are not on the formal content lists for the MCAT, it is still important to be familiar with the jargon of wave motion because sound and light (electromagnetic radiation) *are* on those content lists!

Waves oscillate about a central point called the **equilibrium position**. The **displacement** (**x**) in a wave describes how far a particular point on the wave is from the equilibrium position, expressed as a vector quantity. The maximum magnitude of displacement in a wave is called its **amplitude** (*A*). Be careful with the terminology: note that the amplitude is defined as the maximum displacement from the equilibrium position to the top of a crest or bottom of a trough, not the total displacement between a crest and a trough (which would be double the amplitude). These quantities are shown in Figure 7.2.

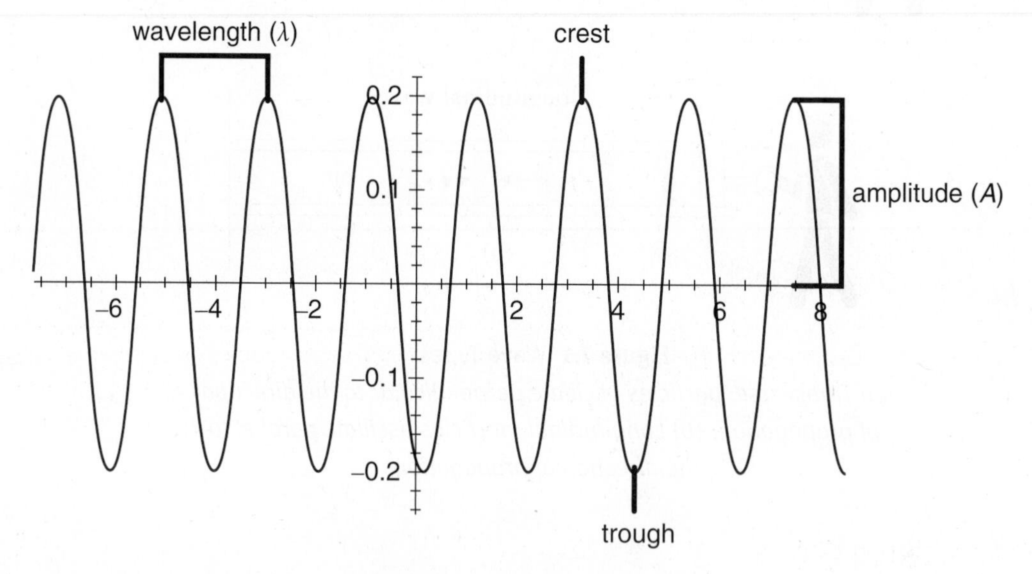

Figure 7.2 Anatomy of a Wave

Phase

When analyzing waves that are passing through the same space, we can describe how "in step" or "out of step" the waves are by calculating the **phase difference**. If we consider two waves that have the same frequency, wavelength, and amplitude and that pass through the same space at the same time, we can say that they are *in phase* if their respective crests and troughs coincide (line up with each other). When waves are perfectly in phase, we say that the phase difference is zero. However, if the two waves travel through the same space in such a way that the crests of one wave coincide with the troughs of the other, then we would say that they are *out of phase*, and the phase difference would be one-half of a wave. This could be expressed as $\frac{\lambda}{2}$ or, if given as an angle, 180° (one cycle = one wavelength = 360°). Of course, waves can be out of phase with each other by any other fraction of a cycle, as well.

Principle of Superposition

The **principle of superposition** states that when waves interact with each other, the displacement of the resultant wave at any point is the sum of the displacements of the two interacting waves. When the waves are perfectly in phase, the displacements always add together and the amplitude of the resultant is equal to the sum of the amplitudes of the two waves. This is called **constructive interference**. When waves are perfectly out of phase, the displacements always counteract each other and the amplitude of the resultant wave is the difference between the amplitudes of the interacting waves. This is called **destructive interference**.

If waves are not perfectly in phase or out of phase with each other, **partially constructive** or **partially destructive** interference can occur. As shown in Figure 7.3a, two waves that are nearly in phase will mostly add together. While the displacement of the resultant is simply the sum of the displacements of the two waves, the waves do not perfectly add together because they are not quite in phase. Therefore, the amplitude of the resultant wave is not quite the sum of the two waves' amplitudes. In contrast, Figure 7.3b shows two waves that are almost perfectly out of phase. The two waves do not quite cancel, but the resultant wave's amplitude is clearly much smaller than that of either of the other waves.

KEY CONCEPT

If two waves are perfectly in phase, the resultant wave has an amplitude equal to the sum of the amplitudes of the two waves. If two equal waves are exactly 180 degrees out of phase, then the resultant wave has zero amplitude.

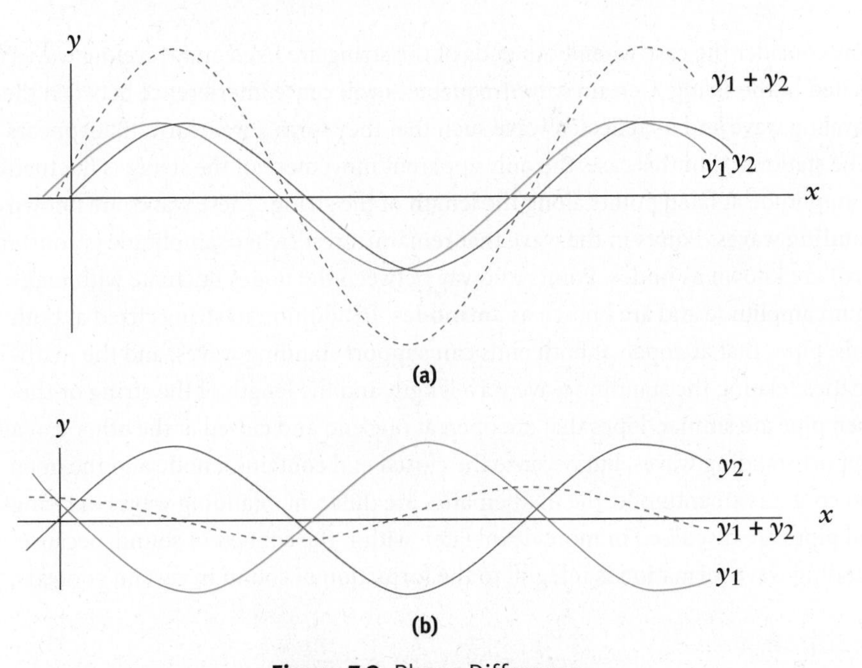

Figure 7.3 Phase Difference
(a) In phase with a difference of almost zero; (b) Out of phase with a difference of almost 180 degrees $\left(\frac{\lambda}{2}\right)$

REAL WORLD

In noise-canceling headphones, pressure waves from noise are canceled by destructive interference. The speaker creates a wave that is 180 degrees out of phase and of similar amplitude. Many frequencies are usually present in the noise, so it is difficult to get perfect noise cancellation.

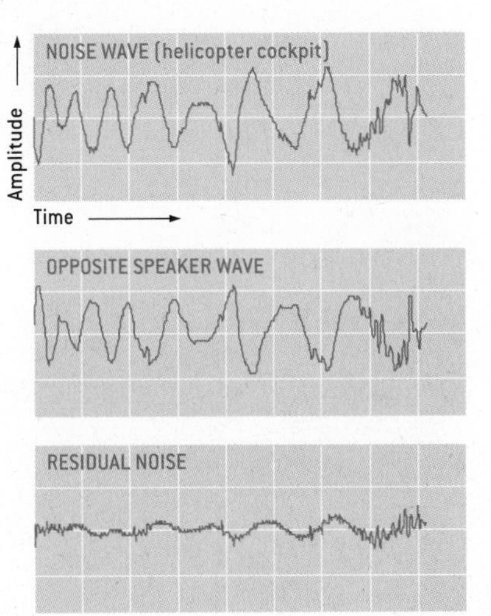

Noise-canceling headphones operate on the principle of superposition. They do not simply muffle sound, but actually capture the environmental noise and, using computer technology, produce a sound wave that is approximately 180 degrees out of phase. The combination of the two waves inside the headset results in destructive interference, thereby canceling—or nearly canceling—the ambient noise.

Traveling and Standing Waves

If a string fixed at one end is moved up and down, a wave will form and travel, or propagate, toward the fixed end. Because this wave is moving, it is called a **traveling wave**. When the wave reaches the fixed boundary, it is reflected and inverted, as shown in Figure 7.4. If the free end of the string is continuously moved up and down, there will then be two waves: the original wave moving down the string toward the fixed end and the reflected wave moving away from the fixed end. These waves will then interfere with each other.

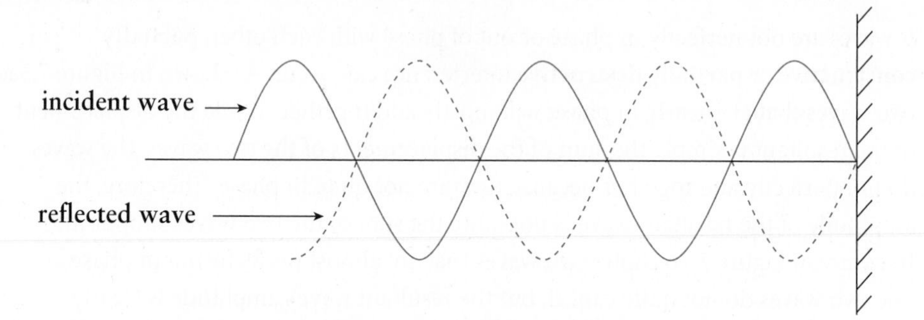

Figure 7.4 Traveling Wave

Now consider the case when both ends of the string are fixed and traveling waves are excited in the string. Certain wave frequencies will cause interference between the traveling wave and its reflected wave such that they form a waveform that appears to be stationary. In this case, the only apparent movement of the string is fluctuation of amplitude at fixed points along the length of the string. These waves are known as **standing waves**. Points in the wave that remain at rest (where amplitude is constantly zero) are known as **nodes**. Points midway between the nodes fluctuate with maximum amplitude and are known as **antinodes**. In addition to strings fixed at both ends, pipes that are open at both ends can support standing waves, and the mathematics relating the standing wave wavelength and the length of the string or the open pipe are similar. Pipes that are open at one end and closed at the other can also support standing waves, but because the closed end contains a node and the open end contains an antinode, the mathematics are different. Standing waves in strings and pipes are discussed in more detail later, within the context of sound, because standing wave formation is integral to the formation of sound in certain contexts.

Resonance

Why are clarinets, pianos, and even half-filled wine glasses considered musical instruments, but not pencils, chairs, or paper? This discrepancy has much to do with the **natural (resonant) frequencies** of these objects. Any solid object, when hit, struck, rubbed, or disturbed in any way will begin to vibrate. Tapping a pencil on a surface will cause it to vibrate, as will hitting a chair or crumpling a piece of paper. Blowing air pressure between a clarinet reed and a mouthpiece, striking a taut piano string, and creating friction on a wine glass's surface will also cause vibration. If the natural frequency is within the frequency detection range of the human ear,

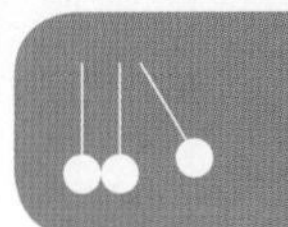

the sound will be audible. The quality of the sound, called **timbre**, is determined by the natural frequency or frequencies of the object. Some objects vibrate at a single frequency, producing a pure tone. Other objects vibrate at multiple frequencies that have no relation to one another. These objects produce sounds that we do not find particularly musical, such as tapping a pencil, hitting a chair, or crumpling paper. These sounds are called **noise**, scientifically. Still other objects vibrate at multiple natural frequencies (a **fundamental pitch** and multiple **overtones**) that are related to each other by whole number ratios, producing a richer, more full tone. The human brain perceives these sounds as being more musical, and all nonpercussion instruments produce such overtones. Of note for the MCAT, the frequencies between 20 Hz and 20,000 Hz are generally audible to healthy young adults, and high-frequency hearing generally declines with age.

The natural frequency of most objects can be changed by changing some aspect of the object itself. For example, a set of eight identical glasses can be filled with different levels of water so that each vibrates at a different natural frequency, producing the eight notes of a diatonic musical scale. Strings have an infinite number of natural frequencies that depend on the length, linear density, and tension of the string.

If a periodically varying force is applied to a system, the system will then be driven at a frequency equal to the frequency of the force. This is known as **forced oscillation**. If the frequency of the applied force is close to that of the natural frequency of the system, then the amplitude of oscillation becomes much larger. This can easily be demonstrated by a child on a swing being pushed by a parent. If the parent pushes the child at a frequency nearly equal to the frequency at which the child swings back toward the parent, the arc of the swinging child will become larger and larger: the amplitude is increasing because the **force frequency** is nearly identical to the swing's natural frequency.

If the frequency of the periodic force is equal to a natural (resonant) frequency of the system, then the system is said to be **resonating**, and the amplitude of the oscillation is at a maximum. If the oscillating system were frictionless, the periodically varying force would continually add energy to the system, and the amplitude would increase indefinitely. However, because no system is completely frictionless, there is always some damping, which results in a finite amplitude of oscillation. In general, **damping** or **attenuation** is a decrease in amplitude of a wave caused by an applied or nonconservative force. Furthermore, many objects cannot withstand the large amplitude of oscillation and will break or crumble. A dramatic demonstration of resonance is the shattering of a wine glass by loudly singing the natural frequency of the glass. This is actually possible with a steady, loud tone—the glass will resonate (oscillate with maximum amplitude) and eventually shatter.

MCAT CONCEPT CHECK 7.1

Before you move on, assess your understanding of the material with these questions.

1. Define the following terms:
   - Wave speed:

     ____

   - Frequency:

     ____

   - Angular frequency:

     ____

   - Period:

     ____

   - Equilibrium position:

     ____

   - Amplitude:

     ____

   - Traveling wave:

     ____

   - Standing wave:

     ____

2. If two waves are out of phase at any interval besides 180 degrees, how does the amplitude of the resultant wave compare to the amplitudes of the two interfering waves?

   ____

3. True or False: Sound waves are a prime example of transverse waves.

4. How does applying a force at the natural frequency of a system change the system?

   ____

7.2 Sound

High-Yield

LEARNING OBJECTIVES

After Chapter 7.2, you will be able to:

- Explain how sound is produced and transmitted
- Connect amplitude and frequency to their corresponding properties in a sound wave
- Predict changes in apparent frequency based on the movement of two objects in relation to one another
- Calculate key values for sound, including intensity, frequency, apparent frequency, and wavelength in Doppler and pipe systems

Sound is a longitudinal wave transmitted by the oscillation of particles in a deformable medium. As such, sound can travel through solids, liquids, and gases, but cannot travel through a vacuum. The speed of sound is given by the equation

$$v = \sqrt{\frac{B}{\rho}}$$

Equation 7.4

where B is the bulk modulus, a measure of the medium's resistance to compression (B increases from gas to liquid to solid), and ρ is the density of the medium. Because the bulk modulus increases disproportionately more than density as one goes from gas to liquid to solid, sound travels fastest through a solid and slowest through a gas. The speed of sound in air at 20°C is approximately 343 $\frac{\text{m}}{\text{s}}$.

KEY CONCEPT

The speed of sound is fastest in a solid with low density, and slowest in a very dense gas.

Production of Sound

Sound is produced by the mechanical disturbance of particles in a material along the sound wave's direction of propagation. Although the particles themselves do not travel along with the wave, they do vibrate or oscillate about an equilibrium position, which causes small regions of compression to alternate with small regions of rarefaction (decompression). These alternating regions of increased and decreased particle density travel through the material, allowing the sound wave to propagate.

Because sound involves vibration of material particles, the source of any sound is ultimately a mechanical vibration of some frequency. They can be produced by the vibration of solid objects or the vibration of fluids, including gases. Solid objects that can vibrate to produce musical sound include strings (on a piano, violin, guitar, and so on), metal (bells), or wood bars (xylophone or marimba). Vibration of air within certain objects, including all woodwinds and brass instruments, pipe organs, and even a soda bottle, can also create musical sound. The pitch (frequency) at which the air column within the instrument vibrates is determined by the length of the air column, which can be changed either by covering holes in the instrument or directly changing its length.

The human voice is no less a musical instrument than any of those listed above. Sound is created by passing air between the vocal cords, which are a pair of thin membranes stretched across the larynx. As the air moves past the cords, they vibrate like the double reed of an oboe or bassoon, causing the air to vibrate at the same frequency. The pitch of the sound is controlled by varying the tension of the cords. Adult male vocal cords are larger and thicker than those of adult females; thus, the male voice is typically lower in pitch.

Frequency and Pitch

We've discussed frequency as the rate at which a particle or wave completes a cycle. Our perception of the frequency of sound is called the **pitch**. Lower-frequency sounds have lower pitch, and higher-frequency sounds have higher pitch. On the MCAT, sound frequencies are usually within the normal range of human hearing, from 20 Hz to 20,000 Hz. Sound waves with frequencies below 20 Hz are called **infrasonic** waves, and those with frequencies above 20,000 Hz are called **ultrasonic** waves. Both dog whistles, which emit frequencies between 20 and 22 kHz, and medical ultrasound machines, which emit frequencies in excess of 2 GHz, are examples of ultrasonic waves.

Doppler Effect

BRIDGE

The Doppler effect applies to all waves, including light. This means that if a source of light is moving toward the detector, the observed frequency will increase. This is called blue shift because blue is at the high-frequency end of the visible spectrum. If the source is moving away from the detector, the observed frequency will decrease, causing red shift. Light and electromagnetic waves are discussed in Chapter 8 of *MCAT Physics and Math Review.*

We've all witnessed the Doppler effect: an ambulance or fire truck with its sirens blaring is quickly approaching from the other lane, and as it passes, one can hear a distinct drop in the pitch of the siren. This phenomenon affecting frequency is called the **Doppler effect**, which describes the difference between the actual frequency of a sound and its perceived frequency when the source of the sound and the sound's detector are moving relative to one another. If the source and detector are moving toward each other, the perceived frequency, f', is greater than the actual frequency, f. If the source and detector are moving away from each other, the perceived frequency is less than the actual frequency. This can be seen from the Doppler effect equation:

$$f' = f\frac{(v \pm v_D)}{(v \mp v_S)}$$

Equation 7.5

where f' is the perceived frequency, f is the actual emitted frequency, v is the speed of sound in the medium, v_D is the speed of the detector, and v_S is the speed of the source. Note the unusual signs in the equation. If memorized in this form, the upper sign should be used when the detector or source is moving *toward* the other object. The lower sign should be used when the detector or source is moving *away* from the other object.

MNEMONIC

Sign convention in the Doppler equation:

- Top sign for "toward"
- Bottom sign for "away"

This sign convention is usually the most confusing part of the Doppler effect equation, so let's take a closer look. Imagine the situation presented earlier: you're driving down the street when you hear an ambulance approaching from behind. In this scenario, you are the detector and the ambulance is the sound source. At this time, you would say that you are driving away from the ambulance; even though the ambulance is moving faster and getting closer to you, the direction in which you are driving is still *away* from the ambulance. By this logic, the lower sign (−) should be used in the numerator, which relates to the detector. The driver of the ambulance, on the other hand, would say that he is driving *toward* you. By this logic, the top sign (−) should be used in the denominator, which relates to the source. In this case, the Doppler effect equation would look like this:

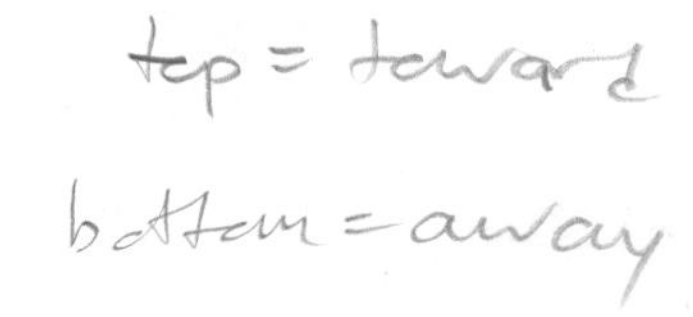

$$f' = f\frac{(v - v_D)}{(v - v_S)}$$

Because $v_S > v_D$, we know that $f' > f$.

Now imagine that the ambulance has passed you and continues to speed down the road. At this point, you would say that you are driving *toward* the ambulance (top sign in the numerator, +), even if you are not going as fast as it is. The ambulance driver would say that he is driving away from you (bottom sign in the denominator, +) and the corresponding Doppler effect equation would be:

$$f' = f\frac{(v + v_D)}{(v + v_S)}$$

Here, because $v_S > v_D$, $f' < f$. This change in f' from being greater than f to being less than f is perceived as a drop in pitch.

The Doppler effect can be visualized by considering the sound waves in front of a moving object as being compressed, while the sound waves behind the object are stretched out, as shown in Figure 7.5.

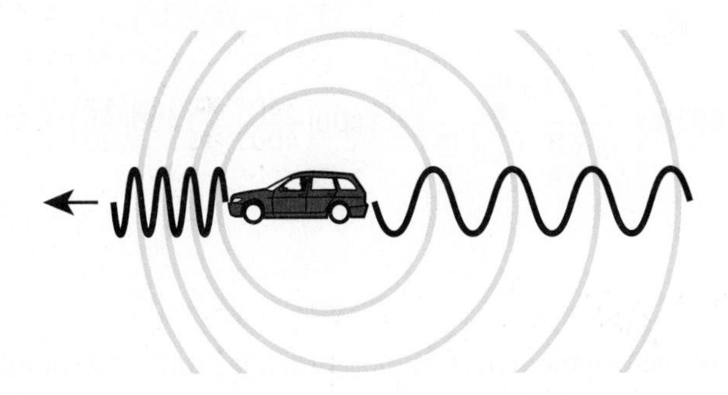

Figure 7.5 The Doppler Effect
The black arrow indicates the direction of motion for the car. In front of the car, crests of the sound wave are compressed together, increasing the frequency (pitch). Behind the car, crests of the sound wave are stretched apart, decreasing the frequency.

The Doppler effect can be used by animals through the process of **echolocation**. In echolocation, the animal emitting the sound (usually a dolphin or bat) serves as both the source and the detector of the sound. The sound bounces off of a surface and is reflected back to the animal. How long it takes for the sound to return, and the change in frequency of the sound, can be used to determine the position of objects in the environment and the speed at which they are moving.

Example: A train traveling south at 216 $\frac{\text{km}}{\text{hr}}$ is sounding its whistle while passing by a stationary observer. The whistle emits sound at a frequency of 1400 Hz. What is the frequency heard by the stationary observer when the train is moving toward the observer, and when the train has passed the observer? (Note: The speed of sound in air is approximately 340 $\frac{\text{m}}{\text{s}}$.)

Solution: To solve this problem, the speed of the train (v_S) must first be converted to $\frac{\text{m}}{\text{s}}$:

$$v_S = 216\ \frac{\text{km}}{\text{hr}}\left[\frac{1\ \text{hr}}{3600\ \text{s}}\right]\left[\frac{1000\ \text{m}}{1\ \text{km}}\right] = 60\ \frac{\text{m}}{\text{s}}$$

When the train is moving toward the stationary observer, the top sign should be used in the denominator. The numerator is simply v because $v_D = 0$. This gives

$$f' = f\,\frac{v}{v - v_S}$$

$$= (1400\ \text{Hz})\left(\frac{340\ \frac{\text{m}}{\text{s}}}{340\ \frac{\text{m}}{\text{s}} - 60\ \frac{\text{m}}{\text{s}}}\right) = 1400\left(\frac{340}{280}\right) = 1400\left(\frac{17}{14}\right) = 1700\ \text{Hz}$$

When the train is moving away from the observer, the sign in the denominator changes. The numerator remains unchanged because the observer is still stationary:

$$f' = f\,\frac{v}{v + v_S}$$

$$= (1400\ \text{Hz})\left(\frac{340\ \frac{\text{m}}{\text{s}}}{340\ \frac{\text{m}}{\text{s}} + 60\ \frac{\text{m}}{\text{s}}}\right) = 1400\left(\frac{340}{400}\right) = 1400\left(\frac{17}{20}\right) = 1190\ \text{Hz}$$

Shock Waves

In a special case of the Doppler effect, an object that is producing sound while traveling at or above the speed of sound allows wave fronts to build upon one another at the front of the object. This creates a much larger amplitude at that point. Because amplitude for sound waves is related to the degree of compression of the medium, this creates a large pressure differential or pressure gradient. This highly condensed wave front is called a **shock wave**, and it can cause physical disturbances as it passes through other objects. The passing of a shock wave creates very high pressure, followed by very low pressure, which is responsible for the phenomenon known as a **sonic boom**. Unlike its depiction in movies and television, a sonic boom can be heard any time that an object traveling at or faster than the speed of sound passes a detector, not just at the point that the speed of sound is exceeded **(Mach 1)**. Once an object moves faster than the speed of sound, some of the effects of the shock wave are mitigated because all of the wave fronts will trail behind the object, destructively interfering with each other.

Intensity and Loudness of Sound

The **loudness** or **volume** of a sound is the way in which we perceive its intensity. Perception of loudness is subjective, and depends not only on brain function, but also physical factors such as obstruction of the ear canal, stiffening of the ossicles, or damage to cochlear hair cells by exposure to loud noises or with age. Sound intensity, on the other hand, is objectively measurable. **Intensity** is the average rate of energy transfer per area across a surface that is perpendicular to the wave. In other words, intensity is the power transported per unit area. The SI units of intensity are therefore watts per square meter $\left(\frac{\text{W}}{\text{m}^2}\right)$. Intensity is calculated using the equation

$$I = \frac{P}{A}$$

Equation 7.6

where P is the power and A is the area. Rearranging this equation, we could consider that the power delivered across a surface, such as the tympanic membrane (eardrum), is equal to the product of the intensity I and the surface area A, assuming the intensity is uniformly distributed.

The amplitude of a sound wave and its intensity are also related to each other: intensity is proportional to the square of the amplitude. Therefore, doubling the amplitude produces a sound wave that has four times the intensity.

Intensity is also related to the distance from the source of the sound wave. As sound waves emanate outward from their source, it is as though the waves are pushing against the interior wall of an ever-expanding spherical balloon. Because the surface area of a sphere increases as a function of the square of the radius ($A = 4\pi r^2$), sound waves transmit their power over larger and larger areas the farther from the source they travel. Intensity, therefore, is inversely proportional to the square of the distance from the source. For example, sound waves that have traveled 2 meters from their source have spread their energy out over a surface area that is four times larger than that for identical sound waves that have traveled 1 meter from their source.

The softest sound that the average human ear can hear has an intensity equal to about $1 \times 10^{-12} \frac{\text{W}}{\text{m}^2}$. The mechanical disturbance associated with the threshold of hearing is remarkably small—the displacement of air particles is on the order of one billionth of a centimeter. At the other end of the spectrum, the intensity of sound at the threshold of pain is $10 \frac{\text{W}}{\text{m}^2}$ and the intensity that causes instant perforation of the eardrum is approximately $1 \times 10^4 \frac{\text{W}}{\text{m}^2}$. This is a huge range, which would be unmanageable to express on a linear scale. To make this range easier to work with, we use a logarithmic scale, called the **sound level** (β), measured in **decibels (dB)**:

$$\beta = 10 \log \frac{I}{I_0}$$

Equation 7.7

where I is the intensity of the sound wave and I_0 is the threshold of hearing $\left(1 \times 10^{-12} \frac{\text{W}}{\text{m}^2}\right)$, which is used as a reference intensity. When the intensity of a sound is changed by some factor, one can calculate the new sound level by using the equation

$$\beta_f = \beta_i + 10 \log \frac{I_f}{I_i}$$

Equation 7.8

where $\frac{I_f}{I_i}$ is the ratio of the final intensity to the initial intensity.

BRIDGE

We use logarithms with scales that have an extremely large range. The MCAT will mostly deal with base-ten logarithms (common logarithms). As an example, $\log 1000 = 3$ because $10^3 = 1000$. As another example, $\log 1 = 0$ because $10^0 = 1$. Logarithms are discussed in Chapter 10 of *MCAT Physics and Math Review*.

The sound levels and relative intensities of several sound sources and thresholds are shown in Table 7.1.

Sound Source	Sound Level (dB)	Intensity $\left(\frac{\text{W}}{\text{m}^2}\right)$
(Threshold of Hearing)	0	1×10^{-12}
Rustling Leaves	10	1×10^{-11}
Whisper	20	1×10^{-10}
Quiet Room at Night	30	1×10^{-9}
Quiet Library	40	1×10^{-8}
Moderate Rainfall	50	1×10^{-7}
Conversational Speech at 1 m	60	1×10^{-6}
Vacuum Cleaner at 1 m	70	1×10^{-5}
Door Slamming	80	1×10^{-4}
Lawn Mower at 1 m	90	1×10^{-3}
Jackhammer at 1 m	100	1×10^{-2}
Loud Rock Concert	110	1×10^{-1}
Thunder	120	1×10^{0}
(Threshold of Pain)	130	1×10^{1}
Rifle at 1 m	140	1×10^{2}
Jet Engine at 30 m	150	1×10^{3}
(Eardrum Perforation)	160	1×10^{4}

Table 7.1 Sound Level and Intensity of Sound Sources and Important Thresholds

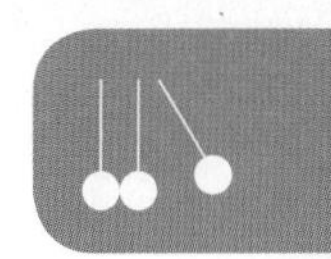

Example: A detector with a surface area of 1 square meter is placed 1 meter from a blender. It measures the average power of the blender's sound as being 10^{-3} W. Find the intensity and sound level of the blender, and the ratio of the intensities of the blender and a jet engine. (Note: Assume $\beta_{jet} = 150$ dB.)

Solution: Intensity is defined as the power per area:

$$I = \frac{P}{A} = \frac{10^{-3}\ \text{W}}{1\ \text{m}^2} = 10^{-3}\ \frac{\text{W}}{\text{m}^2}$$

The sound level can then be calculated from the intensity:

$$\beta = 10 \log \frac{I}{I_0} = 10 \log \left(\frac{10^{-3}\ \frac{\text{W}}{\text{m}^2}}{10^{-12}\ \frac{\text{W}}{\text{m}^2}} \right) = 10 \log 10^9 = 90\ \text{dB}$$

Finally, the ratio of two sound intensities can be found from the difference of their sound levels:

$$\beta_{jet} = \beta_{blender} + 10 \log \frac{I_{jet}}{I_{blender}}$$

$$150\ \text{dB} = 90\ \text{dB} + 10 \log \frac{I_{jet}}{I_{blender}}$$

$$6 = \log \frac{I_{jet}}{I_{blender}}$$

$$10^6 = \frac{I_{jet}}{I_{blender}}$$

Thus, the jet engine's sound is 1,000,000 times more intense than the blender's sound.

Attenuation

Sound is not transmitted undiminished. Even after the decrease in intensity associated with distance, real world measurements of sound will be lower than those expected from calculations. This is a result of **damping**, or **attenuation**. Oscillations are a form of repeated linear motion, so sound is subject to the same nonconservative forces as any other system, including friction, air resistance, and viscous drag.

The presence of a nonconservative force causes the system to decrease in amplitude during each oscillation. Because amplitude, intensity, and sound level (loudness) are related, there is a corresponding gradual loss of sound. Note that damping does not have an effect on the frequency of the wave, so the pitch will not change. This phenomenon, along with reflection, explains why it is more difficult to hear in a confined or cluttered space than in an empty room: friction from the surfaces of the objects in the room actually decreases the sound waves' amplitudes. Over small distances, attenuation is usually negligible.

MCAT EXPERTISE

Like nonconservative forces, attenuation is generally negligible on Test Day. If it is important for answering a question, the MCAT will make it clear that you should consider the effects of damping (attenuation) on an oscillating system.

Beat Frequency

Sound volume can also vary periodically due to interference effects. When two sounds of slightly different frequencies are produced in proximity, as when tuning a pair of instruments next to one another, volume will vary at a rate based on the difference between the two pitches being produced. The frequency of this periodic increase in volume can be calculated by the equation:

$$f_{beat} = |f_1 - f_2|$$

Equation 7.9

where f_1 and f_2 represent the two frequencies that are close in pitch, and f_{beat} represents the resulting beat frequency.

Standing Waves

Remember that standing waves are produced by the constructive and destructive interference of a traveling wave and its reflected wave. More broadly, we can say that a standing wave will form whenever two waves of the same frequency traveling in opposite directions interfere with one another as they travel through the same medium. Standing waves appear to be standing still—that is, not propagating—because the interference of the wave and its reflected wave produce a resultant that fluctuates only in amplitude. As the waves move in opposite directions, they interfere to produce a new wave pattern characterized by alternating points of maximum displacement (amplitude) and points of no displacement. The points in a standing wave with no fluctuation in displacement are called **nodes**. The points with maximum fluctuation are called **antinodes**.

MNEMONIC

Nodes are places of **No D**isplacement

Not every frequency of traveling wave will result in standing wave formation. The length of the medium dictates the wavelengths (and, by extension, the frequencies) of traveling waves that can establish standing waves. Objects that support standing waves have boundaries at both ends. **Closed boundaries** are those that do not allow oscillation and that correspond to nodes. The closed end of a pipe and the secured ends of a string are both considered closed boundaries. **Open boundaries** are those that allow maximal oscillation and correspond to antinodes. The open end of a pipe and the free end of a flag are both open boundaries.

Strings

Consider a string, such as a guitar or violin string, or a piano wire, fixed rigidly at both ends. Because the string is secured at both ends and is therefore immobile at these points, they are considered nodes. If a standing wave is set up such that there is only one antinode between the two nodes at the ends, the length of the string corresponds to one-half the wavelength of this standing wave, as shown in Figure 7.6a. This is because on a sine wave, the distance from one node to the next node is one-half of a wavelength. If a standing wave is set up such that there are two antinodes between the ends, there must be a third node located between the antinodes, as shown in Figure 7.6b. In this case, the length of the string corresponds to the wavelength of this standing wave. Again, the distance on a sine wave from a node to the second consecutive node is exactly one wavelength. This pattern suggests that the length L of a string must be equal to some multiple of half-wavelengths $\left(L = \frac{\lambda}{2}, \frac{2\lambda}{2}, \frac{3\lambda}{2}\right.$, and so on).

The equation that relates the wavelength λ of a standing wave and the length L of a string that supports it is:

$$\lambda = \frac{2L}{n}$$

Equation 7.10

where n is a positive nonzero integer ($n = 1, 2, 3$, and so on) called the **harmonic**. The harmonic corresponds to the number of half-wavelengths supported by the string. From the relationship that $f = \frac{v}{\lambda}$ where v is the wave speed, the possible frequencies are:

$$f = \frac{nv}{2L}$$

Equation 7.11

The lowest frequency (longest wavelength) of a standing wave that can be supported in a given length of string is known as the **fundamental frequency** (first harmonic). The frequency of the standing wave given by $n = 2$ is known as the first overtone or second harmonic. This standing wave has one-half the wavelength and twice the frequency of the first harmonic. The frequency of the standing wave given by $n = 3$ is known as the second overtone or third harmonic, as shown in Figure 7.6c. All the possible frequencies that the string can support form its **harmonic series**.

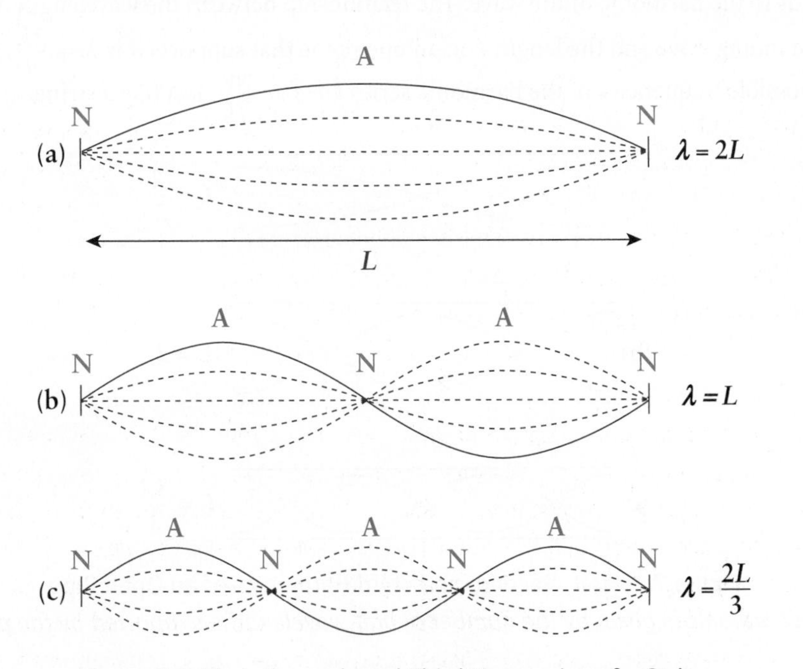

Figure 7.6 First, Second, and Third Harmonics of a String
The harmonic is given by the number of half-wavelengths supported by the string. N = node; A = antinode.

MCAT EXPERTISE

As a shortcut, for strings attached at both ends, the number of antinodes present will tell you which harmonic it is.

Open Pipes

Pipes can support standing waves and produce sound as well. Many musical instruments are straight or curved tubes within which air will oscillate at particular frequencies to set up standing waves. The end of a pipe can be open or closed. If the end of the pipe is open, it will support an antinode. If it is closed, it will support a node. One end of the pipe must be open at least slightly to allow for the entry of air, but sometimes these openings are small and covered by the musician's mouth—in these cases, they function as a closed end. Pipes that are open at both ends are called **open pipes**, while those that are closed at one end (and open at the other) are called **closed pipes**. The flute functions as an open pipe instrument, while the clarinet and brass instruments are closed pipe instruments. If you are a musician, this may be counterintuitive. The distal end of a flute is open, but the proximal end is closed; however, the mouthpiece of a flute is close enough to this closed end for it to function as an open end. Similarly, while air must pass through the mouthpiece of a reed or brass instrument, the opening is sufficiently small to function as a closed end.

An open pipe, being open at both ends, has antinodes at both ends. If a standing wave is set up such that there is only one node between the two antinodes at the ends, the length of the pipe corresponds to one-half the wavelength of this standing wave, as shown in Figure 7.7a. This is analogous to a string except that the ends are both antinodes instead of nodes. The analogy continues throughout: the second harmonic (first overtone) has a wavelength equal to the length of the pipe, as shown in Figure 7.7b. The third harmonic (second overtone) has a wavelength equal to two-thirds the length of the pipe, as shown in Figure 7.7c. Again, an open pipe can contain any multiple of half-wavelengths; the number of half-wavelengths corresponds to the harmonic of the wave. The relationship between the wavelength λ of a standing wave and the length L of an open pipe that supports it is $\lambda = \frac{2L}{n}$, and the possible frequencies of the harmonic series are $f = \frac{nv}{2L}$, just like a string.

MCAT EXPERTISE

As a shortcut, for open pipes, the number of nodes present will tell you which harmonic it is.

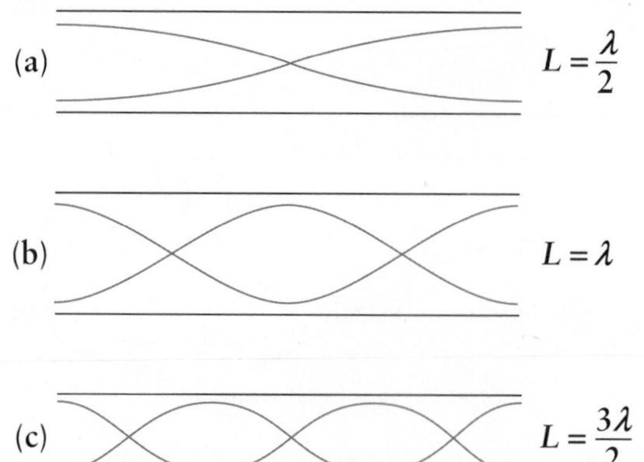

Figure 7.7 First, Second, and Third Harmonics of an Open Pipe
The harmonic is given by the number of half-wavelengths supported by the pipe.

It is worthwhile to note that Figure 7.7 is really a symbolic representation of the first three harmonics in an open pipe. We use the term symbolic because the conventional way of diagramming standing waves is to represent sound waves as transverse, rather than longitudinal, waves (which are much harder to draw).

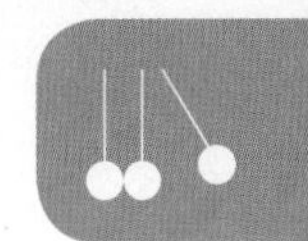

Closed Pipes

In the case of a closed pipe, the closed end will correspond to a node, and the open end will correspond to an antinode. The first harmonic in a closed pipe consists of only the node at the closed end and the antinode at the open end, as shown in Figure 7.8a. In a sinusoidal wave, the distance from a node to the following antinode is one-quarter of a wavelength. Indeed, unlike strings or open pipes, the harmonic in a closed pipe is equal to the number of quarter-wavelengths supported by the pipe. Because the closed end must always have a node and the open end must always have an antinode, there can only be odd harmonics. This is because an even number of quarter-wavelengths would be an integer number of half-wavelengths—which would necessarily have either two nodes or two antinodes at the ends. The first harmonic has a wavelength that is four times the length of the closed pipe. The third harmonic (first overtone) has a wavelength that is four-thirds the length of the closed pipe, as shown in Figure 7.8b. The fifth harmonic (second overtone) has a wavelength that is four-fifths the length of the closed pipe, as shown in Figure 7.8c. The equation that relates the wavelength λ of a standing wave and the length L of a closed pipe that supports it is:

$$\lambda = \frac{4L}{n}$$

Equation 7.12

where n can only be an odd integers ($n = 1, 3, 5$, and so on). The frequency of the standing wave in a closed pipe is:

$$f = \frac{nv}{4L}$$

Equation 7.13

where v is the wave speed.

MCAT EXPERTISE

Unlike strings and open pipes, one cannot simply count the number of nodes or antinodes to determine the harmonic of the wave in closed pipes. Therefore, when presented with a closed pipe, make sure to actually count the number of quarter-wavelengths contained in the pipe to determine the harmonic.

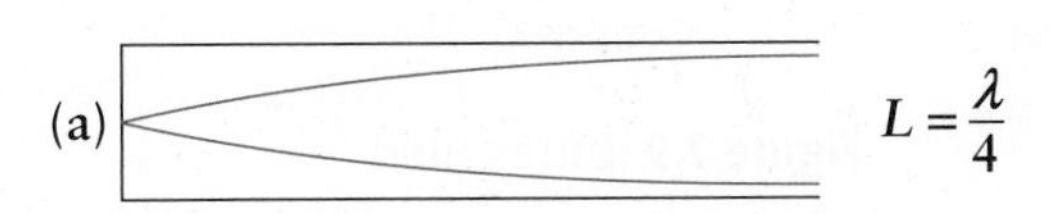

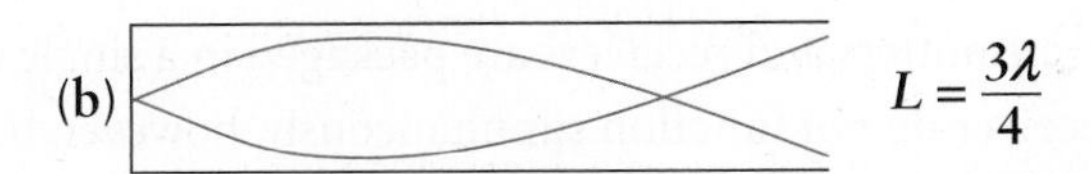

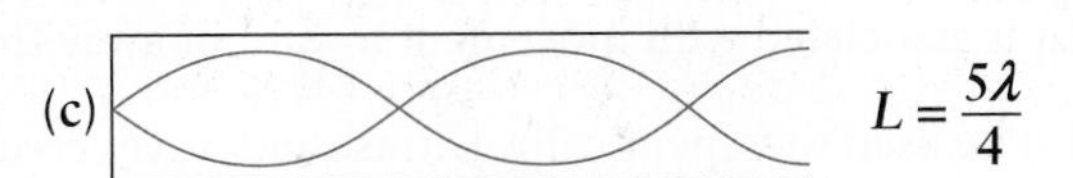

Figure 7.8 First, Third, and Fifth Harmonics of a Closed Pipe
The harmonic is given by the number of quarter-wavelengths supported by the pipe.

Ultrasound

Until this point we've focused on sound in the audible range; however, in medicine we can also use sound waves to visualize organs, anatomy, and pathology. This imaging modality can be used for prenatal screening, or to diagnose gallstones and breast or thyroid masses, or for needle guidance in a biopsy. **Ultrasound** uses high frequency sound waves outside the range of human hearing to compare the relative densities of tissues in the body. An ultrasound machine consists of a transmitter that generates a pressure gradient, which also functions as a receiver that processes the reflected sound, as seen in Figure 7.9. Because the speed of the wave and travel time is known, the machine can generate a graphical representation of borders and edges within the body by calculating the traversed distance. Note that ultrasound ultimately relies on reflection; thus, an interface between two objects is necessary to visualize anything. Reflection will be discussed further in the next chapter.

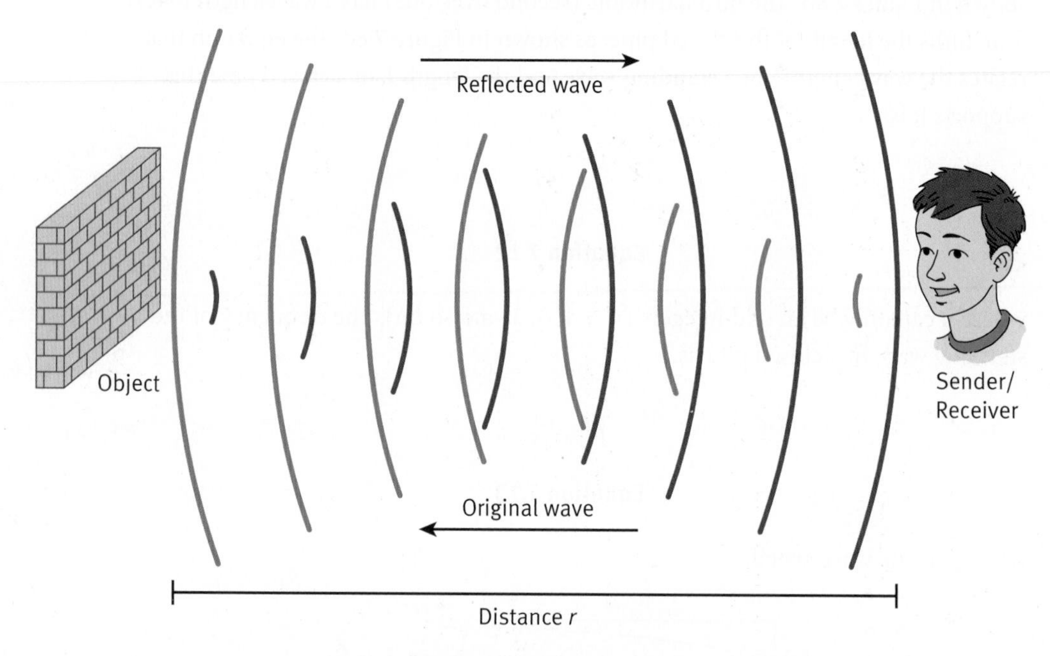

Figure 7.9 Ultrasound
The transmitter (sender) generates a wave, which reflects off of an object and returns to the transmitter (which also functions as a receiver).

Most ultrasound transmitters and receivers are packaged in a single unit. The transmitter and receiver do not function simultaneously, however, because one of the objectives of the system is to reduce interference. In addition to the standard ultrasound, most modern ultrasound machines also have a Doppler mode. **Doppler ultrasound** is used to determine the flow of blood within the body by detecting the frequency shift that is associated with movement toward or away from the receiver.

Ultrasound can also be used therapeutically. Ultrasound waves create friction and heat when they act on tissues, which can increase blood flow to a site of injury in deep tissues and promote faster healing. Focused ultrasound also has a range of applications. Focusing a sound wave using a parabolic mirror causes constructive interference at the focal point of the mirror. This creates a very high-energy wave exactly at that point, which can be used to noninvasively break up a kidney stone (*lithotripsy*) or ablate (destroy) small tumors. Ultrasound can also be used for dental cleaning and destruction of cataracts (*phacoemulsification*). In each case, the ultrasound waves are applied for a sufficient time period to achieve the desired effect.

MCAT CONCEPT CHECK 7.2

Before you move on, assess your understanding of the material with these questions.

1. How is sound produced and transmitted?

___

2. To which properties of a sound wave do amplitude and frequency correspond?
   - Amplitude:

___

   - Frequency:

___

3. If two objects are traveling toward each other, how does the apparent frequency differ from the original frequency? What if two objects are traveling away from each other? What if one object is following the other?
   - Toward each other:

___

   - Away from each other:

___

   - One object follows the other:

___

4. What phenomena can be detected or treated using ultrasound?

___

5. For each of the following diagrams, label the type of pipe or string it represents, a node and antinode, and the relevant equation relating λ and L:

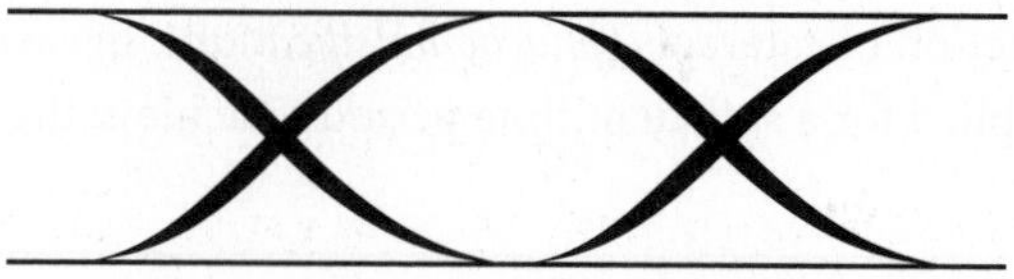

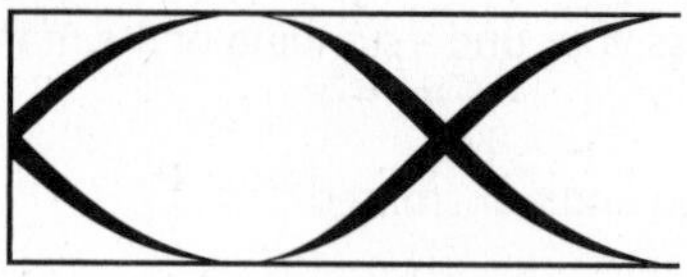

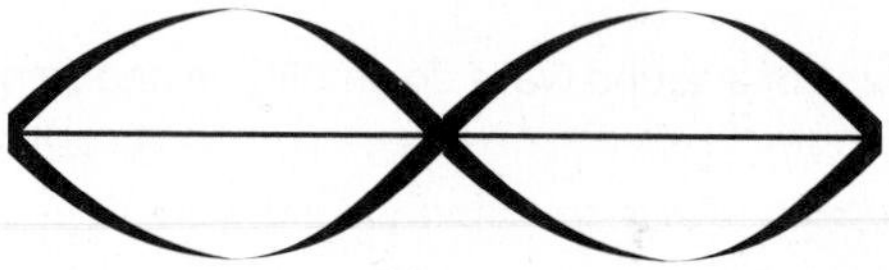

Conclusion

In this chapter, we reviewed the general characteristics of waves, including the phenomena of interference and resonance, and analyzed the characteristics and behaviors of sound as an example of a longitudinal waveform. Sound is the mechanical disturbance of particles creating oscillating regions of compression and rarefaction parallel to the direction of wave movement. The intensity of a sound wave is perceived as the sound level (loudness) of the sound and is measured in decibels. The decibel scale is a logarithmic scale used to describe the ratio of a sound's intensity to a reference intensity (the threshold of human hearing). We also reviewed the Doppler effect and a special case with shock waves. We then reviewed the mathematics governing the formation of standing waves, which are important in the formation of musical sounds in strings, open pipes, and closed pipes. Finally, we discussed a medical application of sound that incorporates many of these topics: ultrasound.

Continue to review these MCAT topics—it's easy to think about sound if you listen to music when you study! Whether you turn on Top 40, smooth jazz, or rococo fugues, the principles of sound production and propagation are key to your enjoyment of these harmonious sounds. Sound, of course, is not the only waveform tested on the MCAT. Light waves (and electromagnetic radiation in general) are heavily tested topics on Test Day—we'll review them in the next chapter.

You've reviewed the content, now test your knowledge and critical thinking skills by completing a test-like passage set in your online resources!

Concept Summary

General Wave Characteristics

- **Transverse waves** have oscillations of wave particles perpendicular to the direction of wave **propagation** (e.g., "The Wave", electromagnetic waves).
- **Longitudinal waves** have oscillations of wave particles parallel to the direction of wave propagation (e.g., sound waves).
- **Displacement** (x) in a wave refers to how far a point is from the **equilibrium position**, expressed as a vector quantity.
- The **amplitude** (A) of a wave is the magnitude of its maximal displacement.
- The maximum point of a wave (point of most positive displacement) is called a **crest**.
- The minimum point of a wave (point of most negative displacement) is called a **trough**.
- The **wavelength** (λ) of a wave is the distance between two crests or two troughs.
- The **frequency** (f) of a wave is the number of cycles it makes per second. It is expressed in **hertz (Hz)**.
- The **angular frequency** (ω) is another way of expressing frequency and is expressed in radians per second.
- The **period** (T) of a wave is the number of seconds it takes to complete a cycle. It is the inverse of frequency.
- **Interference** describes the ways in which waves interact in space to form a **resultant wave**.
  - **Constructive interference** occurs when waves are exactly **in phase** with each other. The amplitude of the resultant wave is equal to the sum of the amplitudes of the two interfering waves.
  - **Destructive interference** occurs when waves are exactly **out of phase** with each other. The amplitude of the resultant wave is equal to the difference in amplitude between the two interfering waves.
  - **Partially constructive** and **partially destructive interference** occur when two waves are not quite perfectly in or out of phase with each other. The displacement of the resultant wave is equal to the sum of the displacements of the two interfering waves.
- **Traveling waves** have continuously shifting points of maximum and minimum displacement.
- **Standing waves** are produced by the constructive and destructive interference of two waves of the same frequency traveling in opposite directions in the same space.
  - **Antinodes** are points of maximum oscillation.
  - **Nodes** are points where there is no oscillation.

- **Resonance** is the increase in amplitude that occurs when a periodic force is applied at the **natural (resonant) frequency** of an object.
- **Damping** is a decrease in amplitude caused by an applied or nonconservative force.

Sound

- **Sound** is produced by mechanical disturbance of a material that creates an oscillation of the molecules in the material.
- Sound propagates through all forms of matter (but not a vacuum).
  - Sound propagates fastest through solids, followed by liquids, and slowest through gases.
  - Within a medium, as density increases, the speed of sound decreases.
- The **pitch** of a sound is related to its frequency.
- The **Doppler effect** is a shift in the perceived frequency of a sound compared to the actual frequency of the emitted sound when the source of the sound and its detector are moving relative to one another.
  - The apparent frequency will be higher than the emitted frequency when the source and detector are moving toward each other.
  - The apparent frequency will be lower than the emitted frequency when the source and detector are moving away from each other.
  - The apparent frequency can be higher, lower, or equal to the emitted frequency when the two objects are moving in the same direction, depending on their relative speeds.
  - When the source is moving at or above the speed of sound, **shock waves (sonic booms)** can form.
- Loudness or volume of sound (**sound level**) is related to its **intensity**.
  - Intensity is related to a wave's amplitude.
  - Intensity decreases over distance and some energy is lost to **attenuation** (damping) from frictional forces.
- Strings and **open pipes** (open at both ends) support standing waves, and the length of the string or pipe is equal to some multiple of half-wavelengths.
- **Closed pipes** (closed at one end) also support standing waves, and the length of the pipe is equal to some odd multiple of quarter-wavelengths.
- Sound is used medically in **ultrasound** machines for both imaging (diagnostic) and treatment (therapeutic) purposes.

Answers to Concept Checks

7.1

1. Wave speed is the rate at which a wave transmits the energy or matter it is carrying. Wave speed is the product of frequency and wavelength. Frequency is a measure of how often a waveform passes a given point in space. It is measured in Hz. Angular frequency is the same as frequency, but is measured in radians per second. Period is the time necessary to complete a wave cycle. The equilibrium position is the point with zero displacement in an oscillating system. Amplitude is the maximal displacement of a wave from the equilibrium position. Traveling waves have nodes and antinodes that move with wave propagation. Standing waves have defined nodes and antinodes that do not move with wave propagation.
2. If two waves are perfectly in phase, the amplitude of the resulting wave is equal to the sum of the amplitudes of the interfering waves. If two waves are perfectly out of phase, the amplitude of the resulting wave is the difference of the amplitudes of the interfering waves. Therefore, if the two waves are anywhere between these two extremes, the amplitude of the resulting wave will be somewhere between the sum and difference of the amplitudes of the interfering waves.
3. False. Sound waves are the most common example of longitudinal waves on the MCAT.
4. The object will resonate because the force frequency equals the natural (resonant) frequency. The amplitude of the oscillation will increase.

7.2

1. Sound is produced by mechanical vibrations. These are usually generated by solid objects like bells or vocal cords, but occasionally can be generated by fluids. Sound is propagated as longitudinal waves in matter, so it cannot propagate in a vacuum.
2. The amplitude of a wave is related to its sound level (volume). The frequency of a wave is related to its pitch.
3. When two objects are traveling toward each other, the apparent frequency is higher than the original frequency $\left(f' = f\,\frac{(v + v_D)}{(v - v_S)}\right)$. When two objects are traveling away from each other, the apparent frequency is lower than the original frequency $\left(f' = f\,\frac{(v - v_D)}{(v + v_S)}\right)$. When one object follows the other, the apparent frequency could be higher, lower, or equal to the original frequency depending on the relative speeds of the detector and the source $\left(f' = f\,\frac{(v + v_D)}{(v + v_S)} \text{ or } f' = f\,\frac{(v - v_D)}{(v - v_S)}\right)$.

4. Ultrasound can be used for prenatal screening or to diagnose gallstones, breast and thyroid masses, and blood clots. It can be used for needle guidance in a biopsy, for dental cleaning, and for treating deep tissue injury, kidney stones, certain small tumors, and cataracts, among many other applications.
5.

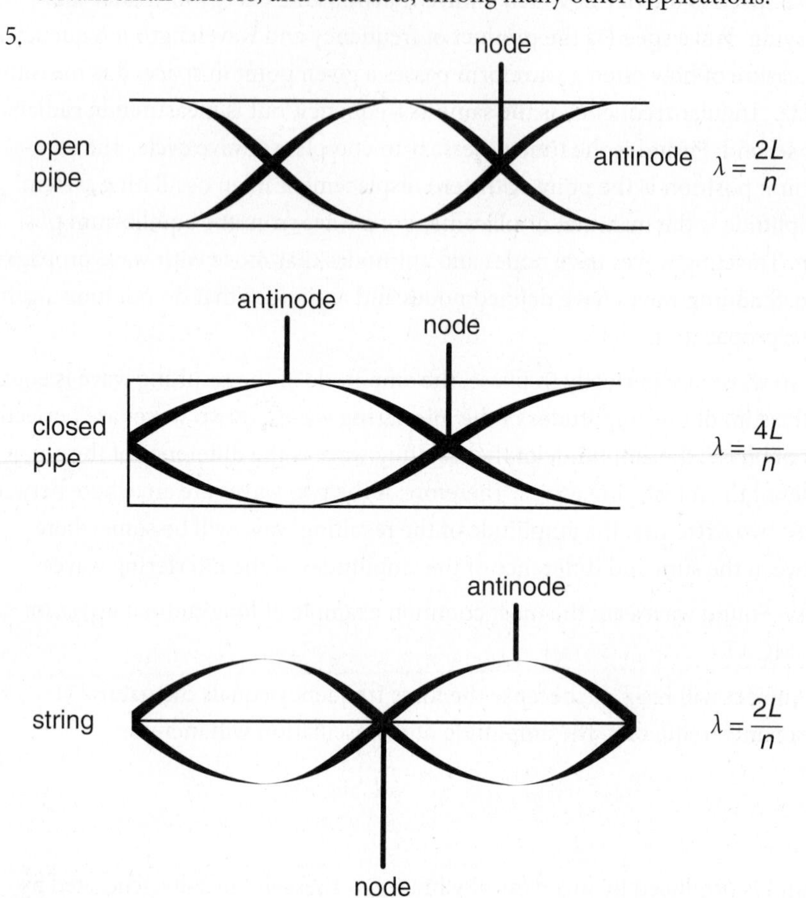

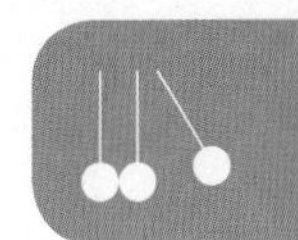

Science Mastery Assessment Explanations

1. **C**

If these two glasses are perfectly identical, then the fact that the first glass shattered at 808 Hz tells us that this is very close (if not identical) to the natural (resonant) frequency of the glass. If she produces a frequency that is not equal (or very close) to the natural frequency, then the applied frequency will not cause the glass to resonate, and there will not be the increase in wave amplitude associated with resonating objects. Attenuation will increase with increased frequency because there is more motion over which non-conservative forces can damp the sound wave; however, even if sound level was matched to that which shattered the first glass when accounting for attenuation, the glass would still not shatter for the reasons described above, eliminating **(D)**.

2. **B**

This question is testing our understanding of pipes open at one or both ends. To begin, remember that high-frequency sounds have a high pitch and low-frequency sounds have a low pitch. The pipe in this example begins as one that is open on both ends, and then one end is closed off. Our task, therefore, is to determine how the frequency of the second harmonic differs between a pipe that is open at both ends from one of equal length that is open at only one end. For a pipe of length L open at both ends, the wavelength for the second harmonic (first overtone) is equal to L:

$$\lambda = \frac{2L}{n} = \frac{2L}{2} = L$$

In contrast, for a pipe open at one end and closed at the other, the wavelength is equal to $\frac{4L}{3}$:

$$\lambda = \frac{4L}{n} = \frac{4L}{3}$$

Keep in mind that the first overtone for a closed pipe corresponds to the third harmonic, not the second. Thus, when the brother covers one end of the flute, the wavelength increases. Given that the wavelength and the frequency of a sound are inversely proportional, an increase in wavelength corresponds to a decrease in frequency. Therefore, when the brother covers one end of the flute, the sound produced by the instrument will be slightly lower in pitch than the original sound.

3. **C**

The angular frequency is related to the frequency through the equation $\omega = 2\pi f$. Therefore, the magnitude of the angular frequency will always be larger than the magnitude of the frequency. The magnitude of the angular frequency may or may not be larger than the magnitude of period; these variables are inversely proportional, eliminating **(A)**. The product of the frequency and the period is always 1 because these two are inverses of each other, eliminating **(B)**. Finally, the product of the angular frequency and period will always be 2π because $\omega = 2\pi f = \frac{2\pi}{T}$, eliminating **(D)**.

4. **B**

Although intensity, **(A)**, could be used to measure distance, time of travel is an easier indication and most commonly used by ultrasound machines. Apparent frequency, **(D)**, is only used in Doppler ultrasound, but is not used to calculate distance. Angle of incidence, **(C)**, can be used to position various structures on the screen of an ultrasound, but is not used to calculate distance.

5. **A**

Period is inversely related to frequency. Because the perceived frequency is doubled, the perceived period must be halved, from 34 ms to 17 ms. While either condition I or II would cause a doubling of the perceived frequency, neither condition must necessarily be true because the opposite could be true instead.

6. **B**

This question is testing our understanding of traveling waves. We know that frequency and wavelength are related through the equation $v = f\lambda$. Frequency and period are inverses of each other, so this equation could be rearranged to solve for period:

$$v = \left(\frac{1}{T}\right)\lambda \rightarrow T = \frac{\lambda}{v} = \frac{(0.1\ \text{m})}{\left(3\ \frac{\text{m}}{\text{s}}\right)} = 0.03\ \text{s}$$

7. **D**

The angular frequency is related to the frequency of a wave through the formula $\omega = 2\pi f$. Thus, our initial task is to calculate the frequency of the wave. Knowing its speed, we determine the frequency of the wave by first calculating wavelength ($v = f\lambda$). For the third harmonic of a standing wave in a pipe with one closed end, the wavelength is

$$\lambda = \frac{4L}{n} = \frac{4(1.5 \text{ m})}{3} = 2 \text{ m}$$

The frequency of the wave is therefore

$$f = \frac{v}{\lambda} = \frac{340 \frac{\text{m}}{\text{s}}}{2 \text{ m}} = 170 \text{ Hz}$$

Finally, obtain the angular frequency by multiplying the frequency of the wave by 2π:

$$\omega = 2\pi f = 340\pi \text{ radians per second}$$

8. **C**

Let I_i be the intensity before the increase and I_f be the intensity after the increase. Using the equation that relates sound level to intensity, obtain the ratio of I_i to I_f:

$$\beta_f = \beta_i + 10 \log \frac{I_f}{I_i} \rightarrow \beta_f - \beta_i = 10 \log \frac{I_f}{I_i}$$

$$20 \text{ dB} = 10 \log \frac{I_f}{I_i}$$

$$2 = \log \frac{I_f}{I_i}$$

$$100 = \frac{I_f}{I_i}$$

9. **B**

Saying that the stapedial footplate has limited displacement during vibration is another way of stating that the amplitude of the vibration has been decreased. Because amplitude is related to intensity, and intensity is related to sound level, the perceived sound level (volume) will be decreased as well. Pitch, described in **(C)** and **(D)**, is related to the frequency of a sound, not its amplitude.

10. **A**

When two waves are out of phase by 180°, the resultant amplitude is the difference between the two waves' amplitudes. In this case, the resulting wave will have an amplitude of 5 cm − 3 cm = 2 cm.

11. **D**

This question is testing you on your understanding of the Doppler effect. A difference of zero between the perceived and the emitted frequencies implies that the source of the sound is not moving relative to the student. If the car in **(D)** is moving at the same speed as the student, then the relative motion between them could be 0. In all of the other cases, the student and the sound source are necessarily moving relative to each other.

12. **D**

Sound is a mechanical disturbance propagated through a deformable medium; it is transmitted by the oscillation of particles parallel to the direction of the sound wave's propagation. As such, sound needs matter to travel through, eliminating **(A)**. The speed of propagation is fastest in solid materials, followed by liquids, and slowest in gases.

13. **B**

Shock waves are the buildup of wave fronts as the distance between those wave fronts decreases. This occurs maximally when an object is traveling at exactly the same speed as the wave is traveling (the speed of sound). Once an object moves faster than the speed of sound, some of the effects of the shock wave are mitigated because all of the wave fronts will trail behind the object, destructively interfering with each other.

14. **A**

Here, an observer is moving closer to a stationary source. The applicable version of the Doppler effect equation is $f' = f\,\frac{(v + v_D)}{v}$ where v is the speed of the sound. Because the numerator is greater than the denominator, f' will be greater than f; therefore, statement I is true. The scenario described in statement II will produce a similar, but not identical, frequency for the officer: the frequency formula would be $f' = f\,\frac{v}{(v - v_S)}$. The apparent frequency will increase, but the increase will not be exactly the same as if the officer had been moving. Statement III is false because we already know the frequency increases for the officer—a decrease in velocity would be associated with a decrease in frequency.

15. **D**

Intensity is equal to power divided by area. In this case, area refers to the surface area of concentric spheres emanating out from the source of the sound. This surface area is given by $4\pi r^2$, so as distance (r) doubles, the intensity will decrease by a factor of four.

Consult your online resources for additional practice. GO ONLINE

Equations to Remember

(7.1) **Wave speed:** $v = f\lambda$

(7.2) **Period:** $T = \frac{1}{f}$

(7.3) **Angular frequency:** $\omega = 2\pi f = \frac{2\pi}{T}$

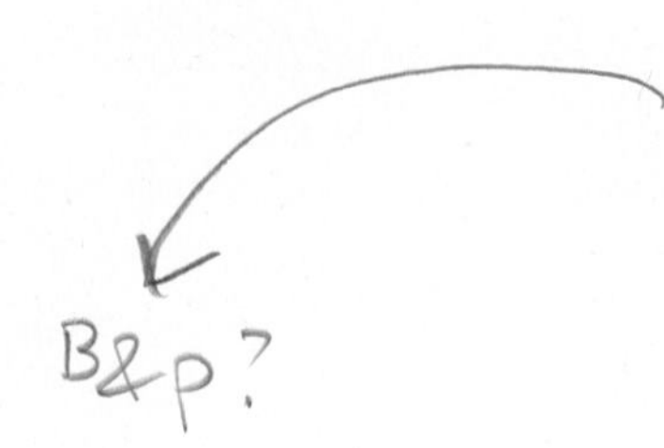

(7.4) **Speed of sound:** $v = \sqrt{\frac{B}{\rho}}$

(7.5) **Doppler effect:** $f' = f\frac{(v \pm v_D)}{(v \mp v_S)}$

(7.6) **Intensity:** $I = \frac{P}{A}$

(7.7) **Sound level:** $\beta = 10 \log \frac{I}{I_0}$

(7.8) **Change in sound level:** $\beta_f = \beta_i + 10 \log \frac{I_f}{I_i}$

(7.9) **Beat frequency:** $f_{beat} = |f_1 - f_2|$

(7.10) **Wavelength of a standing wave (strings and open pipes):** $\lambda = \frac{2L}{n}$

(7.11) **Frequency of a standing wave (strings and open pipes):** $f = \frac{nv}{2L}$

(7.12) **Wavelength of a standing wave (closed pipes):** $\lambda = \frac{4L}{n}$

(7.13) **Frequency of a standing wave (closed pipes):** $f = \frac{nv}{4L}$

Shared Concepts

Behavioral Sciences Chapter 2
Sensation and Perception

General Chemistry Chapter 8
The Gas Phase

Physics and Math Chapter 1
Kinematics and Dynamics

Physics and Math Chapter 2
Work and Energy

Physics and Math Chapter 8
Light and Optics

Physics and Math Chapter 10
Mathematics

CHAPTER 8

Light and Optics

SCIENCE MASTERY ASSESSMENT

Every pre-med knows this feeling: there is so much content I have to know for the MCAT! How do I know what to do first or what's important?

While the high-yield badges throughout this book will help you identify the most important topics, this Science Mastery Assessment is another tool in your MCAT prep arsenal. This quiz (which can also be taken in your online resources) and the guidance below will help ensure that you are spending the appropriate amount of time on this chapter based on your personal strengths and weaknesses. Don't worry though—skipping something now does not mean you'll never study it. Later on in your prep, as you complete full-length tests, you'll uncover specific pieces of content that you need to review and can come back to these chapters as appropriate.

How to Use This Assessment

If you answer 0–7 questions correctly:

Spend about 1 hour to read this chapter in full and take limited notes throughout. Follow up by reviewing **all** quiz questions to ensure that you now understand how to solve each one.

If you answer 8–11 questions correctly:

Spend 20–40 minutes reviewing the quiz questions. Beginning with the questions you missed, read and take notes on the corresponding subchapters. For questions you answered correctly, ensure your thinking matches that of the explanation and you understand why each choice was correct or incorrect.

If you answer 12–15 questions correctly:

Spend less than 20 minutes reviewing all questions from the quiz. If you missed any, then include a quick read-through of the corresponding subchapters, or even just the relevant content within a subchapter, as part of your question review. For questions you got correct, ensure your thinking matches that of the explanation and review the Concept Summary at the end of the chapter.

1. If a light ray has a frequency of 5.0×10^{14} Hz, in which region of the electromagnetic spectrum is it located?
   - **A.** X-ray
   - **B.** UV
   - **C.** Visible
   - **D.** Infrared

2. During the fall, leaves change in color from green to red because chlorophyll breaks down, leaving the secondary anthocyanin pigments. Which of the following best describes the light reflected by anthocyanin?
   - **A.** Has a wavelength of 700 m
   - **B.** Has a wavelength of 580 nm
   - **C.** Has a frequency of 4.2×10^{6} Hz
   - **D.** Has a frequency of 4.2×10^{13} Hz

3. An object is placed at the center of curvature of a concave mirror. Which of the following is true about the image?
   - **A.** It is real and inverted.
   - **B.** It is virtual and inverted.
   - **C.** It is real and upright.
   - **D.** It is virtual and upright.

4. In a double slit experiment, what wavelength of light must be used for the second dark fringe to be at an angle of 30° given that the distance between the two slits is 0.3 mm?
   - **A.** 1×10^{-5} m
   - **B.** 6×10^{-5} m
   - **C.** 3.6×10^{-2} m
   - **D.** 6×10^{-2} m

5. A ray of light ($f = 5 \times 10^{14}$ Hz) travels from air into crystal into chromium. If the indices of refraction of air, crystal, and chromium are 1, 2, and 3, respectively, and the incident angle is 30°, then which of the following describes the frequency and the angle of refraction in the chromium?
   - **A.** 5×10^{14} Hz; 9.6°
   - **B.** 5×10^{14} Hz; 57°
   - **C.** 1.0×10^{10} Hz; 9.6°
   - **D.** 1.0×10^{10} Hz; 57°

6. A source of light ($f = 6.0 \times 10^{14}$ Hz) passes through three plane polarizers. The first two polarizers are in the same direction, while the third is rotated 90° with respect to the second polarizer. What is the frequency of the light that comes out of the third polarizer?
   - **A.** 3.0×10^{14} Hz
   - **B.** 6.0×10^{14} Hz
   - **C.** 9.0×10^{14} Hz
   - **D.** Light will not pass through the third polarizer

7. As part of an organic chemistry lab, a student must determine the specific rotation of a chiral product. He uses a plane polarizer, but accidentally inserts a second polarizer at a 90° angle to the first. What is true of the resulting light if it passes through both polarizers?
   - **A.** It is circularly polarized.
   - **B.** It is twice as plane polarized.
   - **C.** There is no light.
   - **D.** It is the same as when passed through a single polarizer.

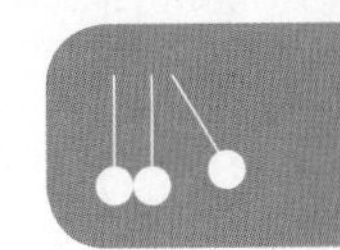

8. Which of the following describes the image formed by an object placed in front of a convex lens at a distance smaller than the focal length?
   - **A.** Virtual and inverted
   - **B.** Virtual and upright
   - **C.** Real and upright
   - **D.** Real and inverted

9. A submarine is inspecting the surface of the water with a laser that points from the submarine to the surface of the water and through the air. At what angle will the laser not penetrate the surface of the water but rather reflect entirely back into the water? (Assume $n_{water} = 1.33$ and $n_{air} = 1$.)
   - **A.** 19°
   - **B.** 29°
   - **C.** 39°
   - **D.** 49°

10. A student is analyzing the behavior of a light ray that is passed through a small opening and a lens and allowed to project on a screen a distance away. What happens to the central maximum (the brightest spot on the screen) when the slit becomes narrower?
   - **A.** The central maximum remains the same.
   - **B.** The central maximum becomes narrower.
   - **C.** The central maximum becomes wider.
   - **D.** The central maximum divides into smaller light fringes.

11. Which of the following are able to produce a virtual image?
   - **I.** Convex lens
   - **II.** Concave lens
   - **III.** Plane mirror

   - **A.** I only
   - **B.** III only
   - **C.** II and III only
   - **D.** I, II, and III

12. Monochromatic red light is allowed to pass between two different media. If the incident angle in medium 1 is 30° and the incident angle in medium 2 is 45°, what is the relationship between the speed of the light in medium 2 compared to that in medium 1?
   - **A.** $\nu_2 = \nu_1\sqrt{2}$
   - **B.** $\nu_2\sqrt{2} = \nu_1$
   - **C.** $\nu_2 = \nu_1\sqrt{3}$
   - **D.** $\nu_2\sqrt{3} = \nu_1$

13. A scientist looks through a microscope with two thin lenses with $m_1 = 10$ and $m_2 = 40$. What is the overall magnification of this microscope?
   - **A.** 0.25
   - **B.** 30
   - **C.** 50
   - **D.** 400

14. Imagine that a beam of monochromatic light originates in air and is allowed to shine upon the flat surface of a piece of glass at an angle of 60° with the normal. The reflected and refracted beams are perpendicular to each other. What is the index of refraction of the glass?
   - **A.** $\frac{\sqrt{3}}{3}$
   - **B.** 1
   - **C.** 2
   - **D.** $\sqrt{3}$

15. Which of the following will not result in the splitting of white light into its component colors?
   - **A.** Dispersion through a prism
   - **B.** Diffraction through a grating
   - **C.** Refraction within a thin film
   - **D.** Reflection from an ideal convex mirror

Answer Key

1. **C** (Ch. 8.1)
2. **D** (Ch. 8.1)
3. **A** (Ch. 8.2)
4. **B** (Ch. 8.3)
5. **A** (Ch. 8.2)
6. **D** (Ch. 8.4)
7. **C** (Ch. 8.4)
8. **B** (Ch. 8.2)
9. **D** (Ch. 8.2)
10. **C** (Ch. 8.3)
11. **D** (Ch. 8.2)
12. **A** (Ch. 8.2)
13. **D** (Ch. 8.2)
14. **D** (Ch. 8.2)
15. **D** (Ch. 8.2)

CHAPTER 8

Light and Optics

In This Chapter

8.1 **Electromagnetic Spectrum** HY
- Electromagnetic Waves 286
- Color and the Visible Spectrum 287

8.2 **Geometrical Optics** HY
- Reflection 289
- Refraction 294
- Lenses 298
- Dispersion 303

8.3 **Diffraction**
- Single Slit 306
- Slit–Lens System 307
- Multiple Slits 307
- X-Ray Diffraction 310

8.4 **Polarization**
- Plane-Polarized Light 311
- Circular Polarization 312

Concept Summary 314

CHAPTER PROFILE

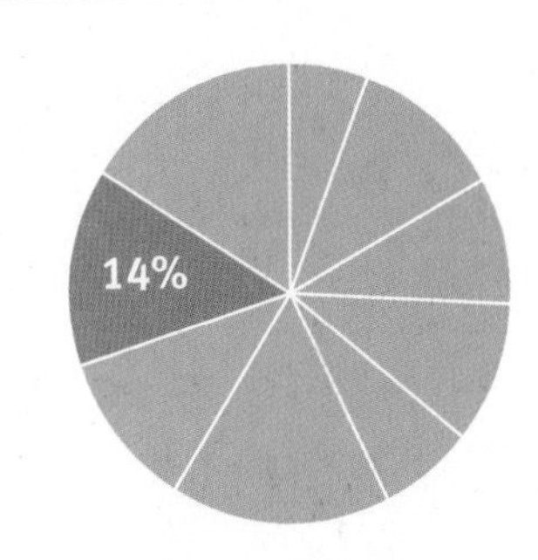

The content in this chapter should be relevant to about 14% of all questions about physics on the MCAT.

This chapter covers material from the following AAMC content category:

4D: How light and sound interact with matter

Introduction

The next time you're browsing your local convenience store, take a look at the security mirrors—the ones that bulge out from the wall, usually above eye level. Looking into one of these mirrors, notice not just that the image you see of the world is distorted but *how* it is distorted: the image is still right-side up, but everything is much smaller than you'd expect, and the curve of the mirror introduces some slopes that are not present in reality. Additionally, you see a much wider field of vision than you would if the mirror were a simple plane mirror. This is why security mirrors are useful: they are a convenient, low-tech solution that allows the cashier to survey the entire store in one glance. All these features result from the fact that the security mirror is a convex, diverging optical system. Parallel light rays that hit the mirror are reflected in multiple directions, which allows observers to see a large field of vision, even if the image is somewhat distorted and the objects in the image are closer than they appear. Indeed, the passenger-side mirror of a car that bears that same message is also a convex mirror, allowing the driver to see a wider view of the cars behind him or her.

This chapter will first complete a topic from Chapter 7 of *MCAT Physics and Math Review* by analyzing the transverse waveform of visible light and other electromagnetic (EM) waves. We will then consider in detail the rules of optics, which describe the behavior of electromagnetic waves as they bounce off of and travel through various shapes and compositions of matter. The optical systems covered are those tested on the MCAT: concave and convex mirrors, which produce images by reflection, and concave and convex lenses, which produce images by refraction. To finish, we will discuss the phenomena of thin-slit experiments (diffraction) and light polarization.

8.1 Electromagnetic Spectrum

High-Yield

LEARNING OBJECTIVES

After Chapter 8.1, you will be able to:

- Order the types of electromagnetic radiation, such as X-rays, microwaves, and visible light, from lowest to highest energy
- Describe the properties of electromagnetic waves
- Compare the visible spectrum to the full electromagnetic spectrum

The full **electromagnetic spectrum** includes **radio waves** on one end (long wavelength, low frequency, low energy) and **gamma rays** on the other (short wavelength, high frequency, high energy). Between the two extremes, we find, in order from lowest energy to highest energy, **microwaves**, **infrared**, visible light, **ultraviolet**, and **X-rays**. This chapter will focus primarily on the range of wavelengths corresponding to the visible spectrum of light (400 nm to 700 nm).

Electromagnetic Waves

A changing magnetic field can cause a change in an electric field, and a changing electric field can cause a change in a magnetic field. Because of the reciprocating nature of these two fields, we can see how electromagnetic waves occur in nature. Each oscillating field causes oscillations in the other field completely independent of matter, so electromagnetic waves can even travel through a vacuum.

Electromagnetic waves are transverse waves because the oscillating electric and magnetic field vectors are perpendicular to the direction of propagation. The electric field and the magnetic field are also perpendicular to each other. This is illustrated in Figure 8.1.

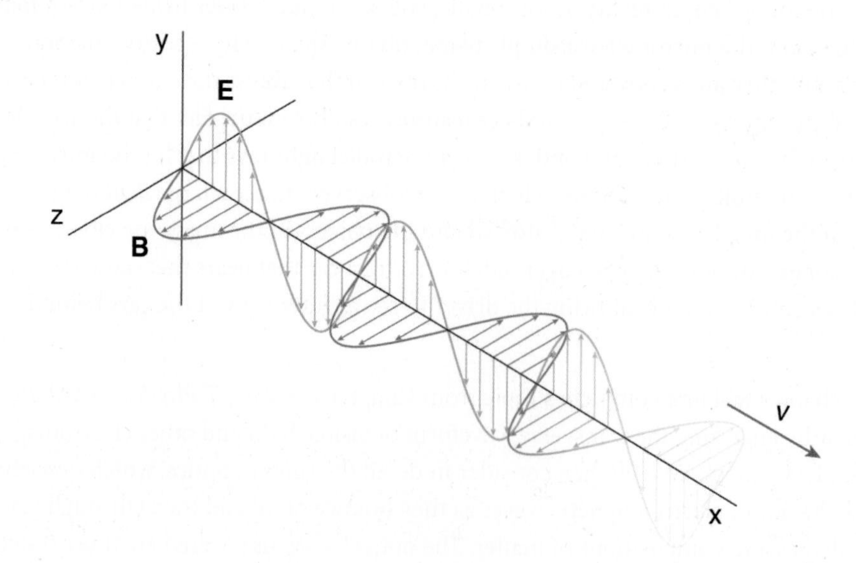

Figure 8.1 Electromagnetic Wave
*The electric field (**E**) oscillates up and down the page; the magnetic field (**B**) oscillates into and out of the page.*

The **electromagnetic spectrum** describes the full range of frequencies and wavelengths of electromagnetic waves. Wavelengths are often given in the following units: mm (10^{-3} m), ∝m (10^{-6} m), nm (10^{-9} m), and Å (ångström, 10^{-10} m). The full spectrum is broken up into many regions, which in descending order of wavelength are radio (10^{9}–1 m), microwave (1 m–1 mm), infrared (1 mm–700 nm), visible light (700–400 nm), ultraviolet (400–50 nm), X-ray (50–10^{-2} nm), and γ-rays (less than 10^{-2} nm). The electromagnetic spectrum is depicted in Figure 8.2.

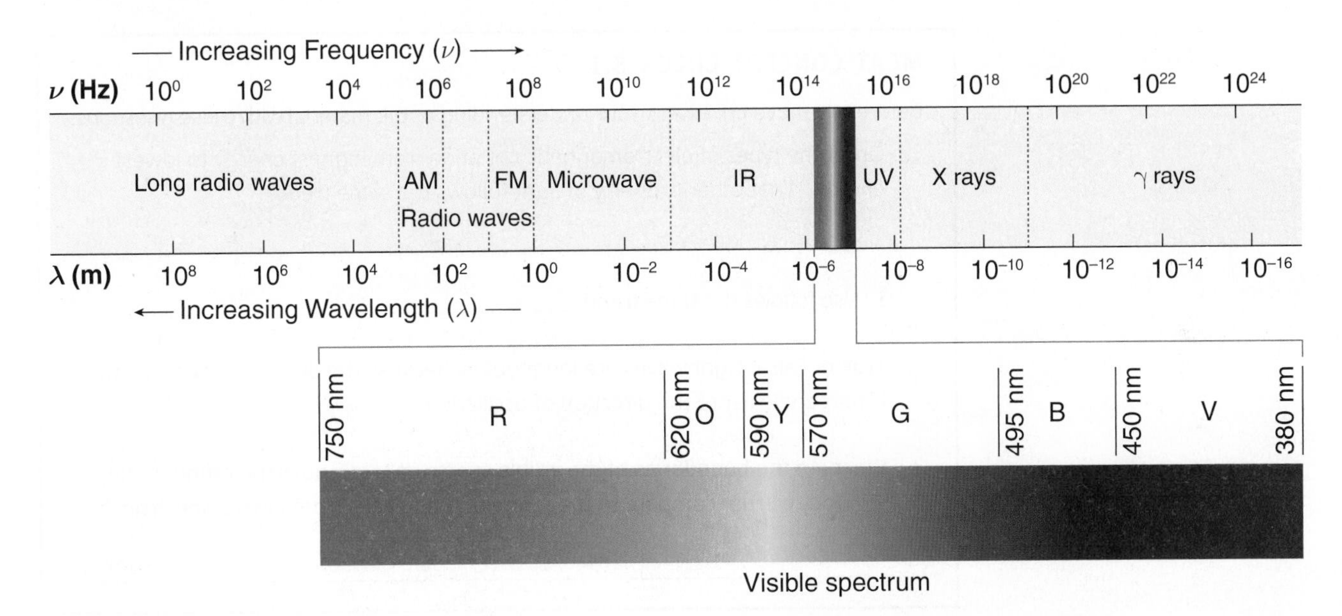

Figure 8.2 The Electromagnetic Spectrum

Electromagnetic waves vary in frequency and wavelength, but in a vacuum, all electromagnetic waves travel at the same speed, called the **speed of light**. This constant is represented by c and is approximately $3.00 \times 10^{8} \frac{\text{m}}{\text{s}}$. To a first approximation—and for the purposes of all MCAT-related equations—electromagnetic waves also travel in air with this speed. In reference to electromagnetic waves, the familiar equation $v = f\lambda$ becomes

$$c = f\lambda$$

Equation 8.1

where c is the speed of light in a vacuum and, to a first approximation, also in air, f is the frequency, and λ is the wavelength.

Color and the Visible Spectrum

The only part of the spectrum that is perceived as light by the human eye is the **visible region**. Within this region, different wavelengths are perceived as different colors, with violet at one end of the visible spectrum (400 nm) and red at the other (700 nm).

MNEMONIC

To recall the order of the colors in the visible spectrum, remember the grade-school "rainbow" of **ROY G. BV** (**r**ed, **o**range, **y**ellow, **g**reen, **b**lue, **v**iolet).

MCAT EXPERTISE

Wavelengths in the visible range are common on the MCAT. Remembering the boundaries of the visible spectrum (about 400–700 nm) will save you time and energy on Test Day.

Light that contains all the colors in equal intensity is perceived as white. The color of an object that does not emit its own light is dependent on the color of light that it reflects. Thus, an object that appears red is one that absorbs all colors of light except red. This implies that a red object under green illumination will appear black because it absorbs the green light and has no light to reflect. The term **blackbody** refers to an ideal absorber of all wavelengths of light, which would appear completely black if it were at a lower temperature than its surroundings.

MCAT CONCEPT CHECK 8.1

Before you move on, assess your understanding of the material with these questions.

1. Order the types of electromagnetic radiation from highest energy to lowest energy. What other property of light follows the same trend?

_______ > _______ > _______ > _______ > _______ > _______ > _______

- Also follows the same trend: ____________________

2. True or False: Light waves are longitudinal because the direction of propagation is perpendicular to the direction of oscillation.

3. What are the boundaries of the visible spectrum? How does the range of the visible spectrum compare to the range of the full electromagnetic spectrum?

____

8.2 Geometrical Optics

High-Yield

LEARNING OBJECTIVES

After Chapter 8.2, you will be able to:

- Apply the sign conventions for mirrors and lenses to optics systems
- Describe the bending of light as it moves between media with different refractive indices
- Explain the impact of dispersion effects and aberrations on the behavior of light
- Recall Snell's law and other key optics equations
- Solve optics and Snell's law problems

When light travels through a homogeneous medium, it travels in a straight line. This is known as **rectilinear propagation**. The behavior of light at the boundary of a medium or interface between two media is described by the theory of geometrical optics. Geometrical optics explains reflection and refraction, as well as the applications of mirrors and lenses.

Reflection

Reflection is the rebounding of incident light waves at the boundary of a medium. Light waves that are reflected are not absorbed into the second medium; rather, they bounce off of the boundary and travel back through the first medium. Figure 8.3 illustrates reflection on a plane mirror.

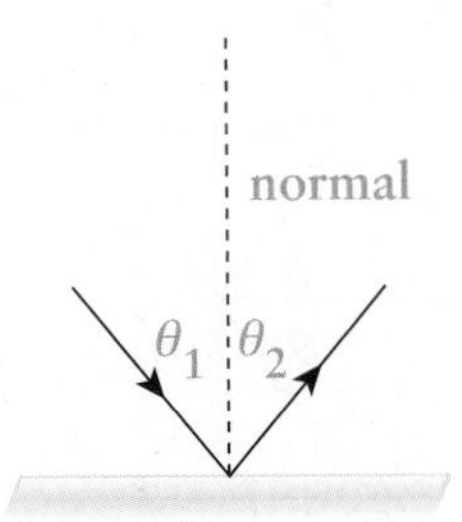

Figure 8.3 Reflection
According to the law of reflection, $\theta_1 = \theta_2$.

The law of reflection is

$$\theta_1 = \theta_2$$

Equation 8.2

where θ_1 is the incident angle and θ_2 is the reflected angle, both measured from the normal. The **normal** is a line drawn perpendicular to the boundary of a medium; all angles in optics are measured from the normal, not the surface of the medium.

Plane Mirrors

In general, images created by a mirror can be either **real** or **virtual**. An image is said to be real if the light actually converges at the position of the image. An image is virtual if the light only *appears* to be coming from the position of the image but does not actually converge there. One of the distinguishing features of real images is the ability of the image to be projected onto a screen.

Parallel incident light rays remain parallel after reflection from a plane mirror; that is, **plane mirrors**—being flat reflective surfaces—cause neither convergence nor divergence of reflected light rays. Because the light does not converge at all, plane mirrors always create virtual images. In a plane mirror, the image appears to be the same distance behind the mirror as the object is in front of it, as shown in Figure 8.4. In other words, plane mirrors create the appearance of light rays originating behind the mirrored surface. Because the reflected light remains in front of the mirror but the image appears behind the mirror, the image is virtual. Plane mirrors include most of the common mirrors found in our homes. To assist in our discussion of spherical mirrors, plane mirrors can be conceptualized as spherical mirrors with an infinite radius of curvature.

plane mirrors always create virtual images

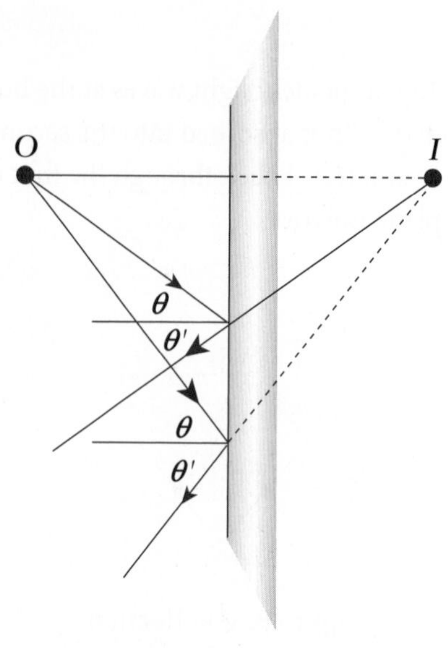

Figure 8.4 Reflection in a Plane Mirror
O is the object and I is the (virtual) image; all incident angles (θ) are equal to their respective reflected angles (θ′).

Spherical Mirrors

Spherical mirrors come in two varieties: concave and convex. The word spherical implies that the mirror can be considered a spherical cap or dome taken from a much larger spherically shaped mirror. Spherical mirrors have an associated center of curvature (*C*) and a radius of curvature (*r*). The **center of curvature** is a point on the optical axis located at a distance equal to the **radius of curvature** from the vertex of the mirror; in other words, the center of curvature would be the center of the spherically shaped mirror if it were a complete sphere.

If we were to look from the inside of a sphere to its surface, we would see a **concave** surface. On the other hand, if we were to look from outside the sphere, we would see a **convex** surface. For a concave surface, the center of curvature and the radius of curvature are located in front of the mirror. For a convex surface, the center of curvature and the radius of curvature are behind the mirror. Concave mirrors are called **converging mirrors** and convex mirrors are called **diverging mirrors** because they cause parallel incident light rays to converge and diverge after they reflect, respectively.

There are several important lengths associated with mirrors, as shown in Figure 8.5. The **focal length** (f) is the distance between the **focal point** (*F*) and the mirror. Note that for all spherical mirrors, $f = \frac{r}{2}$ where the radius of curvature (r) is the distance between C and the mirror. The distance between the object and the mirror is o; the distance between the image and the mirror is i.

REAL WORLD

The passenger-side mirrors in cars are an example of convex mirrors (everything appears smaller and farther away); the small circular mirrors used for applying makeup are an example of concave mirrors (everything appears bigger and closer).

MNEMONIC

Con**cave** is like looking into a **cave.**

KEY CONCEPT

Concave mirrors are converging mirrors. Convex mirrors are diverging mirrors. The reverse is true for lenses.

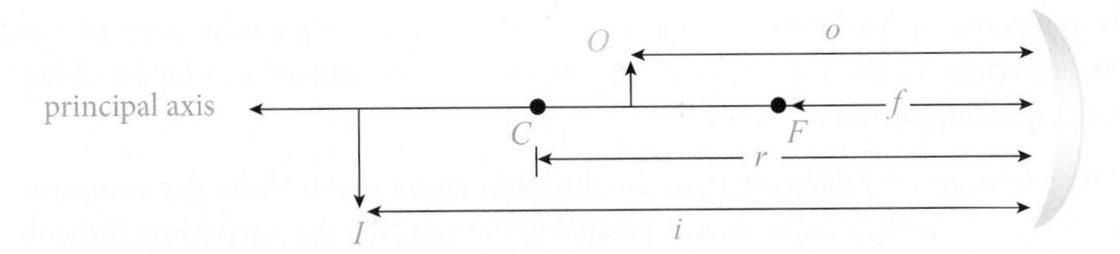

Figure 8.5 Key Variables in Geometrical Optics
The mirror pictured is a concave mirror; light rays are not shown.

There is a simple relationship between these four distances:

$$\frac{1}{f} = \frac{1}{o} + \frac{1}{i} = \frac{2}{r}$$

Equation 8.3

While it is not important which units of distance are used in this equation, it is important that all values used have the *same* units as each other.

On the MCAT, you will most often use this equation to calculate the **image distance** for all types of mirrors and lenses. If the image has a positive distance ($i > 0$), it is a real image, which implies that the image is in front of the mirror. If the image has a negative distance ($i < 0$), it is virtual and thus located behind the mirror. Plane mirrors can be thought of as spherical mirrors with infinitely large focal distances. As such, for a plane mirror, $r = f = \infty$, and the equation becomes $\frac{1}{o} + \frac{1}{i} = 0$ or $i = -o$. This can be interpreted as saying the virtual image is at a distance behind the mirror equal to the distance the object is in front of the mirror.

The **magnification (m)** is a dimensionless value that is the ratio of the image distance to the object distance:

$$m = -\frac{i}{o}$$

Equation 8.4

By extension, the magnification also gives the ratio of the size of the image to the size of the object. Following the sign convention given later in Table 8.1, the orientation of the image (upright or inverted) can be determined: a negative magnification signifies an **inverted** image, while a positive value signifies an **upright** image. If $|m| < 1$, the image is smaller than the object (reduced); if $|m| > 1$, the image is larger than the object (enlarged); and if $|m| = 1$, the image is the same size as the object.

Figure 8.6 shows ray diagrams for a concave spherical mirror with the object at three different points. A **ray diagram** is useful for getting an approximation of where an image is. On Test Day, ray diagrams can be helpful for a quick determination of the type of image that will be produced by an object some distance from the mirror (real *vs.* virtual, inverted *vs.* upright, and magnified *vs.* reduced). Ray diagrams should be used with caution, however: under the pressure of Test Day, it can be easy to draw them incorrectly. Therefore, it is important to practice drawing ray diagrams to avoid careless errors on Test Day, and it is also important to be familiar with how to solve optics questions mathematically.

When drawing a ray diagram, there are three important rays to draw. For a concave mirror, a ray that strikes the mirror parallel to the **axis** (the normal passing through the center of the mirror) is reflected back through the focal point (green lines in Figure 8.6 and 8.7). A ray that passes through the focal point before reaching the mirror is reflected back parallel to the axis (red lines). A ray that strikes the mirror at the point of intersection with the axis is reflected back with the same angle measured from the normal (blue lines). In Figure 8.6a, the object is placed beyond *F*, and the image produced is real, inverted and magnified. In Figure 8.6b, the object is placed at *F*, and no image is formed because the reflected light rays are parallel to each other. In terms of the mirror equation, we say that the image distance $i = \infty$ here. For the scenario in Figure 8.6c, the object is placed between *F* and the mirror, and the image produced is virtual, upright, and magnified.

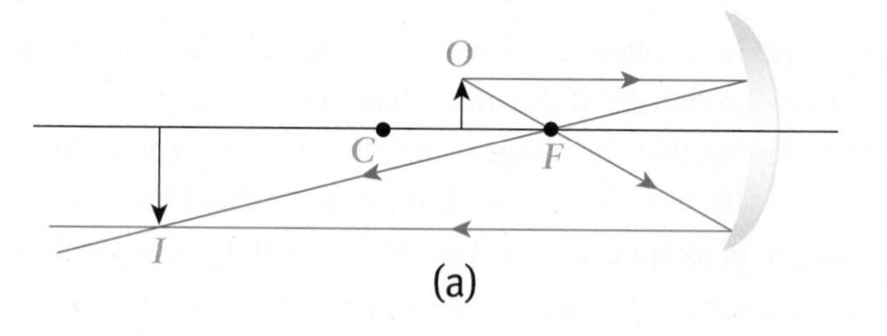

KEY CONCEPT

Any time an object is at the focal point of a converging mirror, the reflected rays will be parallel, and thus, the image will be at infinity.

O
C
F
(b)

I
O
C
F
(c)

Figure 8.6 Ray Diagrams for Concave (Converging) Mirrors
(a) Object is placed beyond F; (b) Object is placed at F; (c) Object is placed between F and the mirror.

A single diverging mirror forms only a virtual, upright, and reduced image, regardless of the position of the object. The farther away the object, the smaller the image will be. To quickly remember these rules, recall the convenience store security mirrors mentioned at the beginning of the chapter. The ray diagram of a diverging mirror is shown in Figure 8.7.

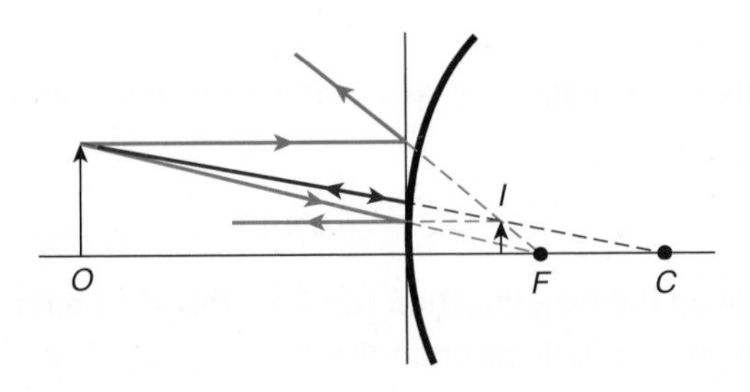

Figure 8.7 Ray Diagrams for Convex (Diverging) Mirrors

Sign Conventions for Mirrors

Table 8.1 provides the sign convention for single mirrors. Note that on the MCAT, for almost all problems involving mirrors, the object will be placed in front of the mirror. Thus, the object distance *o* is almost always positive.

Symbol	Positive	Negative
o	Object is in front of mirror	Object is behind mirror (extremely rare)
i	Image is in front of mirror (real)	Image is behind mirror (virtual)
r	Mirror is concave (converging)	Mirror is convex (diverging)
f	Mirror is concave (converging)	Mirror is convex (diverging)
m	Image is upright (erect)	Image is inverted

Table 8.1 Sign Convention for a Single Mirror

Example: An object is placed 6 cm in front of a concave mirror that has a 10 cm radius of curvature. Determine the image distance, the magnification, whether the image is real or virtual, and whether it is inverted or upright.

Solution: Use the optics equation:

$$\frac{1}{f} = \frac{1}{o} + \frac{1}{i} = \frac{2}{r}$$

$$\frac{1}{i} = \frac{2}{r} - \frac{1}{o}$$

$$\frac{1}{i} = \frac{2}{10\text{ cm}} - \frac{1}{6\text{ cm}} = \frac{12-10}{60} = \frac{2}{60}$$

$$i = \frac{60}{2} = +30\text{ cm}$$

A positive value for *i* signifies that the image is in front of the mirror and is therefore real. For a single lens or mirror with $o > 0$, a real image will always be inverted.

KEY CONCEPT

To find where the image is (for a mirror), draw the following rays and find a point where any two intersect. This point of intersection marks the tip of the image. If the rays you draw do not appear to intersect, extend them to the other side of the mirror, creating a virtual image.

- Ray parallel to axis → reflects back through focal point
- Ray through focal point → reflects back parallel to axis
- Ray to center of mirror → reflects back at same angle relative to normal

KEY CONCEPT

The focal length of converging mirrors (and converging lenses) will always be positive. The focal length of diverging mirrors (and diverging lenses) will always be negative.

MNEMONIC

Image types with a single lens or mirror (assuming *o* is positive): **UV NO IR**

- **U**pright images are always **v**irtual
- **No** image is formed when the object is a focal length away
- **I**nverted images are always **r**eal

After determining i, the magnification m can be calculated as:

$$m = -\frac{i}{o} = -\frac{30\text{ cm}}{6\text{ cm}} = -5$$

The negative sign on the magnification confirms that the image is inverted, and the fact that $|m| > 1$ indicates that the image is enlarged.

REAL WORLD

When a pencil (or any straight object) is dipped into a glass of water at an angle, it looks impossibly bent where it intersects the surface of the water because the light reflecting off of the portion of the pencil under water is refracted.

Refraction

Refraction is the bending of light as it passes from one medium to another and changes speed. The speed of light through any medium is always less than its speed through a vacuum. Remember that the speed of light in a vacuum, c, is equal to $3.00 \times 10^8\ \frac{\text{m}}{\text{s}}$. The speed of light in air is just slightly lower that this value; on the MCAT, it is appropriate to use $3.00 \times 10^8\ \frac{\text{m}}{\text{s}}$ for the speed of light in air.

Snell's Law

When light is in any medium besides a vacuum, its speed is less than c. For a given medium

$$n = \frac{c}{v}$$

Equation 8.5

where c is the speed of light in a vacuum, v is the speed of light in the medium, and n is a dimensionless quantity called the **index of refraction** of the medium. The index of refraction of a vacuum is 1, by definition; for all other materials, the index of refraction will be greater than 1. For air, n is essentially equal to 1 because the speed of light in air is extremely close to c. The indices of refraction for a number of common media are shown in Table 8.2. These values are provided only for reference; they need not be memorized.

Medium	Index of refraction (n)
Vacuum	1 (by definition)
Air	1.0003
Ice	1.31
Water	1.33
Acetone	1.36
Ethanol	1.36
Cornea (human)	1.37–1.40
Lens (human)	1.39–1.41
Glass (various types)	1.48–1.93
Diamond	2.42

Table 8.2 Indices of Refraction of Common Media

Refracted rays of light obey **Snell's law** as they pass from one medium to another:

$$n_1 \sin \theta_1 = n_2 \sin \theta_2$$

Equation 8.6

where n_1 and θ_1 refer to the medium from which the light is coming and n_2 and θ_2 refer to the medium into which the light is entering. Note that θ is once again measured with respect to the normal, as shown in Figure 8.8.

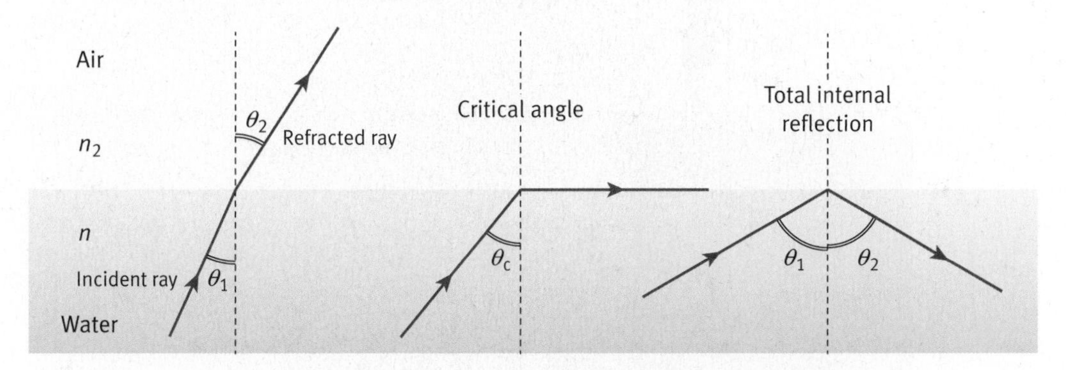

Figure 8.8 Snell's Law

From Snell's law, we can see that when light enters a medium with a higher index of refraction ($n_2 > n_1$), it bends toward the normal ($\sin \theta_2 < \sin \theta_1$; therefore, $\theta_2 < \theta_1$), as shown in Figure 8.9. Conversely, if the light travels into a medium where the index of refraction is smaller ($n_2 < n_1$), the light will bend away from the normal ($\sin \theta_2 > \sin \theta_1$; therefore, $\theta_2 > \theta_1$).

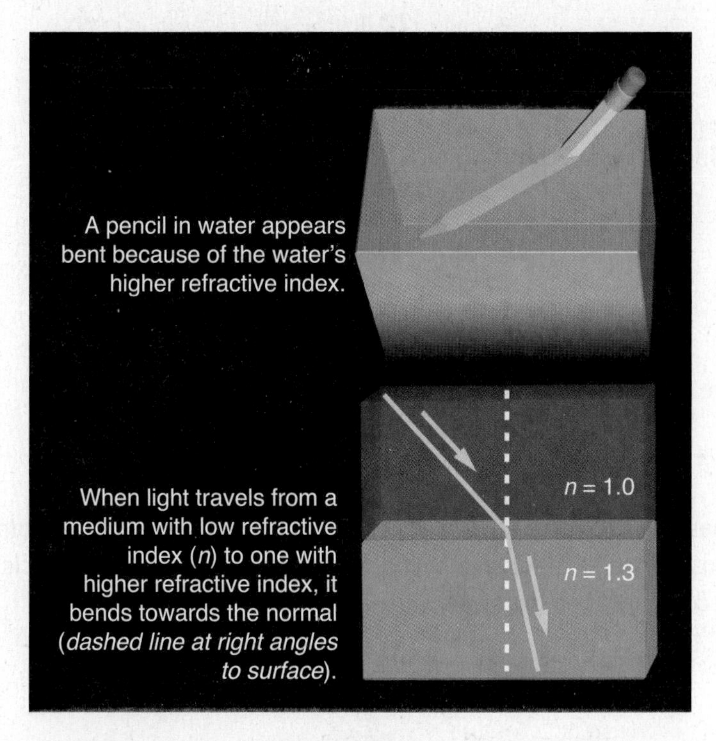

Figure 8.9 Refraction of Light from Air into Water

KEY CONCEPT

Remember that when light enters a medium with a higher index of refraction, it bends toward the normal. When light enters a medium with a lower index of refraction, it bends away from the normal.

Example: A penny sits at the bottom of a pool of water ($n = 1.33$) at a depth of 3.0 m. If an observer 1.8 m tall stands 30 cm away from the edge, how close to the side can the penny be and still be visible?

Solution: First, draw a picture of the situation:

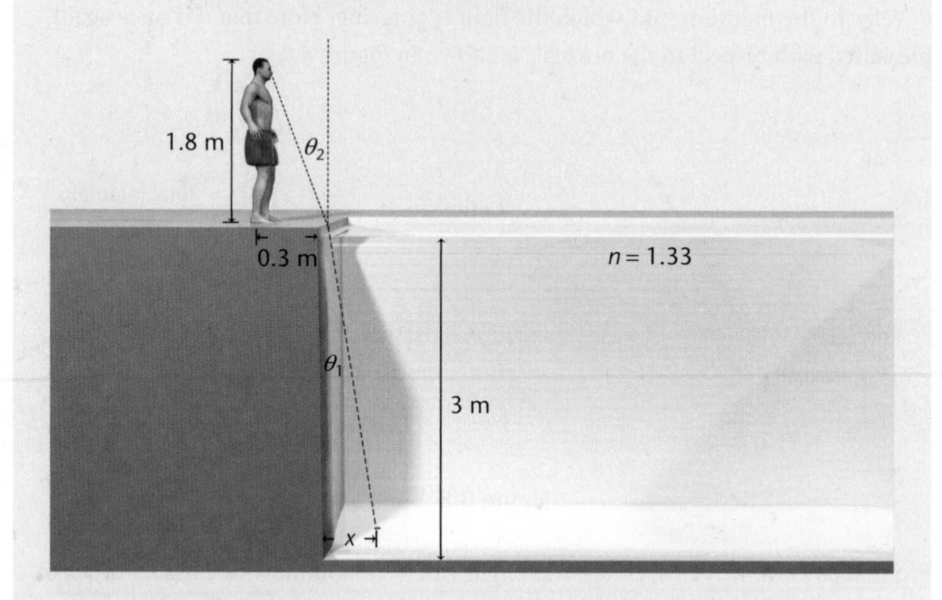

Note that the light is coming from the water ($n_1 = 1.33$) and going into the air ($n_2 \approx 1$), so the light is bent away from the normal ($\theta_2 > \theta_1$). We need to find the angle that the light rays make with the normal to the water's surface:

$$\tan \theta_2 = \frac{0.3 \text{ m}}{1.8 \text{ m}} = 0.167$$

$$\theta_2 = \tan^{-1} 0.167 = 9.5°$$

Using Snell's law, we can solve for θ_1:

$$\sin \theta_1 = \left(\frac{n_2}{n_1}\right) \sin \theta_2 = \left(\frac{1}{1.33}\right) \sin 9.5° = \frac{0.165}{1.33}$$

$$\theta_1 = \sin^{-1}\left(\frac{0.165}{1.33}\right) = 7.1°$$

Now, we can find x using trigonometry:

$$x = (3 \text{ m}) \times \tan \theta_1 = 3 \tan 7.1° = 3 \times 0.124 = 0.37 \text{ m} = 37 \text{ cm}$$

Note that you will not be expected to calculate precise values of trigonometric functions or inverse trigonometric functions on Test Day. This question is provided mainly as an opportunity to see the application of Snell's law.

Total Internal Reflection

When light travels from a medium with a higher index of refraction (such as water) to a medium with a lower index of refraction (such as air), the refracted angle is larger than the incident angle ($\theta_2 > \theta_1$); that is, the refracted light ray bends away from the normal. As the incident angle is increased, the refracted angle also increases, and eventually, a special incident angle called the **critical angle (θ_c)** is reached, for which the refracted angle θ_2 equals 90 degrees. At the critical angle, the refracted light ray passes along the interface between the two media. The critical angle can be derived from Snell's law if $\theta_2 = 90°$, such that

$$\theta_c = \sin^{-1}\left(\frac{n_2}{n_1}\right)$$

Equation 8.7

Total internal reflection, a phenomenon in which all the light incident on a boundary is reflected back into the original material, results with any angle of incidence greater than the critical angle, θ_c, as shown in Figure 8.10.

KEY CONCEPT

Total internal reflection occurs as the light moves from a medium with a higher refractive index to a medium with a lower one.

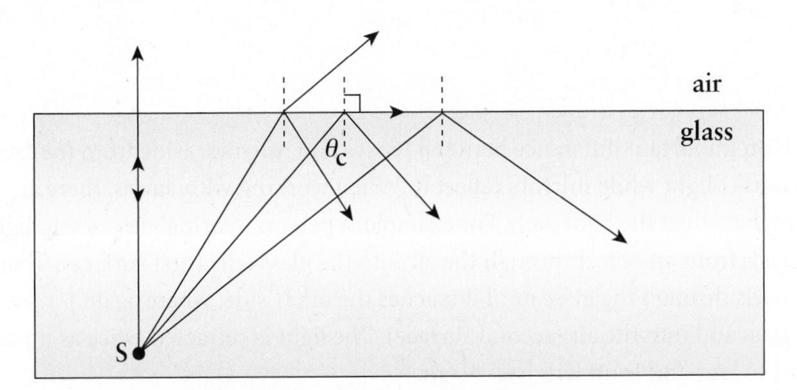

Figure 8.10 Total Internal Reflection
At the incident angle of θ_c, the refracted angle is equal to 90°; at incident angles above 90°, total internal reflection occurs.

Example: From the previous example, suppose another penny is 3 m away from the edge. Will a light ray going from this penny to the edge of the pool emerge from the water?

Solution: The angle made by the second penny's light ray is

$$\theta_1 = \tan^{-1}\left(\frac{\text{opposite}}{\text{adjacent}}\right) = \tan^{-1}\left(\frac{3\text{ m}}{3\text{ m}}\right) = \tan^{-1}(1) = 45°$$

Find the critical angle:

$$\theta_c = \sin^{-1}\left(\frac{n_2}{n_1}\right) = \sin^{-1}\left(\frac{1}{1.33}\right) = \sin^{-1}(0.75) = 48.8°$$

It is not expected on the MCAT that we know the inverse sin of 0.75. However, we do know that $\sin(45°) = \frac{\sqrt{2}}{2} \approx 0.71$. Since $\sin(\theta_c)$ is greater than 0.71, the critical angle must be greater than 45°. Therefore, $\theta_c > \theta_1$ and the light ray will emerge from the pool.

Lenses

There is an important difference between **lenses** and mirrors, aside from the fact that lenses refract light while mirrors reflect it. When working with lenses, there are two surfaces that affect the light path. For example, a person wearing glasses sees light that travels from an object through the air into the glass lens (first surface). Then the light travels through the glass until it reaches the other side, where again it travels out of the glass and into the air (second surface). The light is refracted twice as it passes from air to lens and from lens back to air.

Thin Spherical Lenses

On the MCAT, lenses generally have negligible thickness. Because light can travel from either side of a lens, a lens has two focal points, with one on each side. The focal length can be measured in either direction from the center. For thin spherical lenses, the focal lengths are equal, so we speak of just one focal length for the lens as a whole.

Figure 8.11a illustrates that a converging lens is always thicker at the center, while Figure 8.11b illustrates that a diverging lens is always thinner at the center. The basic formulas for finding image distance and magnification for spherical mirrors also apply to lenses. The object distance *o*, image distance *i*, focal length *f*, and magnification *m*, are related by the equations $\frac{1}{f} = \frac{1}{o} + \frac{1}{i}$ and $m = -\frac{i}{o}$.

REAL WORLD

Converging lenses (reading glasses) are needed by people who are "farsighted." Diverging lenses (standard glasses) are needed by people who are "nearsighted."

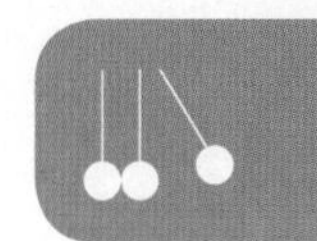

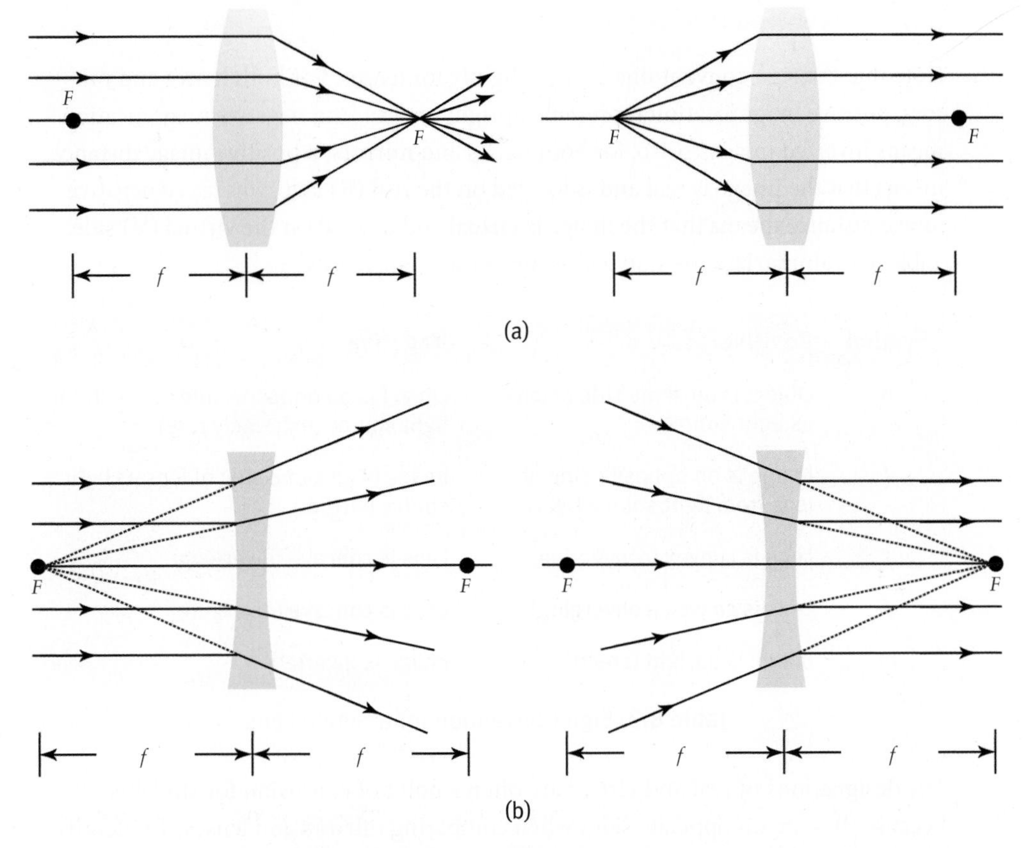

Figure 8.11 Ray Diagrams for Single Lenses
(a) Convex (converging) lenses; (b) Concave (diverging) lenses.

Real Lenses

For lenses where the thickness cannot be neglected, the focal length is related to the curvature of the lens surfaces and the index of refraction of the lens by the **lensmaker's equation**:

$$\frac{1}{f} = (n-1)\left(\frac{1}{r_1} - \frac{1}{r_2}\right)$$

Equation 8.8

where n is the index of refraction of the lens material, r_1 is the radius of curvature of the first lens surface, and r_2 is the radius of curvature of the second lens surface.

The eye is a complex refractive instrument that uses real lenses. The cornea acts as the primary source of refractive power because the change in refractive index from air is so significant. Then, light is passed through an adaptive lens that can change its focal length before reaching the vitreous humor. It is further diffused through layers of retinal tissue to reach the rods and cones. At this point, the image has been focused and minimized significantly, but is still relatively blurry. Our nervous system processes the remaining errors to provide a crisp view of the world.

KEY CONCEPT

To find where the image is (for a lens), draw the following rays and find a point where any two intersect. This point of intersection marks the tip of the image. If the rays you draw do not appear to intersect, extend them to the same side of the lens from which the light came, creating a virtual image.

- Ray parallel to axis → refracts through focal point of front face of the lens
- Ray through or toward focal point before reaching lens → refracts parallel to axis
- Ray to center of lens → continues straight through with no refraction

Sign Conventions for Lenses

Note that the sign conventions change slightly for lenses. For both lenses and mirrors, positive magnification represents upright images, and negative magnification means inverted images. Also, for both lenses and mirrors, a positive image distance means that the image is real and is located on the real (R) side, whereas a negative image distance means that the image is virtual and located on the virtual (V) side. Table 8.3 summarizes the sign conventions for single lenses.

Symbol	Positive	Negative
o	Object is on same side of lens as light source	Object is on opposite side of lens from light source (extremely rare)
i	Image is on opposite side of lens from light source (real)	Image is on same side of lens as light source (virtual)
r	Lens is convex (converging)	Lens is concave (diverging)
f	Lens is convex (converging)	Lens is concave (diverging)
m	Image is upright (erect)	Image is inverted

Table 8.3 Sign Convention for a Single Lens

KEY CONCEPT

It is important to realize that concave mirrors and convex lenses are both converging and thus have similar properties. Convex mirrors and concave lenses are both diverging and also have similar properties.

The designations of real and virtual are often a point of confusion for students because they are on opposite sides when comparing mirrors and lenses. To identify the real side (R), remember that the real side is where light actually goes after interacting with the lens or mirror. For mirrors, light is reflected and, therefore, stays in front of the mirror. Hence, for a mirror, the real side is in front of the mirror, and the virtual side is behind the mirror. For lenses, the convention is different: because light travels through the lens and comes out on the other side, the real side is on the opposite side of the lens from the original light source, and the virtual side is on the same side of the lens as the original light source. Although the object of a single lens is on the virtual side, this does not make the object virtual. Objects are real, with a positive object distance, unless they are placed in certain multiple lens systems in which the image of one lens becomes the object for another (a scenario which is very rarely encountered on the MCAT).

Focal lengths and radii of curvature have a simpler sign convention. For both mirrors and lenses, converging species have positive focal lengths and radii of curvature, and diverging species have negative focal lengths and radii of curvature. Remember that lenses have two focal lengths and two radii of curvature because they have two surfaces. For a thin lens where thickness is negligible, the sign of the focal length and radius of curvature are given based on the first surface the light passes through.

Power

Optometrists often describe a lens in terms of its **power (*P*)**. This is measured in **diopters**, where f (the focal length) is in meters and is given by the equation

$$P = \frac{1}{f}$$

Equation 8.9

P has the same sign as f and is, therefore, positive for a converging lens and negative for a diverging lens. People who are nearsighted (can see near objects clearly) need diverging lenses, while people who are farsighted (can see distant objects clearly) need converging lenses. Bifocal lenses are corrective lenses that have two distinct regions—one that causes convergence of light to correct for farsightedness (**hyperopia**) and a second that causes divergence of light to correct for nearsightedness (**myopia**) in the same lens.

MCAT EXPERTISE

The MCAT does expect students to understand what myopia and hyperopia are, as well as their corresponding ray diagrams and correction strategies.

Multiple Lens Systems

Lenses in contact are a series of lenses with negligible distances between them. These systems behave as a single lens with equivalent focal length given by

$$\frac{1}{f} = \frac{1}{f_1} + \frac{1}{f_2} + \frac{1}{f_3} + \cdots + \frac{1}{f_n}$$

Equation 8.10

Because power is the reciprocal of focal length, the equivalent power is

$$P = P_1 + P_2 + P_3 + \cdots + P_n$$

Equation 8.11

A good example of lenses in contact is a corrective contact lens worn directly on the eye. In this case, the cornea of the eye (a converging lens) is in contact with a contact lens (either converging or diverging, depending on the necessary correction), and their powers would be added.

For lenses not in contact, the image of one lens becomes the object of another lens. The image from the last lens is considered the image of the system. Microscopes and telescopes are good examples of these systems. The magnification for the system is

$$m = m_1 \times m_2 \times m_3 \times \cdots \times m_n$$

Equation 8.12

REAL WORLD

The eye has an optical power of around 60 diopters. Most contact lens wearers have prescriptions between 0.25 and 8 diopters (both positive and negative). Therefore, even at its worst, the human eye can maintain its optical power at about 87% of its maximum.

MCAT EXPERTISE

Because the MCAT is a timed test, you will not be given a massive multiple lens system and be expected to make calculations, although they might test a multiple-lens system conceptually.

Example: An object is 15 cm to the left of a thin diverging lens with a 45 cm focal length as shown below. Find where the image is formed, if it is upright or inverted, and if it is real or virtual. What is the radius of curvature, assuming the lens is symmetrical and is made of glass with a non negligible thickness and an index of refraction of 1.50?

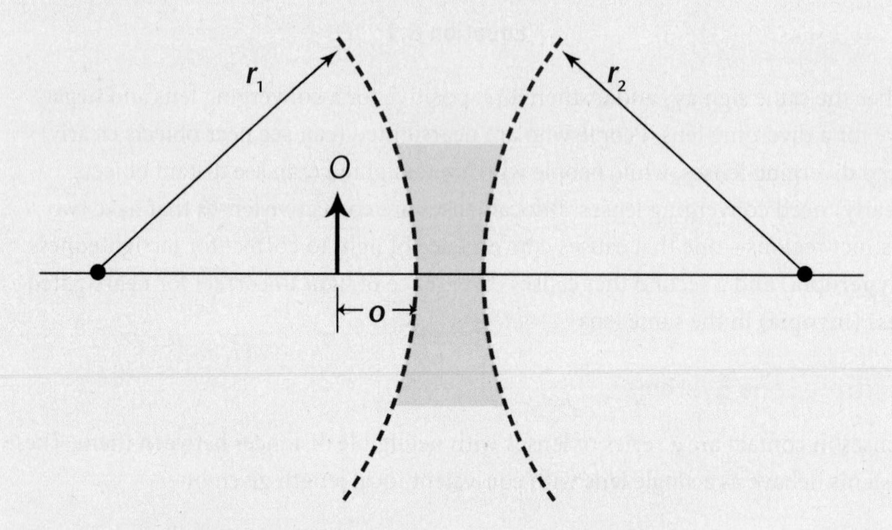

Solution: The image distance (i) is found using the equation

$$\frac{1}{f} = \frac{1}{o} + \frac{1}{i} \rightarrow \frac{1}{i} = \frac{1}{f} - \frac{1}{o}$$

Because the lens is diverging, the focal length has a negative sign ($f = -45$ cm). The object, like any object in a single-lens system, has a positive sign ($o = +15$ cm). Now we can solve for i:

$$\frac{1}{i} = \frac{1}{f} - \frac{1}{o} = \frac{1}{-45\text{ cm}} - \frac{1}{15\text{ cm}} = \frac{-1-3}{45} = \frac{-4}{45}$$
$$i = -\frac{45}{4} = -11.25\text{ cm}$$

The negative sign indicates that the image is on the same side of the light source and is virtual. Remember that for a single lens or mirror, virtual images are always upright.

For a single lens / mirror, virtual images are always upright!

The thickness of a lens is usually negligible, but we are told otherwise in this question. To determine the radii of curvature, we use the lensmaker's equation. Because the lens is symmetrical, the radii are equal but opposite in sign. As the light progresses from left to right, the first surface of the lens is concave ($r_1 < 0$), and the second surface of the lens is convex ($r_2 > 0$).

$$\frac{1}{f} = (n-1)\left(\frac{1}{r_1} - \frac{1}{r_2}\right) = (n-1)\left(-\frac{1}{r} - \frac{1}{r}\right) = (n-1)\left(\frac{-2}{r}\right)$$
$$r = -2f(n-1)$$
$$= -2(-45\text{ cm})(1.5-1) = 2(45)(0.5) = 45\text{ cm}$$

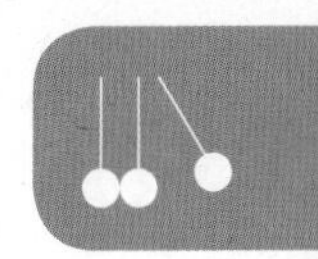

Spherical Aberration

Spherical mirrors and lenses are imperfect. They are therefore subject to specific types of errors or **aberrations**. **Spherical aberration** is a blurring of the periphery of an image as a result of inadequate reflection of parallel beams at the edge of a mirror or inadequate refraction of parallel beams at the edge of a lens. This creates an area of multiple images with very slightly different image distances at the edge of the image, which appears blurry. This phenomenon can be seen in Figure 8.12.

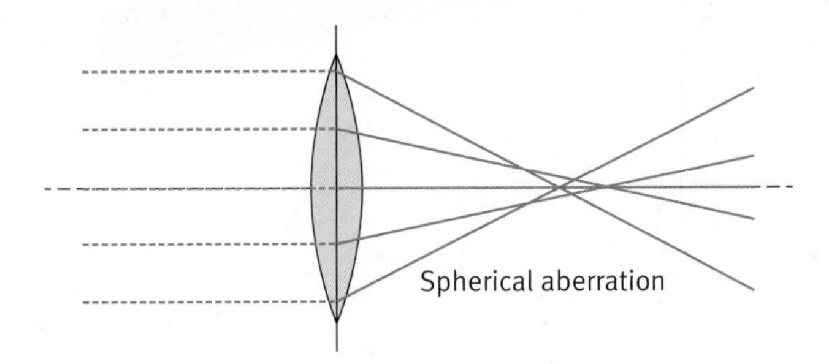

Figure 8.12 Spherical Aberration
Parallel rays are not perfectly reflected or refracted through the focal point, leading to blurriness at the periphery of the image.

Chromatic aberration, discussed below, is predominantly seen in spherical lenses.

REAL WORLD

If you remember back to conic sections from your precalculus class, it should be no surprise that spherical mirrors and lenses do not focus light perfectly. Parabolas are perfect reflectors, meaning that parallel light rays are reflected perfectly through the focal point. This is used in extracorporeal shock wave lithotripsy, in which a parabolic mirror is positioned with a kidney stone at the focal point. Sound waves are reflected off of the mirror and create enough vibration in the kidney stone to shatter it.

Dispersion

As discussed earlier, the speed of light in a vacuum is the same for all wavelengths. However, when light travels through a medium, different wavelengths travel at different speeds. This fact implies that the index of refraction of a medium affects the wavelength of light passing through the medium because the index of refraction is related to the speed of the wave by $n = \frac{c}{v}$. It also implies that the index of refraction itself actually varies with wavelength. When various wavelengths of light separate from each other, this is called **dispersion**. The most common example of dispersion is the splitting of white light into its component colors using a prism.

If a source of white light is incident on one of the faces of a prism, the light emerging from the prism is spread out into a fan-shaped beam, as shown in Figure 8.13. This occurs because violet light has a smaller wavelength than red light and so is bent to a greater extent. Because red experiences the least amount of refraction, it is always on top of the spectrum; violet, having experienced the greatest amount of refraction, is always on the bottom of the spectrum. Note that as light enters a medium with a different index of refraction, the wavelength changes but the frequency of the light does not.

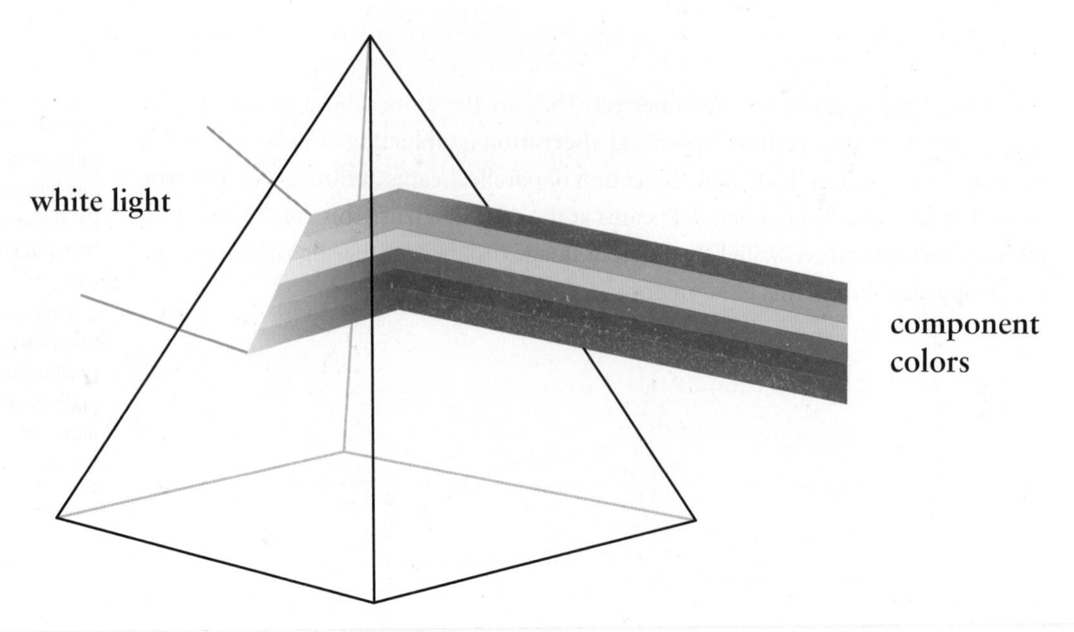

Figure 8.13 Dispersion in a Prism
Due to their different speeds while inside the prism, the various wavelengths of light are refracted to different degrees.

Chromatic Aberration

Chromatic aberration, shown in Figure 8.14, is a dispersive effect within a spherical lens. Depending on the thickness and curvature of the lens, there may be significant splitting of white light, which results in a rainbow halo around images. This phenomenon is corrected for in visual lenses like eyeglasses and car windows with special coatings that have different dispersive qualities from the lens itself.

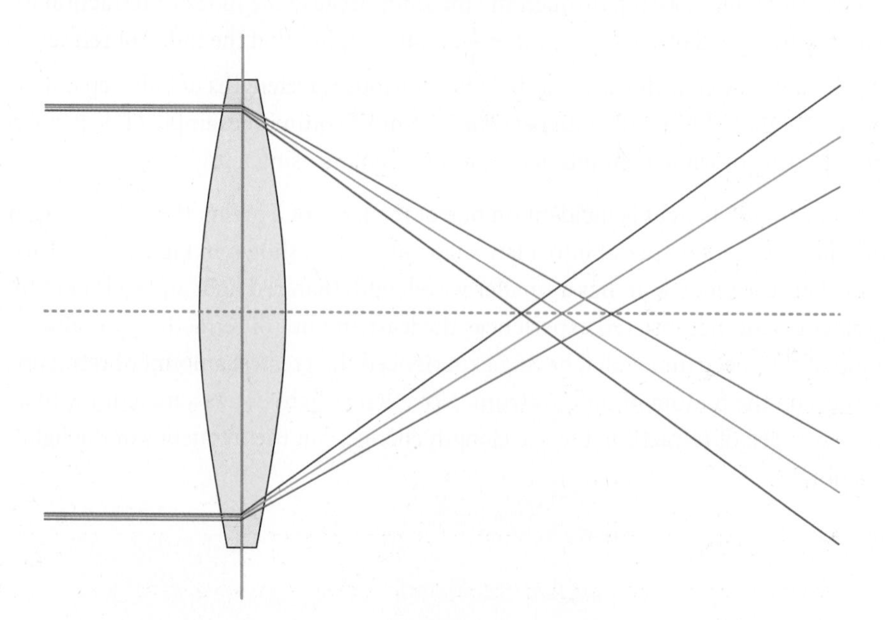

Figure 8.14 Chromatic Aberration
Light dispersion within the glass lens leads to the formation of a rainbow halo at the edge of the image.

MCAT CONCEPT CHECK 8.2

Before you move on, assess your understanding of the material with these questions.

1. Populate the following tables according to the sign conventions for mirrors and lenses:

Mirrors		
Symbol	**Positive**	**Negative**
o		
i		
r		
f		
m		

Lenses		
Symbol	**Positive**	**Negative**
o		
i		
r		
f		
m		

2. True or False: Incident angle is always measured with respect to the normal.

3. Describe the bending of light when moving from a medium with low refractive index to high refractive index, and from a medium with high refractive index to low refractive index:
   - Low *n* to high *n*:
   - High *n* to low *n*:

4. Define the following terms:
   - Dispersion:
   - Aberration:

5. What are the two mathematical relationships between image distance and object distance?

8.3 Diffraction

LEARNING OBJECTIVES

After Chapter 8.3, you will be able to:

- Distinguish between the diffraction patterns in single-slit, double-slit, and slit–lens systems
- Recall the wave phenomena that lead to diffraction fringes
- Describe Young's double-slit experiment:

Diffraction refers to the spreading out of light as it passes through a narrow opening or around an obstacle. Interference between diffracted light rays lead to characteristic fringes in slit–lens and double-slit systems. Diffraction and interference are significant evidence for the wave theory of light.

Single Slit

Although it is usually safe to assume that nonrefracted light travels in a straight line, there are situations where light will not actually travel in a straight-line path. When light passes through a narrow opening (an opening with a size that is on the order of light wavelengths), the light waves seem to spread out (diffract), as is shown in Figure 8.15. As the slit is narrowed, the light spreads out more.

As the slit is narrowed, the light spreads out more!!!

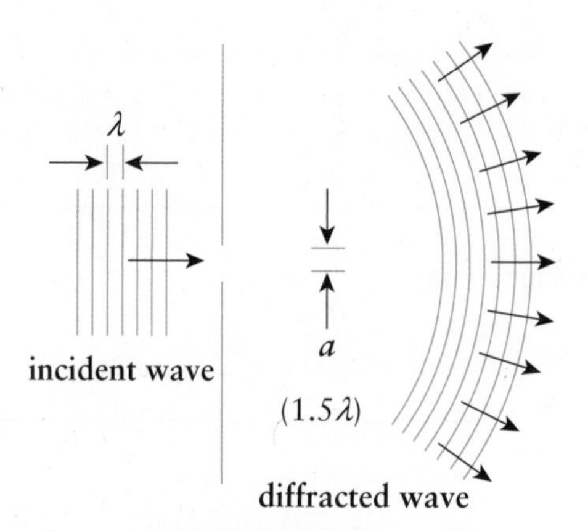

Figure 8.15 Diffraction
Light emerges from a narrow slit in a wide arc, not a narrow beam.

Slit–Lens System

If a lens is placed between a narrow slit and a screen, a pattern is observed consisting of a bright central fringe with alternating dark and bright fringes on each side, as shown in Figure 8.16. The central bright fringe (maximum) is twice as wide as the bright fringes on the sides, and as the slit becomes narrower, the central maximum becomes wider. The location of the dark fringes (minima) is given by the formula

$$a \sin \theta = n\lambda$$

Equation 8.13

where a is the width of the slit, θ is the angle between the line drawn from the center of the lens to the dark fringe and the axis of the lens, n is an integer indicating the number of the fringe, and λ is the wavelength of the incident wave. Note that bright fringes are halfway between dark fringes.

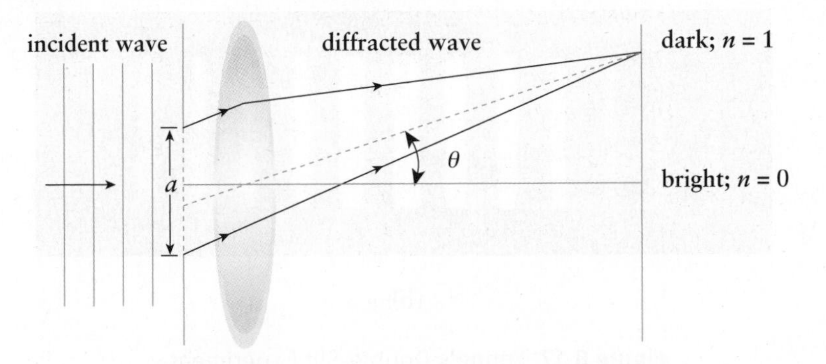

Figure 8.16 Single-Slit Diffraction with Lens

Multiple Slits

When waves interact with each other, the displacements of the waves add together in a process called **interference**, as described in Chapter 7 of *MCAT Physics and Math Review*. In his famous double-slit experiment, Thomas Young showed that the diffracted rays of light emerging from two parallel slits can interfere with one another. This was a landmark finding that contributed to understanding of light as a wave. Figure 8.17 shows the typical setup for Young's double-slit experiment. When monochromatic light (light of only one wavelength) passes through the slits, an interference pattern is observed on a screen placed behind the slits. Regions of constructive interference between the two light waves appear as bright fringes (maxima) on the screen. Conversely, in regions where the light waves interfere destructively, dark fringes (minima) appear.

BRIDGE

Light is similar to other waveforms; it is affected by constructive and destructive interference when light passes through a slit and a lens, and when light passes through multiple slits. Interference also occurs with sound waves, as discussed in Chapter 7 of *MCAT Physics and Math Review*.

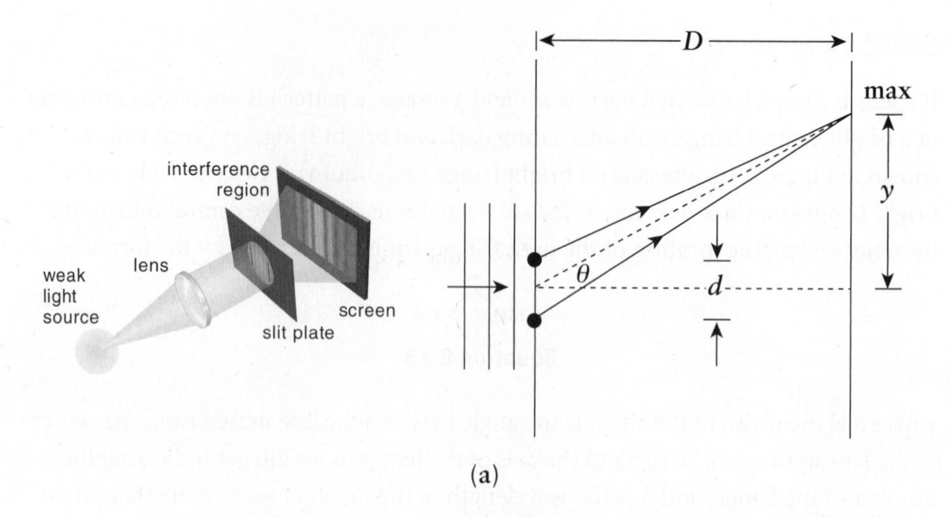

(a)

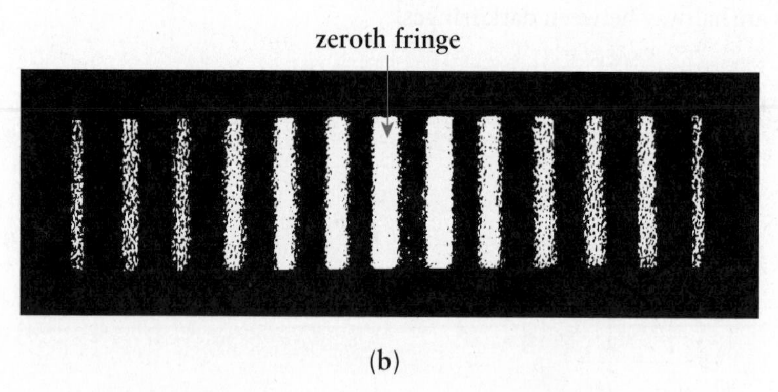

(b)

Figure 8.17 Young's Double-Slit Experiment
(a) Setup for experiment; (b) Interference pattern caused by a double-slit setup.

The positions of dark fringes (minima) on the screen can be found from the equation

$$d \sin \theta = \left(n + \frac{1}{2}\right) \lambda$$

Equation 8.14

where d is the distance between the two slits, θ is the angle between the line drawn from the midpoint between the two slits to the dark fringe and the normal, n is an integer indicating the number of the fringe, and λ is the wavelength of the incident wave. Note that bright fringes are halfway between dark fringes.

Example: In a double-slit experiment, what is the linear distance y between the sixth and eighth minima on the screen? (Note: The wavelength λ is 550 nm, the slits are separated by a distance of 0.14 mm, and the screen is 70 cm from the slits.)

Solution: The position of a dark fringe (minimum) is given by

$$d \sin \theta = \left(n + \frac{1}{2}\right) \lambda$$

We do not know the value of $\sin \theta \left(\frac{\text{opposite}}{\text{hypotenuse}}\right)$. However, for small angles, $\sin \theta \approx \tan \theta$. This is because the length of the hypotenuse is very close to the length of the adjacent side. We do know the value of $\tan \theta \left(\frac{\text{opposite}}{\text{adjacent}}\right)$, so we can substitute it into the equation and still get very close to the correct answer:

$$d \sin \theta = \left(n + \frac{1}{2}\right) \lambda$$

$$d \tan \theta \approx \left(n + \frac{1}{2}\right) \lambda$$

$$d\left(\frac{y}{D}\right) \approx \left(n + \frac{1}{2}\right) \lambda$$

$$y \approx \frac{\left(n + \frac{1}{2}\right) \lambda D}{d}$$

$$y_8 - y_6 \approx \frac{\left(n_8 + \frac{1}{2}\right) \lambda D}{d} - \frac{\left(n_6 + \frac{1}{2}\right) \lambda D}{d}$$

$$\Delta y \approx \frac{(\Delta n)\, \lambda D}{d} = \frac{(2)\left(550 \times 10^{-9}\ \text{m}\right)(0.7\ \text{m})}{0.14 \times 10^{-3}\ \text{m}} = 550 \times 10^{-5}\ \text{m} = 5.5\ \text{mm}$$

Diffraction gratings consist of multiple slits arranged in patterns. Diffraction gratings can create colorful patterns similar to a prism as the different wavelengths interfere in characteristic patterns. For example, the organization of the grooves on a CD or DVD act like a diffraction grating, creating an iridescent rainbow pattern on the surface of the disc. Thin films may also cause interference patterns because light waves reflecting off the external surface of the film interfere with light waves reflecting off the internal surface of the film, as shown in Figure 8.18. Common examples of thin films are soap bubbles or oil puddles in wet parking lots. Note that the interference here is not between diffracted rays, but between reflected rays.

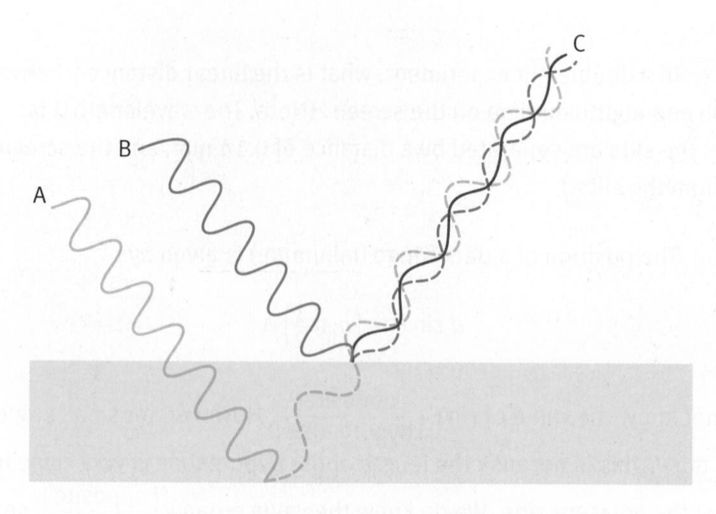

Figure 8.18 Thin Film Interference
Interference patterns, C, occur as light waves reflecting off the external surface of the film, B, interfere with light waves reflecting off the internal surface of the film, A. Note that there would be a small degree of refraction as well, although this is not shown in the image.

BRIDGE

X-ray diffraction and protein crystallography are commonly used to analyze the structure of proteins. These techniques, as well as a number of other protein assays, are discussed in Chapter 3 of *MCAT Biochemistry Review*.

X-Ray Diffraction

X-ray diffraction uses the bending of light rays to create a model of molecules. X-ray diffraction is often combined with protein crystallography during protein analysis. Dark and light fringes do not take on a linear appearance, but rather a complex two dimensional image. An example of an X-ray diffraction pattern is shown in Figure 8.19.

Figure 8.19 X-Ray Diffraction Patterns

MCAT CONCEPT CHECK 8.3

Before you move on, assess your understanding of the material with these questions.

1. How does the diffraction pattern for a single slit differ from a slit with a thin lens?
   - Single slit:

   - Slit–lens system:

   ---

2. What wave phenomenon do diffraction fringes result from?

   ---

3. How does double-slit diffraction and interference differ from single-slit diffraction?
   - Double-slit:

   - Single-slit:

   ---

4. True or False: Maxima in diffraction patterns are always equidistant between two minima.

8.4 Polarization

LEARNING OBJECTIVES

After Chapter 8.4, you will be able to:

- Compare and contrast plane-polarized and circularly polarized light
- Describe how a polarized filter impacts the wavelength and/or frequency of light passing through the filter

Plane-Polarized Light

Plane-polarized (or **linearly polarized**) **light** is light in which the electric fields of all the waves are oriented in the same direction (that is, their electric field vectors are parallel). It follows that their magnetic fields vectors are also parallel, but convention dictates that the plane of the electric field identifies the plane of polarization. Unpolarized light has a random orientation of its electric field vectors; sunlight and light emitted from a light bulb are prime examples. One of the most common applications of plane-polarized light on the MCAT is in the classification of stereoisomers, as discussed in Chapter 2 of *MCAT Organic Chemistry Review*. The optical

REAL WORLD

Plane-polarized light is used to diagnose a number of diseases. Amyloidosis, caused by the buildup of various forms of misfolded proteins, is diagnosed by biopsy and staining the tissue with Congo red stain; a bright "apple green" color is seen under plane-polarized light. Gout (the precipitation of monosodium urate crystals) and pseudogout (the precipitation of calcium pyrophosphate crystals) are differentiated by their precipitate colors under polarized light: monosodium urate appears yellow and calcium pyrophosphate appears blue when the axis of the crystal is aligned with a polarizer.

activity of a compound, due to the presence of chiral centers, causes plane-polarized light to rotate clockwise or counterclockwise by a given number of degrees relative to its concentration (its **specific rotation**). Remember that enantiomers, as nonsuperimposable mirror images, will have opposite specific rotations.

KEY CONCEPT

The electric fields of unpolarized light waves exist in all three dimensions: the direction of the wave's propagation is surrounded by electric fields in every plane perpendicular to that direction. Polarizing light limits the electric field's oscillation to only two dimensions.

There are filters called polarizers, often used in cameras and sunglasses, which allow only light with an electric field pointing in a particular direction to pass through. If one passes a beam of light through a polarizer, it will only let through that portion of the light parallel to the axis of the polarizer. If a second polarizer is then held up to the first, the angle between the polarizers' axes will determine how much light passes through. When the polarizers are aligned, all the light that passes through the first polarizer also passes through the second. When the second polarizer is turned so that its axis is perpendicular, no light gets through at all.

Circular Polarization

Circular polarization is a rarely seen natural phenomenon that results from the interaction of light with certain pigments or highly specialized filters. Circularly polarized light has a uniform amplitude but a continuously changing direction, which causes a helical orientation in the propagating wave, as shown in Figure 8.20. The helix has average electrical field vectors and magnetic field vectors that lie perpendicular to one another, like other waves, with maxima that fall on the outer border of the helix.

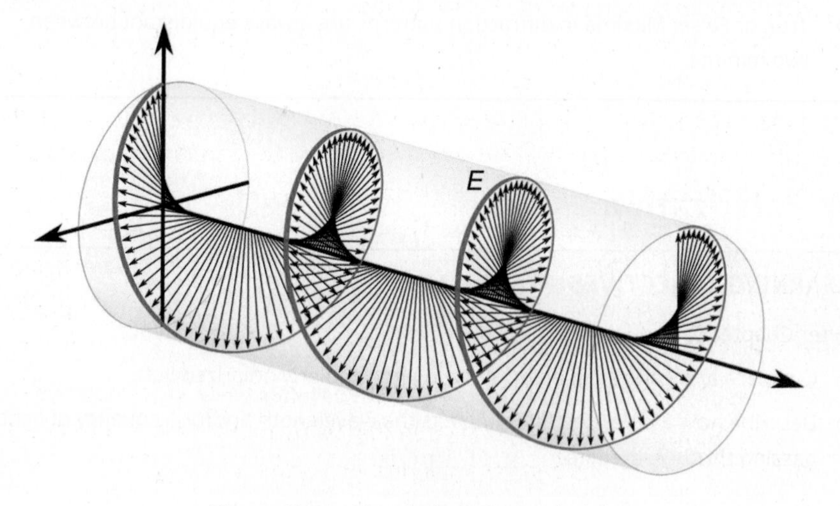

Figure 8.20 Circularly Polarized Light

MCAT CONCEPT CHECK 8.4

Before you move on, assess your understanding of the material with these questions.

1. Contrast plane-polarized and circularly polarized light:
   - Plane-polarized:

     ____

   - Circularly polarized:

     ____

2. How does the application of a polarized filter impact the wavelength of light passing through the filter?

   ____

Conclusion

This chapter illuminated the key behaviors and characteristics of light and optical systems. First, we described the nature of the electromagnetic (EM) wave, noting that we can only perceive light in the visible range (400 nm–700 nm). We then focused on geometrical optics to consider the reflective and refractive behaviors of light, noting the ways in which mirrors reflect light to produce images and lenses refract light to produce images. We acknowledged the fact that light doesn't always travel in straight-line pathways but can bend and spread out through diffraction. We examined the pattern of interference that occurs when light passes through a double slit, as demonstrated in Young's double-slit experiment. Finally, we wrapped up with a discussion on plane-polarized and circularly polarized light. In this chapter, we considered the properties that support the wave theory of light. In the next chapter, we'll explore the photon and properties that support the particle theory of light, as well as other atomic and nuclear phenomena.

GO ONLINE

You've reviewed the content, now test your knowledge and critical thinking skills by completing a test-like passage set in your online resources!

Concept Summary

Electromagnetic Spectrum

- **Electromagnetic waves** are transverse waves that consist of an oscillating electric field and an oscillating magnetic field.
- The two fields are perpendicular to each other and to the direction of propagation of the wave.
- The **electromagnetic spectrum** is the range of frequencies and wavelengths found in EM waves.
- The EM spectrum includes, from lowest to highest energy, **radio waves, microwaves, infrared**, visible light, **ultraviolet**, **X-rays**, and **γ-rays**.
- The **visible spectrum** runs from approximately 400 nm (violet) to 700 nm (red).

Geometrical Optics

- **Reflection** is the rebounding of incident light waves at the boundary of a medium.
- The **law of reflection** states that the incident angle will equal the angle of reflection, as measured from the normal.
- **Spherical mirrors** have **centers** and **radii of curvature**, as well as **focal points**.
  - **Concave** mirrors are **converging** systems and can produce **real, inverted** images or **virtual, upright** images, depending on the placement of the object relative to the focal point.
  - **Convex** mirrors are **diverging** systems and will only produce virtual, upright images.
  - **Plane mirrors** also produce virtual, upright images; these images are always the same size as the object. They may be thought of as spherical mirrors with infinite radii of curvature.
- **Refraction** is the bending of light as it passes from one medium to another.
- The speed of light changes depending on index of refraction of the medium. This speed change causes refraction.
- The amount of refraction depends on the wavelength of the light involved; this behavior causes **dispersion** of light through a prism.
- **Snell's law** (the law of refraction) states that there is an inverse relationship between the index of refraction and the sine of the angle of refraction (measured from the normal).
- **Total internal reflection** occurs when light cannot be refracted out of a medium and is instead reflected back inside the medium.
  - This happens when light moves from a medium with a higher index of refraction to a medium with a lower index of refraction with a high incident angle.
  - The minimum incident angle at which total internal reflection occurs is called the **critical angle**.

- **Lenses** refract light to form images of objects.
    - Thin symmetrical lenses have focal points on each side.
    - Convex lenses are converging systems and can produce real, inverted images or virtual, upright images.
    - Concave lenses are diverging systems and will only produce virtual, upright images.
    - Lenses with non negligible thickness require use of the **lensmaker's equation.**
- The following table summarizes image creation in converging and diverging systems for both mirrors and lenses:

	Converging Systems					Diverging Systems
o relative to f	$o > 2f$	$o = 2f$	$2f > o > f$	$o = f$	$o < f$	all object distances
image	real, inverted, reduced	real, inverted, same	real, inverted, magnified	no image	virtual, upright, magnified	virtual, upright, reduced

Diffraction

- **Diffraction** is the bending and spreading out of light waves as they pass through a narrow slit.
- Diffraction may produce a large central light fringe surrounded by alternating light and dark fringes with the addition of a lens.
- Interference supports the wave theory of light.
- **Young's double-slit experiment** shows the constructive and destructive interference of waves that occur as light passes through parallel slits, resulting in minima (dark fringes) and maxima (bright fringes) of intensity.

Polarization

- In **plane-polarized light**, all of the light rays have electric fields with parallel orientation.
- Plane-polarized light is created by passing unpolarized light through a **polarizer**.
- In **circularly polarized light**, all of the light rays have electric fields with equal intensity but constantly rotating direction.
- Circularly polarized light is created by exposing unpolarized light to special pigments or filters.

Answers to Concept Checks

8.1

1. γ-rays > X-rays > ultraviolet > visible light > infrared > microwaves > radio. Frequency follows the same trend as energy, whereas wavelength follows the opposite trend.
2. False. Light waves are transverse because the direction of propagation is perpendicular to the direction of oscillation.
3. Visible light ranges from wavelengths of about 400 nm to 700 nm. This is in comparison to the entire EM spectrum which ranges from wavelengths of nearly 0 to 10^9 m.

8.2

1.

Mirrors		
Symbol	**Positive**	**Negative**
o	Object is in front of mirror	Object is behind mirror (extremely rare)
i	Image is in front of mirror (real)	Image is behind mirror (virtual)
r	Mirror is concave (converging)	Mirror is convex (diverging)
f	Mirror is concave (converging)	Mirror is convex (diverging)
m	Image is upright (erect)	Image is inverted
Lenses		
Symbol	**Positive**	**Negative**
o	Object is on same side of lens as light source	Object is on opposite side of lens from light source (extremely rare)
i	Image is on opposite side of lens from light source (real)	Image is on same side of lens as light source (virtual)
r	Lens is convex (converging)	Lens is concave (diverging)
f	Lens is convex (converging)	Lens is concave (diverging)
m	Image is upright (erect)	Image is inverted

2. True. In optics, incident angles are always measured relative to the normal.
3. Light will bend toward the normal when going from a medium with low n to high n. Light will bend away from the normal when going from a medium with high n to low n; if the incident angle is larger than the critical angle (θ_c), total internal reflection will occur.

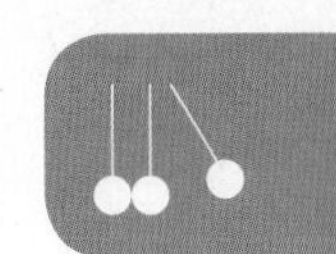

4. Dispersion is the tendency for different wavelengths of light to experience different degrees of refraction in a medium, leading to separation of light into the visible spectrum (a rainbow). Aberration (spherical or chromatic) is the alteration or distortion of an image as a result of an imperfection in the optical system.
5. $\frac{1}{f} = \frac{1}{o} + \frac{1}{i}$ and $m = -\frac{i}{o}$

8.3

1. Diffraction through a single slit does not create characteristic fringes when projected on a screen, although the light does spread out. When a lens is introduced into the system, the additional refraction of light causes constructive and destructive interference, creating fringes.
2. Fringes result from constructive and destructive interference between light rays.
3. The image formed during double-slit diffraction contains fringes because light rays constructively and destructively interfere. A single slit forms an image of a wide band of light, spread out from its original beam.
4. True. Maxima and minima alternate in a diffraction pattern. A maximum is equidistant between two minima, and a minimum is equidistant between two maxima.

8.4

1. Plane-polarized light contains light waves with parallel electric field vectors. Circularly polarized light selects for a given amplitude and has a continuously rotating electric field direction.
2. Plane polarization has no effect on the wavelength (or frequency or speed) of light. Polarization *does* affect the amount of light passing through a medium and light intensity.

Science Mastery Assessment Explanations

1. **C**

It is unnecessary to memorize the entire electromagnetic spectrum for Test Day; however, it is important to know that the visible spectrum runs from 400–700 nm. We can calculate the wavelength of this light ray:

$$c = f\lambda$$

$$\lambda = \frac{c}{f} = \frac{3\times10^{8}\ \frac{\text{m}}{\text{s}}}{5\times10^{14}\ \text{Hz}} = 6\times10^{-7}\ \text{m} = 600\ \text{nm}$$

This wavelength falls within the visible spectrum and has a yellow-orange color.

2. **D**

The color of an object is determined by the wavelength of light reflected by that object. Since anthocyanin pigment appears red, it must reflect red light. The frequency of light can be calculated using the equation $c = f\lambda$. Red light has a wavelength of 700 nm, which can be converted to 700×10^{-9} m. Plugging in values yields: $f = (3 \times 10^{8}) / (700 \times 10^{-9}) = 4.2 \times 10^{13}$ Hz, matching **(D)**. Note that a close eye for units would enable the immediate elimination of **(A)**.

3. **A**

One could solve this question with a ray diagram, but be wary about using ray diagrams on Test Day. It is easy to make small mistakes that cause the light rays not to intersect. Therefore, solve the question using the sign convention. If the object is at the center of curvature, its distance is $2f$. We can plug into the optics equation:

$$\frac{1}{f} = \frac{1}{o} + \frac{1}{i} \rightarrow \frac{1}{i} = \frac{1}{f} - \frac{1}{o} = \frac{1}{f} - \frac{1}{2f} = \frac{1}{2f}$$

$$i = 2f = r$$

Because i is positive, the image is real. For single mirrors or lenses, all real images are inverted.

4. **B**

To calculate the wavelength, use the formula: $d\sin(\theta) = (n + 1/2)\lambda$. Rearrange the formula to solve for wavlength: $\lambda = d\sin(\theta)/(n + 1/2)$. Since the distance is given in millimeters, first convert to meters; $d = 3 \times 10^{-4}$ m. Since the question asks for the second dark fringe, we have $n = 2$. Plugging in the given values yields: $\lambda = (3 \times 10^{-4})\sin(30°)/(2 + 1/2) = 6 \times 10^{-5}$ m, matching choice **(B)**.

5. **A**

This question contains two parts—we have to determine the frequency and the angle of refraction of the light ray. The first part, however, is straightforward because the frequency of a light ray traveling from one medium to another does not change. Because the frequency must be 5×10^{14} Hz, we can eliminate **(C)** and **(D)**. For the angle of refraction, we can either calculate it or determine it using logic. First, the light ray goes from air into crystal; that is, from a low index of refraction to a higher one. According to Snell's law, the angle of refraction will be smaller than the incident angle (closer to the normal). When the light ray moves from crystal to chromium, it again goes from a lower index of refraction into a higher one, thus making the angle of refraction even smaller, eliminating **(B)**. This question could also be answered by calculation using Snell's law, but the calculations are time consuming and unnecessary.

6. **D**

Plane-polarized light is light in which the electric fields of all the waves are oriented in the same direction. Light passing through the first two polarizers will only contain rays with their electric field vectors in the same direction. When it reaches the third polarizer, however, the light will not be able to pass through because all the light rays will be oriented in the direction dictated by the first and second polarizers.

7. **C**

Plane polarizers only allow one specific orientation of the electric field of light to pass through. When plane polarizers are perpendicular to each other, no light can pass through the second polarizer, which supports **(C)** as the correct answer. Note that if two polarizers are aligned, all of the light that passes through the first will also pass through the second; there is no "twice as plane polarized," eliminating **(B)**.

8. **B**

The image produced by a convex lens can be either real or virtual. It is real if the object is placed at a distance greater than the focal point, and virtual if the object is placed at a distance less than the focal point (between the focal point and the lens). Remember that for a single mirror or lens, an image that is real must be inverted and one that is virtual must be upright. In this question, the object is placed in front of the focal point, so the image must be virtual and, therefore, upright. We could also determine this from the optics equation. If $f > o$, then $\frac{1}{f} - \frac{1}{o}$ is negative, and i is therefore negative (virtual).

9. **D**

This question is testing our understanding of total internal reflection. As the laser beam travels from water to air—that is, from a higher to a lower index of refraction—the angle of refraction increases. At the critical angle (θ_c), the angle of refraction becomes 90°; at this point, the refracted ray is parallel to the surface of the water. When the incident angle is greater than the critical angle, all the light is reflected back into the water. The question is asking for the critical angle:

$$\theta_c = \sin^{-1}\frac{n_2}{n_1} = \sin^{-1}\frac{1}{1.33} = \sin^{-1}0.75$$

The inverse sine of 0.75 must be slightly higher than $45°\left(\sin 45° = \frac{\sqrt{2}}{2} = 0.707\right)$. 48.59° is the exact answer.

10. **C**

This question is testing our understanding of diffraction. When light passes through a narrow opening, the light waves spread out; as the slit narrows, the light waves spread out even more. When a lens is placed between the narrow slit and the screen, a pattern consisting of alternating bright and dark fringes can be observed on the screen. As the slit becomes narrower, the central maximum (the brightest and most central fringe) becomes wider. This can also be seen in the equation for the position of dark fringes in a slit–lens setup ($a \sin\theta = n\lambda$). As a, the width of the slit, decreases, $\sin\theta$ must increase because $n\lambda$ is constant for a given fringe. If $\sin\theta$ increases, θ necessarily increases, implying that the fringes are spreading farther apart.

11. **D**

All images produced by plane mirrors will be virtual, so statement III is true. The same goes for diverging species (convex mirrors and concave lenses), so statement II is true. Converging species (concave mirrors and convex lenses) can produce real or virtual images, depending on how far the object is from the species, so statement I is also true.

12. **A**

First, the color of the light is irrelevant here; the ratio would be the same even if the specific color were not mentioned. Second, recall Snell's Law: $n_1 \sin\theta_1 = n_2 \sin\theta_2$. Although we don't know the value of n for either medium, you do know the simple relationship $n = \frac{c}{v}$. Replacing n in Snell's law, and canceling out c from both sides, we get:

$$\frac{c}{v_1}\sin\theta_1 = \frac{c}{v_2}\sin\theta_2$$

$$\frac{\sin 30°}{v_1} = \frac{\sin 45°}{v_2}$$

$$\frac{1}{2v_1} = \frac{\sqrt{2}}{2v_2}$$

$$v_2 = v_1\sqrt{2}$$

13. **D**

The overall magnification of a system of multiple lenses is simply the product of each lens's magnification. In this case, that is $10 \times 40 = 400$.

14. **D**

Drawing a diagram is best here. Because the angle given is with respect to the normal, you know that the incident angle must equal 60°. You know that the reflected beam will have an angle of 60° relative to the normal. Therefore, the reflected beam will make an angle of 30° with the plane of the glass. If the reflected and refracted beams are perpendicular to each other, the refracted beam will make a 60° angle with the plane of the glass. $\theta_{refracted}$ is therefore 30° relative to the normal.

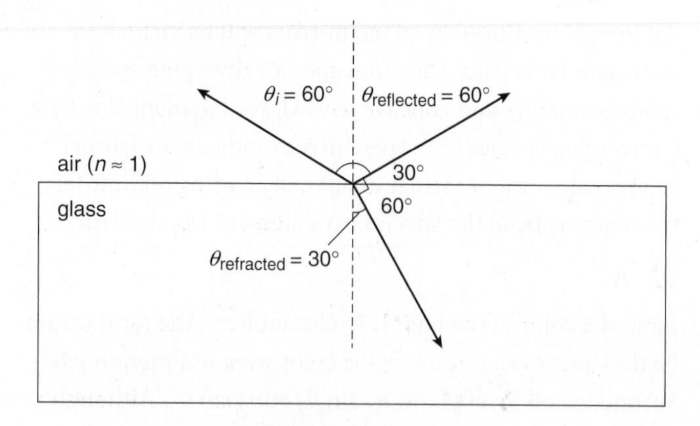

Using $n_1 \sin \theta_1 = n_2 \sin \theta_2$, we have

$$1 \sin 60° = n_2 \sin 30°$$

$$\frac{\sqrt{3}}{2} = n_2 \left(\frac{1}{2}\right)$$

$$\sqrt{3} = n_2$$

15. **D**

Light can be split into its component colors by dispersion, such as that through a prism, eliminating **(A)**. Diffraction by a diffraction grating will also separate colors by their wavelengths, eliminating **(B)**. The refraction of light within a thin film also leads to light dispersion as the different colors are refracted at slightly different angles in the film, eliminating **(C)**. A mirror with significant aberration could lead to a separation of light into its component colors, but we are told that this is an ideal mirror. Thus, **(D)** is the correct answer.

Consult your online resources for additional practice. GO ONLINE

Equations to Remember

(8.1) **Speed of light from frequency and wavelength:** $c = f\lambda$

(8.2) **Law of reflection:** $\theta_1 = \theta_2$

(8.3) **Optics equation:** $\frac{1}{f} = \frac{1}{o} + \frac{1}{i} = \frac{2}{r}$

(8.4) **Magnification:** $m = -\frac{i}{o}$

(8.5) **Index of refraction:** $n = \frac{c}{v}$

(8.6) **Snell's law:** $n_1 \sin \theta_1 = n_2 \sin \theta_2$

(8.7) **Critical angle:** $\theta_c = \sin^{-1}\left(\frac{n_2}{n_1}\right)$

(8.8) **Lensmaker's equation:** $\frac{1}{f} = (n-1)\left(\frac{1}{r_1} - \frac{1}{r_2}\right)$

(8.9) **Power:** $P = \frac{1}{f}$

(8.10) **Focal length of multiple lens system:** $\frac{1}{f} = \frac{1}{f_1} + \frac{1}{f_2} + \frac{1}{f_3} + \cdots + \frac{1}{f_n}$

(8.11) **Power of multiple lens system:** $P = P_1 + P_2 + P_3 + \cdots + P_n$

(8.12) **Magnification of multiple lens system:** $m = m_1 \times m_2 \times m_3 \times \cdots \times m_n$

(8.13) **Positions of dark fringes in slit–lens setup:** $a \sin \theta = n\lambda$

(8.14) **Positions of dark fringes in double-slit setup:** $d \sin \theta = \left(n + \frac{1}{2}\right)\lambda$

Shared Concepts

Behavioral Sciences Chapter 2
Sensation and Perception

Biochemistry Chapter 3
Nonenzymatic Protein Function and Protein Analysis

Organic Chemistry Chapter 2
Isomers

Organic Chemistry Chapter 11
Spectroscopy

Physics and Math Chapter 7
Waves and Sound

Physics and Math Chapter 9
Atomic and Nuclear Phenomena

CHAPTER 9

Atomic and Nuclear Phenomena

SCIENCE MASTERY ASSESSMENT

Every pre-med knows this feeling: there is so much content I have to know for the MCAT! How do I know what to do first or what's important?

While the high-yield badges throughout this book will help you identify the most important topics, this Science Mastery Assessment is another tool in your MCAT prep arsenal. This quiz (which can also be taken in your online resources) and the guidance below will help ensure that you are spending the appropriate amount of time on this chapter based on your personal strengths and weaknesses. Don't worry though—skipping something now does not mean you'll never study it. Later on in your prep, as you complete full-length tests, you'll uncover specific pieces of content that you need to review and can come back to these chapters as appropriate.

How to Use This Assessment

If you answer 0–7 questions correctly:

Spend about 1 hour to read this chapter in full and take limited notes throughout. Follow up by reviewing **all** quiz questions to ensure that you now understand how to solve each one.

If you answer 8–11 questions correctly:

Spend 20–40 minutes reviewing the quiz questions. Beginning with the questions you missed, read and take notes on the corresponding subchapters. For questions you answered correctly, ensure your thinking matches that of the explanation and you understand why each choice was correct or incorrect.

If you answer 12–15 questions correctly:

Spend less than 20 minutes reviewing all questions from the quiz. If you missed any, then include a quick read-through of the corresponding subchapters, or even just the relevant content within a subchapter, as part of your question review. For questions you got correct, ensure your thinking matches that of the explanation and review the Concept Summary at the end of the chapter.

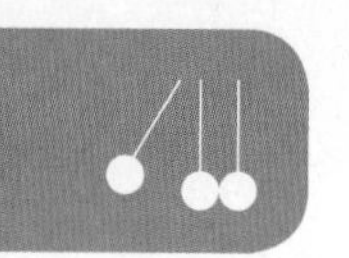

1. If the work function of a metal is 6.622×10^{-20} J and a ray of electromagnetic radiation with a frequency of 1.0×10^{14} Hz is incident on the metal, what will be the speed of the electrons ejected from the metal? (Note: $h = 6.626 \times 10^{-34}$ J·s and $m_{e^-} = 9.1 \times 10^{-31}$ kg)
   A. $2.62 \times 10^{-6} \frac{m}{s}$
   B. $1.07 \times 10^{-4} \frac{m}{s}$
   C. $9.38 \times 10^{3} \frac{m}{s}$
   D. $3.81 \times 10^{5} \frac{m}{s}$

2. What is the wavelength of a photon that causes an electron to be emitted from a metal with a kinetic energy of 50 J? (Note: The work function of the metal is 16 J, and $h = 6.626 \times 10^{-34}$ J·s)
   A. 1.0×10^{-34} m
   B. 3.0×10^{-27} m
   C. 3.0×10^{-26} m
   D. 1.0×10^{35} m

3. Which of the following statements is inconsistent with the Bohr model of the atom?
   A. Energy levels of the electron are stable and discrete.
   B. An electron emits or absorbs radiation only when making a transition from one energy level to another.
   C. To jump from a lower energy to a higher energy orbit, an electron must absorb a photon of precisely the right frequency such that the photon's energy equals the energy difference between the two orbits.
   D. To jump from a higher energy to a lower energy orbit, an electron absorbs a photon of a frequency such that the photon's energy is exactly the energy difference between the two orbits.

4. When a hydrogen atom electron falls to the ground state from the $n = 2$ state, 10.2 eV of energy is emitted. What is the wavelength of this radiation? (Note: 1 eV $= 1.60 \times 10^{-19}$ J, and $h = 6.626 \times 10^{-34}$ J·s)
   A. 5.76×10^{-9} m
   B. 1.22×10^{-7} m
   C. 3.45×10^{-7} m
   D. 2.5×10^{15} m

5. The figure below illustrates an electron with initial energy of –10 eV moving from point A to point B. What change accompanies the movement of the electron?

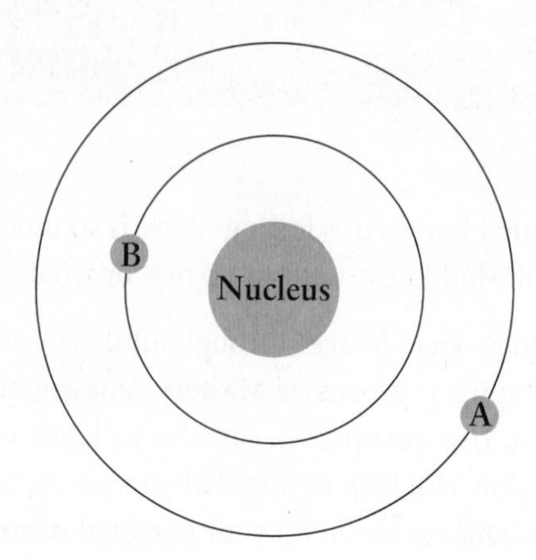

   A. Absorption of a photon
   B. Emission of a photon
   C. Decrease in the atom's work function
   D. Increase in the atom's total energy

6. Which of the following four fundamental forces is primarily responsible for holding the nucleons together?
   A. Binding energy
   B. Strong nuclear force
   C. Electrostatic force
   D. Gravitational force

7. All of the following statements about the photoelectric effect are true EXCEPT:
   A. the intensity of the light beam does not affect the photocurrent.
   B. the kinetic energies of the emitted electrons do not depend on the light intensity.
   C. a weak beam of light of frequency greater than the threshold frequency yields more current than an intense beam of light of frequency lower than the threshold frequency.
   D. for light of a given frequency, the kinetic energy of emitted electrons increases as the value of the work function decreases.

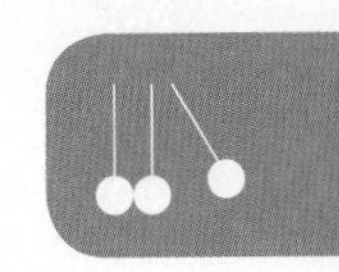

8. What is the binding energy of the argon-40 isotope in MeV? (Note: $m_{proton} = 1.0073$ amu, $m_{neutron} = 1.0087$ amu, $m_{Ar\text{-}40\ nucleus} = 39.9132$ amu, $c^2 = 932\ \frac{MeV}{amu}$)
   - **A.** 0.4096 MeV
   - **B.** 40.3228 MeV
   - **C.** 381.7 MeV
   - **D.** 643.8 MeV

9. Which of the following correctly identifies the following process?

$$^{67}_{31}\text{Ga} + e^- \rightarrow {}^{67}_{30}\text{Zn}$$

   - **A.** β^- decay
   - **B.** β^+ decay
   - **C.** e^- capture
   - **D.** γ decay

10. Consider the following fission reaction.

$$\underset{1.0087}{^{1}_{0}n} + \underset{10.0129}{^{10}_{5}\text{B}} \rightarrow \underset{7.0160}{^{7}_{3}\text{Li}} + \underset{4.0026}{^{4}_{2}\text{He}}$$

The masses of the species involved are given in atomic mass units below each species, and 1 amu can create 932 MeV of energy. What is the energy liberated due to transformation of mass into energy during this reaction?
   - **A.** 0.003 MeV
   - **B.** 1.4 MeV
   - **C.** 2.8 MeV
   - **D.** 5.6 MeV

11. Element X is radioactive and decays via α decay with a half-life of four days. If 12.5 percent of an original sample of element X remains after *n* days, what is the value of *n*?
   - **A.** 4
   - **B.** 8
   - **C.** 12
   - **D.** 16

12. A graph of an exponential decay process is created. The *y*-axis is the natural logarithm of the ratio of the number of intact nuclei at a given time to the number of intact nuclei at time $t = 0$. The *x*-axis is time. What does the slope of such a graph represent?
   - **A.** λ
   - **B.** $-\lambda$
   - **C.** $e^{-\lambda t}$
   - **D.** $\frac{n}{n_0}$

13. A certain carbon nucleus dissociates completely into α-particles. How many particles are formed?
   - **A.** 1
   - **B.** 2
   - **C.** 3
   - **D.** 4

14. The half-life of carbon-14 is approximately 5,730 years, while the half-life of carbon-12 is essentially infinite. If the ratio of carbon-14 to carbon-12 in a certain sample is 25% less than the normal ratio in nature, how old is the sample?
   - **A.** Less than 5,730 years
   - **B.** Approximately 5,730 years
   - **C.** Significantly greater than 5,730 years, but less than 11,460 years
   - **D.** Approximately 11,460 years

15. A nuclide undergoes two alpha decays, two positron decays, and two gamma decays. What is the difference between the atomic number of the parent nuclide and the atomic number of the daughter nuclide?
   - **A.** 0
   - **B.** 2
   - **C.** 4
   - **D.** 6

Answer Key

1. **C** (Ch. 9.1)
2. **B** (Ch. 9.1)
3. **D** (Ch. 9.2)
4. **B** (Ch. 9.2)
5. **B** (Ch. 9.2)
6. **B** (Ch. 9.3)
7. **A** (Ch. 9.1)
8. **C** (Ch. 9.3)
9. **C** (Ch. 9.4)
10. **C** (Ch. 9.4)
11. **C** (Ch. 9.4)
12. **B** (Ch. 9.4)
13. **C** (Ch. 9.4)
14. **A** (Ch. 9.4)
15. **D** (Ch. 9.4)

CHAPTER 9

Atomic and Nuclear Phenomena

In This Chapter

9.1 **The Photoelectric Effect**
- Threshold Frequency 330
- Kinetic Energy of Ejected Electrons 331

9.2 **Absorption and Emission of Light** 334

9.3 **Nuclear Binding Energy and Mass Defect** 336

9.4 **Nuclear Reactions** HY
- Fusion 339
- Fission 340
- Radioactive Decay 341

Concept Summary 349

CHAPTER PROFILE

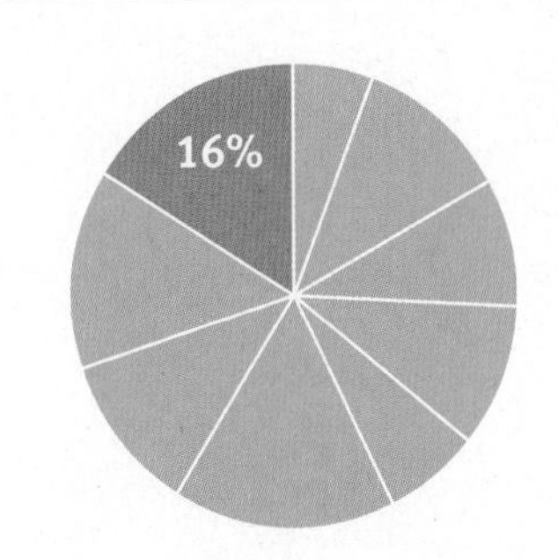

The content in this chapter should be relevant to about 16% of all questions about physics on the MCAT.

This chapter covers material from the following AAMC content categories:

4D: How light and sound interact with matter

4E: Atoms, nuclear decay, electronic structure, and atomic chemical behavior

Introduction

All of life depends on the photoelectric effect. As a photon of light enters the chloroplast in a plant cell, it reacts with chlorophyll, causing the ejection of an electron from certain magnesium-containing dyes. This electron feeds into synthetic pathways that ultimately result in glucose production. While the MCAT does not include photosynthesis in its content lists, its principles are a primary example of the photoelectric effect. It was Albert Einstein who described this effect, and it was this that won him the Nobel Prize—not the theory of relativity. We now use the photoelectric effect in many industrial applications, such as solar panels.

After discussing the photoelectric effect, we will examine nuclear radiation. Nuclear radiation is curiously full of opposites: it can cause life-threatening diseases such as cancer, but it can also be used in the treatment of cancer. It can be used safely for mass power generation, but it can cause untold devastation in meltdowns or weapons of mass destruction. In addition to nuclear radiation, we will examine the strong nuclear force and the equation of mass defect, perhaps the most quoted equation in all of science. At the end of this chapter, we'll have covered all of the physics content tested on the MCAT, and will be ready to move on to mathematics and some skills-based practice.

9.1 The Photoelectric Effect

LEARNING OBJECTIVES

After Chapter 9.1, you will be able to:

- Relate the work function to the energy needed to emit an electron from a metal
- Recall the factor(s) that threshold frequency depends on
- Recognize the phenomena that result from the application of the photoelectric effect:

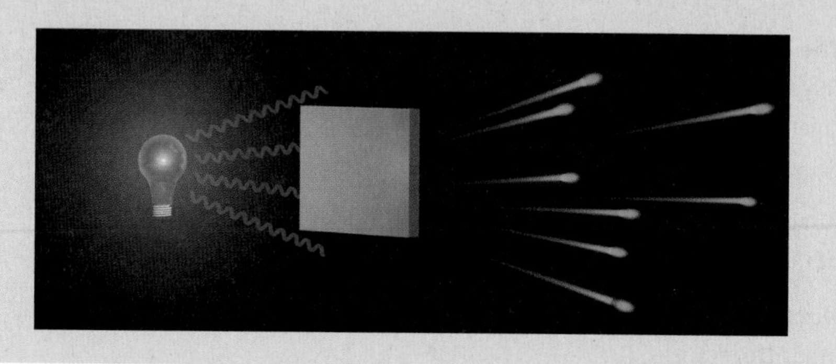

When light of a sufficiently high frequency (typically, blue to ultraviolet light) is incident on a metal in a vacuum, the metal atoms emit electrons. This phenomenon, discovered by Heinrich Hertz in 1887, is called the **photoelectric effect**. As mentioned earlier, Albert Einstein's 1905 explanation of the photoelectric effect won him the Nobel Prize.

Electrons liberated from the metal by the photoelectric effect will produce a net charge flow per unit time, or **current**. Provided that the light beam's frequency is above the threshold frequency of the metal, light beams of greater intensity produce larger current in this way. The higher the intensity of the light beam, the greater the number of photons per unit time that fall on an electrode, producing a greater number of electrons per unit time liberated from the metal. When the light's frequency is above the threshold frequency, the magnitude of the resulting current is directly proportional to the intensity (and amplitude) of the light beam.

Threshold Frequency

The minimum frequency of light that causes ejection of electrons is known as the **threshold frequency,** f_T. The threshold frequency depends on the type of metal being exposed to the radiation. The photoelectric effect is, for all intents and purposes, an "all-or-nothing" response: if the frequency of the incident photon is less than the threshold frequency ($f < f_T$), then no electron will be ejected because the photons do not have sufficient energy to dislodge the electron from its atom. But if the frequency of the incident photon is greater than the threshold frequency ($f > f_T$), then an electron will be ejected, and the maximum kinetic energy of the ejected electron will be equal to the difference between hf and hf_T (also called the

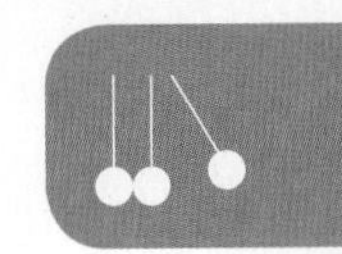

work function). Einstein's explanation of these results was that the light beam consists of an integral number of light quanta called **photons.** The energy of each photon is proportional to the frequency of the light:

$$E = hf$$

Equation 9.1

where E is the energy of the photon of light, h is Planck's constant (6.626×10^{-34} J·s), and f is the frequency of the light. Once we know the frequency, we can easily find the wavelength λ according to the equation $c = f\lambda$, as described in Chapter 8 of *MCAT Physics and Math Review.* According to these equations, waves with higher frequency have shorter wavelengths and higher energy (toward the blue and ultraviolet end of the spectrum); waves with lower frequency have longer wavelengths and lower energy (toward the red and infrared end of the spectrum). In nuclear physics, wavelength is commonly measured in nanometers ($1\text{ nm} = 10^{-9}\text{ m}$) and ångströms ($1\text{ Å} = 10^{-10}\text{ m}$).

KEY CONCEPT

The energy of a photon increases with increasing frequency. The reason that we only discuss electrons being ejected from metals (and not protons or neutrons) is because of the weak hold that metals have on their valence electrons due to their low ionization energies.

Kinetic Energy of Ejected Electrons

If the frequency of a photon of light incident on a metal is at the threshold frequency for the metal, the electron barely escapes from the metal. However, if the frequency of an incident photon is above the threshold frequency of the metal, the photon will have more than enough energy to eject a single electron, and the excess energy will be converted to kinetic energy in the ejected electron. We can calculate the maximum kinetic energy of the ejected electron with the formula:

$$K_{max} = hf - W$$

Equation 9.2

where W is the work function of the metal in question. The **work function** is the minimum energy required to eject an electron and is related to the threshold frequency of that metal by:

$$W = hf_T$$

Equation 9.3

These formulas solve for the maximum kinetic energy of the electron rather than exact kinetic energy because the actual energy can be anywhere between 0 and K_{max}, depending on the specific subatomic interactions between the photon and the metal atom. K_{max} is only achieved when all possible energy from the photon is transferred to the ejected electron.

MCAT EXPERTISE

The photoelectric effect is not frequently tested on the MCAT, but the underlying principles are simple. This is simply another example of energy transfer in which light energy causes an increase in electrical potential energy in the atom—enough to allow the electron to escape. If any energy is "left over," it cannot be destroyed. Rather, it is transferred into kinetic energy in the ejected electron.

BRIDGE

Think of the work function like activation energy, in the sense that it must be matched or exceeded to cause the reaction (escape of an electron) to occur. Activation energy is discussed in Chapter 5 of *MCAT General Chemistry Review.*

Example: If blue light of frequency 6.00×10^{14} Hz is incident on rubidium ($W = 2.26$ eV), will there be photoejection of electrons? If so, what is the maximum kinetic energy that an ejected electron will carry away?
(Note: $h = 6.626 \times 10^{-34}\ \text{J}\cdot\text{s} = 4.14 \times 10^{-15}\ \text{eV}\cdot\text{s}$)

Solution: If the photons have a frequency of 6.00×10^{14} Hz, each photon has an energy of:

$$E = hf = (4.14 \times 10^{-15}\ \text{eV}\cdot\text{s})(6.00 \times 10^{14}\ \text{Hz}) = 2.48\ \text{eV}$$

Clearly then, any given photon has more than enough energy to allow an electron in the metal to overcome the 2.26 eV barrier. In fact, the maximum excess kinetic energy carried away by the electron turns out to be:

$$K = hf - W = 2.48 - 2.26 = 0.22\ \text{eV}$$

In general, the photoelectric effect is strong support for the particle theory of light, which states that light is not a continuous wave but acts as discrete bundles of energy called photons, as shown in Figure 9.1.

MCAT CONCEPT CHECK 9.1

Before you move on, assess your understanding of the material with these questions.

1. How does the work function relate to the energy necessary to emit an electron from a metal?

____

2. What does the threshold frequency depend upon?

____

3. What electrical phenomenon results from the application of the photoelectric effect?

____

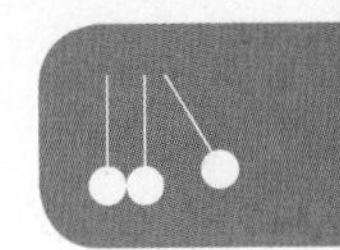

THE PHOTOELECTRIC EFFECT

Making Waves and Particles

The photoelectric effect, exploited in sensors, solar cells, and other electronic light detectors, refers to the ability of light to dislodge electrons from a metal surface. One aspect of the effect is that the speed of ejected electrons depends on the color of the light, not its intensity. Classical physics, which describes light as a wave, cannot explain this feature. By deducing that light could also act as a discrete bundle of energy—that is, a particle—Einstein accounted for the observation.

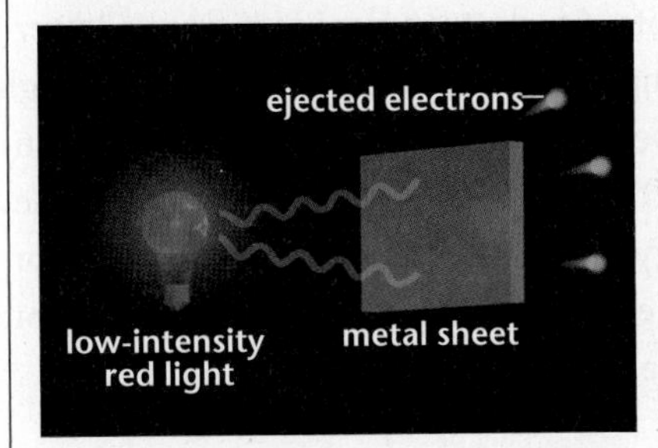

1 Red light sends electrons flying off a piece of metal. In the classical view, light is a continuous wave with energy spread out over the wave.

2 Increasing the brightness ejects more electrons. Classical physics also suggests that ejected electrons should move faster with more waves to ride—but they don't.

3 Changing the light to blue results in much speedier electrons. The reason is that light can behave not just as continuous waves but also as discrete bundles of energy called photons. A blue photon packs more energy than a red photon and essentially acts as a billiard ball with greater momentum, thereby hitting an electron harder (right). The particle view of light also explains why greater intensity increases the number of ejected electrons—with more photons impinging the metal, more electrons are likely to be struck.

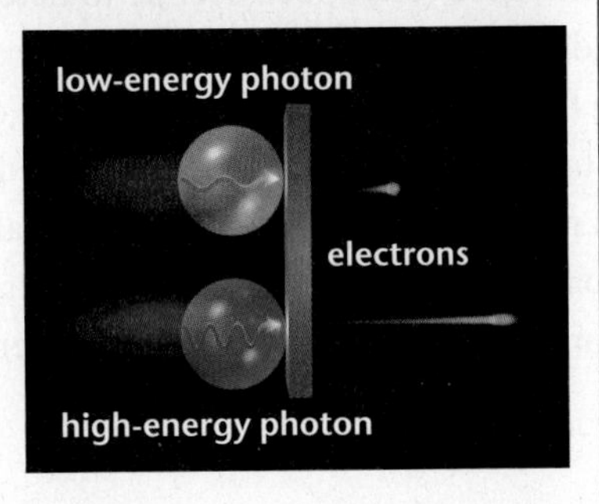

Figure 9.1 The Photoelectric Effect

9.2 Absorption and Emission of Light

LEARNING OBJECTIVES

After Chapter 9.2, you will be able to:

- Describe how the absorption spectrum of a single atom is determined
- Recall when photon emissions are most common during electronic transitions
- Explain the phenomenon of fluorescence

In Chapter 1 of *MCAT General Chemistry Review*, we explored the Bohr model of the atom. As a reminder, the Bohr model states that electron energy levels are stable and discrete, corresponding to specific orbits. An electron can jump from a lower-energy to a higher-energy orbit by absorbing a photon of light of precisely the right frequency to match the energy difference between the orbits ($E = hf$). If a photon does not carry enough energy, then the electron cannot jump to a higher energy level. When an electron falls from a higher-energy level to a lower-energy level, a photon of light is emitted with an energy equal to the energy difference between the two orbits. These processes of **atomic absorption** and **emission** are shown in Figure 9.2.

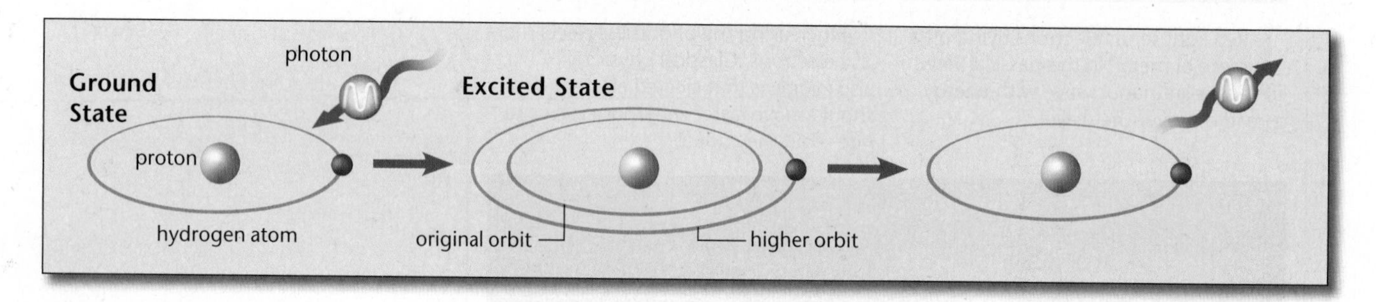

Figure 9.2 Bohr Model: Light Absorption and Emission

BRIDGE

Spectroscopy is discussed in greater detail in Chapter 11 of *MCAT Organic Chemistry Review*.

While information about a single electron is a great foundation for Test Day, in the real world we'll often be handling more complex structures. In organic chemistry, we use **infrared (IR) spectroscopy** to determine chemical structure because different bonds will absorb different wavelengths of light. **UV–Vis spectroscopy** takes this one step further, looking at the absorption of light in the visible and ultraviolet range. Absorption spectra may be represented as a color bar with peak areas of absorption represented by black lines. It can also be shown as a graph with the absolute absorption as a function of wavelength. This is shown in Figure 9.3, which shows the absorption spectrum for the atmosphere across the entire electromagnetic spectrum.

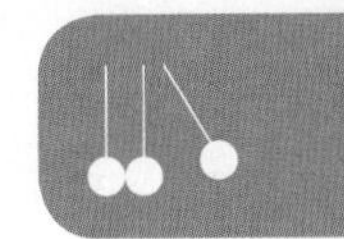

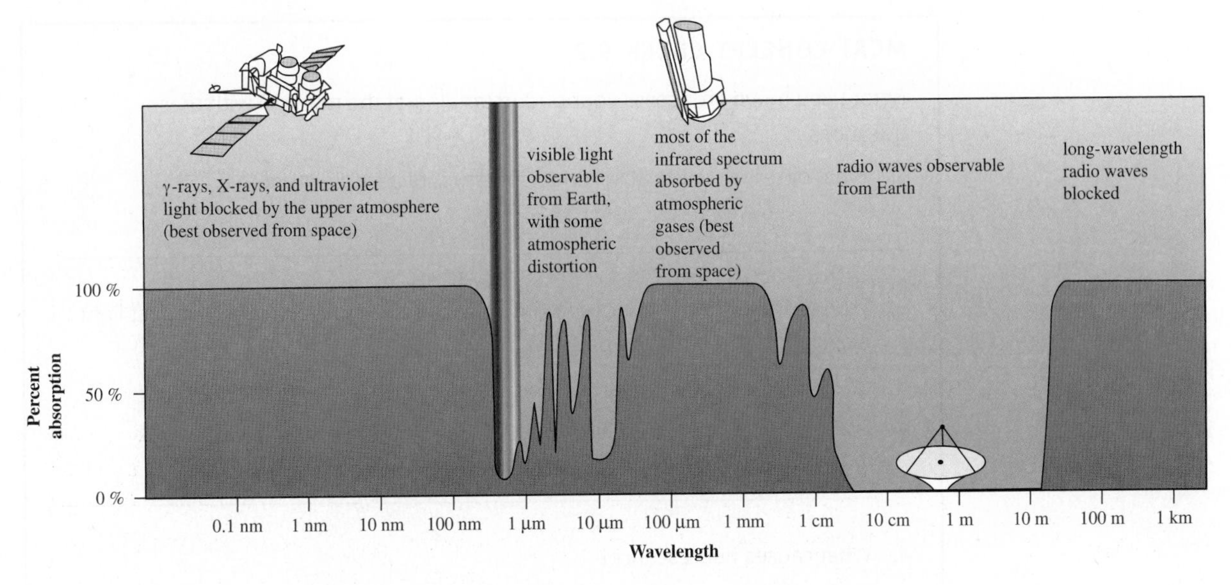

Figure 9.3 Absorption Spectrum of the Atmosphere
The sky is blue because blue light is the least absorbed by atmospheric gases.

Changes in molecular structure can cause dramatic shifts in the absorption patterns of a substance. Consider indicators like phenolphthalein. This indicator has a clear appearance in its acidic state, and thus does not absorb any visible light. In its basic state, it is a bright pink, and thus is absorbing all but the longer wavelengths of visible light—remember that we see the colors that are *not* absorbed. Most indicators contain large organic compounds that have strikingly different absorption patterns based solely on the protonation state of the compound. These compounds often have conjugated double bonds or aromatic ring systems, as this permits the absorption of light from photons in the visible range.

Another phenomenon related to absorption and emission of visible light is **fluorescence**. If one excites a fluorescent substance (such as a ruby, an emerald, or the phosphors found in fluorescent lights) with ultraviolet radiation, it will begin to glow with visible light. Photons of ultraviolet light have relatively high frequencies (short wavelengths). After being excited to a higher energy state by ultraviolet radiation, the electron in the fluorescent substance returns to its original state in two or more steps. By returning in two or more steps, each step involves less energy, so at each step, a photon is emitted with a lower frequency (longer wavelength) than the absorbed ultraviolet photon. If the wavelength of this emitted photon is within the visible range of the electromagnetic spectrum, it will be seen as light of the particular color corresponding to that wavelength. The wide range of colors in fluorescent lights, from the whitish-green of office lighting to the glaring colors of neon signs, is the result of the distinct multi-step emission spectra of different fluorescent materials.

MCAT CONCEPT CHECK 9.2

Before you move on, assess your understanding of the material with these questions.

1. What determines the absorption spectrum of a single atom?

____

2. True or False: Small changes in chemical structure only minimally impact light absorption and emission patterns.

3. During which electronic transitions is photon emission most common?

____

4. What causes fluorescence?

____

9.3 Nuclear Binding Energy and Mass Defect

LEARNING OBJECTIVES

After Chapter 9.3, you will be able to:

- Describe key concepts in nuclear binding energy, including strong nuclear force, mass defect, and binding energy
- Recall the four fundamental forces of nature
- Apply the equation $E = mc^2$

Until this point, we've examined the relationships between electromagnetic radiation and matter—particularly electrons. Now, we'll shift to the energy that is stored in the nucleus, which can be emitted under specific circumstances. While one would assume that the mass of the nucleus is simply the sum of the masses of all of the protons and neutrons within it, the actual mass of every nucleus (other than hydrogen) is slightly smaller than that. This difference is called the **mass defect**. Scientists had difficulty explaining why this mass defect occurred until Einstein characterized the equivalence of matter and energy, embodied by the equation

$$E = mc^2$$

Equation 9.4

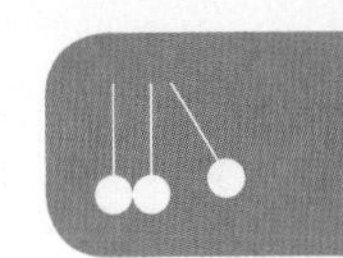

where E is energy, m is mass, and c is the speed of light. The mass defect is a result of matter that has been converted to energy. Because of the large exponent on the speed of light—which is squared in the equation—a very small amount of mass will yield a huge amount of energy. For example, the conversion of one gram of mass to energy will produce 89.9 terajoules ($1 \text{ TJ} = 10^{12}$ joules) or 21.5 billion kilocalories.

When protons and neutrons (**nucleons**) come together to form the nucleus, they are attracted to each other by the **strong nuclear force**, which is strong enough to more than compensate for the repulsive electromagnetic force between the protons. Although the strong nuclear force is the strongest of the four fundamental forces, it only acts over extremely short distances, less than a few times the diameter of a proton or neutron. The nucleons have to get very close together in order for the strong nuclear force to hold them together. The bonded system is at a lower energy level than the unbonded constituents, and this difference in energy must be radiated away in the form of heat, light, or other electromagnetic radiation before the mass defect becomes apparent. This energy, called **binding energy**, allows the nucleons to bind together in the nucleus. Given the strength of the strong nuclear force, the amount of mass that is transformed into the dissipated energy will be a measurable fraction of the initial total mass. The binding energy per nucleon peaks at the element iron, which implies that iron contains the most stable nucleus. In general, intermediate-sized nuclei are more stable than very large or small nuclei.

The **weak nuclear force** also contributes to the stability of the nucleus, but is about one-millionth as strong as the strong nuclear force. The strong and weak nuclear forces constitute two of the four fundamental forces of nature. The other two are electrostatic forces and gravitation.

Example: Measurements of the atomic mass of a neutron and a proton yield these results:

$$\text{proton} = 1.00728 \text{ amu}$$
$$\text{neutron} = 1.00867 \text{ amu}$$

$^{4}_{2}He$ contains two protons and two neutrons, which should theoretically give a helium nucleus a mass of $2 \times 1.00728 + 2 \times 1.00867 = 4.03190$ amu. However, the true mass of the helium nucleus is 4.00260 amu. What is the mass defect and binding energy of this nucleus? (Note: $c^2 = 932 \frac{\text{MeV}}{\text{amu}}$)

Solution: The difference $4.03190 - 4.00260 = 0.02930$ amu is the mass defect for the helium nucleus. This is the mass that contributed to the binding energy of the nucleus:

$$\begin{aligned} E &= mc^2 \\ &= (0.02930 \text{ amu})\left(932 \frac{\text{MeV}}{\text{amu}}\right) \\ &\approx 0.03 \times 900 = 27 \text{ MeV} \,(\text{actual} = 27.3 \text{ MeV}) \end{aligned}$$

MCAT CONCEPT CHECK 9.3

Before you move on, assess your understanding of the material with these questions.

1. Define the following terms:
   - Strong nuclear force:
   - Mass defect:
   - Binding energy:
2. What are the four fundamental forces of nature?
   -
   -
   -
   -
3. How does the mass defect relate to the binding energy?

9.4 Nuclear Reactions

LEARNING OBJECTIVES

After Chapter 9.4, you will be able to:

- Compare and contrast nuclear fission and nuclear fusion reactions
- Recall the emissions, ΔZ, and ΔA of different radioactive processes
- Recall the type of decay that could be detected in an atomic absorption spectrum
- Predict the number of half-lives necessary for decay of some portion of a radioactive sample:

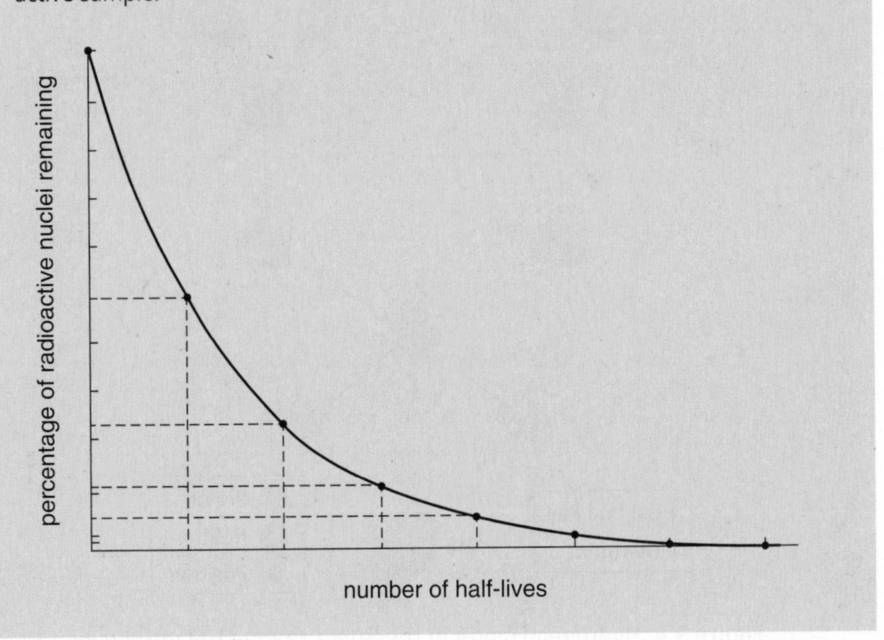

Nuclear reactions, such as fusion, fission, and radioactive decay, involve either combining or splitting the nuclei of atoms. Because the binding energy per nucleon is greatest for intermediate-sized atoms (that is, intermediate-sized atoms are most stable), when small atoms combine or large atoms split, a great amount of energy is released.

When written in **isotopic notation**, elements are preceded by their atomic number as a subscript and mass number as a superscript $\left({}^{A}_{Z}\mathrm{X}\right)$. The **atomic number** (***Z***) corresponds to the number of protons in the nucleus; the **mass number** (***A***) corresponds to the number of protons plus neutrons. When balancing nuclear equations, it is important to balance the number of nucleons on both sides by balancing the atomic numbers and mass numbers.

Fusion

Fusion occurs when small nuclei combine to form a larger nucleus. As an example, many stars (including the Sun) power themselves by fusing four hydrogen nuclei to make one helium nucleus as shown in Figure 9.4. By this method, the Sun produces 3.85×10^{26} joules per second (385 yottawatts), which accounts for the mass defect that arises from the formation of helium nuclei from hydrogen nuclei. Here on Earth, fusion power plants—which are far less common than fission power plants—generate energy from deuterium $\left(^{2}_{1}\text{H}\right)$ and lithium nuclei.

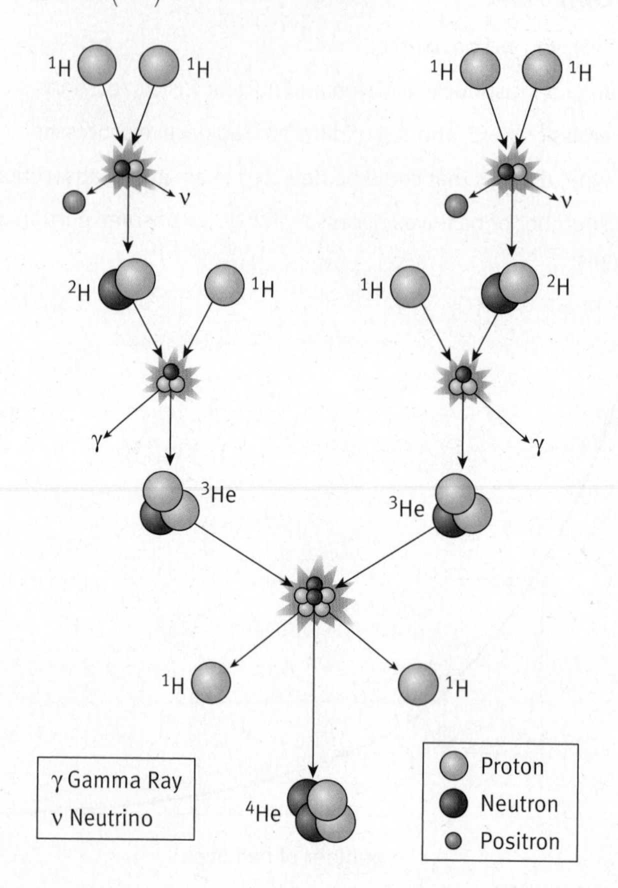

Figure 9.4 Hydrogen Fusion in the Sun, Creating Helium Nuclei

Fission

Fission is a process by which a large nucleus splits into smaller nuclei. Spontaneous fission rarely occurs. However, through the absorption of a low-energy neutron, fission can be induced in certain nuclei. Of special interest are those fission reactions that release more neutrons because these other neutrons will cause a chain reaction in which other nearby atoms can undergo fission. This in turn releases more neutrons, continuing the chain reaction. Such induced fission reactions power most commercial nuclear power plants.

Example: A fission reaction occurs when uranium-235 (U-235) absorbs a low-energy neutron, briefly forming an excited state of U-236, which then splits into xenon-140, strontium-94, and more neutrons. In isotopic notation form, the unbalanced reaction is:

$$^{235}_{92}\text{U} + ^{1}_{0}n \rightarrow ^{236}_{92}\text{U}^* \rightarrow ^{140}_{54}\text{Xe} + ^{94}_{38}\text{Sr} + ^{1}_{0}n$$

When balanced, how many neutrons are produced in the last reaction?

Solution: By treating each arrow as an equals sign, the problem is simply asking us to balance the last equation. The mass numbers (A) on either side of each arrow must be equal. This is an application of nucleon number conservation, which says that the total number of neutrons plus protons remains the same, even if neutrons are converted to protons and vice-versa, as they are in some decays. Because $235 + 1 = 236$, the first arrow is indeed balanced. Looking at the atomic numbers, the number of protons are balanced throughout ($92 + 0 = 92 = 54 + 38 + 0$). To find the number of neutrons, determine how many nucleons remain after accounting for xenon-140 and strontium-94:

$$236 - (140 + 94) = 236 - 234 = 2 \text{ nucleons}$$

Because the protons are balanced, these remaining nucleons are both neutrons. Therefore, two neutrons are produced in this reaction. These neutrons are free to go on and be absorbed by more U-235 and cause more fission reactions. Note that it was not actually necessary to know that the intermediate high-energy state $^{236}_{92}\text{U}^*$ was formed.

Radioactive Decay

Radioactive decay is a naturally occurring spontaneous decay of certain nuclei accompanied by the emission of specific particles. On the MCAT, you should be prepared to answer three general types of radioactive decay problems:

1. The integer arithmetic of particle and isotope species
2. Radioactive half-life problems
3. The use of exponential decay curves and decay constants

Isotope Decay Arithmetic and Nucleon Conservation

Let the letters X and Y represent nuclear isotopes. When the **parent nucleus** X undergoes nuclear decay to form **daughter nucleus** Y, the balanced reaction is:

$$^{A}_{Z}\text{X} \rightarrow ^{A'}_{Z'}\text{Y} + \text{emitted decay particle}$$

Equation 9.5

MCAT EXPERTISE

Whenever you approach radioactive decay problems on the MCAT, start by balancing the number of protons (the atomic numbers). Often, wrong answer choices will simply have an error in the number of protons and can be eliminated before even checking the mass numbers.

KEY CONCEPT

Alpha particles do not have any electrons, so they carry a charge of +2.

When balancing nuclear reactions, the sum of the atomic numbers must be the same on both sides of the equation, and the sum of the mass numbers must be the same on both sides as well.

Alpha Decay

Alpha decay is the emission of an **α-particle**, which is a $^{4}_{2}\text{He}$ nucleus that consists of two protons, two neutrons, and zero electrons. The alpha particle is very massive compared to a beta particle and carries double the charge. Alpha particles interact with matter very easily; hence, they do not penetrate shielding (such as lead sheets) very extensively.

The emission of an α-particle means that the atomic number of the daughter nucleus will be two less than that of the parent nucleus, and the mass number will be four less. This can be expressed in the balanced equation:

$$^{A}_{Z}\text{X} \rightarrow {}^{A-4}_{Z-2}\text{Y} + {}^{4}_{2}\alpha$$

Equation 9.6

Example: Suppose a parent nucleus X alpha decays as follows:

$$^{238}_{92}\text{X} \rightarrow {}^{A'}_{Z'}\text{Y} + \alpha$$

What are the mass number and atomic number of the daughter isotope Y?

Solution: To solve this question, we simply need to balance the atomic numbers and mass numbers:

$$^{A}_{Z}\text{X} \rightarrow {}^{A-4}_{Z-2}\text{Y} + {}^{4}_{2}\alpha$$
$$^{238}_{92}\text{U} \rightarrow {}^{234}_{90}\text{Y} + {}^{4}_{2}\alpha$$

While it is not necessary to identify the elements to answer the question, answers on the MCAT are usually given with the element's symbol. Y must be thorium (Th) because its atomic number is 90. Therefore, the daughter nucleus is $^{234}_{90}\text{Th}$.

Beta Decay

Beta decay is the emission of a **β-particle**, which is an electron and is given the symbol e^- or β^-. Electrons do not reside in the nucleus, but they are emitted by the nucleus when a neutron decays into a proton, a β-particle, and an antineutrino ($\overline{\nu}$). Because an electron is singly charged and 1836 times lighter than a proton, the beta radiation from radioactive decay is more penetrating than alpha radiation. In some cases of induced decay (**positron emission**), a **positron** is released, which has the mass of an electron but carries a positive charge. The positron is given the symbol e^+ or β^+. A neutrino (ν) is emitted in positron decay, as well. Note that neutrinos and antineutrinos are not tested on the MCAT, and are therefore omitted in subsequent discussion.

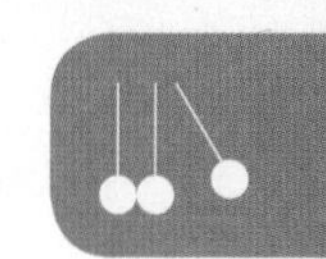

During β^- decay, a neutron is converted into a proton and a β^--particle ($Z = -1$, $A = 0$) is emitted. Hence, the atomic number of the daughter nucleus will be one higher than that of the parent nucleus, and the mass number will not change. This can be expressed in the balanced equation:

$$^{A}_{Z}\mathrm{X} \rightarrow {}^{A}_{Z+1}\mathrm{Y} + \beta^-$$

Equation 9.7

During β^+ decay, a proton is converted into a neutron and a β^+-particle ($Z = +1$, $A = 0$) is emitted. Hence, the atomic number of the daughter nucleus will be one lower than that of the parent nucleus, and the mass number will not change. This can be expressed in the balanced equation:

$$^{A}_{Z}\mathrm{X} \rightarrow {}^{A}_{Z-1}\mathrm{Y} + \beta^+$$

Equation 9.8

KEY CONCEPT

In both types of beta decay, there needs to be conservation of charge. If a negative charge (β^-) is produced, a neutron is converted into a proton to maintain charge. Conversely, if a positive charge (β^+) is produced, a proton is converted into a neutron to maintain charge. Remember that negative beta decay produces a negative β-particle and positive beta decay produces a positive β-particle.

Example: Suppose a promethium-146 nucleus beta-decays as follows:

$$^{146}_{61}\mathrm{Pm} \rightarrow {}^{A'}_{Z'}\mathrm{Y} + \beta^-$$

What are the mass number and atomic number of the daughter isotope Y?

Solution: Again, balance the atomic numbers and mass numbers:

$$^{146}_{61}\mathrm{Pm} \rightarrow {}^{A}_{Z+1}\mathrm{Y} + {}^{0}_{-1}\beta^-$$

$$^{146}_{61}\mathrm{Pm} \rightarrow {}^{146}_{62}\mathrm{Y} + {}^{0}_{-1}\beta^-$$

Y must be samarium (Sm) because its atomic number is 62. Therefore, the daughter nucleus is $^{146}_{62}\mathrm{Sm}$.

Gamma Decay

Gamma decay is the emission of **γ-rays**, which are high-energy (high-frequency) photons. They carry no charge and simply lower the energy of the parent nucleus without changing the mass number or the atomic number. The high-energy state of the parent nucleus may be represented by an asterisk.

This can be expressed in the balanced equation:

$$^{A}_{Z}\mathrm{X}^* \rightarrow {}^{A}_{Z}\mathrm{X} + \gamma$$

Equation 9.9

KEY CONCEPT

Gamma decay questions are the easiest on the MCAT. No changes occur in the mass number or atomic number; only a γ-ray is emitted.

Example: Suppose an excited parent isotope ${}^{A}_{Z}X^*$ gamma decays to ${}^{A'}_{Z'}X$, which then undergoes positron emission to form ${}^{A''}_{Z''}Y$, which in turn alpha decays to ${}^{A'''}_{Z'''}Z$. If Z is americium-241, what is ${}^{A}_{Z}X^*$?

Solution: Because the final daughter nucleus is given, it will be necessary to work backwards through the reactions. The last reaction is the following alpha decay:

$${}^{A''}_{Z''}Y \rightarrow {}^{241}_{95}Am + {}^{4}_{2}\alpha$$

The atomic number of the parent nucleus must be 97, and the mass number is 245. This is berkelium-245. The preceding reaction is the following positron emission:

$${}^{A'}_{Z'}X \rightarrow {}^{245}_{97}Bk + {}^{0}_{+1}\beta^+$$

The atomic number of the parent nucleus must be 98, and the mass number is 245. This is californium-245. Finally, the preceding reaction is the following gamma decay:

$${}^{A}_{Z}X^* \rightarrow {}^{245}_{98}Cf + \gamma$$

The atomic number of the parent nucleus must be 98, and the mass number is 245. This is a higher-energy form of californium-245: ${}^{A}_{Z}X^* = {}^{245}_{98}Cf^*$

Electron Capture

Certain unstable radionuclides are capable of capturing an inner electron that combines with a proton to form a neutron, while releasing a neutrino. The atomic number is now one less than the original but the mass number remains the same. **Electron capture** is a rare process that is perhaps best thought of as the reverse of β^- decay:

$${}^{A}_{Z}X + e^- \rightarrow {}^{A}_{Z-1}Y$$

Equation 9.10

Half-Life

MCAT EXPERTISE

Half-life problems are common on the MCAT. Make sure you write them out; it's easy to lose your place when doing them in your head.

In a sample of radioactive particles, the **half-life** $\left(T_{\frac{1}{2}}\right)$ of the sample is the time it takes for half of the sample to decay. In each subsequent half-life, one-half of the remaining sample decays so that the remaining amount asymptotically approaches zero.

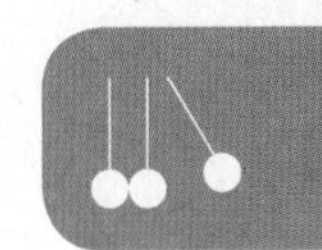

Example: If the half-life of a certain isotope is 4 years, what fraction of a sample of that isotope will remain after 12 years?

Solution: If 4 years is one half-life, then 12 years is 3 half-lives. During the first half-life—the first 4 years—half of the sample will decay. During the second half-life (years 4 to 8), half of the remaining half will decay, leaving one-fourth of the original. During the third and final half-life (years 8 to 12), half of the remaining fourth will decay, leaving one-eighth of the original sample. Thus, the fraction remaining after 3 half-lives is $\left(\frac{1}{2}\right)^3 = \frac{1}{8}$.

Exponential Decay

Let n be the number of radioactive nuclei that have not yet decayed in a sample. It turns out that the rate at which the nuclei decay, $\frac{\Delta n}{\Delta t}$, is proportional to the number that remain (n). This suggests the equation

$$\frac{\Delta n}{\Delta t} = -\lambda n$$

Equation 9.11

where λ is known as the decay constant. The solution of this equation tells us how the number of radioactive nuclei changes with time. This is known as an **exponential decay**:

$$n = n_0 e^{-\lambda t}$$

Equation 9.12

where n_0 is the number of undecayed nuclei at time $t = 0$. The decay constant is related to the half-life by

$$\lambda = \frac{\ln 2}{T_{\frac{1}{2}}} = \frac{0.693}{T_{\frac{1}{2}}}$$

Equation 9.13

A typical exponential decay curve is shown in Figure 9.5.

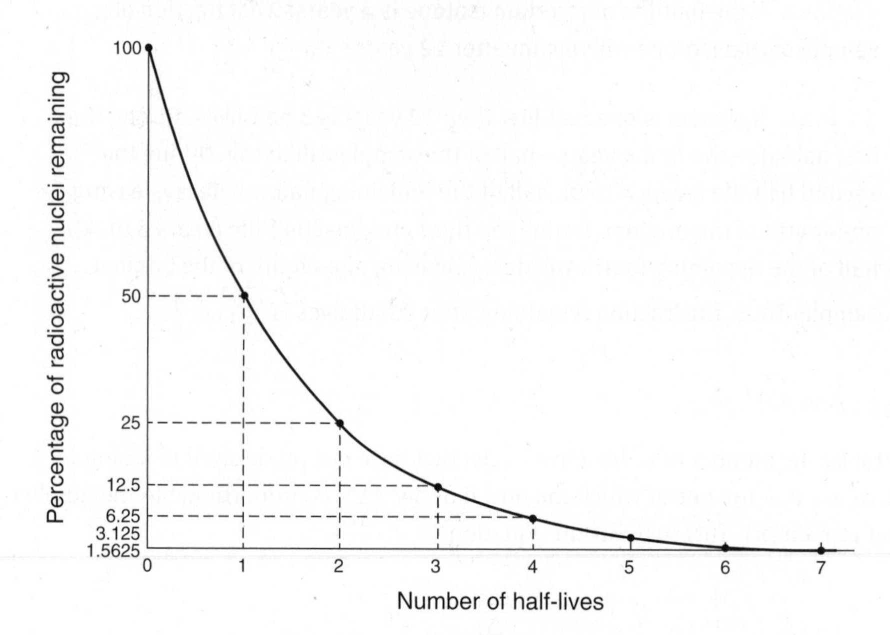

Figure 9.5 Exponential Decay

Example: If at time $t = 0$, there is a 2 mole sample of radioactive isotopes, how many nuclei remain after 45 minutes, assuming a decay constant of 2 hr^{-1}? (Hint: $e^{-\frac{3}{2}} = 0.22$).

Solution: This question is asking for an application of the exponential decay equation:

$$n = n_0 e^{-\lambda t} = n_0 e^{-\left(2 \text{ hr}^{-1}\right)\left(\frac{3}{4} \text{ hr}\right)} = n_0 e^{-\frac{3}{2}}$$

Raising Euler's number (*e*) to an exponent—especially a fractional exponent—is beyond the scope of the math on the MCAT, but the value of $e^{-\frac{3}{2}}$ is 0.22. Thus, 22% of the original 2 mole sample remains. This constitutes 0.44 mol, which, if multiplied by Avogadro's number, gives us the number of nuclei remaining:

$$0.44 \text{ mol} \times 6.02 \times 10^{23} \frac{\text{nuclei}}{\text{mol}} = 2.64 \times 10^{23} \text{ nuclei}$$

MCAT CONCEPT CHECK 9.4

Before you move on, assess your understanding of the material with these questions.

1. True or False: Nuclear fission and nuclear fusion both release energy.
2. Compare and contrast nuclear fission and nuclear fusion reactions:

Nuclear Reaction	Size of Reactant Particles	Change in Nuclear Mass during Reaction (Increase or Decrease)
Fission		
Fusion		

3. Complete the following chart:

Nuclear Reaction	Emits . . .	ΔZ	ΔA
Alpha decay			
Beta-negative decay			
Beta-positive decay			
Gamma decay			
Electron capture			

4. How many half-lives are necessary for the complete decay of a radioactive sample?

____

5. Which type of nuclear decay could be detected in an atomic absorption spectrum?

____

Conclusion

Congratulations! You've finished the physics material that will be needed for Test Day. Our last topic was the interaction of energy and matter on the atomic level. We began by examining the photoelectric effect. Further, we took a look at Bohr's model of the hydrogen ion, and made some generalizations about electronic structure and the permissible states in regards to absorption and emission of light energy. We also noted that on a molecular level, small changes in structure can lead to significant shifts in absorption. We studied the interactions of the nucleus with energy, including the prototypical nuclear reactions of fusion and fission. We finished our discussion of nuclear reactions by examining the most common forms of nuclear decay and some of the mathematics for determining half-life or sample remaining. In the next few chapters, we'll focus on building Test Day skills, including MCAT math shortcuts that will make many of these concepts more rewarding.

You've reviewed the content, now test your knowledge and critical thinking skills by completing a test-like passage set in your online resources!

Concept Summary

The Photoelectric Effect

- The **photoelectric effect** is the ejection of an electron from the surface of a metal in response to light.
- The **threshold frequency** is the minimum light frequency necessary to eject an electron from a given metal.
  - The **work function** is the minimum energy necessary to eject an electron from a given metal. Its value depends on the metal used and can be calculated by multiplying the threshold frequency by **Planck's constant**.
  - The greater the energy of the incident photon above the work function, the more kinetic energy the ejected electron can possess.
- The ejected electrons create a current; the magnitude of this current is proportional to the intensity of the incident beam of light.

Absorption and Emission of Light

- The **Bohr model of the atom** states that electron energy levels are stable and discrete, corresponding to specific orbits.
  - An electron can jump from a lower-energy to a higher-energy orbit by **absorbing** a photon of light of the same frequency as the energy difference between the orbits.
  - When an electron falls from a higher-energy to a lower-energy orbit, it **emits** a photon of light of the same frequency as the energy difference between the orbits.
- **Absorption spectra** may be impacted by small changes in molecular structure.
- **Fluorescence** occurs when a species absorbs high-frequency light and then returns to its **ground state** in multiple steps. Each step has less energy than the absorbed light and is within the visible range of the electromagnetic spectrum.

Nuclear Binding Energy and Mass Defect

- **Nuclear binding energy** is the amount of energy that is released when **nucleons** (protons and neutrons) bind together.
  - The more binding energy per nucleon released, the more stable the nucleus.
  - The four fundamental forces of nature are the **strong** and **weak nuclear force**, which contribute to the stability of the nucleus, electrostatic forces, and gravitation.
- The **mass defect** is the difference between the mass of the unbonded nucleons and the mass of the bonded nucleons within the nucleus.
  - The unbonded constituents have more energy and, therefore, more mass than the bonded constituents.
  - The mass defect is the amount of mass converted to energy during nuclear fusion.

Nuclear Reactions

- **Fusion** occurs when small nuclei combine into larger nuclei.
- **Fission** occurs when a large nucleus splits into smaller nuclei.
- Energy is released in both fusion and fission because the nuclei formed in both processes are more stable than the starting nuclei.
- **Radioactive** decay is the loss of small particles from the nucleus.
  - **Alpha (α) decay** is the emission of an alpha particle $\left(\alpha, {}_{2}^{4}\alpha, {}_{2}^{4}\text{He}\right)$, which is a helium nucleus.
  - **Beta-negative (β^-) decay** is the decay of a neutron into a proton, with emission of an electron (e^-, β^-) and an antineutrino ($\overline{\nu}$).
  - **Beta-positive (β^+) decay**, also called **positron emission**, is the decay of a proton into a neutron, with emission of a **positron** (e^+, β^+) and a neutrino (ν).
  - **Gamma (γ) decay** is the emission of a gamma ray, which converts a high-energy nucleus into a more stable nucleus.
  - **Electron capture** is the absorption of an electron from the inner shell that combines with a proton in the nucleus to form a neutron.
- **Half-life** is the amount of time required for half of a sample of radioactive nuclei to decay.
- In **exponential decay**, the rate at which radioactive nuclei decay is proportional to the number of nuclei that remain.

Answers to Concept Checks

9.1

1. The work function describes the minimum amount of energy necessary to emit an electron. Any additional energy from a photon will be converted to excess kinetic energy during the photoelectric effect.
2. The threshold frequency depends on the chemical composition of a material (that is, the identity of the metal).
3. The accumulation of moving electrons creates a current during the photoelectric effect.

9.2

1. The energy differences between ground-state electrons and higher-level electron orbits determine the frequencies of light a particular material absorbs (its absorption spectrum).
2. False. Small changes, such as protonation and deprotonation, change in oxidation state or bond order, and others may cause dramatic changes in light absorption in a material.
3. When electrons transition from a higher-energy state to a lower-energy state, they will experience photon emission.
4. Fluorescence is a special stepwise photon emission in which an excited electron returns to the ground state through one or more intermediate excited states. Each energy transition releases a photon of light. With smaller energy transitions than the initial energy absorbed, these materials can release photons of light in the visible range.

9.3

1. The strong nuclear force is one of the four primary forces and provides the adhesive force between the nucleons (protons and neutrons) within the nucleus. Mass defect is the apparent loss of mass when nucleons come together, as some of the mass is converted into energy. That energy is called the binding energy.
2. The four fundamental forces of nature are the strong and weak nuclear forces, electrostatic forces, and gravitation.
3. Mass defect is related to the binding energy such that there is a transformation of nuclear matter to energy with a resultant loss of matter. They are related by the equation $E = mc^2$.

9.4

1. True. While they may seem like inverses of each other, both nuclear fusion and nuclear fission reactions release energy.

2.

Nuclear Reaction	Size of Reactant Particles	Change in Nuclear Mass during Reaction (Increase or Decrease)
Fission	Large (actinides, lanthanides)	Decrease
Fusion	Small (hydrogen, helium)	Increase

3.

Nuclear Reaction	Emits...	ΔZ	ΔA
Alpha decay	Alpha particle $\left(\alpha, {}_2^4\alpha, {}_2^4\text{He}\right)$	−2	−4
Beta-negative decay	Electron (e^-, β^-) and antineutrino ($\bar{\nu}$)	+1	0
Beta-positive decay	Positron (e^+, β^+) and neutrino (ν).	−1	0
Gamma decay	Gamma ray (γ)	0	0
Electron capture	Nothing (absorbs an electron from inner shell)	−1	0

4. Because the amount remaining is cut in half after each half-life, the portion remaining will never quite reach zero. This is mostly a theoretical consideration; "all" of a sample is considered to have decayed after 7 to 8 half-lives.

5. Because gamma radiation produces electromagnetic radiation (rather than nuclear fragments), it can be detected on an atomic absorption spectrum.

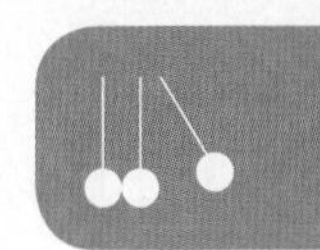

Science Mastery Assessment Explanations

1. **C**

To determine the speed of the electrons ejected, we must first calculate their kinetic energy:

$$\begin{aligned}K &= hf - W \\ &= \left(6.626\times10^{-34}\ \text{J}\cdot\text{s}\right)\left(1.0\times10^{14}\ \text{Hz}\right) - 6.622\times10^{-20}\ \text{J} \\ &= (6.626-6.622)\times10^{-20} = 0.004\times10^{-20} = 4\times10^{-23}\ \text{J}\end{aligned}$$

Now with a value for kinetic energy, we can calculate the speed of the ejected electrons:

$$\begin{aligned}K &= \frac{1}{2}mv^2 \rightarrow v = \sqrt{\frac{2K}{m}} \\ &= \sqrt{\frac{2\left(4\times10^{-23}\ \text{J}\right)}{9.1\times10^{-31}\ \text{kg}}} \approx \sqrt{\frac{8\times10^{-23}}{9\times10^{-31}}} \approx \sqrt{10^{8}} = 10^{4}\ \frac{\text{m}}{\text{s}}\end{aligned}$$

Notice the wide range in the exponents for the answer choices. While the math in this question may seem complex, this allows us to round significantly.

2. **B**

To determine the wavelength of the light ray, first calculate its frequency from the photoelectric effect equation:

$$\begin{aligned}K &= hf - W \rightarrow f = \frac{K+W}{h} \\ &= \frac{50\ \text{J} + 16\ \text{J}}{6.626\times10^{-34}\ \text{J}\cdot\text{s}} \approx \frac{6.6\times10^{1}}{6.6\times10^{-34}} = 10^{35}\ \text{Hz}\end{aligned}$$

In this case, estimation of Planck's constant makes our calculation much simpler without leading us to a nonviable answer. It is worth attempting an estimation first to avoid doing more work than necessary. Now we can determine the wavelength of the incident ray of light by relating the frequency to the speed of light:

$$\begin{aligned}c &= f\lambda \rightarrow \lambda = \frac{c}{f} \\ &= \frac{3.00\times10^{8}\ \frac{\text{m}}{\text{s}}}{10^{35}\ \text{Hz}} = 3\times10^{-27}\ \text{m}\end{aligned}$$

3. **D**

The Bohr model is based on a set of postulates originally put forward to discuss the behavior of electrons in hydrogen. In summary, these postulates state that the energy levels of the electron are stable and discrete, and they correspond to specific orbits, eliminating **(A)**. They also state that an electron emits or absorbs radiation only when making a transition from one energy level to another, eliminating **(B)**. Specifically, when an electron jumps from a lower-energy orbit to a higher-energy one, it must absorb a photon of light of precisely the right frequency such that the photon's energy equals the energy difference between the two orbits, eliminating **(C)**. When falling from a higher-energy orbit to a lower-energy one, an electron emits a photon of light with a frequency that corresponds to the energy difference between the two orbits, This is the opposite of **(D)**, which makes it the right answer.

4. **B**

To solve this question correctly, one must be careful with the units. First, convert 10.2 eV to joules:

$$10.2\ \text{eV}\left(1.60\times10^{-19}\ \frac{\text{J}}{\text{eV}}\right) \approx 1.6\times10^{-18}\ \text{J}$$

Next, to determine the wavelength of the radiation, we can combine the formulas $E = hf$ and $c = f\lambda$:

$$\begin{aligned}E &= hf = \frac{hc}{\lambda} \rightarrow \lambda = \frac{hc}{E} \\ &= \frac{\left(6.626\times10^{-34}\ \text{J}\cdot\text{s}\right)\left(3.00\times10^{8}\ \frac{\text{m}}{\text{s}}\right)}{\left(1.6\times10^{-18}\ \text{J}\right)} \\ &\approx \frac{(4)(3)\left(10^{-26}\right)}{10^{-18}} = 12\times10^{-8} = 1.2\times10^{-7}\ \text{m}\end{aligned}$$

5. **B**

The electron moves from a higher energy level to a lower energy level; this can only occur if the extra energy is dissipated through the emission of a photon. If the electron moved from B to A, it would absorb a photon and increase the atom's total energy; however, the opposite is occurring, so **(A)** and **(D)** can be eliminated. The work function is the amount of energy required to eject an electron from a material; when moving from A to B, the electrical potential energy of the atom decreases, meaning that more energy will be required to free the electron from the atom, eliminating **(C)**.

6. **B**

The strong nuclear force is the attractive force that holds protons and neutrons together in the nucleus, supporting choice **(B)**. This force is greater than the electrostatic repulsion between protons. Note that binding energy, **(A)**, is not one of the four fundamental forces.

7. **A**

The greater the intensity, the greater the number of incident photons and, therefore, the greater the number of electrons that will be ejected from the metal surface (provided that the frequency of the light remains above the threshold). This means a larger current. Remember that the frequency of the light (assuming it is above the threshold frequency) will determine the kinetic energy of the ejected electrons; the intensity of the light determines the number of electrons ejected per time (the current).

8. **C**

To determine the binding energy, we must first determine the mass defect. The mass defect is simply the masses of each of the protons and neutrons in the unbound state added together minus the mass of the formed argon-40 nucleus (which contains 18 protons and 40 − 18 = 22 neutrons):

$$\begin{aligned}\text{mass defect} &= (18 \times 1.0073\text{ amu}) + (22 \times 1.0087\text{ amu}) \\ &\quad - 39.9132\text{ amu} \\ &\approx (18 \times 1.007) + (22 \times 1.009) - 39.9132 \\ &= 18.126 + 22.198 - 39.9132 \\ &= 40.324 - 39.9132 \\ &\approx 0.4\ (\text{actual} = 0.4096\text{ amu})\end{aligned}$$

This math was difficult without a calculator, but by rounding one value down (proton) and one value up (neutron) by similar amounts we ended up very near the actual value. Calculating $18 \times 7 = 126$ and $22 \times 9 = 198$ for the decimal values is more manageable than the exact numbers and the spacing of the answer choices allows for our estimation. The binding energy can then be determined from this mass defect:

$$E = mc^2 = 0.4096\text{ amu} \times 932\,\frac{\text{MeV}}{\text{amu}} \approx 0.4 \times 900 = 360\text{ MeV}$$

The closest answer is **(C)**.

9. **C**

This process can be described as electron capture. Certain unstable radionuclides are capable of capturing an inner electron that combines with a proton to form a neutron. The atomic number becomes one less than the original, but the mass number remains the same. Electron capture is a relatively rare process and can be thought of as the reverse of β^- decay. Notice that the equation is similar to that of β^+ decay but not identical because a particle is absorbed, not emitted.

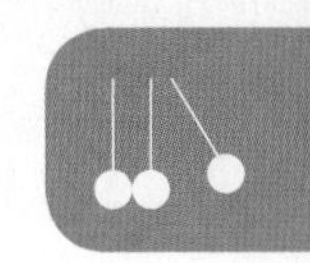

10. **C**

This problem presents a reaction and asks for the energy liberated due to transformation of mass into energy. To convert mass into energy, we are told that 1 amu can be converted into 932 MeV of energy. All we need to do now is calculate how much mass, in amu, is converted in the reaction. Because we are given the atomic mass for each of the elements in the reaction, this is simply a matter of balancing the equation:

$$\begin{aligned}\text{mass defect} &= (1.0087\text{ amu} + 10.0129\text{ amu}) \\ &\quad - (7.0160\text{ amu} + 4.0026\text{ amu}) \\ &= 11.0216 - 11.0186 \\ &= 0.0030\text{ amu}\end{aligned}$$

Given both the small magnitude of this value and the small difference of the answer choices, it is best to not round at this point of the calculation. This is the amount of mass that has been converted into energy. To obtain energy from mass, we have to multiply by the conversion factor (1 amu = 932 MeV):

$$E = 0.003 \times 932 \approx 0.003 \times 900 = 2.7\text{ MeV}$$

At this point we were able to round for an easier calculation that keeps us very near the correct answer choice.

11. **C**

Because the half-life of element X is four days, 50 percent of an original sample remains after four days, 25% remains after eight days, and 12.5% remains after 12 days. Therefore, $n = 12$ days. Another approach is to set $\left(\frac{1}{2}\right)^x = 0.125$, where x is the number of half-lives that have elapsed. Solving for x gives $x = 3$. Thus, 3 half-lives have elapsed, and because each half-life is four days, we know that $n = 12$ days.

12. **B**

The expression $n = n_0 e^{-\lambda t}$ is equivalent to $\frac{n}{n_0} = e^{-\lambda t}$. Taking the natural logarithm of both sides, we get:

$$\ln\left(\frac{n}{n_0}\right) = -\lambda t$$

From this expression, it is clear that plotting $\ln\left(\frac{n}{n_0}\right)$ *vs.* t will give a straight line with a slope of $-\lambda$.

13. **C**

A typical carbon nucleus contains 6 protons and 6 neutrons. An α-particle contains 2 protons and 2 neutrons. Therefore, one carbon nucleus can dissociate into $\frac{6}{2} = 3$ α-particles.

14. **A**

Because the half-life of carbon-12 is essentially infinite, a 25% decrease in the ratio of carbon-14 to carbon-12 means the same as a 25% decrease in the amount of carbon-14. If less than half of the carbon-14 has deteriorated, then less than one half-life has elapsed. Therefore, the sample is less than 5,730 years old. Be careful with the wording here—the question states that the ratio is 25% *less* than the ratio in nature, not 25% *of* the ratio in nature, which would correspond to **(D)**.

15. **D**

In alpha decay, an element loses two protons. In positron decay, a proton is converted into a neutron. Gamma decay has no impact on the atomic number of the nuclide. Therefore, two alpha decays and two positron decays will yield a daughter nuclide with six fewer protons than the parent nuclide.

Consult your online resources for additional practice. GO ONLINE

Equations to Remember

(9.1) Energy of a photon of light: $E = hf$

(9.2) Maximum kinetic energy of an electron in the photoelectric effect: $K_{max} = hf - W$

(9.3) Work function: $W = hf_T$

(9.4) Mass defect and energy: $E = mc^2$

(9.5) Nuclear decay (general form): ${}^{A}_{Z}X \rightarrow {}^{A'}_{Z'}Y +$ emitted decay particle

(9.6) Alpha decay: ${}^{A}_{Z}X \rightarrow {}^{A-4}_{Z-2}Y + {}^{4}_{2}\alpha$

(9.7) Beta-negative decay: ${}^{A}_{Z}X \rightarrow {}^{A}_{Z+1}Y + \beta^-$

(9.8) Beta-positive decay (positron emission): ${}^{A}_{Z}X \rightarrow {}^{A}_{Z-1}Y + \beta^+$

(9.9) Gamma decay: ${}^{A}_{Z}X^* \rightarrow {}^{A}_{Z}X + \gamma$

(9.10) Electron capture: ${}^{A}_{Z}X + e^- \rightarrow {}^{A}_{Z-1}Y$

(9.11) Rate of nuclear decay: $\frac{\Delta n}{\Delta t} = -\lambda n$

(9.12) Exponential decay: $n = n_0 e^{-\lambda t}$

(9.13) Decay constant: $\lambda = \frac{\ln 2}{T_{\frac{1}{2}}} = \frac{0.693}{T_{\frac{1}{2}}}$

Shared Concepts

General Chemistry Chapter 1
Atomic Structure

General Chemistry Chapter 2
The Periodic Table

Organic Chemistry Chapter 11
Spectroscopy

Physics and Math Chapter 1
Kinematics and Dynamics

Physics and Math Chapter 2
Work and Energy

Physics and Math Chapter 8
Light and Optics

CHAPTER 10

Mathematics

SCIENCE MASTERY ASSESSMENT

Every pre-med knows this feeling: there is so much content I have to know for the MCAT! How do I know what to do first or what's important?

While the high-yield badges throughout this book will help you identify the most important topics, this Science Mastery Assessment is another tool in your MCAT prep arsenal. This quiz (which can also be taken in your online resources) and the guidance below will help ensure that you are spending the appropriate amount of time on this chapter based on your personal strengths and weaknesses. Don't worry though—skipping something now does not mean you'll never study it. Later on in your prep, as you complete full-length tests, you'll uncover specific pieces of content that you need to review and can come back to these chapters as appropriate.

How to Use This Assessment

If you answer 0–7 questions correctly:

Spend about 1 hour to read this chapter in full and take limited notes throughout. Follow up by reviewing **all** quiz questions to ensure that you now understand how to solve each one.

If you answer 8–11 questions correctly:

Spend 20–40 minutes reviewing the quiz questions. Beginning with the questions you missed, read and take notes on the corresponding subchapters. For questions you answered correctly, ensure your thinking matches that of the explanation and you understand why each choice was correct or incorrect.

If you answer 12–15 questions correctly:

Spend less than 20 minutes reviewing all questions from the quiz. If you missed any, then include a quick read-through of the corresponding subchapters, or even just the relevant content within a subchapter, as part of your question review. For questions you got correct, ensure your thinking matches that of the explanation and review the Concept Summary at the end of the chapter.

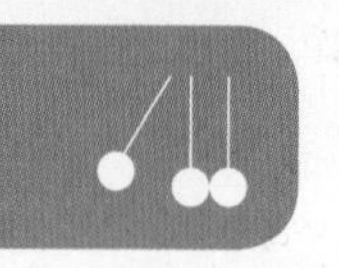

1. How would the number 17,060 be written in scientific notation?

A. 1706×10^1
B. 1.706×10^4
C. 1.7060×10^4
D. 0.17060×10^5

2. How does the number of significant digits differ between 14,320,010 and 3.618000?

A. 14,320,010 has more significant digits than 3.618000
B. 14,320,010 has fewer significant digits than 3.618000
C. 14,320,010 has the same number of significant digits as 3.618000
D. A comparison cannot be made because the numbers are not both in scientific notation.

3. Using the appropriate number of significant digits, what is the answer to the following math problem? (Note: Assume all numbers are the results of measurements.)

$$3.060 \times 4.10 + 200 =$$

A. 210
B. 213
C. 212.5
D. 212.55

4. Which of the following would be the most appropriate setup for estimating the value 3.6×4.85 for questions in which answer choices differ by a small margin?

A. 3.5×5
B. 3.5×4.5
C. 4×4
D. 4×5

5. The value of $200^{0.25}$ is closest to:

A. 4
B. 14
C. 50
D. 800

6. If A and B are real numbers, which of the following equations is INCORRECT?

A. $A^3 \times B^3 = (AB)^3$
B. $A^5 \div A^7 = A^{-2}$
C. $(A^{0.5})^4 + A^2 = 2A^2$
D. $(A^3)^2 = A^9$

7. How can the value of a natural logarithm be converted to the value of a common logarithm?

A. The natural logarithm is divided by a constant.
B. A constant is added to or subtracted from the natural logarithm.
C. The natural logarithm is raised to an exponent.
D. The inverse of the natural logarithm is taken.

8. What is the minimum value of $2 \cos \theta - 1$?

A. -3
B. -2
C. -1
D. 0

9. Which of the following relationships is INCORRECT?

A. $|\sin \theta \times \cos \theta| < |\sin \theta| + |\cos \theta|$
B. $\sin \theta \div \cos \theta = \tan \theta$
C. $\tan 90°$ is undefined
D. $\sin \theta = \sin (90° - \theta)$

10. What is the approximate pH of a solution with a pK_a of 3.6, [HA] = 100 m*M*, and $[A^-]$ = 0.1 *M*?

(Note: $pH = pK_a + \log \frac{[A^-]}{[HA]}$)

A. 1.6

B. 3.6

C. 5.6

D. 7.6

11. At what temperature do the Fahrenheit and Celsius scales give equal values?

A. 0 K

B. 233 K

C. 313 K

D. 273 K

12. In a certain rigid container, pressure and temperature are directly proportional. If the pressure is changed from 540 torr to 180 torr via a temperature change, by what factor has the temperature changed?

A. 360

B. 3

C. 1

D. $\frac{1}{3}$

13. A 150 pound man must be given a drug that is dosed at $1.5\ \frac{\text{mg drug}}{\text{kg body mass}}$. Approximately how many milligrams of the drug should be administered per dose? (Note: 1 lb = 4.45 N)

A. 33 mg

B. 67 mg

C. 100 mg

D. 225 mg

14. The rate of a reaction is calculated as a change in concentration per time. What are the units of the rate constant, *k*, in a reaction that is second order overall with respect to one species? (Note: A second-order reaction of this type has a rate law with the form rate = $k[A]^2$, where [A] is the concentration of the species.)

A. $\frac{1}{s}$

B. $\frac{M}{s}$

C. $\frac{L}{mol \cdot s}$

D. $\frac{L^2}{mol^2 \cdot s}$

15. Middle-aged men require a base level of 900 Calories per day plus an additional 12 Calories per kilogram of body mass per day. Young adult women require a base level of 500 Calories per day, plus 15 Calories per kilogram of body mass per day. At what mass do middle-aged men and young adult women have the same caloric needs?

A. 26 kg

B. 67 kg

C. 133 kg

D. 266 kg

Answer Key

1. **B** (Ch. 10.1)
2. **C** (Ch. 10.1)
3. **B** (Ch. 10.1)
4. **A** (Ch. 10.1)
5. **A** (Ch. 10.2)
6. **D** (Ch. 10.2)
7. **A** (Ch. 10.2)
8. **A** (Ch. 10.3)
9. **D** (Ch. 10.3)
10. **B** (Ch. 10.2)
11. **B** (Ch. 10.4)
12. **D** (Ch. 10.4)
13. **C** (Ch. 10.4)
14. **C** (Ch. 10.4)
15. **C** (Ch. 10.4)

CHAPTER 10

Mathematics

In This Chapter

10.1 **Arithmetic and Significant Figures**
- Scientific Notation 364
- Significant Figures 364
- Estimation 366

10.2 **Exponents and Logarithms**
- Exponents 368
- Rules of Logarithms 371
- Common vs. Natural Logarithms ... 371

10.3 **Trigonometry**
- Definitions and Relationships 373
- Common Values 374

10.4 **Problem-Solving**
- Use of Relationships............ 376
- Conversions 376
- Unit Analysis.................. 378
- Algebraic Systems 379

Concept Summary 383

Introduction

Going to the grocery store is not so different from solving an MCAT multiple choice question. You begin the process by determining how much of each item you need in the near future. Once you know what you need, you check what you already have in order to determine the quantity you need to buy to reach your goal. When you get to the store, you compare that amount to the containers on the shelves. Often they won't match exactly. Say, for example, that you need a total of 16 ounces of peas for a recipe. You already have five at home, so you only need 11 more. Packages of peas, however, may only come in 10- or 16-ounce packages. At that point, you choose the best one for your needs—the 16-ounce package; better to have a little extra than to run short!

If you've ever shopped in an international grocery store, this process can become even more elaborate because the packaging sizes and currency may not match the units with which you are familiar. You may have never taken the time to consider how intensive one's critical thinking must be to efficiently navigate the grocery store, but recognize that it's the same process you need to use on Test Day. First, figure out what you want (what is the question looking for), what you have (information in a passage, question stem, or outside knowledge), and what's needed (calculations and critical thinking), and then make a decision (by matching your answer, eliminating wrong answer choices, or guessing strategically). In this chapter, we'll be focusing on the calculations and critical thinking of mathematics. The math required for the MCAT is on the level of precalculus. You won't need any derivatives or integrals on Test Day, but rapid application of arithmetic, exponent and logarithm rules, trigonometry, statistics, and graphical analysis may be necessary to navigate the MCAT efficiently. In this chapter, you won't see very much new content, but consider this an opportunity to hone your mathematics skills.

MCAT EXPERTISE

You may note that this chapter (and the chapters following) do not contain the typical chapter profile, as they aren't directly related to Physics or AAMC content categories. However, this chapter covers fundamental math content required for calculations across all three science test sections. As such, Chapter 10 should be considered a must-review chapter!

10.1 Arithmetic and Significant Figures

LEARNING OBJECTIVES

After Chapter 10.1, you will be able to:

- Convert values to and from scientific notation
- Determine the significant digits within a number, such as: 1,547,200
- Apply rounding techniques in multiplication and division problems

KEY CONCEPT

It is essential to use judicious rounding and math strategies on Test Day to get through the MCAT efficiently. This is particularly true in the *Chemical and Physical Foundations of Biological Systems* section. Do not try to solve for an exact answer—do only as much as you need to be able to choose the right answer choice!

The MCAT often uses numbers that aren't particularly "nice" looking, especially considering that calculators cannot be used on the test. However, the testmakers *also* know that calculators aren't allowed, so even the most complex math still has to be solvable in a reasonable amount of time. We reconcile these two opposing concepts by using a few Test Day tricks: scientific notation, which can help us narrow down the exponent of our answer choice and often gives the answer directly; and judicious estimation, which will differentiate between otherwise similar answers. While significant figures won't lead us to an answer in the way that the other MCAT skills will, it is a testable topic on the MCAT.

Scientific Notation

Scientific notation is a method of writing numbers that takes advantage of **powers of ten**. In scientific notation, a number is written with a **significand** and an **exponent**. This is much easier to conceptualize with an example. Consider the number 217. The math using this number can be somewhat cumbersome. By transforming it into scientific notation (2.17×10^2), the number becomes easier to manipulate because the power of 10 has been pulled out. In this case, 2.17 is the significand (also called the **coefficient** or **mantissa**), and the 2 in 10^2 is the exponent.

The significand must be a number with an absolute value in the range [1,10). This means that it is any real number between -10 and -1 (not including -10) or between 1 and 10 (not including 10). By extension, the significand cannot begin with a 0, nor can it begin with two digits before the decimal point. The exponent, on the other hand, can be any whole number—positive, negative, or 0.

If at any time your calculations are not in scientific notation, consider adjusting them. While there is a small time investment converting to scientific notation, the time saved on subsequent calculations usually makes up for—and often exceeds—this time investment. This is especially true for questions in which the answers differ by powers of ten. The only exception to maintaining scientific notation is in the calculation of square roots, which are discussed later in this chapter.

Significant Figures

Significant figures provide an indication of our certainty of a measurement, and help us to avoid exceeding that certainty when performing calculations. Significant figures are determined by the precision of the instrument being used for measurement. For example, imagine that you are measuring the width of a block of wood

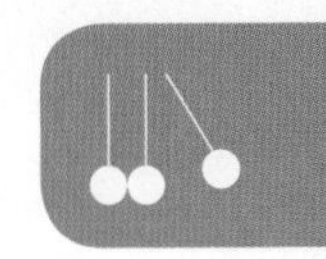

with a ruler. The ruler has markings for centimeters and millimeters; you could state with confidence the width of the block in millimeters—say, 55 millimeters.

However, on this ruler, there are no markings smaller than millimeters; you'd be forced to estimate where within the interval between two millimeter markings the block reaches—say 55.2 millimeters. You cannot be 100 percent confident about this decimal, but some information is better than none, and writing it down lets you know that you *were* confident about the first two digits.

In the situation we just described, only the first two digits would be considered significant because we know that they were measured accurately. We can hold on to the third digit during calculations, but by the time we reach a final answer, we need to reduce the answer to an appropriate number of significant figures. To determine the number of significant figures in a number:

- Count all numbers between the first nonzero digit on the left and the last nonzero digit on the right. Any digit between these two markers (including 0) is significant.
- Any zeroes to the left of the first nonzero digit are considered **leading zeroes** and are not significant.
- If there are zeroes to the right of the last nonzero digit *and* there is a decimal point in the number, then those zeroes are significant figures. If there is no decimal point, they are not significant. For example, 3,490 has three significant figures, while 3,490.0 has five.
- For measurements, the last digit is usually an estimation and is not considered significant (as in the example above).

Scientific notation can clarify significant figures when it contains a decimal point. When converting between standard numbers and scientific notation, be sure to maintain the number of significant figures. 100.0 is written in scientific notation as 1.000×10^2, while 100 is written as 1×10^2 because the **trailing zeroes** in the first example are significant while in the second example they are not.

Math with Significant Figures

Significant figure estimations are most important in the laboratory sciences, particularly analytical chemistry. For multiplication and division, maintain as many digits as possible throughout the calculations so that there is very little rounding error, then round to the number of significant digits that is the same as the *least number of significant digits* in any of the factors, divisors, or dividends. With addition and subtraction, *decimal points are maintained* rather than maintaining significant figures. The convention for decimal points is the same as for significant figures: the answer may have only as many decimal digits as the initial number with the fewest decimal digits.

BRIDGE

Significant figures are important because they give an indication of the accuracy of a measurement. Inaccurate measurements can bias research or lead to faulty conclusions. When presented with data, look for accuracy of the measurements in two ways: identifying the number of significant digits in a number, and looking for error margins or statistical significance in graphs. These latter topics are discussed in Chapter 12 of *MCAT Physics and Math Review*.

MCAT EXPERTISE

Most Test Day math (and, by extension, this *Kaplan MCAT Review* series) neglects significant figures in the answer choices. These calculations are only necessary when specified by the question stem or passage.

Example: Determine the volume of a cylinder with a radius that is measured as 7.45 m and a height of 8.323 m. (Note: Use 3.14159 as π, and round the answer to the correct number of significant digits.)

Solution:

$$\begin{aligned} V &= A_{\text{base}} \times h \\ &= \left(\pi r^2\right) \times h \\ &= (3.14159)(7.45 \text{ m})^2(8.323 \text{ m}) \\ &= 1{,}451.249 \end{aligned}$$

Because all of the factors are multiplied, the answer should have the same number of significant digits as the factor with the fewest number of significant digits. In this case, that is the radius, which has only two significant digits (remember that, in the case of measurements, the last digit is an estimate and is not considered significant). Therefore, the correct answer is 1,500 or 1.5×10^3.

Estimation

On Test Day, much of your math will be determined by the answer choices provided. If the answer choices are very close together, there will be minimal opportunity for rounding; when they are far apart, rough estimations are all that are necessary. While estimation of addition and subtraction are relatively simple rounding choices, we'll review a few tricks for multiplication and division.

Multiplication

Consider the following multiplication problem: $(3.17 \times 10^4) \times (4.53 \times 10^5)$. To three significant digits, the answer to this multiplication problem is 1.44×10^{10}, but this precise calculation is beyond the scope of mental math. However, even if the answer choices are close, it is generally acceptable to round to one decimal place, or $(3.2 \times 10^4) \times (4.5 \times 10^5)$. When rounding numbers in multiplication, keep in mind whether the rounded number is larger or smaller than the original number. If one number is rounded up, it is best to round the other number down slightly to compensate. Even with this rounding, the answer still comes out as 1.44×10^{10}.

If the answer choices are *very* far apart—differing by, say, powers of ten—we can adjust the numbers so that one contains only one significant digit, further simplifying the math. In this example, the calculations could be adjusted to $(3 \times 10^4) \times (4.5 \times 10^5)$, or 1.35×10^{10}. This represents an error of 6.25%, which is still close enough to choose the correct answer for most questions on Test Day.

Division

Let's also consider division as an avenue for estimation. While in our multiplication example we adjusted each number in an opposite direction, with division we are attempting to make proportional adjustments in the same direction. Consider the following example:

Example: Estimate the value of $15.4 \div 3.80$.

Solution: Estimations in division should be made by shifting both numbers in the same direction. It is often easier to adjust the divisor first to simplify calculations. If we round the divisor up to 4, we should round the dividend up accordingly. In this case, it makes sense to round the dividend up to 16—which is not only a whole number, but also a multiple of 4. Our estimate is $16 \div 4 = 4$. Note that, despite this very rough adjustment, we are still very close to the true value of 4.05.

KEY CONCEPT

When rounding numbers to be multiplied, round one number up and one number down to compensate. When rounding numbers to be divided, round both numbers in the same direction to compensate.

MCAT CONCEPT CHECK 10.1

1. Describe the process for converting a number into scientific notation. What values are possible for the significand?

____

____

____

2. Highlight or circle the significant digits in the following numbers:
   - 34,600.
   - 0.0003201
   - 1.10
   - 525,600

3. When rounding two numbers containing decimals, in which direction(s) should each number go for multiplication? For division?
   - Multiplication

   ____

   - Division

   ____

10.2 Exponents and Logarithms

LEARNING OBJECTIVES

After Chapter 10.2, you will be able to:

- Estimate the square root of a given value, like $\sqrt{1942}$
- Estimate the log value of a given number
- Simplify expressions such as: $(a + 2b)^3$

For many students, exponents and logarithms are topics filed away in the depths of memory. While exponential and logarithmic functions are uncommon in everyday life, a number of science topics and equations regularly tested on the MCAT require use of these concepts, as shown in Table 10.1.

Topic	Equation	Location in *Kaplan MCAT Review* Series
Sound level	$\beta = 10 \log \frac{I}{I_0}$	Chapter 7 of *MCAT Physics and Math Review*
Exponential decay	$n = n_0 e^{-\lambda t}$	Chapter 9 of *MCAT Physics and Math Review*
Arrhenius equation for activation energy	$k = Ae^{\frac{-E_a}{RT}}$	Chapter 5 of *MCAT General Chemistry Review*
Gibbs free energy	$\Delta G^\circ_{rxn} = -RT \ln K_{eq}$	Chapter 7 of *MCAT General Chemistry Review*
p scales (pH, pOH, pK_a, pK_b)	$pH = -\log [H^+]$	Chapter 10 of *MCAT General Chemistry Review*
Henderson–Hasselbalch equation	$pH = pK_a + \log \frac{[A^-]}{[HA]}$	Chapter 10 of *MCAT General Chemistry Review*

Table 10.1 Common Exponential and Logarithmic Equations on the MCAT

Exponents

In addition to exponential equations, exponents appear frequently on the MCAT in the context of scientific notation, discussed earlier. Here, we look at the rules of arithmetic with exponents.

Exponent Identities

Only a basic understanding of exponents is necessary for the MCAT, although it can be helpful to know a few values and basic rules. First, any number to the zeroth power is equal to 1:

$$X^0 = 1$$

Equation 10.1

When adding or subtracting numbers with exponents, the true value must be calculated before the addition or subtraction can be performed. For example, $3^2 + 3^2 \neq 6^2$; rather, $3^2 + 3^2 = 9 + 9 = 18$. However, if the base and exponent are the same, we can add the coefficients: $3^2 + 3^2 = (1 + 1) \times 3^2 = 2 \times 3^2 = 18$.

In cases of multiplication and division, the exponents can be manipulated directly, as long as the base number is the same. When multiplying two numbers with the same base, the exponents are added to determine the new number:

$$X^A \times X^B = X^{(A+B)}$$

Equation 10.2

In division, we subtract the exponent of the denominator from the exponent in the numerator to find the exponent in the quotient, as long as all bases are the same:

$$\frac{X^A}{X^B} = X^{(A-B)}$$

Equation 10.3

For a number that is raised to an exponent and then raised again to another exponent, the two exponents are multiplied:

$$(X^A)^B = X^{(A \times B)}$$

Equation 10.4

When a fraction is raised to an exponent, the exponent is distributed to the numerator and denominator:

$$\left(\frac{X}{Y}\right)^A = \frac{X^A}{Y^A}$$

Equation 10.5

Negative exponents represent inverse functions:

$$X^{-A} = \frac{1}{X^A}$$

Equation 10.6

KEY CONCEPT

When adding, subtracting, multiplying, or dividing numbers with exponents, the base must be the same.

For fractional exponents, the numerator can be treated as the exponent, and the denominator represents the root of the number:

$$X^{\frac{A}{B}} = \sqrt[B]{X^A}$$

Equation 10.7

Estimating Square Roots

On Test Day, you may be expected to calculate approximate square roots. To do so, it is useful to be familiar with the values in Table 10.2.

X	X²	X	X²	X	X²	X	X²
1	1	**6**	36	**11**	121	**16**	256
2	4	**7**	49	**12**	144	**17**	289
3	9	**8**	64	**13**	169	**18**	324
4	16	**9**	81	**14**	196	**19**	361
5	25	**10**	100	**15**	225	**20**	400

Table 10.2 Square Values of Integers from 1 to 20

If you are asked to calculate the square root of any number less than 400, you can approximate its value by determining which two perfect squares it falls between. As an alternative method, you can divide the number given to you by known squares to attempt to reduce it:

$$\sqrt{180} = \sqrt{4} \times \sqrt{9} \times \sqrt{5} = 2 \times 3 \times \sqrt{5} = 6\sqrt{5}$$

MCAT EXPERTISE

Estimation of square roots and logarithms is generally sufficient to the first decimal place; don't struggle to become more precise because it won't be necessary on Test Day.

One can estimate this value by considering that the square root of five is somewhere between 2 and 3 ($2^2 = 4$ and $3^2 = 9$), and is closer to 2 than 3. If we estimate $\sqrt{5}$ to be about 2.2, then $6\sqrt{5} \approx 13.2$, which is congruent with our knowledge that the square root of 180 will be between 13 and 14. The true value of $\sqrt{180}$ is approximately 13.4.

If you are using a number in scientific notation, adjust the decimal by one place if necessary so that the exponent is easily divisible by two:

$$\sqrt{4.9 \times 10^{-7}} = \sqrt{49 \times 10^{-8}} = 7 \times 10^{-4}$$

Finally, it is useful to know the values of $\sqrt{2}$ and $\sqrt{3}$:

$$\sqrt{2} \approx 1.414 \text{ (use 1.4)}$$
$$\sqrt{3} \approx 1.732 \text{ (use 1.7)}$$

Equation 10.8

Rules of Logarithms

Logarithms follow many of the same rules as exponents because they are inverse functions. The logarithmic rules are described below:

$$\log_A 1 = 0$$
$$\log_A A = 1$$
$$\log A \times B = \log A + \log B$$
$$\log \frac{A}{B} = \log A - \log B$$
$$\log A^B = B \log A$$
$$\log \frac{1}{A} = -\log A$$

Equations 10.9 to 10.14

It is also useful to know that "p" can be shorthand for $-\log$; thus, $\text{pH} = -\log [\text{H}^+]$, $\text{p}K_\text{a} = -\log K_\text{a}$, and so on.

Example: Derive the Henderson–Hasselbalch equation from the expression for K_a.

Solution:

$$K_\text{a} = \frac{[\text{H}^+][\text{A}^-]}{[\text{HA}]}$$
$$\log K_\text{a} = \log \frac{[\text{H}^+][\text{A}^-]}{[\text{HA}]}$$
$$\log K_\text{a} = \log [\text{H}^+] + \log [\text{A}^-] - \log [\text{HA}]$$
$$-\log [\text{H}^+] = -\log K_\text{a} + \log [\text{A}^-] - \log [\text{HA}]$$
$$\text{pH} = \text{p}K_\text{a} + \log \frac{[\text{A}^-]}{[\text{HA}]}$$

Common vs. Natural Logarithms

Logarithms can use any base, but the most common are base ten, as in our decimal system, and base ***e*** (**Euler's number**, about 2.718). Base-ten logarithms ($\log_{10}$) are called **common logarithms**, whereas those based on Euler's number ($\log_e$ or ln) are called **natural logarithms**. Both common and natural logarithms obey the rules discussed above, but it can be easier to estimate common logarithms because of our familiarity with the decimal number system. Therefore, it is useful to be able to convert between natural logarithms and common logarithms:

$$\log x \approx \frac{\ln x}{2.303}$$

Equation 10.15

KEY CONCEPT

e is Euler's number, which is 2.718281828459045. . . . It is also the base for the natural logarithm.

Estimating Logarithms

When estimating the logarithm of a number, use scientific notation. An exact logarithmic calculation of a number that is not an integer power of 10 is unnecessary on the MCAT. The testmakers are interested, however, in testing your ability to apply mathematical concepts appropriately in solving certain problems. Fortunately, there is a simple method of approximation that can be used on Test Day. If a value is written in proper scientific notation, it will be in the form $n \times 10^m$, where n is a number between 1 and 10. From this fact, we can use logarithm rules to approximate the value:

$$\begin{aligned}\log\left(n \times 10^m\right) &= \log(n) + \log\left(10^m\right) \\ &= m + \log(n)\end{aligned}$$

Because n is a number between 1 and 10, its logarithm will be a decimal between 0 and 1 ($\log 1 = 0$ and $\log 10 = 1$). The closer n is to 1, the closer $\log n$ will be to 0; the closer n is to 10, the closer $\log n$ will be to 1. As a reasonable approximation, one can say that

$$\log\left(n \times 10^m\right) \approx m + 0.n$$

Equation 10.16

where $0.n$ represents sliding the decimal point of n one position to the left (dividing n by ten). For example, $\log\left(9.2 \times 10^8\right) \approx 8 + 0.92 = 8.92$ (actual $= 8.96$).

BRIDGE

A similar concept for estimating logarithms is used in calculations of pH, as described in Chapter 10 of *MCAT General Chemistry Review*. The shortcut is slightly different because we are working with negative logarithms and a negative exponent in the case of pH:

$$-\log\left(n \times 10^{-m}\right) \approx m - 0.n.$$

MCAT CONCEPT CHECK 10.2

Before you move on, assess your understanding of the material with these questions.

1. Simplify the following expressions:
   - $(a + b)^2 =$ ____________________
   - $\frac{a^2 + 2a^2}{5a^3} =$ ____________________
   - $\log_a(a) =$ ____________________
   - $\log(a^3) - \log(a) =$ ____________________

2. Estimate $\sqrt{392}$:

____

____

____

3. Estimate log 7,426,135,420:

____

10.3 Trigonometry

LEARNING OBJECTIVES

After Chapter 10.3, you will be able to:

- Explain the appropriate way to orient vectors in vector addition
- Calculate the value of sine, cosine, or tangent for a given right triangle
- Recall the sine, cosine, and tangent values of key angles

Very little trigonometry is required for the MCAT, but a basic understanding of definitions and a strong knowledge of two special right triangles is essential for strong performance, especially on physics material.

Definitions and Relationships

For any given right triangle and angle, there are characteristic values of sine, cosine, and tangent that depend on the lengths of the legs of the triangle and of the hypotenuse, as shown in Figure 10.1.

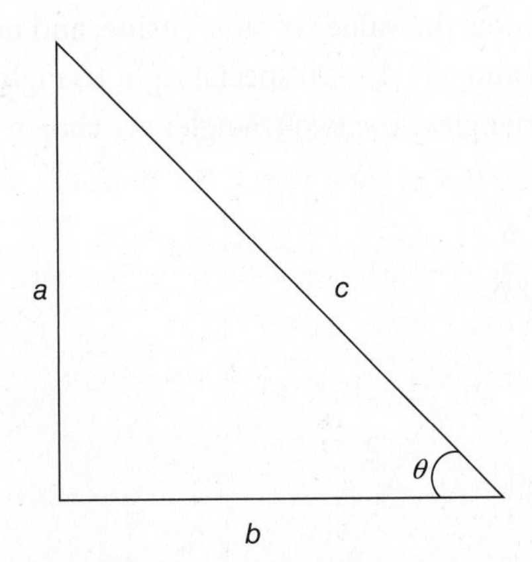

Figure 10.1 Right Triangle and Sides

Sine is calculated as the ratio between the side opposite the angle of interest and the hypotenuse:

$$\sin \theta = \frac{\text{opposite}}{\text{hypotenuse}} = \frac{a}{c}$$

Equation 10.17

Cosine is calculated as the ratio between the side adjacent to the angle of interest and the hypotenuse:

$$\cos \theta = \frac{\text{adjacent}}{\text{hypotenuse}} = \frac{b}{c}$$

Equation 10.18

MNEMONIC

Trigonometric ratios: **SOH CAH TOA:**

- **Sine = Opposite ÷ Hypotenuse**
- **Cosine = Adjacent ÷ Hypotenuse**
- **Tangent = Opposite ÷ Adjacent**

KEY CONCEPT

Trigonometric functions are useful for splitting a vector into its components; inverse trigonometric functions are useful for determining the direction of a resultant from its components.

Tangent is calculated as the ratio between the side opposite the angle of interest and the side adjacent to the angle of interest:

$$\tan\theta = \frac{\text{opposite}}{\text{adjacent}} = \frac{a}{b}$$

Equation 10.19

The values of both sine and cosine range from -1 to 1. The values of tangent, however, range from $-\infty$ to ∞.

Each trigonometric function also has an inverse function: **inverse sine ($\sin^{-1}$ or arcsin)**, **inverse cosine ($\cos^{-1}$ or arccos)**, and **inverse tangent ($\tan^{-1}$ or arctan)**. These functions use the calculated value of sine, cosine, or tangent, and yield a numerical value for the angle of interest. For the triangle in Figure 10.1, $\sin^{-1}\left(\frac{a}{c}\right) = \theta$. Inverse trigonometric functions are most likely to appear in questions asking for the direction of a resultant in vector addition or subtraction.

Common Values

On Test Day, you must know the values of sine, cosine, and tangent for all of the angles in the 30–60–90 and 45–45–90 special right triangles, either by memorization or by drawing the triangles. The two triangles are shown in Figure 10.2.

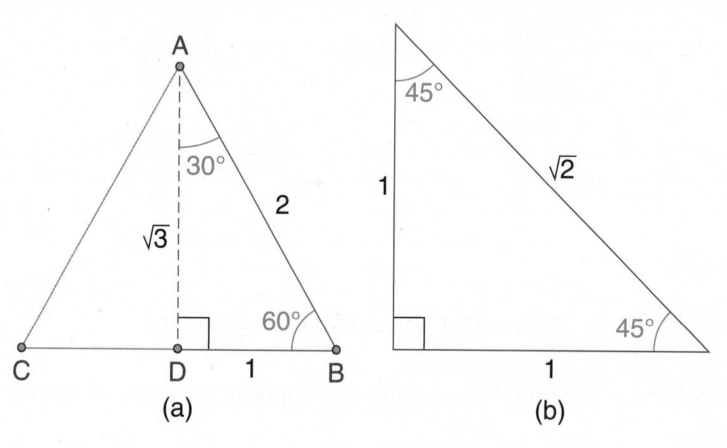

Figure 10.2 Special Right Triangles
(a) 30–60–90; (b) 45–45–90.

Important values of the trigonometric ratios at these angles are shown in Table 10.3.

θ	$\sin\theta$	$\cos\theta$	$\tan\theta$
0°	0	1	0
30°	$\frac{1}{2}$	$\frac{\sqrt{3}}{2}$	$\frac{\sqrt{3}}{3}$
45°	$\frac{\sqrt{2}}{2}$	$\frac{\sqrt{2}}{2}$	1
60°	$\frac{\sqrt{3}}{2}$	$\frac{1}{2}$	$\sqrt{3}$
90°	1	0	*undefined*
180°	0	−1	0

Table 10.3 Common Trigonometric Ratios on the MCAT

MCAT CONCEPT CHECK 10.3

Before you move on, assess your understanding of the material with these questions.

1. During vector addition, how is the angle of the resultant calculated?

2. How are sine, cosine, and tangent calculated when given the dimensions of a right triangle?
   - Sine:
   - Cosine:
   - Tangent:

3. True or False: Only angles in right triangles have characteristic values of the trigonometric functions.

4. At what angle between 0° and 180° does each trigonometric function have a value of 1?

10.4 Problem-Solving

LEARNING OBJECTIVES

After Chapter 10.4, you will be able to:

- Distinguish between direct and inverse relationships
- Convert between metric prefixes
- Solve a system of equations using substitution, setting equations equal, or elimination, such as: $3x + 4y = 17$ and $5x - 2y = 11$

Now that we've examined some individual mathematical skills, let's explore some common problem-solving strategies to attack MCAT questions. The use of relationships and proportionality is especially important in passage-based questions, while unit analysis can help determine which formulas are appropriate for a given question. The use of conversion factors is ubiquitous on the MCAT, as answer choices are often given in different units than the information presented. Algebraic systems are less often required, but may underlie passage interpretation and the approach to some questions.

Use of Relationships

Relationships are generally indicated in MCAT passages by formulas or the use of proportionality constants. In other cases they may be implied and require a bit more work on our part to calculate ratios. Calculations of this type are based on multiplication and division, but explaining the relationship in words—rather than math—may make it challenging to decode the connections between the variables. In **direct relationships**, increasing one variable proportionately increases the other; as one decreases, the other decreases by the same proportion. In **inverse relationships**, an increase in one variable is associated with a proportional decrease in the other.

BRIDGE

According to Boyle's law, pressure and volume have an inverse relationship: as one is doubled, the other is cut in half (keeping all else constant). On the other hand, according to Gay-Lussac's law, pressure and temperature have a direct relationship: as one is doubled, so is the other (keeping all else constant).

Conversions

The MCAT routinely increases the difficulty of a question by requiring the use of conversion factors. Equations may require that variables be in certain formats, or answer choices may differ in units from those given in the question stem. In both cases, it is necessary to convert units. The simplest conversions to perform maintain the same base unit. For example, conversion between grams, kilograms, and milligrams only requires multiplication by an appropriate power of ten. Metric prefixes and their associated powers of ten are found in Table 10.4.

Factor	Prefix	Prefix Abbreviation
10^{12}	*tera–*	T
10^{9}	*giga–*	G
10^{6}	*mega–*	M
10^{3}	*kilo–*	k
10^{2}	*hecto–*	h
10^{1}	*deka–*	da
10^{-1}	*deci–*	d
10^{-2}	*centi–*	c
10^{-3}	*milli–*	m
10^{-6}	*micro–*	μ
10^{-9}	*nano–*	n
10^{-12}	*pico–*	p

Table 10.4 Metric Prefixes

In addition to the conversions that are necessary for changes in prefixes, we must often convert between units, particularly between the British system and SI units. Table 10.5 shows several important conversion factors to recognize on Test Day. Conversion factors (except those for time) should not be memorized; the MCAT will provide them as necessary.

Base Unit	Equivalent Units
1 mile	5280 feet (ft)
1 ft	12 inches (in)
1 inch (in)	2.54 cm
1 Calorie (Cal)	1000 cal
1 calorie (cal)	4.184 J
1 electron–volt (eV)	1.602×10^{-19} J
1 L	33.8 ounces (oz)
1 pound (lb)	4.45 N
1 atomic mass unit (amu)	1.661×10^{-27} kg

Table 10.5 Common Conversion Factors on the MCAT

Example: A car's speedometer registers a speed of 33 miles per hour. What was its speed in meters per second?

Solution: First convert distance measurements, being careful to cancel them out by arranging numerators and denominators.

$$33\ \frac{\text{miles}}{\text{hour}}\left(\frac{5280\ \text{feet}}{1\ \text{mile}}\right)\left(\frac{12\ \text{inches}}{1\ \text{foot}}\right)\left(\frac{2.54\ \text{cm}}{1\ \text{inch}}\right)\left(\frac{1\ \text{m}}{100\ \text{cm}}\right)$$
$$\approx \frac{(33)(5280)(30)}{100} \approx \frac{(1000)(5280)}{100} = 52{,}800\ \frac{\text{m}}{\text{hour}}\left(\text{actual} = 53{,}108\ \frac{\text{m}}{\text{hour}}\right)$$

Then, repeat the procedure with the time measurements.

$$53{,}108\ \frac{\text{m}}{\text{hour}}\left(\frac{1\ \text{hour}}{3600\ \text{s}}\right) \approx \frac{540}{36} = \frac{90}{6} = 15\ \frac{\text{m}}{\text{s}}\left(\text{actual} = 14.8\ \frac{\text{m}}{\text{s}}\right)$$

REAL WORLD

It is important to be able to convert between Fahrenheit and Celsius scales in medicine as different hospitals (and different medical records) may use different units. Body temperature is 98.6°F or 37°C. A fever is usually defined as a temperature above 100.4°F or 38°C. Hypothermia is usually defined as a temperature below 95.0°F or 35°C.

One special case of conversions occurs with temperature. Rather than simply multiplying by a conversion factor, there is also a component of addition or subtraction. The following formulas relate the Fahrenheit, Celsius, and Kelvin systems:

$$F = \frac{9}{5}C + 32$$
$$K = C + 273$$

Equation 10.20

where *F*, *C*, and *K* are the temperatures in degrees Fahrenheit, degrees Celsius, and kelvins, respectively.

Unit Analysis

Unit analysis, also called **dimensional analysis**, may help determine the correct answer even if you forget a relevant formula on Test Day. It can also serve as a double check on one's calculations because the units of the calculated answer must match the units of the answer choices. For example, consider a question in which we are given two quantities: one in $\frac{\text{N}}{\text{C}}$, and the other in volts. The answer choices for the question are all in meters. Even without remembering the equation $V = Ed$, we can infer that we must divide the voltage $\left(1\ \text{V} = 1\ \frac{\text{J}}{\text{C}} = 1\ \frac{\text{N}\cdot\text{m}}{\text{C}}\right)$ by the electric field to get a distance in meters:

$$\frac{\text{N}\cdot\text{m}}{\text{C}} \div \frac{\text{N}}{\text{C}} = \text{m}$$

Dimensional analysis is not a foolproof strategy; it is always better to know the true relationships between variables than to infer them based on units. Still, this strategy can be effective for narrowing down (or even choosing) answer choices on Test Day.

Example: The ejection fraction is the proportion of the blood volume in the left ventricle expelled with each contraction of the heart. A patient is known to have an ejection fraction of 0.6, a cardiac output of 5 $\frac{\text{L}}{\text{min}}$ and a heart rate of 80 $\frac{\text{beats}}{\text{min}}$. What is the volume of blood in this person's left ventricle just prior to contraction?

Solution: A formula was not provided in this question, but we can recognize that the desired answer is a volume. We can start with the cardiac output and heart rate terms to determine the volume ejected per beat.

$$\frac{5\,\frac{\text{L}}{\text{min}}}{80\,\frac{\text{beats}}{\text{min}}} = \frac{1}{16} = 0.0625\,\frac{\text{L}}{\text{beat}}$$

The question also explains that only 60 percent of the blood volume is expelled from the left ventricle per heartbeat. From this, we can determine the volume of blood prior to contraction.

$$0.0625\,\frac{\text{L}}{\text{beat}} = \frac{0.6 \times \text{ventricular volume}}{\text{beat}}$$

$$\frac{0.0625}{0.6}\,\text{L} = \text{ventricular volume} \approx 0.1\text{ L}\ (\text{actual} = 0.104\text{ L})$$

Algebraic Systems

The last key mathematical skill for Test Day is the ability to solve systems of linear equations. In order to solve a system of equations, there must be at least as many equations as there are variables. Where there is only one variable (which does not truly constitute a system), only one equation is necessary; for example, $6 - x = 1$ reduces to $x = 5$. In contrast, with an equation like $3x + 4y = 17$, there is insufficient data to solve for either variable with only the one equation. If a second equation is introduced, such as $5x - 2y = 11$, then we can solve for both variables using one of three methods: substituting one variable in terms of the other, setting equations equal to each other, or manipulating the equations to eliminate one of the variables.

Substitution

In substitution, we solve for one variable in one of the equations, and then insert this term into the other equation. The steps of this method are listed below.

- Solve for one of the variables in one of the equations:

$$5x - 2y = 11$$
$$y = \frac{5x - 11}{2}$$

- Insert the expression into the other equation:

$$3x + 4\left(\frac{5x - 11}{2}\right) = 17$$

- Isolate the variable and solve the resulting equation:

$$3x + 10x - 22 = 17$$
$$13x = 39$$
$$x = 3$$

- Solve for the other variable using this value:

$$3(3) + 4y = 17$$
$$4y = 8$$
$$y = 2$$

Setting Equations Equal

Setting equations equal to one another is a specialized case of substitution. In this method, we solve for the same variable in both equations and then set the two equations equal to each other. The steps of this method are listed below.

- Solve for the same variable in both equations:

$$y = \frac{17 - 3x}{4}$$
$$y = \frac{5x - 11}{2}$$

- Set the equations equal to each other, isolate the variable, and solve for the variable:

$$\frac{17 - 3x}{4} = \frac{5x - 11}{2}$$
$$17 - 3x = 10x - 22$$
$$39 = 13x$$
$$3 = x$$

- Solve for the other variable using this value:

$$3(3) + 4y = 17$$
$$4y = 8$$
$$y = 2$$

Elimination

In elimination, multiply or divide one (or both) of the equations to get the same coefficient in front of one of the variables in both equations. Then, add or subtract the equations as necessary to eliminate one of the variables. The steps of this method are listed below.

- Multiply or divide one (or both) of the equations by a constant so that the coefficient in front of one of the variables in both equations is the same:

$$3x + 4y = 17$$
$$2 \times [5x - 2y = 11]$$

- If the sign of both coefficients is the same, subtract one equation from the other. If the sign is opposite, add the two equations together:

$$\begin{aligned} 4y + 3x &= 17 \\ -\ 4y + 10x &= 22 \\ \hline 13x &= 39 \\ x &= 3 \end{aligned}$$

- Solve for the other variable using this value:

$$\begin{aligned} 3(3) + 4y &= 17 \\ 4y &= 8 \\ y &= 2 \end{aligned}$$

Note that each method results in the same answer despite slight differences in the steps taken. As a matter of convention, the answers for systems of equations with the variables x and y are reported as coordinates on the Cartesian plane (x,y); thus, our answer for this system would be (3,2). Systems of equations can have many variables, but it is unlikely that you will encounter a system with more than three variables (x,y,z) on the MCAT.

MCAT CONCEPT CHECK 10.4

Before you move on, assess your understanding of the material with these questions.

1. How are conversions between metric prefixes accomplished? How would a distance be converted from millimeters to kilometers?

____

____

2. What does it mean for two variables to have a direct relationship? An inverse relationship?
   - Direct:

____

   - Inverse:

____

3. Each of the three methods for solving systems of equations discussed in this chapter solve for one variable, and then use this value to solve for the other. How does each method solve for the first variable?
   - Substitution:

   ___

   - Setting equations equal:

   ___

   - Elimination:

   ___

Conclusion

In this chapter, we reviewed many of the skills that are necessary for successful performance on the MCAT science sections. We began by examining relevant arithmetic calculations for Test Day, including scientific notation and significant figures. We continued our review by examining logarithms and exponents before discussing the most common trigonometric functions and their values. We finished our math review by working on problem-solving skills that will be valuable in your studying and during the MCAT itself. In the next two chapters, we'll review Test Day skills in experimental design and data analysis so that you'll be ready to answer all of the Test Day questions.

You've reviewed the content, now test your knowledge and critical thinking skills by completing a test-like passage set in your online resources!

GO ONLINE

Concept Summary

Arithmetic and Significant Figures

- **Scientific notation** is a method of writing numbers in a way that improves the ease of calculations and the comparability of significant digits.
  - Scientific notation takes the format [significand] $\times 10^{[\text{exponent}]}$.
  - The **significand** must be greater than or equal to 1 and less than 10.
  - The **exponent** must be an integer.
- **Significant figures** include all nonzero digits and any trailing zeroes in a number with a decimal point.
  - Measurements are an exception, in that the last digit provided is not significant.
  - In addition and subtraction, reduce the answer to have the same number of decimal places as the number with the fewest number of decimal places.
  - In multiplication and division, reduce the answer to have the same number of significant digits as the number with the fewest number of significant digits.
  - The entire number should be maintained throughout calculations to minimize rounding error.
- Estimation of multiplication and division should be done logically.
  - In multiplication, if one number is rounded up, the other should be rounded down in proportion.
  - In division, if one number is rounded up, the other should also be rounded up in proportion.

Exponents and Logarithms

- **Exponents** are a notation for repeated multiplication. They may be manipulated mathematically, especially when the bases are the same.
- **Logarithms** are the inverse of exponents and are subject to similar mathematical manipulations.
- **Natural logarithms**, which use base ***e*** (**Euler's number**) can be converted into **common logarithms**, which use base 10.

Trigonometry

- Trigonometric relationships can be calculated based on the lengths of the sides of right triangles.
- **Sine** is the ratio of the length of the side opposite an angle to the length of the hypotenuse.
- **Cosine** is the ratio of the length of the side adjacent to an angle to the length of the hypotenuse.
- **Tangent** is the ratio of the side opposite an angle to the side adjacent to it.
- **Inverse trigonometric functions** use the calculated value from a ratio of side lengths to calculate the angle of interest.

Problem-Solving

- In **direct relationships**, as one variable increases, the other increases in proportion.
- In **inverse relationships**, as one variable increases, the other decreases in proportion.
- Conversions between metric prefixes require multiplication or division by corresponding powers of ten.
- Conversions between units of different scales require multiplication or division, and may require addition or subtraction.
- Unit analysis (**dimensional analysis**) can determine the appropriate computation based on given information.
- Algebraic systems may be solved by substitution, setting equations equal, or elimination. The general ideas are the same in each—solve for one variable, and then substitute the variable into an equation to solve for the other—although the specific methods are different.

Answers to Concept Checks

10.1

1. First, determine which digits are significant, as these will be preserved in scientific notation. Then, move the decimal point until the significand is greater than or equal to 1 and less than 10. Finally, determine what power of 10 is necessary for multiplication to restore the original number.
2. 34,600.; 0.0003201; 1.10; 525,600
3. In multiplication, adjust the two decimals in opposite directions. In division, adjust the two decimals in the same direction.

10.2

1. • $(a+b)^2 = a^2 + 2ab + b^2$
   • $\frac{a^2+2a^2}{5a^3} = \frac{3a^2}{5a^3} = \frac{3}{5a}$
   • $\log_a(a) = 1$
   • $\log(a^3) - \log(a) = \log\frac{a^3}{a} = \log a^2 = 2\log a$
2. $\sqrt{392}$ is between $\sqrt{361}$ and $\sqrt{400}$, so the value is between 19 and 20. We can also simplify this radical:

$$\sqrt{392} = \sqrt{4} \times \sqrt{49} \times \sqrt{2} \approx 14 \times 1.4 = 19.6 \text{ (actual} = 19.8)$$

3. $\log 7{,}426{,}135{,}420 \approx \log(7.4 \times 10^9) \approx 9 + 0.74 = 9.74$ (actual $= 9.87$). Note that—even with an absurdly large number—we can still get relatively accurate estimations by following basic logarithm rules.

10.3

1. The value of a trigonometric function calculated from the dimensions of the resultant vector is used in the inverse tangent function to calculate the resultant vector angle. Inverse trigonometric ratios, in general, can be used to calculate angles.
2. The sine of an angle is equal to the ratio of the side opposite the angle to the hypotenuse. Cosine is the ratio of the side adjacent to the angle to the hypotenuse. Tangent is the ratio of the side opposite the angle to the side adjacent to the angle.
3. False. Even through calculating the values of sine, cosine, and tangent is more complicated in a triangle that does not contain a right angle, all possible angles do still have characteristic trigonometric values.
4. Sin is equal to 1 at 90°, cos at 0°, and tan at 45°. Many MCAT questions utilize the trend that sin increases going from 0° to 90°, whereas cos decreases over this range.

10.4

1. Conversion between metric prefixes is accomplished by multiplication or division by the appropriate power of ten. To convert from millimeters (10^{-3}) to kilometers (10^{3}), it is necessary to multiply by 10^{-6}. It's wise to double-check your work when converting: a kilometer is a larger unit of distance than a millimeter, so the number of kilometers should be smaller than the number of millimeters.
2. In direct relationships, as one quantity increases, the other also increases in proportion. In inverse relationships, as one quantity increases, the other decreases in proportion.
3. In substitution, solve one equation for one variable in terms of the other; then, substitute this expression into the other equation. In setting equations equal (a modified version of substitution), solve both equations for the same variable and set them equal to each other. In elimination, multiply or divide one (or both) equations so that the coefficient in front of one of the variables is the same in both equations; then, add or subtract the equations to eliminate one of the variables.

Science Mastery Assessment Explanations

1. **B**

This question, while overtly testing the ability to use scientific notation, is also checking on the appropriate use of significant digits. Because there is no decimal point, the last zero is not significant and should not be used in scientific notation. The significand in scientific notation should always be between one and ten.

2. **C**

Significant digits include all nonzero digits, all zeroes that are between nonzero digits, and trailing zeroes in any number with a decimal point. In 14,320,010 there is no decimal point; thus, the last zero is insignificant and there are seven significant digits. In 3.618000, all of the digits are significant; thus there are also seven significant digits.

3. **B**

While all digits are preserved during calculations, the final determination of the number of digits is made by both significant figures and decimal places. During multiplication, the answer is maintained to the smallest number of significant digits. During addition, it is maintained to the smallest number of decimal places. By following the order of operations, addition is the last operation; thus we cannot have a decimal in our answer choice. Because multiplication occurred earlier, the result of that multiplication may be shortened according to the two significant figures in 4.10, but not the entire answer.

4. **A**

When estimating the product of two numbers, it is best to round one up while rounding the other down, as in **(A)**. **(B)** and **(D)** each round both numbers in the same direction, which would increase the amount of error in the answer. **(C)** rounds the numbers in opposite directions, but the degree of rounding is significantly larger than in **(A)** and too extreme for answer choices that differ by small amounts.

5. **A**

The fourth root of a number, or a number raised to the one-quarter power, is the square root of the square root of that number:

$$200^{0.25} = 200^{\frac{1}{2}\times\frac{1}{2}} = \sqrt{\sqrt{200}} = \sqrt[4]{200}$$

The square root of 200 should be a bit larger than 14 ($14^2 = 196$); therefore, the fourth root of 200 should be a bit less than 4.

6. **D**

Raising an exponent to another exponent requires multiplying the exponents. Thus, $(A^3)^2 = A^6$.

7. **A**

The relationship between the natural logarithm of a number and the common logarithm of a number is $\log x = \frac{\ln x}{2.303}$. Therefore, the natural logarithm of a number must be divided by the constant 2.303 to obtain the common logarithm of the same number.

8. **A**

The minimum value of the cosine function is -1 ($\cos 180° = -1$). Therefore, the minimum value of $2 \cos \theta - 1$ is $2 \times (-1) - 1 = -3$.

9. **D**

$\sin \theta \neq \sin (90° - \theta)$, although $\sin \theta = \cos (90° - \theta)$. The other statements must all be true. Because sine and cosine values are always between -1 and 1, the product of sine and cosine will always have a magnitude less than 1. The sum of the absolute value of sine and the absolute value of cosine, on the other hand, will always be greater than 1. Therefore, **(A)** can be eliminated. Because sine is the ratio of opposite to hypotenuse and cosine is the ratio of adjacent to hypotenuse, the quotient between the two is the ratio of opposite to adjacent, or the tangent of the angle. Therefore, **(B)** can be eliminated. By the same logic, because $\sin 90° = 1$ and $\cos 90° = 0$, $\tan 90°$ is undefined, eliminating **(C)**.

10. **B**

This question involves both a unit conversion between millimolar values and molar values, and calculation of a logarithm. The relationship between pH and pK_a is described by the Henderson–Hasselbalch equation given in the question stem. 100 $mM = 0.1\ M$, so

$$\text{pH} = \text{p}K_a + \log \frac{[\text{A}^-]}{[\text{HA}]} = 3.6 + \log(1) = 3.6$$

11. **B**

This question requires not only unit conversions, but algebra as well. Given that $F = \frac{9}{5}C + 32$, the temperature T can calculated as:

$$T = \frac{9}{5}T + 32 \rightarrow -\frac{4}{5}T = 32 \rightarrow T = -\frac{5}{4}(32) = -40$$

However, the answers are given in kelvin. $-40°\text{C} + 273 = 233$ K.

12. **D**

In a direct relationship, a change in one of the variables will be associated with a proportional change in the other. Because the pressure was multiplied by $\frac{1}{3}$, the temperature must also be multiplied by $\frac{1}{3}$. Note that the fractional relationships can only be used with temperatures in kelvin.

13. **C**

Because grams are a unit of mass and pounds are a unit of force, we must first convert pounds to newtons, and then divide by the acceleration due to gravity to find kilograms. The weight of the person in newtons is

$$150 \text{ lb} \times 4.45 \frac{\text{N}}{\text{lb}} \approx 150 \times 4.4 = 600 + 60$$
$$= 660 \text{ (actual} = 667.5 \text{ N)}.$$

This corresponds to a mass of

$$\frac{667.5 \text{ N}}{9.8 \frac{\text{m}}{\text{s}^2}} = \frac{667.5 \frac{\text{kg} \cdot \text{m}}{\text{s}^2}}{9.8 \frac{\text{m}}{\text{s}^2}} \approx 67 \text{ kg}.$$

Now, we can determine the dose:

$$67 \text{ kg}\left(\frac{1.5 \text{ mg drug}}{\text{kg body mass}}\right) \approx 100 \text{ mg}.$$

14. **C**

According to the question stem, the rate of a reaction is measured as a change in concentration over time, and thus has the units $\frac{M}{\text{s}}$, where M (molarity) is measured in moles per liter. However, the rate of the reaction is equal to a rate constant times the concentrations of certain reactants squared. In this case, we know the units of everything except the rate constant and must solve for its units:

$$\text{rate} = k[\text{A}]^2$$
$$\frac{M}{\text{s}} = k[M]^2 \rightarrow k = \frac{1}{M \cdot \text{s}} = \frac{\text{L}}{\text{mol} \cdot \text{s}}$$

15. **C**

This is a system of equations couched in data. From this information, we can construct two equations:

$$C = 900 + 12m$$
$$C = 500 + 15m$$

These equations can be solved by setting them equal:

$$900 + 12m = 500 + 15m$$
$$400 = 3m$$
$$133 \text{ kg} = m$$

Consult your online resources for additional practice. GO ONLINE

Equations to Remember

(10.1) **Zero exponent identity:** $X^0 = 1$

(10.2) **Multiplying like bases with exponents:** $X^A \times X^B = X^{(A+B)}$

(10.3) **Dividing like bases with exponents:** $\frac{X^A}{X^B} = X^{(A-B)}$

(10.4) **Raising an exponent to another exponent:** $(X^A)^B = X^{(A \times B)}$

(10.5) **Raising fractions to exponents:** $\left(\frac{X}{Y}\right)^A = \frac{X^A}{Y^A}$

(10.6) **Raising bases to negative exponents:** $X^{-A} = \frac{1}{X^A}$

(10.7) **Raising bases to fractional exponents:** $X^{\frac{A}{B}} = \sqrt[B]{X^A}$

(10.8) **Square root approximations:** $\sqrt{2} \approx 1.414$ (use 1.4)

$\sqrt{3} \approx 1.732$ (use 1.7)

(10.9) **Logarithm of 1 identity:** $\log_A 1 = 0$

(10.10) **Logarithm of base identity:** $\log_A A = 1$

(10.11) **Logarithm of product:** $\log A \times B = \log A + \log B$

(10.12) **Logarithm of quotient:** $\log \frac{A}{B} = \log A - \log B$

(10.13) **Logarithm of exponent-containing expression:** $\log A^{\mathrm{B}} = B \log A$

(10.14) **Logarithm of inverse:** $\log \frac{1}{A} = -\log A$

(10.15) **Conversion of natural to common logarithm:** $\log x \approx \frac{\ln x}{2.303}$

(10.16) **Scientific notation logarithm approximation:** $\log (n \times 10^m) \approx m + 0.n$

(10.17) **Definition of sine:** $\sin \theta = \frac{\text{opposite}}{\text{hypotenuse}} = \frac{a}{c}$

(10.18) **Definition of cosine:** $\cos \theta = \frac{\text{adjacent}}{\text{hypotenuse}} = \frac{b}{c}$

(10.19) **Definition of tangent:** $\tan \theta = \frac{\text{opposite}}{\text{adjacent}} = \frac{a}{b}$

(10.20) **Temperature conversions:** $F = \frac{9}{5}C + 32$

$K = C + 273$

Shared Concepts

General Chemistry Chapter 5
Chemical Kinetics

General Chemistry Chapter 7
Thermochemistry

General Chemistry Chapter 10
Acids and Bases

Physics and Math Chapter 7
Waves and Sound

Physics and Math Chapter 9
Atomic and Nuclear Phenomena

Physics and Math Chapter 12
Data-Based and Statistical Reasoning

CHAPTER 11

Reasoning About the Design and Execution of Research

SCIENCE MASTERY ASSESSMENT

Every pre-med knows this feeling: there is so much content I have to know for the MCAT! How do I know what to do first or what's important?

While the high-yield badges throughout this book will help you identify the most important topics, this Science Mastery Assessment is another tool in your MCAT prep arsenal. This quiz (which can also be taken in your online resources) and the guidance below will help ensure that you are spending the appropriate amount of time on this chapter based on your personal strengths and weaknesses. Don't worry though—skipping something now does not mean you'll never study it. Later on in your prep, as you complete full-length tests, you'll uncover specific pieces of content that you need to review and can come back to these chapters as appropriate.

How to Use This Assessment

If you answer 0–7 questions correctly:

Spend about 1 hour to read this chapter in full and take limited notes throughout. Follow up by reviewing **all** quiz questions to ensure that you now understand how to solve each one.

If you answer 8–11 questions correctly:

Spend 20–40 minutes reviewing the quiz questions. Beginning with the questions you missed, read and take notes on the corresponding subchapters. For questions you answered correctly, ensure your thinking matches that of the explanation and you understand why each choice was correct or incorrect.

If you answer 12–15 questions correctly:

Spend less than 20 minutes reviewing all questions from the quiz. If you missed any, then include a quick read-through of the corresponding subchapters, or even just the relevant content within a subchapter, as part of your question review. For questions you got correct, ensure your thinking matches that of the explanation and review the Concept Summary at the end of the chapter.

1. An experimenter is attempting to investigate the effect of a new antibiotic on *E. coli*. He plates cells and administers one milliliter of the antibiotic. Which of the following is an appropriate negative control in this experiment?
   - **A.** A plate with no cells that was coated with one milliliter of antibiotic.
   - **B.** A plate with *E. coli* and no additional treatment.
   - **C.** A plate with *E. coli* and one milliliter of isotonic saline.
   - **D.** A plate of epithelial cells treated with one milliliter of antibiotic.

2. Which of the following would best establish a causal link?
   - **A.** A cross-sectional study using survey data for hand-washing and colds.
   - **B.** A case–control study of an exposure during childhood and development of a certain disease later in life.
   - **C.** A randomized clinical controlled trial of a new antipyretic drug.
   - **D.** An IQ test where the results are later segregated by gender.

3. An experimenter is attempting to determine the internal energy of a well-known compound. He cleans his glassware, completes the synthesis, calibrates a bomb calorimeter, and then uses it to measure the appropriate thermodynamic values. Which of the following errors did he make?
   - **A.** He did not determine if the compound was novel or if the information has already been determined.
   - **B.** He did not have a specific goal at the beginning of his research.
   - **C.** He should not be involved in both the synthesis of the compound and later testing.
   - **D.** He should have calibrated the calorimeter before the synthesis of the compound.

4. A researcher wishes to generate a parameter for American women's mean weight. Which of the following is the most significant concern?
   - **A.** Measuring a person's weight may have psychological consequences and is unethical.
   - **B.** Gathering all of the necessary study participants would be prohibitive.
   - **C.** Knowing the average weight of women does not provide any useful information.
   - **D.** Enough studies have already been conducted on this topic to render it unnecessary.

5. A cross-sectional study in which current smoking status and cancer history are assessed simultaneously cannot satisfy which of Hill's criteria?
   - **A.** Strength
   - **B.** Coherence
   - **C.** Plausibility
   - **D.** Temporality

6. After randomization, it is discovered that one group in a study has almost twice as many women as the other. Which of the following is an appropriate response?
   - **A.** Move men and women between groups manually so that they have the same gender profile.
   - **B.** Check the randomization algorithm; if it is fair, continue with the research.
   - **C.** Eliminate all of these subjects because of potential bias and randomize a new cohort.
   - **D.** Keep the current cohort and continue randomizing subjects until the gender profiles are equal.

7. An experimenter is attempting to determine the effects of smoking on very low birth weight (VLBW) and of VLBW on IQ. Which of the following statements is correct?
   - **I.** Smoking is an independent variable.
   - **II.** Smoking is a dependent variable.
   - **III.** VLBW is an independent variable.
   - **IV.** VLBW is a dependent variable.

   - **A.** I only
   - **B.** II and IV only
   - **C.** I, III, and IV only
   - **D.** II, III, and IV only

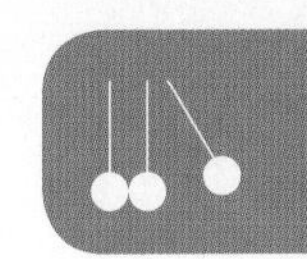

8. A study is performed on a new medication. Subjects in the experimental group are told about the potential side effects of the medication, while subjects in the placebo group are not. The subjects have no contact with each other and do not know in which group they are placed. The side effects end up being significantly more severe in the treatment group, when seen by the same assessor physician. This is most likely caused by which of the following?
   A. Physician unblinding only
   B. Patient unblinding only
   C. Both physician and patient unblinding
   D. Both physician and patient blinding

9. As part of a graduate program entrance exam, a student must submit a grant proposal for a scientific investigation of her choosing. While her hypothesis is well formulated and she has some preliminary data, her proposal is rejected, with the committee reviewing her proposal citing several articles that support her hypothesis. What mistake did she make in drafting her proposal?
   A. She did not formulate a testable hypothesis.
   B. She did not spend sufficient time reviewing existing studies.
   C. Her proposed study did not adequately test her hypothesis.
   D. She did not have enough preliminary data.

10. A new study of a weight loss drug uses a radio advertisement to generate study participation. What type of error is most likely to result?
   A. Hawthorne effect
   B. Selection bias
   C. Confounding
   D. Detection bias

11. A researcher designing a study has paid for it to be professionally translated into several languages. She discusses the potential risks and benefits with each participant and allows them to bring documentation home for review before committing to the study. This researcher has put special focus on:
   A. justice by explaining potential risks.
   B. beneficence by describing the potential benefits of the study.
   C. respect for persons by acknowledging the subject's perspective and rights.
   D. selection bias, by making the recruitment documents inclusive.

12. Which of the following methods would be most appropriate for an initial assessment of hemoglobin saturation during an experiment about breath holding?
   A. A pulse oximeter, which uses a small light on an adhesive bandage.
   B. An arterial cannula, which permits repeated blood draws with a single puncture.
   C. Repeated venipuncture, because a single puncture would cause data overlap.
   D. A Swan–Ganz catheter inserted through the femoral artery, which can measure saturation nearest the heart.

13. A medical student attempting to impress her attending physician refers to a recent article that says that there is a statistically significant difference in pregnancy length with a new therapy. Which of the following is most likely to be a valid criticism of the article?
   A. Medical students usually reference articles in the newest journals, which may be unreliable.
   B. The effect didn't change patient outcomes, only a secondary measure.
   C. There is a lack of internal validity in the results, despite significance.
   D. Selection bias is inherent in the scientific process.

14. Which sample would be the most appropriate participants for a study on hormone replacement therapy for postmenopausal symptoms?

- **A.** Prepubescent girls
- **B.** Premenopausal adult women
- **C.** Pregnant women
- **D.** Postmenopausal women

15. Use of a colorimetric assay to determine protein concentration may be subject to all of the following EXCEPT:

- **A.** the use of standards.
- **B.** measurement error.
- **C.** the Hawthorne effect.
- **D.** systematic error.

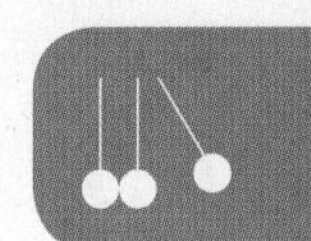

Answer Key follows on next page.

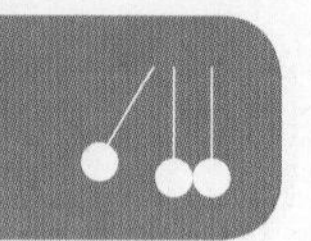

Answer Key

1. **C** (Ch. 11.2)
2. **C** (Ch. 11.3)
3. **A** (Ch. 11.1)
4. **B** (Ch. 11.5)
5. **D** (Ch. 11.3)
6. **B** (Ch. 11.3)
7. **C** (Ch. 11.2)
8. **A** (Ch. 11.3)
9. **B** (Ch. 11.1)
10. **B** (Ch. 11.3)
11. **C** (Ch. 11.4)
12. **A** (Ch. 11.4)
13. **B** (Ch. 11.5)
14. **D** (Ch. 11.5)
15. **C** (Ch. 11.3)

CHAPTER 11

Reasoning About the Design and Execution of Research

In This Chapter

11.1 **The Scientific Method**
The FINER Method 401

11.2 **Basic Science Research** HY
Controls 403
Causality............405
Error Sources 406

11.3 **Human Subjects Research**
Experimental Approach............408
Observational Approach 409
Error Sources 410

11.4 **Ethics**
Respect for Persons............413
Justice............413
Beneficence 414

11.5 **Research in the Real World**
Populations *vs.* Samples............416
Generalizability 416
Support for Interventions 417

Concept Summary 419

Introduction

In the modern world, there are standard places to search for the answers to questions we have: encyclopedias, academic journals, online databases, and other publications. By perusing current research on a topic, we can figure out whether someone else has already asked and answered our question. This is just as true for scientists and other professionals. In many cases, our searches may take us in different directions. This may be as simple as finding additional resources that confirm the answers to our questions; however, research is often more complicated than that. We may find conflicting answers and have to perform critical analysis to determine which data set (and conclusion) was obtained in the most legitimate, unbiased way. Other times, we'll have to find the answers on our own through experiment or observation and data analysis.

In order to generate a consistent body of knowledge, the ways in which scientists generate and search for information must be orderly and uniform. In this chapter, we will discuss the premise of basic science, biomedical, and social sciences research. Our examination will include the necessary criteria for causality, different types of error, and the ethical implications of human subjects research. These research methods are core skills that the MCAT tests—in fact, they constitute one of the four *Scientific Inquiry and Reasoning Skills* on the exam. In addition, their importance will last for the rest of your medical career. As a physician, you will constantly be seeking answers in research to determine prognoses, assess the appropriateness of a

MCAT EXPERTISE

As with the previous chapter, this chapter does not contain any content that falls directly within an AAMC content category. That said, the AAMC has confirmed that a full 10% of the science questions on every MCAT will touch on material in this chapter—and many of those questions will require only information from this chapter, without any other supportive science content. That makes it, point for point, one of the most important chapters in the entire *MCAT Review* series!

treatment modality for a given patient, and answer patients' questions. You may be a researcher yourself in a basic science, clinical, or translational setting. Regardless of your path, evaluating research is critical to the progress of all fields of medicine and will be a key component of your life as a physician.

11.1 The Scientific Method

LEARNING OBJECTIVES

After Chapter 11.1, you will be able to:

- Determine the relative value of a research question by applying the FINER method
- Identify the stages of the scientific method, and evaluate whether they have been appropriately completed
- Evaluate the quality and testability of a hypothesis

The basic paradigm for all scientific inquiry is the scientific method. The **scientific method** is a set of steps that defines the appropriate order of events to structure and carry out an experiment. As such, the scientific method is the established protocol for transitioning from a question to a new body of knowledge. The steps in the scientific method are:

1. **Generate a testable question:** This usually occurs after observing something anomalous in another scientific inquiry or in daily life.
2. **Gather data and resources:** Think back to our introduction; this is the phase of journal and database searches and compiling information. At this step, we as scientists must be careful to look for all information, not just that consistent with our expectations.
3. **Form a hypothesis:** A **hypothesis** is the proposed explanation or proposed answer to our testable question. It is often in the form of an **if–then statement**, which will be tested in subsequent steps.
4. **Collect new data:** This step results from either **experimentation**, which involves manipulating and controlling variables of interest; or **observation**, which often involves no changes in the subject's environment.
5. **Analyze the data:** Look for trends and perform mathematical manipulations to solidify the connections between variables.
6. **Interpret the data and existing hypothesis:** Consider whether the data analysis is consistent with the original hypothesis. If the data is inconsistent, consider alternative hypotheses.
7. **Publish:** Publication provides an opportunity for **peer review**; a summary of what was done during all six prior steps should be included in the publication.
8. **Verify results:** Most experiments are repeated to verify the results under new conditions.

BRIDGE

It is easy to focus on research that agrees with our expectations or opinions, and to ignore research that goes against them. This is an example of confirmation bias at work. Specific types of biases are discussed in Chapter 4 of *MCAT Behavioral Sciences Review* and later in this chapter.

MCAT EXPERTISE

The MCAT will most often test experimental or logical errors during research. Pay particular attention to the scientific method now in order to recognize each area in which an error may occur.

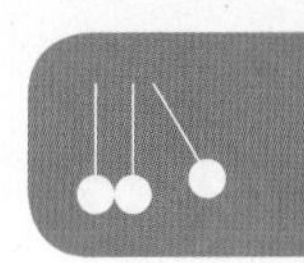

Formulating a testable question often presents students with a challenge because we tend to overreach, creating all-encompassing broad questions; however, in order to form a good testable question, it must be restricted to a relatively narrow area. The same is true of our hypotheses. One might wonder *Why do hot objects cause injury?* This is not a testable question. In fact, most questions that begin with *Why* are too broad to be testable through a single experiment. In this case, a better, more testable question could be *How do epithelial cells respond to heat* in vivo*?* Even this question is likely too broad for a single experiment, but it is testable, and we can form related hypotheses. One possible hypothesis would be: *if heat is applied to* in vivo *epithelial cells, then those cells will lyse*. Pay attention to the format of the hypothesis, as the if–then format ensures that it is testable. We will examine other components of the scientific method as they pertain to basic science research, biochemical and biomedical research, and social science research in the relevant sections.

The FINER Method

The **FINER method** for evaluating a research question is a method to determine whether the answer to one's question will add to the body of scientific knowledge in a practical way and within a reasonable time period. The FINER method asks five questions to make this determination.

- Is the necessary research study going to be **feasible**? A question about the response of chemosynthetic bacteria to a particular antibiotic requires access to chemosynthetic bacteria (which are often associated with harsh and difficult environments to access). If the scientist cannot obtain the necessary supplies, then the research is not feasible. Financial or time constraints, or the inability to gather enough subjects are also feasibility concerns.
- Do other scientists find this question **interesting**? This is somewhat subjective, but if there is little interest in the outcome of a particular research question, then the research will have little utility.
- Is this particular question **novel**? If someone has asked this question before, and answered it to the satisfaction of a peer-reviewed journal, then it's now in the confirmatory stages of the scientific method. Barring any anomalies, asking this question again isn't likely to gain new knowledge.
- Would the study obey **ethical** principles? Just because we are capable of carrying out a research study does not mean it is ethically or morally acceptable. If there's an ethical or moral reason not to perform a study, this should dissuade researchers from carrying out the study just as much as an inability to secure funding.
- Is the question **relevant** outside the scientific community? The more people that the research will impact in everyday life, the more important it usually is. There are exceptions, of course—many people might agree that curing a rare fatal illness is more important than improving the odor of a popular perfume, although a much larger group may be impacted by the latter study.

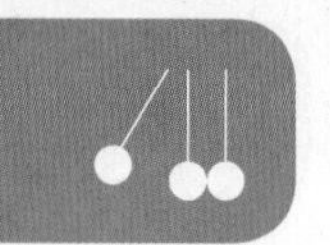

MCAT CONCEPT CHECK 11.1

Before you move on, assess your understanding of the material with these questions.

1. Rank the following research questions from 1 (best) to 3 (worst) using the FINER method and explain your rationale:

   - How long does it take for the Earth to complete one revolution around the Sun?
   - How do medical errors relate to sleep deprivation of medical residents?
   - What is the average lifespan of bacteria in Martian rocks?

   1. ______________________________

   ____

   2. ______________________________

   ____

   3. ______________________________

   ____

2. Errors or biases during publication of results are most likely to affect which stages of the scientific method?

   ____

   ____

   ____

3. True or False: *Most people with hepatitis C acquired it through IV drug use* is an example of a well-formatted hypothesis.

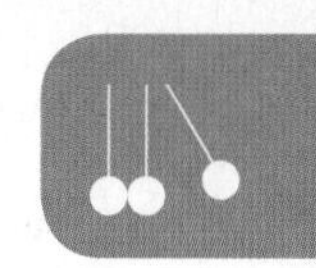

11.2 Basic Science Research

High-Yield

LEARNING OBJECTIVES

After Chapter 11.2, you will be able to:

- Identify common types and sources of error
- Recognize independent and dependent variables and how they are typically displayed graphically
- Explain the importance of the different types of control, including positive and negative controls
- Distinguish between accuracy and precision:

Unreliable & Unvalid

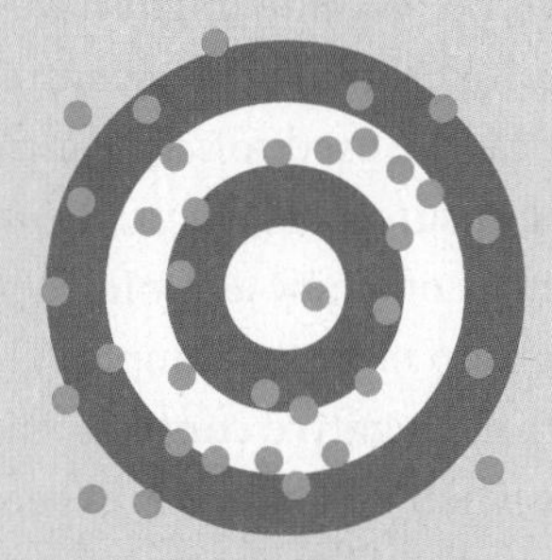

Unreliable, But Valid

Reliable, Not Valid

Both Reliable & Valid

Basic science research—the kind conducted in a laboratory, and not on people—is generally the easiest to design because the experimenter has the most control. Often a causal relationship is being examined because the hypothesis generally states a condition and an outcome. In order to make generalizations about our experiments, we must make sure that the outcome of interest would not have occurred without our intervention, and therefore, we use controls. We must also demonstrate causality, which is relatively simple in basic science research, but less so in other research areas.

Controls

In basic science research, conditions can be applied to multiple trials of the same experiment that are as near to identical as possible. In this way, a **control** or **standard** acts as a method of verifying results. Consider the following experiment: a scientist has an unknown concentration of a basic ammonia solution and wishes to determine the concentration experimentally. He takes a standardized

REAL WORLD

The use of controls also allows investigators to check for contamination of reagents.

solution of hydrochloric acid (made by comparison to a potassium hydrogen phthalate [KHP] standard) and titrates the basic solution in the presence of the same calibrated pH meter he used for the hydrochloric acid standardization. He then determines the ammonia concentration from the results of the titration. Because the concentration of the acid used to determine the ammonia concentration was verified against a standard, he can be confident that the calculated ammonia concentration is accurate.

Controls can also be separate experimental conditions altogether. For example, when testing the reaction of a tissue culture to an antibiotic, a separate culture is generally grown and administered an equal quantity of a compound known to be inert, like water or saline. The control corrects for any impact that the simple addition of volume might have had on the experiment. Some experiments have both positive and negative controls for points of comparison or a group of controls that can be used to create a curve of known values. **Positive controls** are those that ensure a change in the dependent variable when it is expected. In the development of a new assay for detection of HIV, for example, administering the test to a group of blood samples known to contain HIV could constitute a positive control. **Negative controls**, in contrast, ensure no change in the dependent variable when no change is expected. With the same assay, administering the test to a group of samples known not to contain the HIV virus could constitute a negative control. In drug trials, a negative control group is often used to assess for the **placebo effect**—an observed or reported change when an individual is given a sugar pill or sham intervention.

Example: A clinical trial is devised to quantify the effectiveness of a retinal scanning system for the detection of Alzheimer's disease. The trial will include a positive control group, a negative control group, and an experimental group. The experimental group is a large sample of 70-year-olds showing symptoms of Alzheimer's disease. Assuming the retinal scanning system is effective at detecting Alzheimer's disease, rank the groups in decreasing order of percentage of expected disease detection.

Solution: Members of the positive control group should consist of patients who have previously been diagnosed with Alzheimer's disease, whereas members of the negative control group should consist of patients who have previously tested negative for Alzheimer's disease. Therefore, the positive control group and negative control groups should establish the upper and lower bounds of percent detection, respectively. The experimental group is likely to fall between those two extremes. The predicted order is: positive, experimental, and negative.

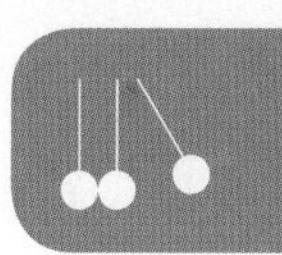

Causality

The other big advantage to being able to manipulate all of the relevant experimental conditions is that basic science researchers can often establish causality. Causality is an if–then relationship, and is often the hypothesis being tested. In basic science research, we manipulate an **independent variable**, and measure or observe a **dependent variable**. When there is a theoretical or known mechanism that links the independent and dependent variables, a causal relationship can be investigated. If the change in the independent variable always precedes the change in the dependent variable, and the change in the dependent variable does not occur in the absence of the experimental intervention, the relationship is said to be causal.

Example: A test was conducted to determine if water consumption impacts systolic blood pressure in mice. The water consumption and systolic blood pressure of eight mice was tracked for three weeks. The average daily water consumption and the average systolic blood pressure of each mouse is listed in the following table.

Mouse	Average Daily Water Consumption (mL)	Average Systolic Blood Pressure (mmHg)
1	3.9	122
2	3.3	128
3	1.5	146
4	4.7	117
5	6.1	108
6	0.8	153
7	10.3	103
8	8.6	105

The researchers plotted pressure and consumption and obtained the following graph. Identify the independent and dependent variables, and determine if a causal relationship exists between the two.

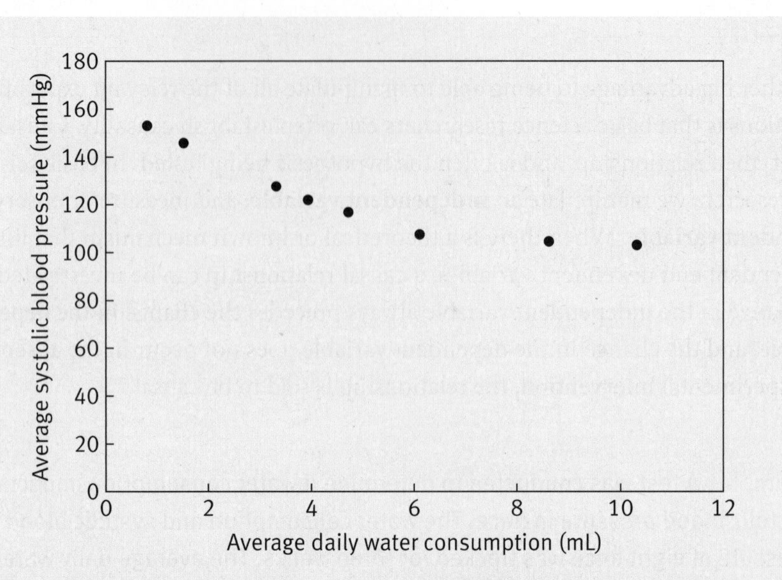

Solution: The independent variable is the "Average Daily Water Consumption" and the dependent variable is the "Average Systolic Blood Pressure (mmHg)."

Not enough data is given to determine if a causal relationship exists. More information is needed to determine if the differences in systolic blood pressure occur due to differences in water consumption. It's possible that the mice could have the given systolic blood pressure values even when consuming equal amounts of water.

KEY CONCEPT

The independent variable is the one that the experimenter is manipulating, and the dependent or outcome variable is the one that is being observed. On a graph the independent variable belongs on the *x*-axis and the dependent variable belongs on the *y*-axis.

Error Sources

In basic science research, experimental bias is usually minimal. The most likely way for an experimenter's personal opinions to be incorporated is through the generation of a faulty hypothesis from incomplete early data and resource collection. However, there can be manipulation of the results by eliminating trials without appropriate background, or by failing to publish works that contradict the experimenter's own hypothesis.

The low levels of bias introduced by the experimenter do not eliminate all error from basic science research. Measurements are especially important in the laboratory sciences, and the instruments may give faulty readings. Instrument error may affect accuracy, precision, or both. **Accuracy**, also called **validity**, is the ability of an instrument to measure a true value. For example, an accurate scale should register a 170-pound person's weight as 170 pounds. **Precision**, also called **reliability**, is the ability of the instrument to read consistently, or within a narrow range.

The same person standing on a scale that is accurate but imprecise may get readings between 150 and 190 pounds. The same person standing on a scale that is inaccurate but precise may get readings between 129 and 131 pounds, a relatively narrow range. Accuracy and precision are represented in Figure 11.1. Because bias is a **systematic error** in data, only an inaccurate tool will introduce bias, but an imprecise tool will still introduce error. Random chance can also introduce error into an experiment; while **random error** is difficult to avoid, it is usually overcome by using a large sample size.

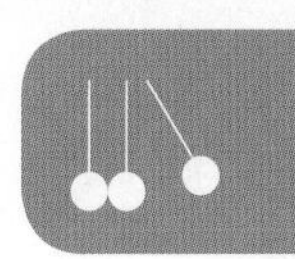

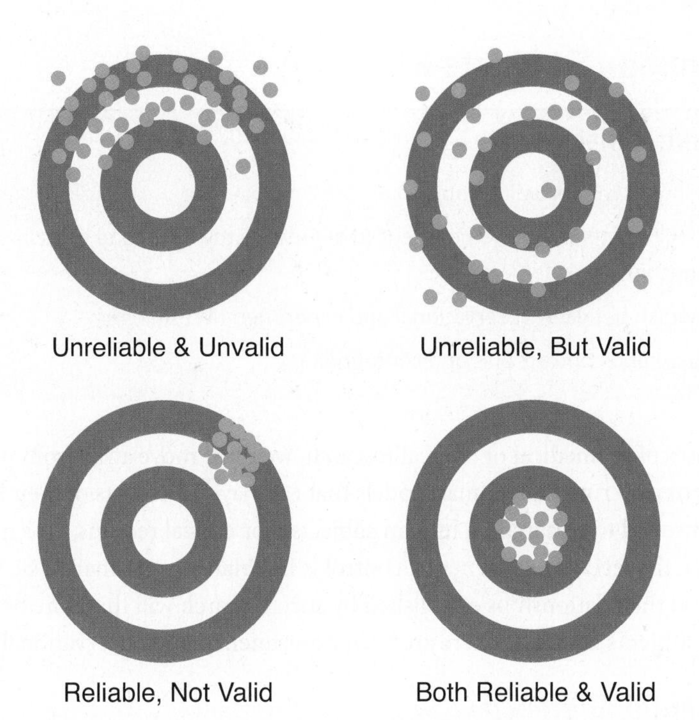

Figure 11.1 Accuracy (Validity) and Precision (Reliability) of Measurements

MCAT CONCEPT CHECK 11.2

Before you move on, assess your understanding of the material with these questions.

1. An experiment with an improperly tared (zeroed) mass balance would suffer from what type of error?

____

2. Label the axes and provide representative data for the following situation: An experimenter adds sodium hydroxide to an experimental solution and records the pH. He finds that the relationship is sigmoidal and that the pH is rising.

3. What is the purpose of a control during experiments? What characteristic of experimental research would be reduced in the absence of a control?

____

11.3 Human Subjects Research

LEARNING OBJECTIVES

After Chapter 11.3, you will be able to:

- Apply Hill's criteria to an experiment to determine the likelihood of a causal relationship
- Distinguish between observational and experimental research
- Compare and contrast bias and confounding

In some cases of biomedical or clinical research, we must move away from petri dishes full of cells or experimental animal models that can have all aspects of their living conditions controlled to research on human subjects. For ethical reasons, which we will discuss later, the level of experimental control is invariably lower than basic science research, and the relationships established by such research will therefore be weaker. In **human subjects research**, there are both experimental and observational studies.

Experimental Approach

In biomedical research, it is possible to perform experiments in which an independent variable is manipulated and an outcome is observed. In these experiments, we are still attempting to elicit a causal relationship. Because subjects are in less-controlled conditions, the data analysis phase is more complicated than in laboratory studies. In clinical and social sciences research, it is often still possible to conduct experiments by manipulating the environment or circumstances of the subject.

Randomization

Randomization is the method used to control for differences between subject groups in biomedical research. Randomization uses an algorithm to determine the placement of each subject into either a control group that receives no treatment or a sham treatment, or one or more treatment groups. A proper randomization algorithm will be equivalent to a coin toss or die roll. Once each individual is assigned to a group, the intervention is performed and the results are measured. Ideally, each group is perfectly matched on conditions such as age and gender; however, as long as there is an appropriate randomization algorithm, the collected data may be analyzed without concern.

Blinding

REAL WORLD

Blinding isn't only useful in drug trials; even sham treatments of acupuncture have been used to blind subjects in randomized controlled trials focusing on the use of acupuncture for musculoskeletal pain.

Because many of the measures in biomedical research are subjective, the perception of the subject and the investigator may be biased by knowing which group the subject is in. To remove this bias, the subjects and/or investigators may be **blinded**, which means they do not have information about which group the subject is in. In **single-blind experiments**, only the patient or the **assessor** (the person who makes measurements on the patient or performs subjective evaluations) is blinded. In **double-blind experiments,** the investigator, subject, and assessor all do not know the subject's group. Without blinding, the **placebo effect** would be greatly reduced in the control group, but still be present in the treatment group.

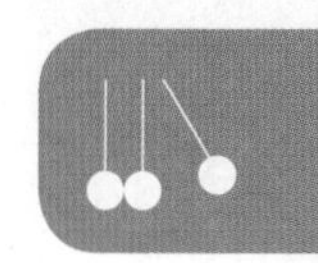

Data Analysis

In biomedical research, data analysis must account for variables outside of the independent and dependent variables considered. Most often, these include gender and age; lifestyle variables, such as smoking status and body mass index (BMI), and other factors that may affect the measured outcomes. Some of these other factors can be inferred from the initial literature review, although other unexpected **confounding variables** may exist. Software programs can use **binary** (yes *vs.* no, better *vs.* worse), as well as **continuous** (amount of weight lost, percent improvement in cardiac output) or **categorical variables** (state of residence, socioeconomic status) to create a regression model. **Regression analysis** may demonstrate linear, parabolic, exponential, logarithmic, or other relationships, as we will discuss in Chapter 12 of *MCAT Physics and Math Review*.

KEY CONCEPT

Research methods that generate numerical data are quantitative, while those that generate non-numerical data are qualitative; mixed-method research utilizes both.

Observational Approach

We may wish to study certain causal associations for which an experiment cannot be performed for ethical or practical reasons. In such a case, we must draw on the available data and analyze it. **Observational studies** in medicine fit into one of three categories: cohort studies, cross-sectional studies, and case–control studies. These studies often look for the connections between exposures and outcomes. Observational studies do not demonstrate causality, although the tendency toward causality may be demonstrated by Hill's criteria, which we will examine later.

REAL WORLD

Ethnographic studies are observational studies utilized by sociologists. These studies attempt to understand cultures by looking at the complete social environment.

Cohort studies are those in which subjects are sorted into groups based on differences in risk factors (**exposures**), and then assessed at various intervals to determine how many subjects in each group had a certain **outcome**. For example, a study in which 100 smokers and 100 nonsmokers are followed for 20 years while counting the number of subjects who develop lung cancer in each group would be an example of a cohort study.

KEY CONCEPT

A longitudinal study is an observational research method that follows the same subjects over time. Therefore, a cohort study is a form of longitudinal study.

Cross-sectional studies attempt to categorize patients into different groups at a single point in time. For example, a study to determine the prevalence of lung cancer in smokers and nonsmokers at a given point in time would be an example of a cross-sectional study.

Case-control studies start by identifying the number of subjects with or without a particular outcome, and then look backwards to assess how many subjects in each group had exposure to a particular risk factor. For example, a study in which 100 patients with lung cancer and 100 patients without lung cancer are assessed for their smoking history would be an example of a case–control study.

Hill's Criteria

Hill's criteria describe the components of an observed relationship that increase the likelihood of causality in the relationship. While only the first criterion is necessary for the relationship to be causal, it is not sufficient. The more criteria that are satisfied by a relationship, the likelier it is that the relationship is causal. Hill's criteria do not provide an absolute guideline on whether a relationship is causal; thus, for any observational study, the relationship should be described as a **correlation**.

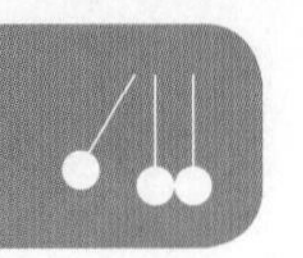

- **Temporality:** The exposure (independent variable) *must* occur before the outcome (dependent variable).
- **Strength:** As more variability in the outcome variable is explained by variability in the study variable, the relationship is more likely to be causal.
- **Dose–response relationship:** As the study or independent variable increases, there is a proportional increase in the response. The more consistent this relationship, the more likely it is to be causal.
- **Consistency:** The relationship is found to be similar in multiple settings.
- **Plausibility:** There is a reasonable mechanism for the independent variable to impact the dependent variable supported by existing literature.
- **Consideration of alternative explanations:** If all other plausible explanations have been eliminated, the remaining explanation is more likely.
- **Experiment:** If an experiment can be performed, a causal relationship can be determined conclusively.
- **Specificity:** The change in the outcome variable is only produced by an associated change in the independent variable.
- **Coherence:** The new data and hypothesis are consistent with the current state of scientific knowledge.

Error Sources

In addition to the measurement error found in basic science research, we must be aware of bias and error introduced by using human subjects as part of an experimental or observational model. As mentioned earlier, bias is a systematic error. As such, it generally does not impact the precision of the data, but rather skews the data in one direction or another. **Bias** is a result of flaws in the data collection phase of an experimental or observational study. **Confounding** is an error during analysis.

Selection Bias

The most prevalent type of bias is **selection bias**, in which the subjects used for the study are not representative of the target population. People who volunteer for a study in a particular area may be significantly different from people who do not volunteer. For example, someone volunteering for a drug trial that requires clinical visits may be healthier or more likely to benefit from the study than someone who does not volunteer because they cannot make it to the hospital.

Selection bias may also apply in cases where one gender is more prevalent in a study than another, or where there are differences in the age profile of the experiment group and the population. Measurement and assessment of selection bias occurs before any intervention.

Detection Bias

Detection bias results from educated professionals using their knowledge in an inconsistent way. Because prior studies have indicated that there is a correlation between two variables, finding one of them increases the likelihood that the

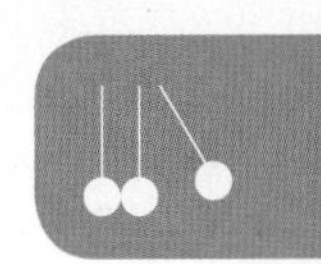

researcher will search for the second. For example, high blood pressure (hypertension) and diabetes mellitus are more common in the obese population; thus, a physician may screen obese patients for hypertension and diabetes at a higher rate than healthy-weight patients, inflating the true value of the secondary measurement (although, as described in Chapter 12 of *MCAT Behavioral Sciences Review*, other biases against obese individuals actually tend to lead to lower rates of screening and preventative care).

Observation Bias

The **Hawthorne effect**, or **observation bias**, posits that the behavior of study participants is altered simply because they recognize that they are being studied. Often these lifestyle alterations improve the health of the sample population. For example, patients in a study for a given weight loss drug may begin exercising more frequently or may make healthier diet choices, thus artificially increasing the perceived effect of the drug. Because the change in data is systematic and occurs before data analysis, this is an example of bias.

Confounding

Confounding, sometimes inaccurately called confounding bias or omitted variable bias, is a data analysis error. The data may or may not be flawed, but an incorrect relationship is characterized. For example, consider the statement *Having natural red hair leads to a decreased pain tolerance and higher opiate tolerance.* There are two flaws with this statement. First, the statement implies a causal relationship as a result of what would almost certainly be an observational study. Second, consider whether or not this is realistic. How could red hair cause the findings described? According to current research, there is no likely causality between these two. However, a third variable, such as a gene mutation, could potentially cause both parts of this statement. If one measured the degree of red hair pigment and the degree of pain intolerance, there might be a very strong statistical relationship, but there is no causal relationship between the two. These "third-party" variables are called **confounding variables** or **confounders**, as illustrated schematically in Figure 11.2.

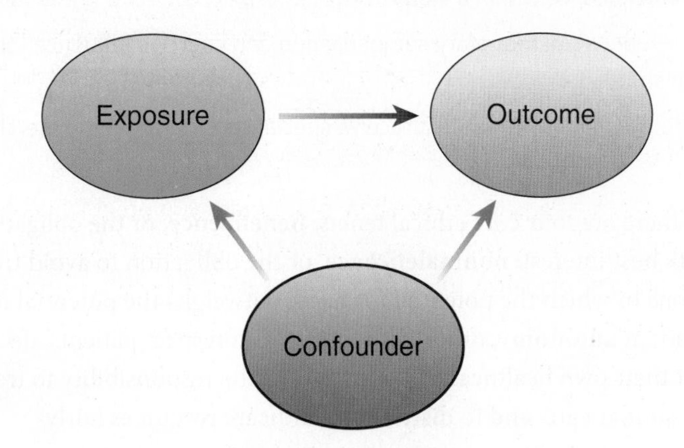

Figure 11.2 Confounding

MCAT CONCEPT CHECK 11.3

Before you move on, assess your understanding of the material with these questions.

1. True or False: A researcher who fails to demonstrate temporality can still provide evidence for a causal relationship by satisfying the rest of Hill's criteria.
2. How does observational research differ from experimental research?
    - Observational research:

    ___

    - Experimental research:

    ___

3. What is the difference between bias and confounding?
    - Bias:

    ___

    - Confounding:

    ___

11.4 Ethics

LEARNING OBJECTIVES

After Chapter 11.4, you will be able to:

- Distinguish between autonomy (medical ethics) and respect for persons (research ethics)
- Predict the ethical issues regarding respect for persons, justice, beneficence, and nonmaleficence within a study
- Distinguish between monetary compensation and coercive influence for a research study
- Recall the populations that must receive special consideration for coercion

BRIDGE

The key ethical tenets of medicine are also discussed in Chapter 11 of *MCAT Behavioral Sciences Review* in conjunction with an analysis of major institutions in the United States.

In medicine, there are four core ethical tenets: **beneficence**, or the obligation to act in the patient's best interest; **nonmaleficence**, or the obligation to avoid treatments or interventions in which the potential for harm outweighs the potential for benefit; respect for patient **autonomy**, or the responsibility to respect patients' decisions and choices about their own healthcare; and **justice**, or the responsibility to treat similar patients with similar care, and to distribute healthcare resources fairly.

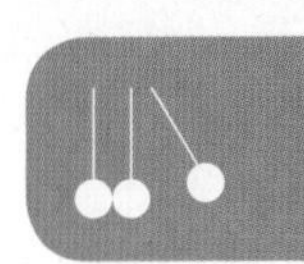

In research, these four principles are replaced by a slightly modified set. The *Belmont Report*, a landmark document published by the National Commission for the Protection of Human Subjects of Biomedical and Behavioral Research in 1979, delineates the three necessary pillars of research ethics: respect for persons, justice, and a slightly more inclusive version of beneficence.

Respect for Persons

Respect for persons includes the need for honesty between the subject and the researcher, and generally—but not always—prohibits deception. Respect for persons also includes the process of **informed consent**, in which a patient must be adequately counseled on the procedures, risks and benefits, and goals of a study to make a knowledgeable decision about whether or not to participate in the study. Further, the investigator cannot exert a coercive influence over the subjects, or they will not be acting autonomously. This coercive influence may be the result of a natural power imbalance, such as that between a teacher and a student, or may be the result of an extreme financial incentive or the inability to otherwise receive treatment for a condition. Respect for persons also includes the need to respect the subjects' wishes to continue with or cease participation in a study. The subject may withdraw consent that was previously granted at any time.

Many older studies did not abide by respect for persons. For example, sentinel studies on the severity of untreated syphilis were conducted without the participants' knowledge or consent. Many early psychological and sociological studies involved significant deception, which was not always disclosed after the fact. In current practice, hospital or university **institutional review boards** have put into place systematic protections against unethical studies. **Vulnerable persons**, which include children, pregnant women, and prisoners, require special protections above and beyond those taken with the general population. **Confidentiality** is also generally considered as part of respect for persons during research.

REAL WORLD

The Tuskegee syphilis experiment was a notorious forty-year study (1932–1972) by the United States Public Health Service that was fraught with extreme violations of the ethical principle of respect for persons. In this study, poor African-American men were enrolled into a study on the natural progression of syphilis. These men were given sham treatments, barred from accessing appropriate healthcare, and repeatedly deceived by investigators—including the fact that they were never told they had syphilis! This study was so significant in bioethical history that it is actually considered the primary impetus for the writing of the Belmont Report.

Justice

Justice in research applies to both the selection of a research topic and the execution of the research. In a world where all individuals and all questions are ethically the same, the only way to determine the selection of a research question to maintain justice is through random chance, in theory. Thankfully, we live in a world with morally relevant differences as established by our cultures. **Morally relevant differences** are defined as those differences between individuals that are considered an appropriate reason to treat them differently. For example, age is a significant moral difference in ethical deliberations: all else being equal, a transplant that is as likely to benefit a young child or an elderly adult might be given to the child because of a longer life expectancy. Population size is often morally relevant in study design because a study that impacts a large population will generally have more potential to do good than

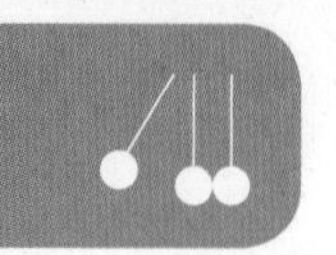

one that impacts a small population. In contrast, race, ethnicity, sexual orientation, and financial status are generally *not* considered morally relevant differences. It should be noted that religion may or may not be a valid moral criterion, depending on the context; for example, certain interventions that are prohibited by a given religion are grounds to avoid that treatment in an individual of that religion—this is in keeping with patient autonomy.

Justice is also important in the selection of subjects and the execution of research. When there is risk associated with a study, it must be fairly distributed so as not to unduly harm any group. This generally corresponds to seeking a diverse group for a study. Note that this also has a side benefit of increasing external validity, which will be discussed later in the chapter. However, the population that is most likely to benefit from the study may be required to bear a greater proportion of the risk. This apparent discrepancy—that all individuals should equally share the burden of risk, and that the target population may assume a higher proportion of risk—is reflective of the fact that likelihood of benefit is a morally relevant difference between individuals. Therefore, in studies in which there is no perceived difference in the likelihood of benefit between individuals, all individuals should assume equal risk; in those for which a particular population is most likely to benefit, that population should assume a higher proportion of risk. Finally, be aware that in some cases, such as drug trials, it may necessary to test the intervention in healthy individuals unaffected by the illness for which the drug has been designed. In this case, the burden of risk falls on a secondary population. This is permissible as long as the potential risks and benefits for the subjects have been addressed through informed consent and respect for persons has been maintained.

Beneficence

Beneficence is the easiest concept to understand in the context of research ethics. It must be our intent to cause a net positive change for both the study population and general population, and we must do our best to minimize any potential harms. This benefit may be as intangible as a feeling of personal satisfaction, and it may be unrelated to the original purpose of the study, such as small financial incentives. It may also be a future benefit if the participant is a member of the target population.

Research should be conducted in the least invasive, painful, or traumatic way possible. For example, a measurement that could be taken with either a finger stick or an indwelling catheter should be taken with a finger stick because it is far less painful and invasive. In addition, in studies comparing two potential treatment options, one cannot approach the research with the knowledge that one treatment is superior to the other. This is termed **equipoise**. If it becomes evident that one treatment option is clearly superior before a study is scheduled to finish, the trial must be stopped because providing an inferior treatment is a net harm.

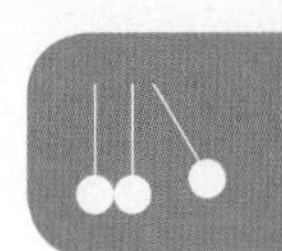

MCAT CONCEPT CHECK 11.4

Before you move on, assess your understanding of the material with these questions.

1. What is the difference between autonomy in medical ethics and respect for persons in research ethics?

 ____

 ____

 ____

2. During study design, a company wishing to market a drug to severe diabetics proposes to enroll only mild diabetics. Which principle of research ethics is the company violating? Are there any research concerns in this proposed study besides ethics?

 ____

 ____

 ____

3. What is the difference between a coercive influence and monetary compensation for a research study?

 ____

 ____

 ____

4. What are some populations that must receive special consideration for coercion?

 ____

 ____

11.5 Research in the Real World

LEARNING OBJECTIVES

After Chapter 11.5, you will be able to:

- Compare and contrast internal and external validity
- Explain the impact of sample size on generalizability
- Recall the qualities a study must have to justify an intervention

Until this point, we've been discussing research in a vacuum but our goals with any research are application-based. In order to apply the data generated, there are practical concerns that we must consider. For example, we must take into account the statistical strengths and weaknesses of a study, especially those that are related to the differences between the target population and the study sample. We also consider ways in which bias impacts the ability to use study conclusions in the real world, and whether there is any true justification for an intervention.

Populations *vs.* Samples

In statistics and research, we generally work with a sample rather than an entire population. A **population** is the complete group of every individual that satisfies the attributes of interest. Populations may be very large; for example, the population of humans is over seven billion people. In contrast, a population with a large number of qualifiers—for example, the population of American females between 18 and 30 years old who have Darier's disease, a rare skin condition—will be much smaller (in this case, about 100 people). Information that is calculated using every person in a population is called a **parameter**.

Working with a population is generally not feasible, even for smaller groups. Therefore, we make generalizations about populations based on sample data. A **sample** is any group taken from a population that does not include all individuals from the population. Ideally, samples will be representative of the population, and there are several methods of ensuring this. Random samples are generally considered the gold standard, although selecting for certain small subgroups may also be used. Information about a sample is called a **statistic**. With comparatively large or repeated samples, statistics can be used to estimate population parameters. If only a single small sample is taken, then very little information can be gleaned about the population.

Generalizability

When analyzing a study, we also look for markers of **internal validity** (or support for causality as discussed earlier) and **external validity**, or **generalizability**. Studies with low generalizability have very narrow conditions for sample selection that do not reflect the target population, whereas studies with high generalizability have samples that are representative of the target population. For example, a psoriasis study with low generalizability might have only participants who were diagnosed within the last year, while a study with high generalizability would have participants with a distribution of time since diagnosis that is similar to the population of all psoriatic patients.

REAL WORLD

Drugs undergo continuous evaluation in part because of poor preclinical generalizability. Some marketing changes or additional warnings may become necessary, or a drug may even be taken off the market. These are unforeseen risks or outcomes that only become apparent when the drug becomes available to the entire population.

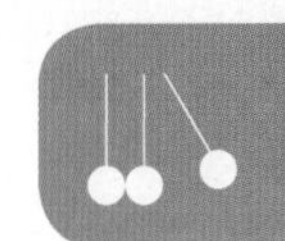

Support for Interventions

As future doctors, we are interested in applying research to our patients. To do so, we'll need to consider whether the data is sufficient for the recommendation or exclusion of any therapy or treatment plan.

Statistical *vs.* Clinical Effect

In research, the primary marker of success is being able to generate results that are **statistically significant**—that is, not the result of random chance. However, even the smallest difference between two treatments may be significant mathematically. For example, a decrease in systolic blood pressure of one millimeter of mercury could be statistically significant; however, it is not likely to change patient outcomes. In this way, we must assess whether there is **clinical significance**—a notable or worthwhile change in health status as a result of our intervention.

MCAT CONCEPT CHECK 11.5

Before you move on, assess your understanding of the material with these questions.

1. What is the difference between internal validity and external validity?
   - Internal validity:

   ____

   - External validity:

   ____

2. Why might small samples provide insufficient information about a population?

   ____

   ____

3. What qualities must a study have to provide justification for an intervention?

   ____

   ____

Conclusion

In this chapter, we focused on one of the four *Scientific Inquiry and Reasoning Skills* that will be tested on the MCAT: reasoning about the design and execution of research. We began by reviewing the scientific method and the value of historical data in the formulation of a research question. We then compared the methodology for both basic sciences research and human subjects research, especially with regard to error. We finished our investigation by examining the ethical and practical concerns in research design. The questions for this chapter are designed to allow you to practice this new skill, rather than to test your memorization of this content. In the next chapter, we'll specifically work with data and graphical analysis, another of the *Scientific Inquiry and Reasoning Skills* that will be essential on Test Day.

You've reviewed the content, now test your knowledge and critical thinking skills by completing a test-like passage set in your online resources!

GO ONLINE

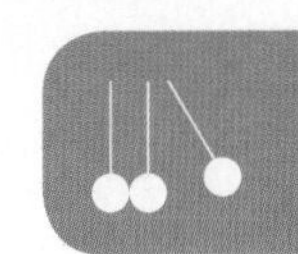

Concept Summary

The Scientific Method

- The **scientific method** is a series of eight steps for the generation of new knowledge.
  - The initial steps (generate a testable question, gather data and resources, form a hypothesis) focus on generating a **hypothesis**.
  - The intermediate steps (collect new data, analyze the data, interpret the data and existing hypothesis) focus on testing that hypothesis.
  - The final steps (publish and verify results) relate to providing the results for further testing of the hypothesis.
- The **FINER method** assesses the value of a research question on the basis of whether or not it is feasible, interesting, novel, ethical, and relevant.

Basic Science Research

- Basic science research uses chemicals, cell cultures, or animal subjects and is experiment-based.
- During research, we manipulate **independent variables** and observe changes in the **dependent variable**.
- **Controls** are used to correct for any influences of an intervention that are not part of the model. Controls may be positive or negative.
  - **Positive controls** ensure that a change in the dependent variable occurs when expected.
  - **Negative controls** ensure that no change in the dependent variable occurs when none is expected.
- Basic science research is often the best type for demonstrating causality because the experimenter has the highest degree of control over the experimental conditions.
- Error in basic science research most often results from errors in measurement.
  - **Accuracy (validity)** is the quality of approximating the true value.
  - **Precision (reliability)** is the quality of being consistent in approximations.

Human Subjects Research

- Human subjects research is subject to ethical constraints that are generally absent in basic science research.
- Experiments may still be performed, but causal conclusions are harder to determine because circumstances are harder to control.

- Much of human subjects research is **observational**.
  - **Cohort studies** record exposures throughout time and then assess the rate of a certain outcome.
  - **Cross-sectional studies** assess both exposure and outcome at the same point in time.
  - **Case–control studies** assess outcome status and then assess for exposure history.
  - Causality in observational studies is supported by **Hill's criteria**, which include temporality, strength, dose–response relationships, consistency, plausibility, consideration of alternative explanations, experiments, specificity, and coherence.
- Error may be in the form of bias, confounding, or random error.
- **Bias** is systematic and results from a problem during data collection.
  - **Selection bias**, in which the sample differs from the population, is most common in human subjects research.
  - **Detection bias** arises from educated professionals using their knowledge in an inconsistent way by searching for an outcome disproportionately in certain populations.
  - The **Hawthorne effect** results from changes in behavior—by the subject, experimenter, or both—that occur as a result of the knowledge that the subject is being observed.
- **Confounding** is an error in data analysis that results from a common connection of both the dependent and independent variables to a third variable.

Ethics

- Medical ethics generally refers to the four principles of **beneficence**, **nonmaleficence**, respect for patient **autonomy**, and **justice**.
- Research ethics were established by the **Belmont Report**.
  - **Respect for persons** includes autonomy, informed consent, and confidentiality.
  - **Justice** dictates which study questions are worth pursuing and which subjects to use.
  - **Beneficence** requires us to do the most good with the least harm. We cannot perform an intervention without **equipoise**—a lack of knowledge about which arm of the research study is better for the subject.

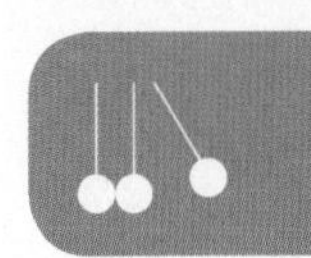

Research in the Real World

- **Populations** are all of the individuals who share a set of characteristics. Population data are called **parameters**.
- **Samples** are a subset of a population that are used to estimate population data. Sample data are called **statistics.**
- **Internal validity** refers to the identification of causality in a study between the independent and dependent variables. **External validity** refers to the ability of a study to be **generalized** to the population that it describes.
- In order to be supported, an intervention must display both statistical and clinical significance.
  - **Statistical significance** refers to the low likelihood of the experimental findings being due to chance.
  - **Clinical significance** refers to the usefulness or importance of experimental findings to patient care or patient outcomes.

Answers to Concept Checks

11.1

1. 1. *How do medical errors relate to sleep deprivation of medical residents?* This is a current topic of investigation and a consensus in the scientific community is still being reached. Medical residents are available for interview, and the research has very relevant outcomes.
   2. *What is the average lifespan of bacteria in Martian rocks?* While it is not very feasible to acquire the Martian rocks, the results would be both novel and interesting.
   3. *How long does it take the Earth to complete one revolution around the Sun?* This question has been asked and answered to the satisfaction of the scientific community. It is neither novel nor interesting (in terms of further research).
2. Errors during publication of current studies adversely affect the quality of future experimentation by providing an incomplete or flawed research base. Without accurate resources, subsequent hypotheses are likely to be flawed.
3. False. While the statement may or may not be true, this is not an easily testable hypothesis. While not required, if–then formatting of a hypothesis necessarily implies a testable relationship between ideas.

11.2

1. This experiment would likely have inaccuracy error but not imprecision error. In other words, the scale would reliably read the same mass or weight, but the mass or weight it reads is not correct. This would lead to bias in the results.
2. 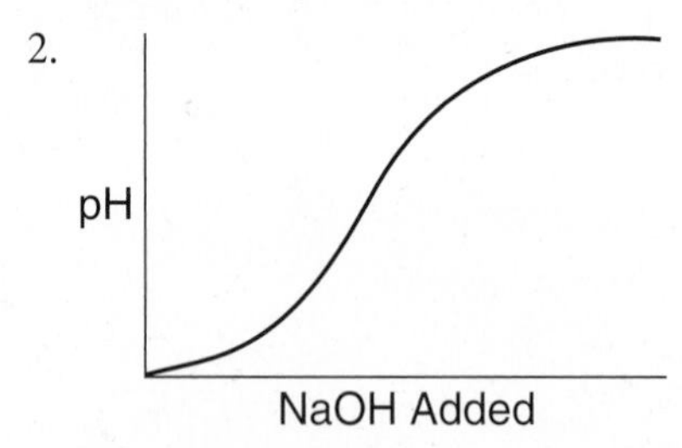

3. Controls in experiments help to establish causality by demonstrating that the outcome does not occur in the absence of an intervention. Controls are used to keep the manipulations of different systems as similar as possible, or as a known standard against which to judge an experimental manipulation. Without controls, it is far more difficult to establish causality.

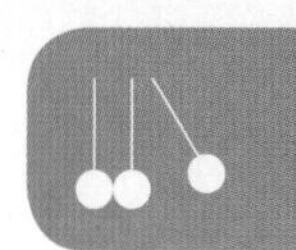

11.3

1. False. Temporality is the only necessary criterion from Hill's criteria. If temporality is not satisfied, the relationship cannot be said to be causal. The addition of other criteria increases the probability of a causal relationship, assuming that temporality has not been invalidated.
2. Observational research does not involve manipulation of the subjects' environment. It is generally less conclusive and more subjective than experimental research, which does involve manipulation of the subject or environment.
3. Bias is a systematic (unidirectional) error that occurs during the selection of subjects or the measurement and collection of data. Confounding is an error that occurs during data analysis, in which an association is erroneously drawn between two variables because of a shared connection to a third variable.

11.4

1. Autonomy is simply the right of an individual to make decisions on his or her own behalf and to have those decisions be respected. Respect for persons also requires honesty, confidentiality, informed consent, and freedom from coercion.
2. The company is violating the principle of justice by choosing participants that are not part of the target population. The company is also introducing selection bias.
3. The line between a coercive influence and a compensatory influence is often debated. In general, a compensatory influence is one that does not impact the decision to participate, while a coercive influence is one in which the subject loses autonomy to make the decision to participate.
4. Children, pregnant women, and prisoners are considered especially at risk for coercion and thus are granted special protections.

11.5

1. Internal validity is the tendency of the same experiment to produce the same results when repeated, and provides support for causality. External validity is the ability to take the information generated during research and apply it to a larger group. External validity is also called generalizability.
2. Small samples are subject to more random variation than large samples. If only one person is selected, he or she may be an outlier, but if a much larger sample is selected, an outlier will have less of an effect on the results.
3. A study must have both statistical significance and clinical significance to provide justification for an intervention. A study without statistical significance may be the result of random chance, whereas one without clinical significance will not impact patients.

Science Mastery Assessment Explanations

1. **C**

The purpose of a control is to keep the conditions of two experiments as close as possible to establish causality. In this case, the one milliliter volume addition might have impacted the growth of *E. coli*; thus, we must control for this by administering an equal volume of a theoretically inert compound to a plate of *E. coli*.

2. **C**

An experiment will always establish a clearer causal link than an observational study. **(A)**, **(B)**, and **(D)** are all examples of observational data.

3. **A**

The experimenter has not completed the initial phases of research. There was no data acquisition or refinement, and there was no indication that the question required an experiment to be answered. Were the experimenter doubtful of the validity of the reported value, an experiment could be appropriate—but there is no information to indicate that this is so. Based on the question stem, it is clear that the experimenter had a clear goal, eliminating **(B)**. In human subjects research, tasks may be divided to facilitate blinding, but this is generally unnecessary in basic sciences research, eliminating **(C)**. As long as the calorimeter was calibrated prior to its use, it does not matter when this calibration occurred relative to the synthesis of the compound, eliminating **(D)**.

4. **B**

A parameter is a population measure, so to calculate it, every single member must be measured. Identifying, measuring, and recording data for a population that large—over 160 million—is essentially impossible. Common biometric measures, if not misused, generally neither cause significant psychological harm nor are unethical, eliminating **(A)**. Knowing a mean weight could have major ramifications, including public health measures, medical recommendations, and shifting of body image, eliminating **(C)**. While a number of studies on weight have been performed, there is not yet a parameter describing the entire population, eliminating **(D)**.

5. **D**

Because both the exposure and outcome are measured at the same time, we cannot make any conclusions about temporality. The cancer patient may have only begun smoking after diagnosis, and this type of study doesn't examine that possibility.

6. **B**

Randomization is based on the idea that the results will only vary as a result of random chance as long as the assignment is proper. The appropriate response to a fair algorithm that assigns groups in an unexpected way is to proceed with the research. Participants should never be assigned by the researcher, nor should one continue randomizing samples to achieve a desired outcome—these are likely to introduce more error than leaving unequal groups, eliminating **(A)** and **(D)**. It is unnecessary to drop this entire cohort, assuming the randomization algorithm was fair, eliminating **(C)**.

7. **C**

Two relationships are being assessed. In the relationship between smoking and very low birth weight (VLBW), smoking is the independent variable and VLBW is the dependent variable. In the second relationship, VLBW is being compared to IQ. VLBW is the independent variable here, while IQ is the dependent variable.

8. **A**

Because the same physician sees both the control groups and the experimental groups, there is the potential for the physician to realize which of the groups is receiving which treatment—especially if the subject mentions expected side effects. In this study, patients were not told which group they were in, but if they were assigned to the medication group, they were told about its side effects. If the patients talked to each other they could experience patient unblinding, but we are told they have no communication with each other.

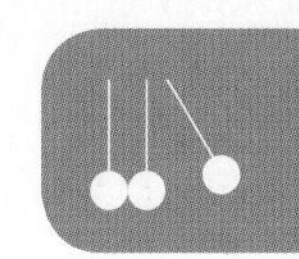

9. **B**

As part of the scientific method, after formulating a testable question, one must search through journals and databases to review the available information. This student likely did not spend sufficient time reviewing existing studies because her review committee was able to cite several studies that had already tested and affirmed her hypothesis, meaning that her hypothesis is not scientifically interesting. This observation is consistent with **(B)**. On the other hand, the question stem indicates that her hypothesis was well formulated and that she had some preliminary data, which eliminates **(A)** and **(D)**. There is no stated criticism of her research methods, which eliminates **(C)**.

10. **B**

Requiring subjects to volunteer for a study and to seek the study out will introduce selection bias. The people who end up volunteering listen to the radio, which the general population may not, and are interested in the topic and willing to volunteer. Most studies suffer from selection bias and it is the most common impediment to generalizability.

11. **C**

The behaviors described in the question stem (informing the patient, providing time to make a decision) are consistent with informed consent and autonomy, which are both part of respect for persons.

12. **A**

While it may appear that this question is asking to determine which method is the most accurate or reliable, this is an ethics question. All of these methods (with the exception of venipuncture, **(C)**) measure oxygen saturation. According to the principle of beneficence, we must minimize potential harms associated with our investigations; thus, the noninvasive pulse oximeter should be greatly favored over other measurements for an initial assessment.

13. **B**

Statistical significance is not the same as clinical significance. There are medications that increase the length of pregnancy in preterm labor, but some only do so for a few hours. Because this may not impact patient outcomes, it should not inform treatment decisions.

14. **D**

Samples should always be taken from the target population or population of interest. Given that the target population is postmenopausal women, this group should be used for the sample. Both prepubescent girls and pregnant women are populations with special precautions against coercion, and are not appropriate for this study, eliminating **(A)** and **(C)**. Premenopausal women are unlikely to require hormone replacement therapy unless they have a condition that specifically necessitates it, eliminating **(B)**.

15. **C**

The Hawthorne effect—a change in behavior as a result of the knowledge that one is being observed—is only present with human subjects. Basic science research generally does not suffer from the Hawthorne effect.

Consult your online resources for additional practice.

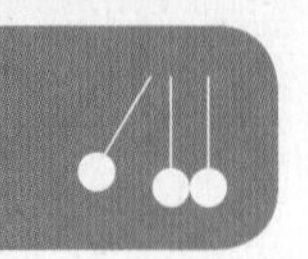

Shared Concepts

Behavioral Sciences Chapter 4
Cognition, Consciousness, and Language

Behavioral Sciences Chapter 8
Social Processes, Attitudes, and Behavior

Behavioral Sciences Chapter 11
Social Structure and Demographics

Behavioral Sciences Chapter 12
Social Stratification

Physics and Math Chapter 12
Data-Based and Statistical Reasoning

CHAPTER 12

Data-Based and Statistical Reasoning

SCIENCE MASTERY ASSESSMENT

Every pre-med knows this feeling: there is so much content I have to know for the MCAT! How do I know what to do first or what's important?

While the high-yield badges throughout this book will help you identify the most important topics, this Science Mastery Assessment is another tool in your MCAT prep arsenal. This quiz (which can also be taken in your online resources) and the guidance below will help ensure that you are spending the appropriate amount of time on this chapter based on your personal strengths and weaknesses. Don't worry though—skipping something now does not mean you'll never study it. Later on in your prep, as you complete full-length tests, you'll uncover specific pieces of content that you need to review and can come back to these chapters as appropriate.

How to Use This Assessment

If you answer 0–7 questions correctly:

Spend about 1 hour to read this chapter in full and take limited notes throughout. Follow up by reviewing **all** quiz questions to ensure that you now understand how to solve each one.

If you answer 8–11 questions correctly:

Spend 20–40 minutes reviewing the quiz questions. Beginning with the questions you missed, read and take notes on the corresponding subchapters. For questions you answered correctly, ensure your thinking matches that of the explanation and you understand why each choice was correct or incorrect.

If you answer 12–15 questions correctly:

Spend less than 20 minutes reviewing all questions from the quiz. If you missed any, then include a quick read-through of the corresponding subchapters, or even just the relevant content within a subchapter, as part of your question review. For questions you got correct, ensure your thinking matches that of the explanation and review the Concept Summary at the end of the chapter.

1. The inclusion of an outlier in statistical analysis will most greatly affect which of the following measurements?
   - **A.** Median
   - **B.** Mode
   - **C.** Mean
   - **D.** All three are equally affected

2. In a sample of hospital patients, the mean age is found to be significantly lower than the median. Which of the following best describes this distribution?
   - **A.** Skewed right
   - **B.** Skewed left
   - **C.** Normal
   - **D.** Bimodal

3. What is the median of the following data set?

   7, 17, 53, 23, 4, 2, 4
   - **A.** 4
   - **B.** 7
   - **C.** 15.7
   - **D.** 23

4. On Santa Cruz island, the available food sources led to disruptive selection in the beak sizes of finches, in which large and small beak sizes were favored over intermediate sized beaks. As a result, few finches have an intermediate sized beak. If a sample of finches were analyzed, what distribution would best describe the distribution of beak size?
   - **A.** Normal
   - **B.** Bimodal
   - **C.** Skewed left
   - **D.** Skewed right

5. A 95% confidence interval will fall within what distance from the mean?
   - **A.** $\pm\sigma$
   - **B.** $\pm2\sigma$
   - **C.** $\pm3\sigma$
   - **D.** $\pm4\sigma$

6. Approximately 20% of Americans eat fast food at least once per week. A new study conducted by the National Heart Association showed that 48% of Americans have some form of cardiovascular disease. Which of the following hypothetical findings could be used to show that cardiovascular disease is independent of weekly fast food consumption?
   - **A.** Individuals who do not consume fast food do not have cardiovascular disease.
   - **B.** Of the 48% of those who had cardiovascular disease, only half of them ate fast food on a weekly basis.
   - **C.** 9.6% of Americans have cardiovascular disease and consume fast food on a weekly basis.
   - **D.** None of the individuals who consume fast food on a weekly basis develop cardiovascular disease.

7. Which of the following outliers would most likely be the easiest to correct?
   - **A.** A typographical error in data transfer
   - **B.** A measurement error in instrument calibration
   - **C.** A heavily skewed distribution
   - **D.** A correctly measured anomalous result

8. Assume that blonde hair and blue eyes are independent recessive traits. If one parent is a carrier for each gene while the other parent is homozygous recessive for both genes, what is the probability that the first two offspring will both have blonde hair and blue eyes?
   - **A.** 6.25%
   - **B.** 25%
   - **C.** 43.75%
   - **D.** 50%

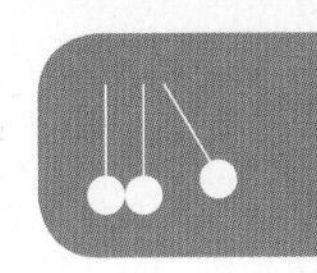

9. Based on the county-level map below, which of the following statements best represents the data about elderly individuals? (Note: The darker the shade of green, the higher the percentage of elderly persons in the county.)

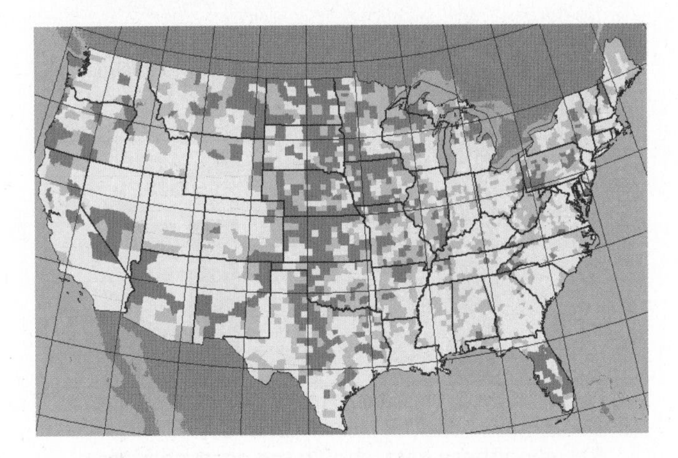

A. Most of the elderly people in the United States live in the center of the country.
B. Most of the people living in the center of the United States are elderly.
C. The center of the United States tends to have a larger proportion of elderly people.
D. There are more elderly people moving to the center of the country than elsewhere.

10. As the confidence level increases, a confidence interval:

A. becomes wider.
B. becomes thinner.
C. shifts to higher values.
D. shifts to lower values.

11. Which of the following measures of distribution is most useful for determining probabilities?

A. Range
B. Average distance from mean
C. Interquartile range
D. Standard deviation

12. Are there any outliers on the following box plot?

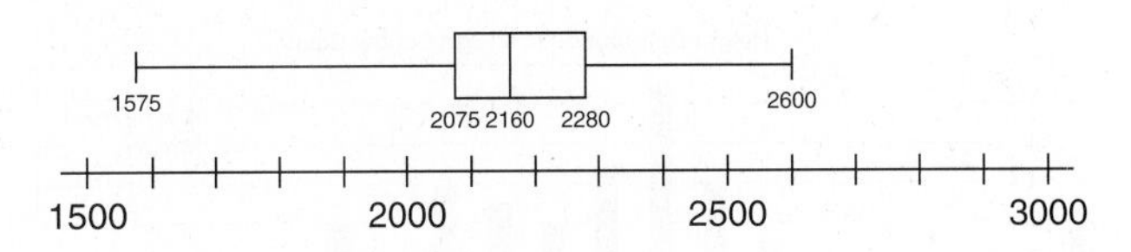

A. Yes; 1575 is an outlier.
B. Yes; 2600 is an outlier.
C. Yes; both 1575 and 2600 are outliers.
D. No; there are no outliers.

13. A gas station attendant notices that there are more assaults near his store on the days that he sells a lot of ice cream and hypothesizes that the ingestion of excess glucose leads to an increase in violent behavior. He begins to tally how much ice cream he sells on a daily basis and checks the daily crime statistics to generate the graph below.

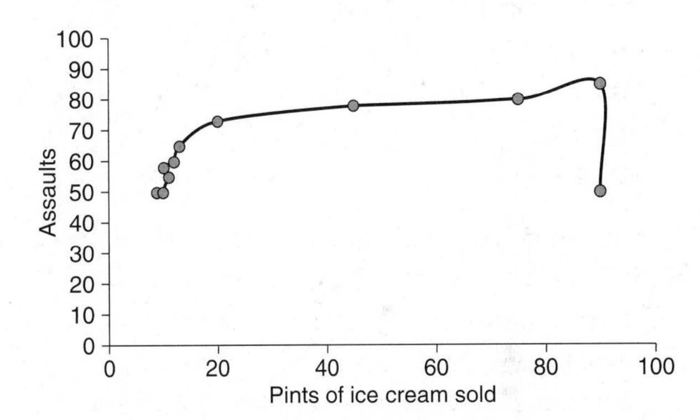

He analyzes the data and concludes that his hypothesis is true. Are there any flaws in his conclusion?

A. No, the number of assaults increase as the pints of ice cream sold increase.
B. No, the study is appropriately controlled.
C. Yes, the study is missing a positive control.
D. Yes, he has proved correlation, but not a causal link.

14. The following histogram:

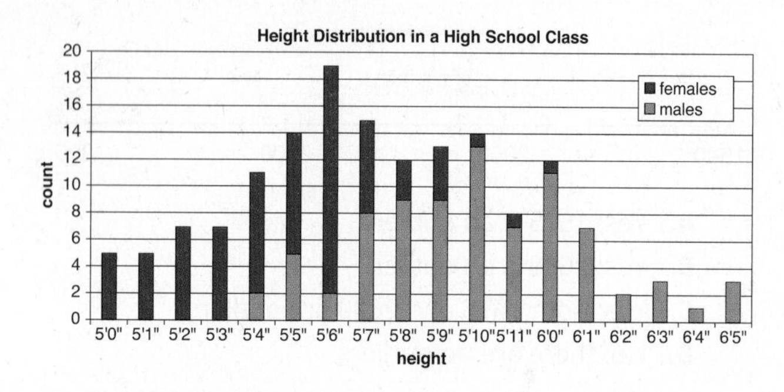

I. contains a bimodal distribution.
II. should be analyzed as two separate distributions.
III. contains one mode.

A. II only
B. I and II only
C. I and III only
D. I, II, and III

15. A cardiologist is investigating a new drug therapy for the treatment of hypertension. He measures his patients' baseline blood pressure, then treats his patients with the drug for ten weeks, after which time he again measures their blood pressure.

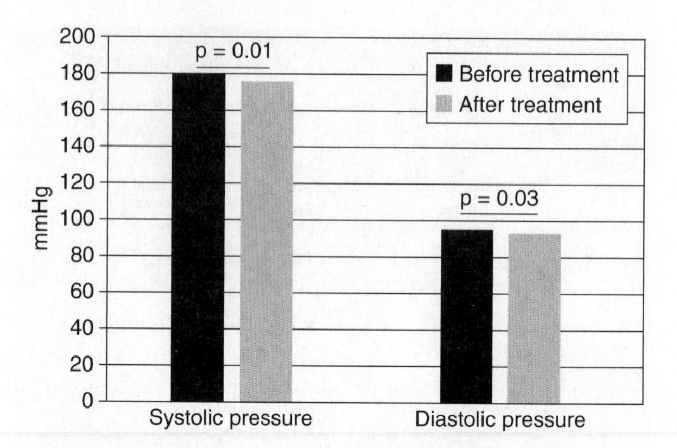

Based on these data, should the doctor advocate for making this drug the standard treatment for hypertension?

A. Yes, there is a statistically significant decrease in blood pressure.
B. Yes, the drug is effective at decreasing blood pressure.
C. No, while statistically significant, the results are not clinically significant.
D. No, the data is not statistically significant.

Answer Key follows on next page.

Answer Key

1. **C** (Ch. 12.1)
2. **B** (Ch. 12.2)
3. **B** (Ch. 12.1)
4. **B** (Ch. 12.2)
5. **B** (Ch. 12.5)
6. **C** (Ch. 12.4)
7. **A** (Ch. 12.3)
8. **A** (Ch. 12.4)
9. **C** (Ch. 12.6)
10. **A** (Ch. 12.5)
11. **D** (Ch. 12.3)
12. **C** (Ch. 12.3)
13. **D** (Ch. 12.7)
14. **D** (Ch. 12.6)
15. **C** (Ch. 12.7)

CHAPTER 12

Data-Based and Statistical Reasoning

In This Chapter

12.1 **Measures of Central Tendency**
Mean........................436
Median......................437
Mode........................438

12.2 **Distributions**
Normal Distributions............440
Skewed Distributions............440
Bimodal Distributions...........441

12.3 **Measures of Distribution**
Range.......................442
Interquartile Range.............443
Standard Deviation..............444
Outliers....................446

12.4 **Probability**
Independence, Mutual Exclusivity, and Exhaustiveness..........447
Calculations.................448

12.5 **Statistical Testing**
Hypothesis Testing..............449
Confidence Intervals............450

12.6 **Charts, Graphs, and Tables** HY
Types of Charts.................452
Graphs and Axes................455
Interpreting Tables.............459

12.7 **Applying Data**
Correlation and Causation........460
In the Context of Scientific Knowledge..................460

Concept Summary................463

Introduction

Academic papers are extremely predictable. They generally begin with an abstract that reflects the major points of the rest of the paper. The authors then provide an expanded introduction, materials and methods, data, and discussion. The key to a high-quality research paper is making this discussion unnecessary—any scientists, when given the prior sections, should be led to the same conclusions as those given by the author. The testmakers are keenly aware of this fact. On Test Day, you may be presented with research in the form of an experiment-based passage and part of your task will be inferring the important conclusions that can be supported by the findings of the study.

This chapter covers the last of the *Scientific Inquiry and Reasoning Skills* tested on the MCAT: the statistical analysis of raw data, interpretation of visual representations of this data, and application of data to answer research questions. We'll begin by examining basic statistical principles like distribution types, measures of central tendency, and measures of distribution. We'll also discuss probability, and the semantics of

MCAT EXPERTISE

Chapter 12 does not contain a chapter profile as it does not directly cover any AAMC content categories. That said, just like with Chapter 11 of this book, the AAMC has confirmed that 10% of the science questions on every MCAT will require material in this chapter—and many of those questions will require only information from this chapter, without any other supportive science content. With 10% from Chapter 11 and 10% from Chapter 12, there will be more than 30 questions(!) on your exam that will test one of these two skills.

this branch of mathematics. We'll conclude our discussion of probability and statistics with an exploration of statistical significance in basic hypothesis testing and confidence intervals. Then, we'll move on to the interpretation of charts and graphs. Finally, we'll link all of this information with the skills we gained in the last chapter and assess the future use and validity of studies.

12.1 Measures of Central Tendency

LEARNING OBJECTIVES

After Chapter 12.1, you will be able to:

- Calculate mean, median, and mode for a data set
- Predict the best measure of central tendency for a given data set

Measures of central tendency are those that describe the middle of a sample. How we define *middle* can vary. Is it the mathematical average of the numbers in the data set? Is it the result in a data set that divides the set into two—with half the sample values above this result and half the sample values below? Both of these data can be important, and the difference between them can also provide useful information on the shape of a distribution.

Mean

The **mean** or **average** of a set of data (more accurately, the **arithmetic mean**) is calculated by adding up all of the individual values within the data set and dividing the result by the number of values:

$$\bar{x} = \frac{\sum_{i=1}^{n} x_i}{n}$$

Equation 12.1

where x_i to x_n are the values of all of the data points in the set and n is the number of data points in the set. As we discussed in the last chapter, the mean may be a parameter or a statistic (as is true of all of the measures of central tendency) depending on whether we are discussing a population or a sample. Mean values are a good indicator of central tendency when all of the values tend to be fairly close to one another. Having an **outlier**—an extremely large or extremely small value compared to the other data values—can shift the mean toward one end of the range. For example, the average income in the United States is about $70,000, but half of the population makes less than $50,000. In this case, the small number of extremely high-income individuals in the distribution shifts the mean to the high end of the range.

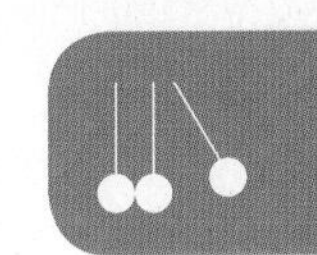

Example: The following data were collected on the ages of attendees at Ray's birthday party:

23, 22, 25, 22, 22, 24, 36, 20

What is the mean age of the attendees? Is this an appropriate measure for this data?

Solution: The mean is the sum of the data points divided by the number of data points:

$$\bar{x} = \frac{23+22+25+22+22+24+36+20}{8} = \frac{194}{8} = 24.25$$

Because the mean is relatively near most of the values collected for this data set, it may be appropriate. Keep in mind, though, that the presence of an outlier and the fact that the mean is greater than all but two of the values collected indicates that the mean has been shifted toward the high end of the range. The presence of a single outlier does not invalidate the mean, but it does make interpretation in context necessary.

Median

The **median** value for a set of data is its midpoint, where half of data points are greater than the value and half are smaller. In data sets with an odd number of values, the median will actually be one of the data points. In data sets with an even number of values, the median will be the mean of the two central data points. To calculate the median, a data set must first be listed in increasing fashion. The position of the median can be calculated as follows:

$$\text{median position} = \frac{(n+1)}{2}$$

Equation 12.2

where n is the number of data values. In a data set with an even number of data points, this equation will solve for a noninteger number; for example, in a data set with 18 points, it will be $\frac{18+1}{2} = 9.5$. The median in this case will be the arithmetic mean of the ninth and tenth items in the data set when sorted in ascending order. The median tends to be the least susceptible to outliers, but may not be useful for data sets with very large ranges (the distance between the largest and smallest data point, as discussed later in this chapter) or multiple modes.

Example: Using the same data from the last question, find the median age of the attendees. Comparing this value to the mean, is the median a better or worse indicator of central tendency in this sample?

Solution: The first step in finding the median is to order the data from smallest to largest. Our original data was:

23, 22, 25, 22, 22, 24, 36, 20

Reordered, this becomes:

20, 22, 22, 22, 23, 24, 25, 36

n, the number of data points, is 8, so the median will be the average of the fourth and fifth data points. The median is therefore $\frac{22+23}{2} = 22.5$. The median is a better indicator of central tendency for this data than the mean of 24.25. The median is unaffected by the outlier and lies close to most of the values in the data set. One could improve the representativeness of the mean by excluding 36 from the data set, in which case the mean would be 22.6 while the median would be 22.

KEY CONCEPT

The median divides the data set into two groups with 50% of values higher than the median and 50% of values lower than it.

If the mean and the median are far from each other, this implies the presence of outliers or a skewed distribution, as discussed later in this chapter. If the mean and median are very close, this implies a symmetrical distribution.

Mode

The **mode**, quite simply, is the number that appears the most often in a set of data. There may be multiple modes in a data set, or—if all numbers appear equally—there can even be no mode for a data set. When we examine distributions, the peaks represent modes. The mode is not typically used as a measure of central tendency for a set of data, but the number of modes, and their distance from one another, is often informative. If a data set has two modes with a small number of values between them, it may be useful to analyze these portions separately or to look for other variables that may be responsible for dividing the distribution into two parts.

MCAT CONCEPT CHECK 12.1

Before you move on, assess your understanding of the material with these questions.

1. What types of data sets are best analyzed using the mean as a measure of central tendency?

____

____

2. Calculate the mean, median, and mode of the following data set:

25, 23, 23, 6, 9, 21, 4, 4, 2

- Mean: ______________________
- Median: ______________________
- Mode: ______________________

12.2 Distributions

LEARNING OBJECTIVES

After Chapter 12.2, you will be able to:

- Assess whether data without a normal distribution can be analyzed with measures of central tendency and distribution
- Distinguish between normal, skewed, and bimodal distributions
- Describe the relationship between mean, median, and mode in different types of distributions:

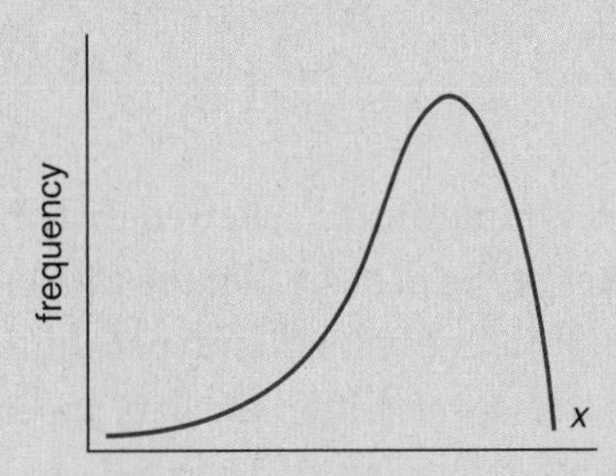

Often a single statistic for a data set is insufficient for a detailed or relevant analysis. In this case, it is useful to look at the overall shape of the distribution as well as specifics about how that shape impacts our interpretation of the data. The shape of a distribution will impact all of the measures of central tendency that we have already discussed, as well as some measures of distribution, which we will examine later.

KEY CONCEPT

The normal distribution and its counterpart, the standard distribution, are the basis of most statistical testing on the MCAT. In the normal distribution, all of the measures of central tendency are the same.

Normal Distributions

In statistics, we most often work with a **normal distribution**, shown in Figure 12.1. Even when we know that this is not quite the case, we can use special techniques so that our data will approximate a normal distribution. This is very important because the normal distribution has been "solved" in the sense that we can transform any normal distribution to a **standard distribution** with a mean of zero and a standard deviation of one, and then use the newly generated curve to get information about probability or percentages of populations. The normal distribution is also the basis for the **bell curve** seen in many scenarios, including exam scores on the MCAT.

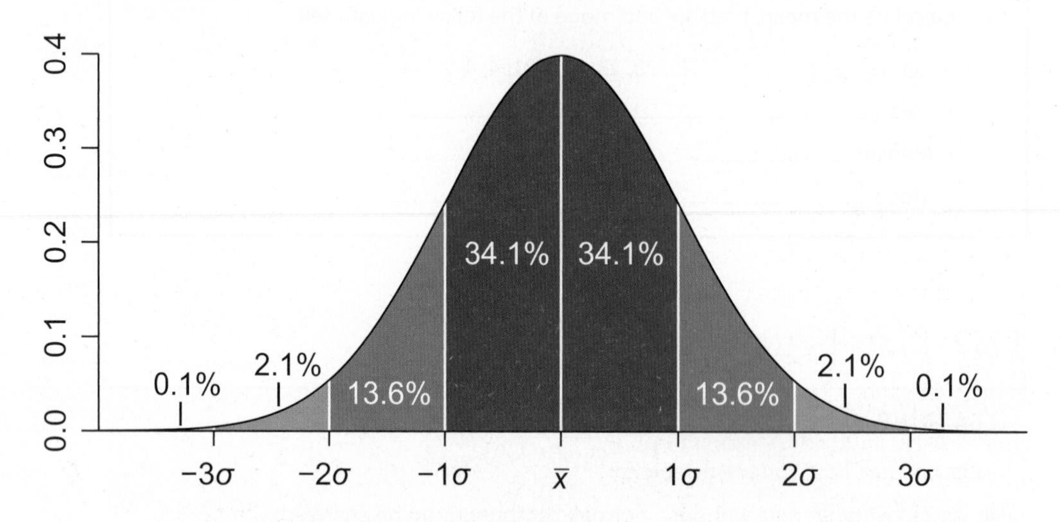

Figure 12.1 The Normal Distribution
The mean, median, and mode are at the center of the distribution. Approximately 68% of the distribution is within one standard deviation of the mean, 95% within two, and 99% within three.

Skewed Distributions

Distributions are not always symmetrical. A **skewed distribution** is one that contains a **tail** on one side or the other of the data set. On the MCAT, skewed distributions are most often tested by simply identifying their type. This is often an area of confusion for students because the *visual* shift in the data appear *opposite* the direction of the skew. A negatively skewed distribution has a tail on the left (or negative) side, whereas a positively skewed distribution has a tail on the right (or positive) side. Because the mean is more susceptible to outliers than the median, the mean of a negatively skewed distribution will be lower than the median, while the mean of a positively skewed distribution will be higher than the median. These distributions, and their measures of central tendency, are shown in Figure 12.2.

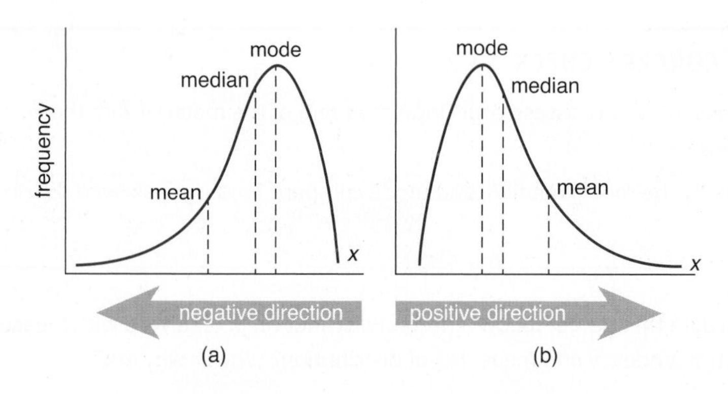

Figure 12.2 Skewed Distributions
(a) Negatively skewed distribution, with mean lower than median;
(b) Positively skewed distribution, with mean higher than median.

KEY CONCEPT

The direction of skew in a sample is determined by its tail, *not* the bulk of the distribution.

Bimodal Distributions

Some distributions have two or more peaks. A distribution containing two peaks with a valley in between is called **bimodal**, as shown in Figure 12.3. It is important to note that a bimodal distribution, strictly speaking, might have only one mode if one peak is slightly higher than the other. However, even when the peaks are of two different sizes, we still call the distribution bimodal. If there is sufficient separation of the two peaks, or a sufficiently small amount of data within the valley region, bimodal distributions can often be analyzed as two separate distributions. On the other hand, bimodal distributions do not *have* to be analyzed as two separate distributions either; the same measures of central tendency and measures of distribution can be applied to them as well.

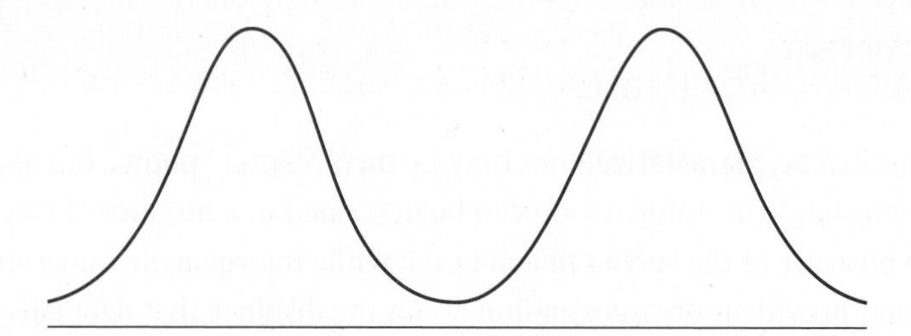

Figure 12.3 Bimodal Distribution

MCAT CONCEPT CHECK 12.2

Before you move on, assess your understanding of the material with these questions.

1. How do the mean, median, and mode compare for a right-skewed distribution?

____

2. Can data that do not follow a normal distribution be analyzed with measures of central tendency and measures of distribution? Why or why not?

____

3. What is the difference between normal or skewed distributions, and bimodal distributions?

____

12.3 Measures of Distribution

LEARNING OBJECTIVES

After Chapter 12.3, you will be able to:

- Identify outliers using interquartile range or standard deviation
- Describe the relationship between range and standard deviation
- Justify whether certain measures of distribution are or are not appropriate for a given situation

Distributions can be characterized not only by their "center" points, but also by the spread of their data. This information can be described in a number of ways. Range is an absolute measure of the spread of a data set, while interquartile range and standard deviation provide more information about the distance that data falls from one of our measures of central tendency. We can use these quantities to determine if a data point is truly an outlier in our data set.

Range

The **range** of a data set is the difference between its largest and smallest values:

$$\text{range} = x_{\text{max}} - x_{\text{min}}$$

Equation 12.3

Range does not consider the number of items of the data set, nor does it consider the placement of any measures of central tendency. Range is therefore heavily affected by the presence of data outliers. In cases where it is not possible to calculate the standard deviation for a normal distribution because the entire data set is not provided, it is possible to approximate the standard deviation as one-fourth of the range.

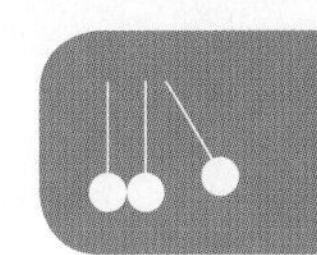

Interquartile Range

Interquartile range is related to the median, first, and third quartiles. **Quartiles,** including the median (Q_2), divide data (when placed in ascending order) into groups that comprise one-fourth of the entire set. There is some debate over the most appropriate way to calculate quartiles; for the purposes of the MCAT, we will use the most common (and simplest) method:

1. To calculate the position of the first quartile (Q_1) in a set of data sorted in ascending order, multiply n by $\frac{1}{4}$.
2. If this is a whole number, the quartile is the mean of the value at this position and the next highest position.
3. If this is a decimal, round *up* to the next whole number, and take that as the quartile position.
4. To calculate the position of the third quartile (Q_3), multiply the value of n by $\frac{3}{4}$. Again, if this is a whole number, take the mean of this position and the next. If it is a decimal, round up to the next whole number, and take that as the quartile position.

The interquartile range is then calculated by subtracting the value of the first quartile from the value of the third quartile:

$$\text{IQR} = Q_3 - Q_1$$

Equation 12.4

The interquartile range can be used to determine outliers. Any value that falls more than 1.5 interquartile ranges below the first quartile or above the third quartile is considered an **outlier**.

KEY CONCEPT

One definition of an outlier is any value lower than $1.5 \times \text{IQR}$ below Q_1 or any value higher than $1.5 \times \text{IQR}$ above Q_3.

Example: Using the interquartile range, determine whether the 36-year-old from Ray's party is an outlier. The ages are provided in numerical order below for convenience:

20, 22, 22, 22, 23, 24, 25, 36

Solution: In order to determine whether this point is an outlier, we must first determine the interquartile range. To do so, we must determine the first and third quartiles. This data set contains eight values. Multiplying 8 by $\frac{1}{4}$ gives us 2, so the first quartile is the mean of the second and third values in the ordered data set:

$$Q_1 = \frac{22 + 22}{2} = 22$$

Multiplying 8 by $\frac{3}{4}$ gives us 6, so the third quartile is the mean of the sixth and seventh values in the ordered data set:

$$Q_3 = \frac{24+25}{2} = 24.5$$

The interquartile range is the difference between these:

$$\text{IQR} = Q_3 - Q_1 = 24.5 - 22 = 2.5$$

Outliers are data values more than 1.5 interquartile ranges below Q_1 or above Q_3. Thus, any value above $24.5 + 1.5 \times 2.5 = 24.5 + 3.75 = 28.25$ or below $22 - 1.5 \times 2.5 = 22 - 3.75 = 18.25$ will be an outlier. 36 is well above 28.25, so it is an outlier in this data set.

Standard Deviation

Standard deviation is the most informative measure of distribution, but it is also the most mathematically laborious. It is calculated relative to the mean of the data. **Standard deviation** is calculated by taking the difference between each data point and the mean, squaring this value, dividing the sum of all of these squared values by the number of points in the data set minus one, and then taking the square root of the result. Expressed mathematically

$$\sigma = \sqrt{\frac{\sum_{i=1}^{n}(x_i - \bar{x})^2}{n-1}}$$

Equation 12.5

where σ is the standard deviation, x_i to x_n are the values of all of the data points in the set, $\bar{x}$ is the mean, and n is the number of data points in the set. The use of $n - 1$ instead of n is mathematically—but not practically—important and the reason for doing so is beyond the scope of the MCAT.

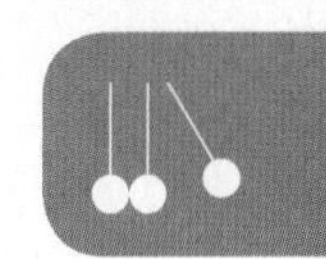

Example: Calculate the standard deviation for the following data set:

1, 2, 3, 9, 10

Solution: First, determine the value of the mean:

$$\bar{x} = \frac{\sum_{i=1}^{n} x_i}{n} = \frac{1+2+3+9+10}{5} = \frac{25}{5} = 5$$

Then, find the difference between each data point and the mean, and square this value. This is a rather tedious project, but is best solved with the use of a table as seen below:

x_i	$x_i - \bar{x}$	$(x_i - \bar{x})^2$
1	−4	16
2	−3	9
3	−2	4
9	4	16
10	5	25

Now we can determine the standard deviation:

$$\sigma = \sqrt{\frac{\sum_{i=1}^{n} (x_i - \bar{x})^2}{n-1}} = \sqrt{\frac{16+9+4+16+25}{4}} = \sqrt{\frac{70}{4}}$$
$$= \sqrt{17.5} \approx 4 \text{ (actual} = 4.18)$$

Keep in mind that when calculating the mean, we use n as the denominator, but when calculating standard deviation, we use $n - 1$.

The standard deviation can also be used to determine whether a data point is an outlier. If the data point falls more than three standard deviations from the mean, it is considered an outlier. The standard deviation relates to the normal distribution as well. On a normal distribution, approximately 68% of data points fall within one standard deviation of the mean, 95% fall within two standard deviations, and 99% fall within three standard deviations, as shown in Figure 12.1 earlier. Integration or specialized software can be used to determine percentages falling within other intervals.

KEY CONCEPT

Another definition of outlier is any value that lies more than three standard deviations from the mean.

Outliers

While we have already discussed methods for determining if a data point is an outlier, it is useful to know how to approach data with outliers. Outliers typically result from one of three causes:

1. A true statistical anomaly (e.g., a person who is over seven feet tall).
2. A measurement error (for example, reading the centimeter side of a tape measure instead of inches).
3. A distribution that is not approximated by the normal distribution (e.g., a skewed distribution with a long tail).

When an outlier is found, it should trigger an investigation to determine which of these three causes applies. If there is a measurement error, the data point should be excluded from analysis. However, the other two situations are less clear.

If an outlier is the result of a true measurement, but is not representative of the population, it may be weighted to reflect its rarity, included normally, or excluded from the analysis depending on the purpose of the study and preselected protocols. The decision should be made before a study begins—not once an outlier has been found. When outliers are an indication that a data set may not approximate the normal distribution, repeated samples or larger samples will generally demonstrate if this is true.

MCAT EXPERTISE

The existence of outliers is key for determining whether or not the mean is an appropriate measure of central tendency. It may also indicate a measurement error on the part of the investigators.

MCAT CONCEPT CHECK 12.3

Before you move on, assess your understanding of the material with these questions.

1. Compare the method of determining outliers from the interquartile range and from the standard deviation:
   - From interquartile range:

   ____

   - From standard deviation:

   ____

2. How do range and standard deviation generally relate to one another mathematically? Is this relationship accurate for the data set used earlier in this section (1, 2, 3, 9, 10; $\sigma = 4.18$)?

   ____

   ____

3. Why would the average difference from the mean be an inappropriate measure of distribution?

   ____

   ____

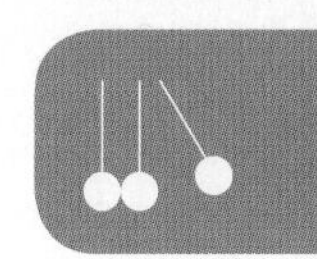

12.4 Probability

LEARNING OBJECTIVES

After Chapter 12.4, you will be able to:

- Define independence, mutual exclusivity, and exhaustiveness
- Calculate the probability of an event, or of co-occurrence of multiple independent events

Probability is usually tested on the MCAT in the context of a science question, rather than being tested on its own. In particular, genetics questions involving the Hardy-Weinberg equilibrium and Punnett squares are common applications of probability. Probability also underlies statistical testing, which we will investigate in the next section.

Independence, Mutual Exclusivity, and Exhaustiveness

In probability problems, we must first determine the relationship between events and outcomes. For events, we are most interested in **independence** or **dependence**. Conceptually, **independent events** have no effect on one another. If you roll a die and get a 3, then pick it up and roll it again, the probability of getting a 3 on the second roll is no different than it was before the first roll. Independent events can occur in any order without impacting one another.

KEY CONCEPT

Independent events do not impact each other, so their probabilities are never expected to change.

Dependent events do have an impact on one another, such that the order changes the probability. Consider a container with five red balls and five blue balls. The probability that one will choose a red ball is $\frac{5}{10}$. If a red ball is indeed chosen, then the probability of drawing another red ball is $\frac{4}{9}$. If, however, a blue ball is chosen, then the probability of drawing a red ball is $\frac{5}{9}$. In this way, the probability of the second event (getting a red ball on the second draw) is indeed dependent on the result of the first event.

We are also concerned with whether events are mutually exclusive or not. This term applies to outcomes, rather than events. **Mutually exclusive outcomes** cannot occur at the same time. One cannot flip both heads and tails in one throw, or be both ten and twenty years old. The probability of two mutually exclusive outcomes occurring together is 0%.

Finally, we must consider if a set of outcomes is exhaustive or not. A group of outcomes is said to be **exhaustive** if there are no other possible outcomes. For example, flipping heads or tails are said to be exhaustive outcomes of a coin flip; these are the only two possibilities.

Calculations

For independent events, the probability of two or more events occurring at the same time is the product of their probabilities alone

$$P(\text{A} \cap \text{B}) = P(\text{A and B}) = P(\text{A}) \times P(\text{B})$$

Equation 12.6

For example, the probability of getting heads on a coin flip twice in a row is the same as the probability of getting heads the first time times the probability of getting heads the second time, or $0.5 \times 0.5 = 0.25$. The probability of two independent events co-occurring is shown diagrammatically in Figure 12.4.

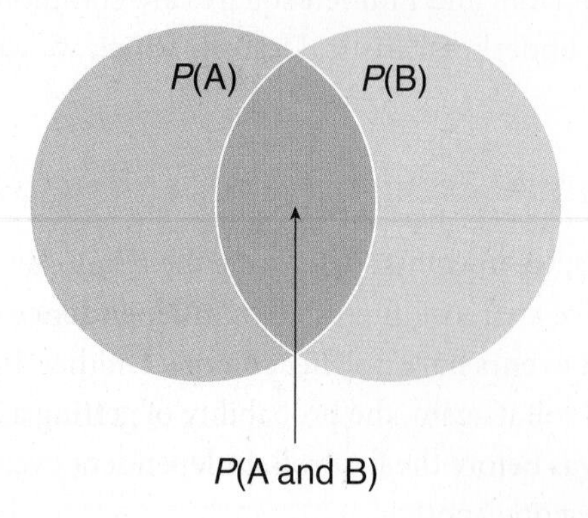

Figure 12.4 Probability of Two Independent Events Co-Occurring
$P(A \text{ and } B) = P(A) \times P(B)$

KEY CONCEPT

In probability, when using the word:

- *and*—multiply the probabilities
- *or*—add the probabilities (and subtract the probability of both happening together)

The probability of at least one of two events occurring is equal to the sum of their initial probabilities, minus the probability that they will both occur.

$$P(\text{A} \cup \text{B}) = P(\text{A or B}) = P(\text{A}) + P(\text{B}) - P(\text{A and B})$$

Equation 12.7

Example: In a certain population, 10% of the population has diabetes and 30% is obese. If 7% of the population has both diabetes and obesity, are these events independent? If one chose an individual at random from this population, what would be the probability of that patient having at least one of the two conditions?

Solution: With the numbers given, these events cannot be independent. For independent events, $P(\text{A and B}) = P(\text{A}) \times P(\text{B}) = P(\text{having diabetes}) \times P(\text{being obese}) = 0.1 \times 0.3 = 0.03$. In this population, the probability of having diabetes and being obese is 0.07.

To determine the probability of the individual having at least one of the conditions, we use the "or" equation:

$$P(\text{A or B}) = P(\text{A}) + P(\text{B}) - P(\text{A and B}) = 0.1 + 0.3 - 0.07 = 0.33 \text{ or } 33\%$$

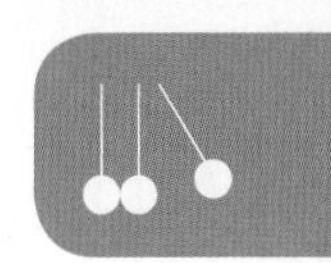

MCAT CONCEPT CHECK 12.4

Before you move on, assess your understanding of the material with these questions.

1. Assume the likelihood of having a male child is equal to the likelihood of having a female child. In a series of ten live births, the probability of having at least one boy is equal to:

 ____

2. Define the following terms:
   - Independence:

 ____

   - Mutual exclusivity:

 ____

   - Exhaustiveness:

 ____

12.5 Statistical Testing

LEARNING OBJECTIVES

After Chapter 12.5, you will be able to:

- Distinguish between hypothesis tests and confidence intervals
- Recall how *p*-values are calculated during a hypothesis test
- Predict the outcome of a test given its *p*- and α-values
- Explain the importance of power in statistical testing

Hypothesis testing and confidence intervals allow us to draw conclusions about populations based on our sample data. Both are interpreted in the context of probabilities, and what we deem to be an acceptable risk of error.

Hypothesis Testing

Hypothesis testing begins with an idea about what may be different between two populations. We have a **null hypothesis,** which is always a hypothesis of equivalence. In other words, the null hypothesis says that two populations are equal, or that a single population can be described by a parameter equal to a given value. The **alternative hypothesis** may be **nondirectional** (that the populations are not equal) or **directional** (for example, that the mean of population A is greater than the mean of population B).

The most common hypothesis tests are *z*- or ***t*-tests**, which rely on the standard distribution or the closely related *t*-distribution. From the data collected, a **test statistic** is calculated and compared to a table to determine the likelihood that that statistic was obtained by random chance (under the assumption that our null hypothesis is true). This is our ***p*-value**. We then compare our *p*-value to a **significance level** (α); 0.05 is commonly used. If the *p*-value is greater than α, then we fail to reject the null hypothesis, which means that there is not a statistically significant difference between the two populations. If the *p*-value is less than α, then we reject the null hypothesis and state that there is a statistically significant difference between the two groups. Again, when the null hypothesis is rejected, we state that our results are statistically significant.

The value of α is the level of risk that we are willing to accept for incorrectly rejecting the null hypothesis. This is also called a **type I error**. In other words, a type I error is the likelihood that we report a difference between two populations when one does not actually exist. A **type II error** occurs when we incorrectly fail to reject the null hypothesis. In other words, a type II error is the likelihood that we report no difference between two populations when one actually exists. The probability of a type II error is sometimes symbolized by β. The probability of correctly rejecting a false null hypothesis (reporting a difference between two populations when one actually exists) is referred to as **power**, and is equal to $1 - \beta$. Finally, the probability of correctly failing to reject a true null hypothesis (reporting no difference between two populations when one does not exist) is referred to as **confidence**. These conditions are summarized in Table 12.1.

		Truth About the Population	
		H_0 true (no difference)	H_a true (difference exists)
Conclusion Based on Sample	Reject H_0	Type I error (α)	Power ($1 - \beta$)
	Fail to reject H_0	Confidence	Type II error (β)

Table 12.1 Results of Hypothesis Testing

Confidence Intervals

Confidence intervals are essentially the reverse of hypothesis testing. With a confidence interval, we determine a range of values from the sample mean and standard deviation. Rather than finding a *p*-value, we begin with a desired confidence level (95% is standard) and use a table to find its corresponding *z*- or *t*-score. When we multiply the *z*- or *t*-score by the standard deviation, and then add and subtract this number from the mean, we create a range of values. For example, consider a population for which we wish to know the mean age. We draw a sample from that population and find that the mean of the sample is 30, with a standard deviation of 3. If we wish to have 95% confidence, the corresponding *z*-score (which would be provided on Test Day) is 1.96. Thus, the range is $30 - (3)(1.96)$ to $30 + (3)(1.96) = 24.12$ to

35.88. We can then report that we are 95% confident that the true mean age of the population from which this sample is drawn is between 24.12 and 35.88.

MCAT CONCEPT CHECK 12.5

Before you move on, assess your understanding of the material with these questions.

1. How do hypothesis tests and confidence intervals differ?
   - Hypothesis tests:

   ____

   - Confidence intervals:

   ____

2. If the *p*-value is greater than α in a given statistical test, what is the outcome of the test?

   ____

3. How is the *p*-value calculated during a hypothesis test?

   ____

   ____

4. True or False: Power is the probability of correctly rejecting the null hypothesis.

12.6 Charts, Graphs, and Tables

LEARNING OBJECTIVES

After Chapter 12.6, you will be able to:

- Recognize when data relationships call for transformation into semilog or log-log plots
- Recall the pros and cons of different types of visual data representation, including pie charts, bar graphs, box plots, maps, graphs, and tables
- Distinguish between exponential and parabolic curves

Because your career will be filled with evidence-based medicine, it is important to be able to recognize and interpret data in multiple forms. We have already considered the mathematical side of statistics; now, let's take a look at the visual side. On the MCAT, anticipate that most passages in the sciences will be accompanied by a visual aid in some way—frequently, this will be a chart, graph, or data table.

Types of Charts

Charts present information in a visual format and are frequently used for categorical data.

Pie or Circle Charts

Pie or **circle charts** are used to represent relative amounts of entities and are especially popular in demographics. They may be labeled with raw numerical values or with percent values. The primary downside to pie charts is that as the number of represented categories increases, the visual representation loses impact and becomes confusing. For example, in Figure 12.5, the population of each of the 50 states and the District of Columbia is presented on a pie chart, but the large number of entities makes the graph incoherent.

MCAT EXPERTISE

Questions about pie charts are likely to be qualitative, asking for the smallest or largest group, or the percentage occupied by one or more groups combined. These questions are unlikely to require additional analysis because pie charts are not dense with information.

BRIDGE

Pie charts are frequently used to present demographic information. Demographics is the statistical arm of sociology and is discussed in Chapter 11 of *MCAT Behavioral Sciences Review.*

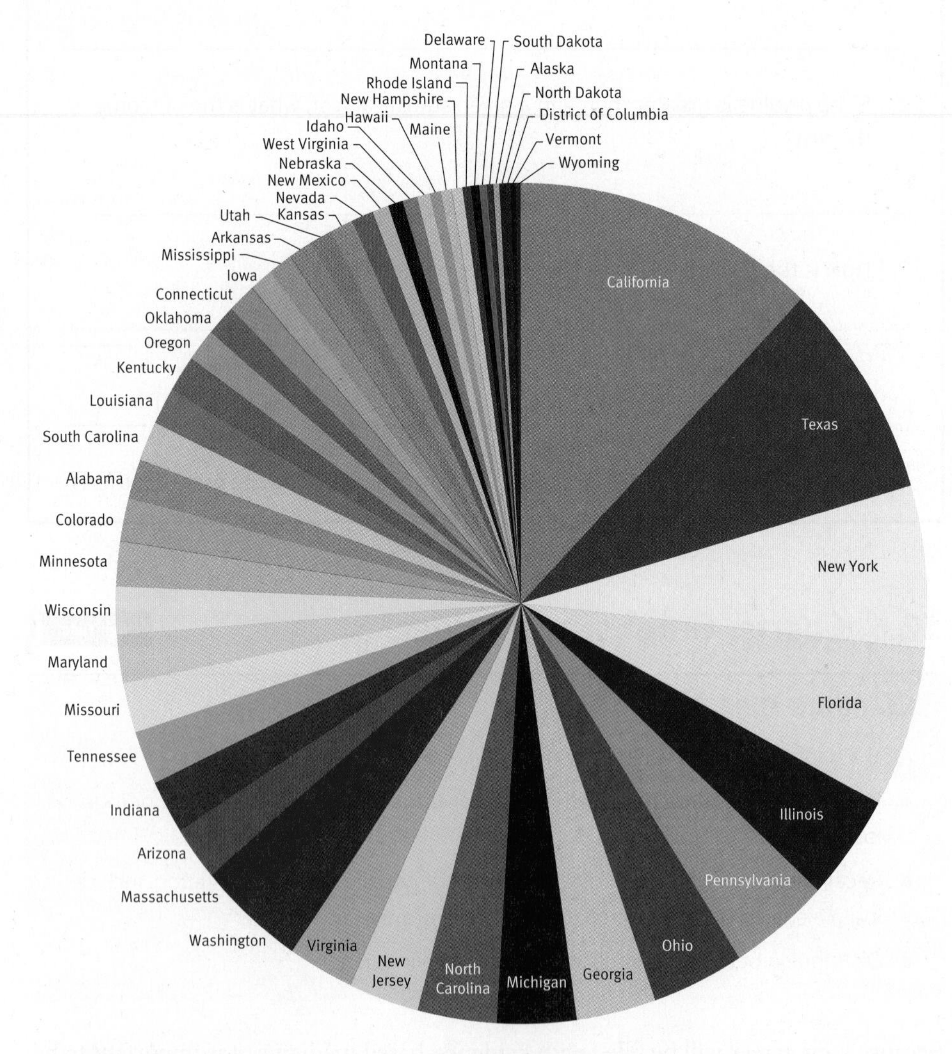

Figure 12.5 Pie Chart of United States Population by State, 2010 Census
Pie charts become difficult to interpret when too many categories are included.

Bar Charts and Histograms

Bar charts and histograms are likely to contain significantly more information than a pie chart for the same amount of page space. **Bar charts** are used for categorical data, which sort data points based on predetermined categories. The bars may then be sorted by increasing or decreasing bar length. The length of a bar is generally proportional to the value it represents. Wherever possible, breaks should be avoided in the chart because of the potential to distort scale. To that end, be wary of graphs that contain breaks; they may be enlarging the difference between bars. Figure 12.6 shows a representative bar graph for causes of cancer death in the United States in 2010.

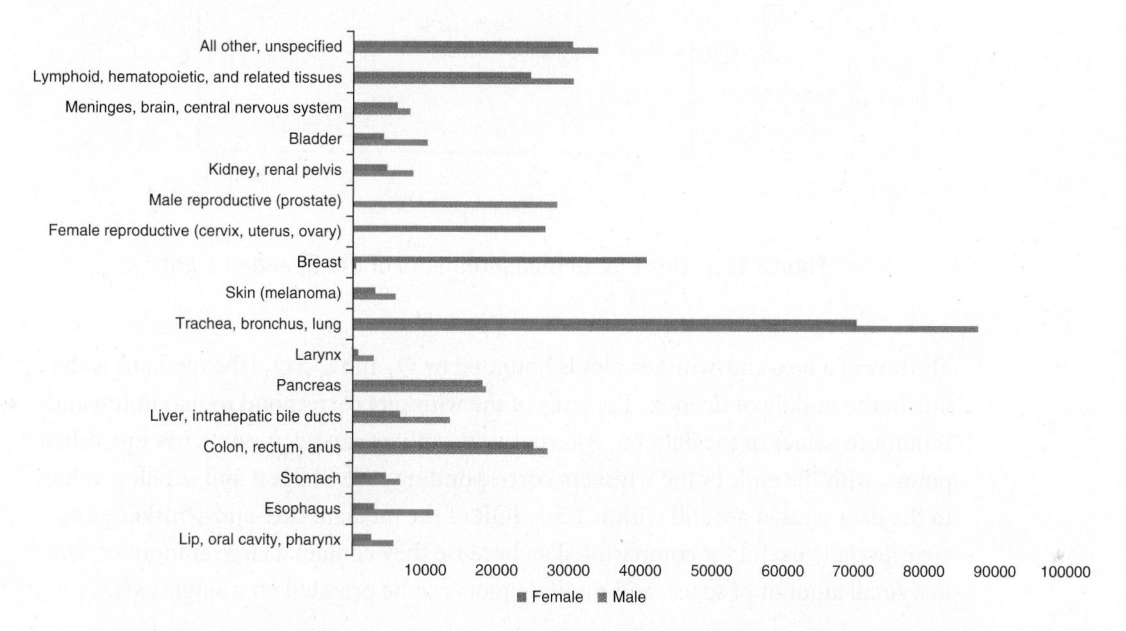

Figure 12.6 Causes of Cancer Death by Type, 2010
Source: *Centers for Disease Control and Prevention National Vital Statistics Reports*

Histograms present numerical data rather than discrete categories. Histograms are particularly useful for determining the mode of a data set because they are used to display the distribution of a data set.

Box Plot

Box plots are used to show the range, median, quartiles and outliers for a set of data. A labeled box plot, also called a **box-and-whisker**, is shown in Figure 12.7.

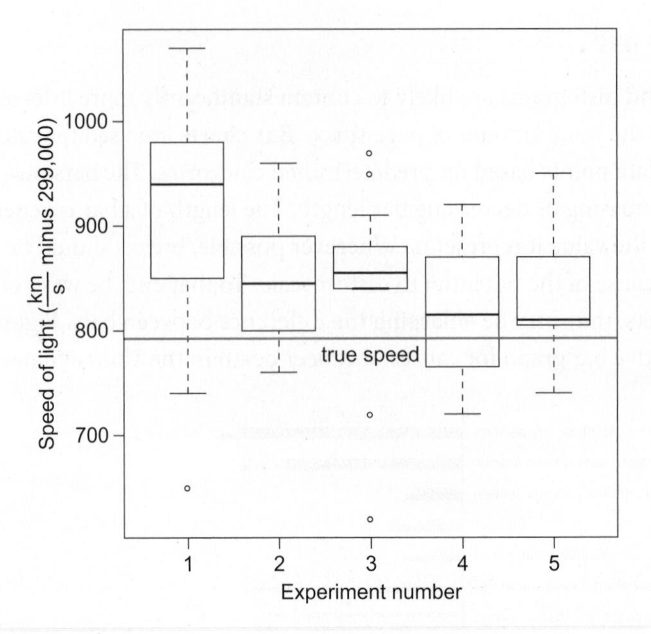

Figure 12.7 Box Plot of Measurements of the Speed of Light

The box of a box-and-whisker plot is bounded by Q_1 and Q_3; Q_2 (the median) is the line in the middle of the box. The ends of the whiskers correspond to maximum and minimum values of the data set. Alternatively, outliers can be presented as individual points, with the ends of the whiskers corresponding to the largest and smallest values in the data set that are still within 1.5 × IQR of the median. Box-and-whisker plots are especially useful for comparing data because they contain a large amount of data in a small amount of space, and multiple plots can be oriented on a single axis.

Maps

In addition to the other forms of charts, data can be illustrated geographically. Maps of health conditions, population density, political districts, and ethnicity are relatively easy to comprehend and may show geographic clustering for some data. The best map data will examine one or at most two pieces of information simultaneously. Any further data may inhibit clarity. A map of population density in each country of the world is shown in Figure 12.8.

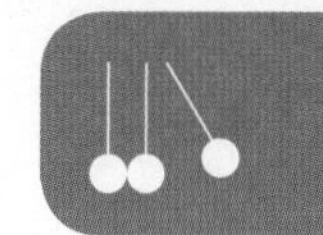

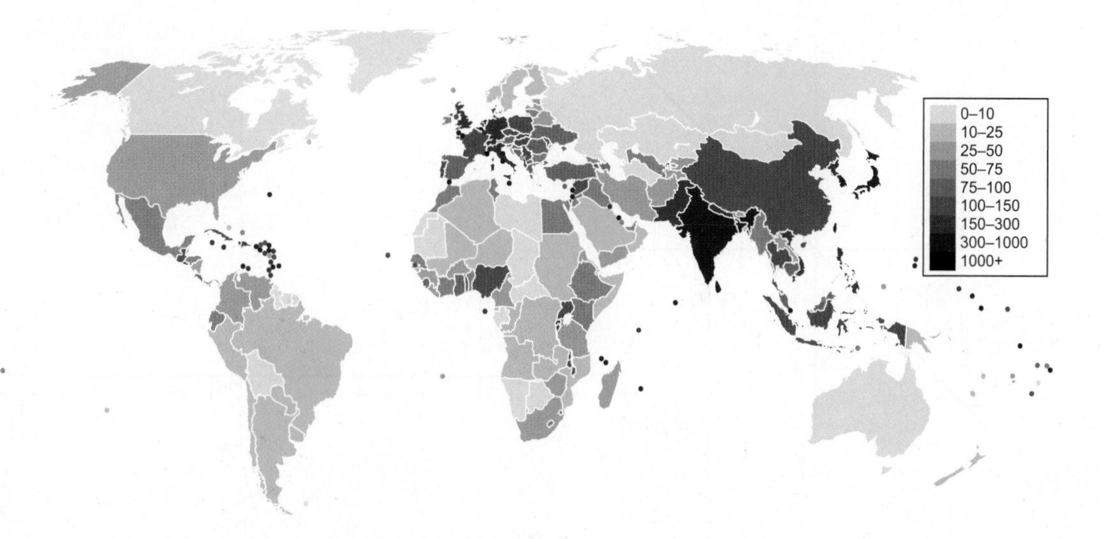

Figure 12.8 Population Density by Country, 2006

Graphs and Axes

While we're all familiar with constructing graphs—especially scatter plots and line graphs—it is important to know some important features and potential stumbling blocks of graphs as we move toward Test Day. When presented with a graph, you should attempt to draw rough conclusions immediately but should not spend time analyzing all of the details of the graph unless asked to do so by a question. The first thing to do when you encounter a graph on Test Day is to look at the axes.

Linear Graphs

Linear graphs show the relationships between two variables. They generally involve two direct measurements and, strictly speaking, do not have to be a straight line. The shape of the curve on this type of graph may be **linear**, **parabolic**, **exponential**, or **logarithmic**. These are shown in Figure 12.9. On Test Day, you should be able to recognize at least these four shapes of graphs.

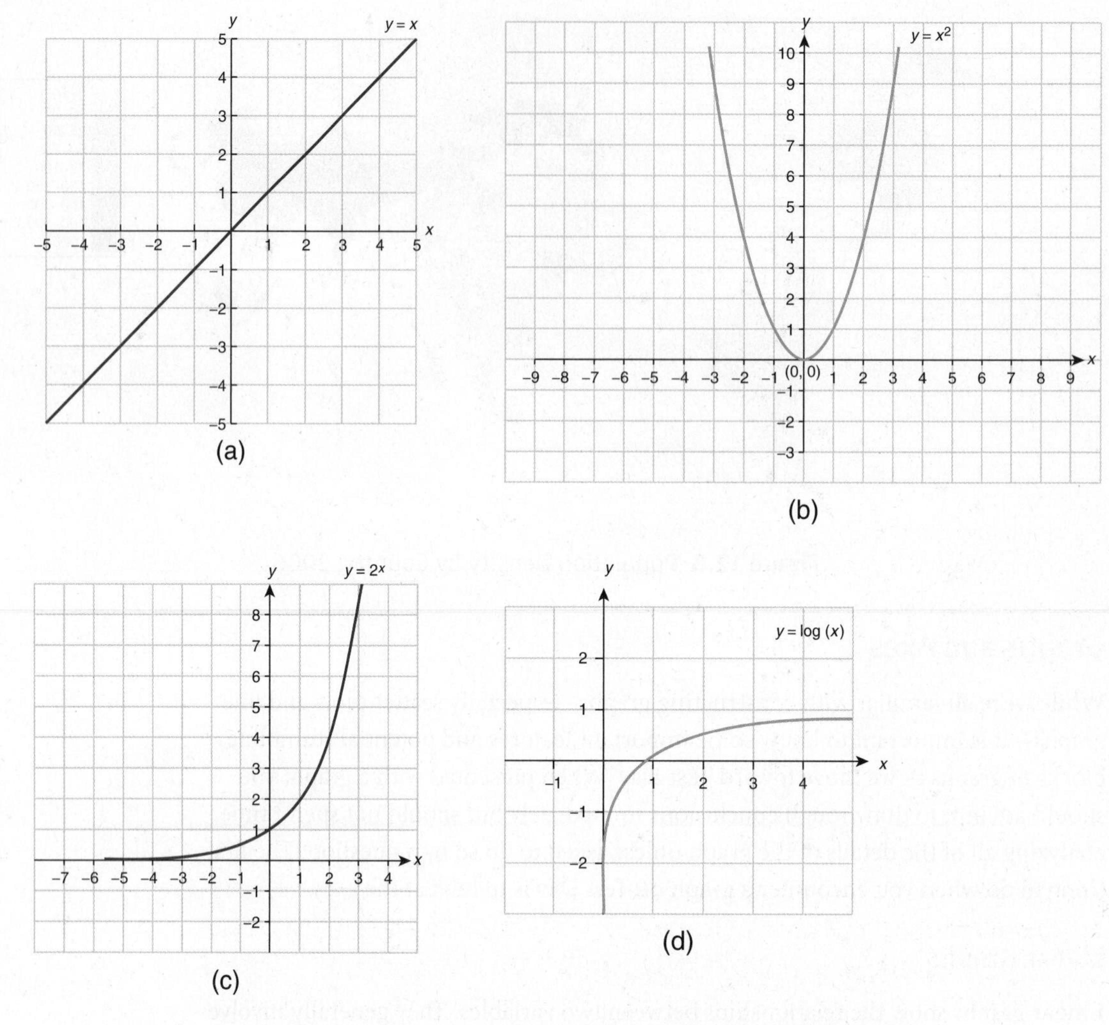

Figure 12.9 Shapes of Common Relationships on a Linear Graph
(a) Linear; (b) Parabolic; (c) Exponential; (d) Logarithmic

MNEMONIC

Slope is like waking up in the morning: Slope is always rise (vertical) over run (horizontal) because you have to get up from bed (rise) before you get moving (run).

The axes of a linear graph will be consistent in the sense that each unit will occupy the same amount of space (the distance from 1 to 2 to 3 to 4 on each axis remains the same size). As with bar graphs, be wary of scale and breaks in axes. Where both the shape of the graph and the graph type are linear, we should be able to calculate the slope of the line. **Slope (m)** is the change in the y-direction divided by the change in the x-direction for any two points:

$$m = \frac{\text{rise}}{\text{run}} = \frac{\Delta y}{\Delta x}$$

Equation 12.8

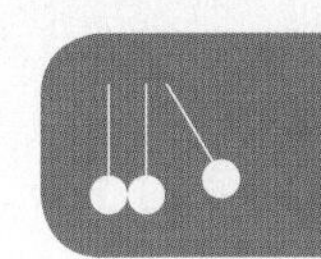

Example: Calculate the slope of the line in the graph shown below.

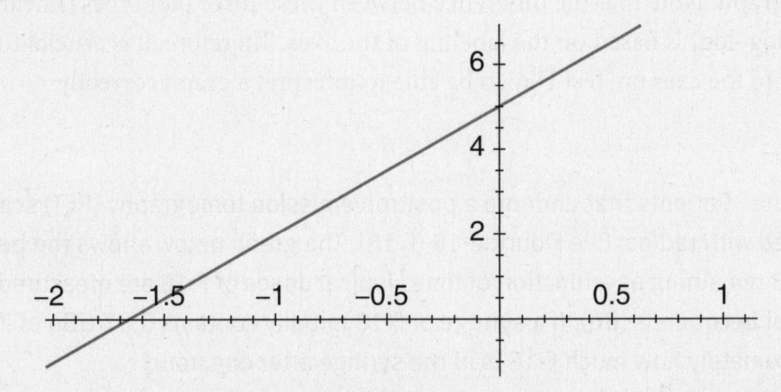

Solution: The slope of a line is equal to the difference in the values of two points in the *y*-direction divided by the difference in the values of the same two points in the *x*-direction. The *x*- and *y*-intercepts are generally good choices because one of the values will be zero for each point:

$$m = \frac{y_2 - y_1}{x_2 - x_1} = \frac{0 - 5}{-1.66 - 0} = \frac{-5}{-1.66} = \frac{5}{\frac{5}{3}} = 3$$

Semilog and Log–Log Graphs

Semilog graphs are a specialized representation of a logarithmic data set. They can be easier to interpret because the otherwise curved nature of the logarithmic data is made linear by a change in the **axis ratio**. In semilog graphs, one axis (usually the *x*-axis) maintains the traditional unit spacing. The other axis assigns spacing based on a ratio, usually 10, 100, 1000, and so on. The multiples may be of any number as long as there is consistency in the ratio from one point on the axis to the next. Figure 12.10 shows an example of a semilog plot.

KEY CONCEPT

The axes on a graph will determine which type of plot is being used, and provide key information about the underlying relationship between the relevant variables.

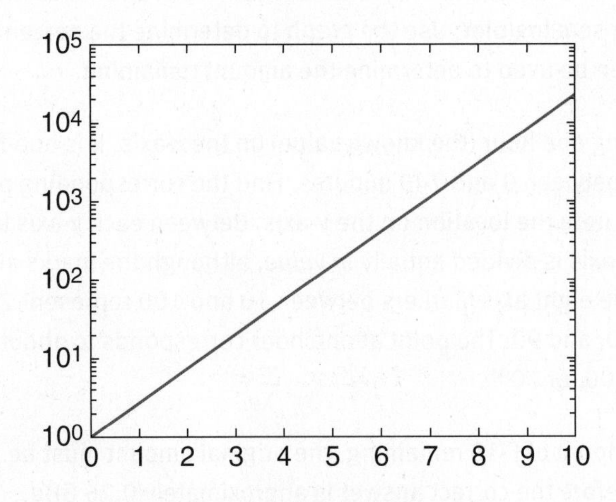

Figure 12.10 A Semilog Plot

In some cases, both axes can be given a different axis ratio to create a linear plot. When both axes use a constant ratio from point to point on the axis, this is termed a **log–log graph**. Note that the difference between these three plot types (linear, semi-log, and log–log) is based on the labeling of the axes. Therefore, it is crucial to pay attention to the axes on Test Day to be able to interpret a graph correctly.

Example: Patients that undergo a positron emission tomography (PET) scan are injected with radioactive Flouride-18 (F-18). The graph below shows the percent of F-18 remaining as a function of time. Typical doses of F-18 are measured in units of becquerels (Bq). If a syringe of F-18 initially contains 0.37 GBq of F-18, approximately how much F-18 is in the syringe after one hour?

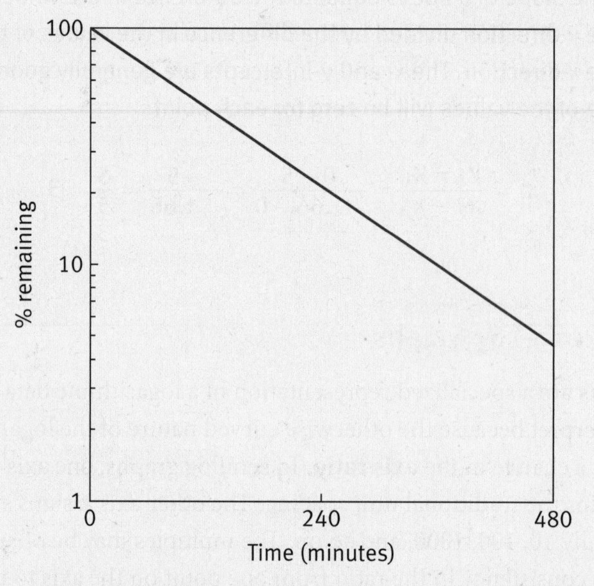

Solution: Notice that the values on the *y*-axis are equally spaced and multiples of ten. That means the *y*-axis is logarithmic. The *x*-axis is linear so the graph is a semilog plot. Use the graph to determine the percent remaining, which can then be used to determine the amount remaining.

Start by finding one hour (the known value) on the *x*-axis. It is one-fourth of the distance between 0 and 240 minutes. Find the corresponding point on the line and then note the location on the *y*-axis. Between each *y*-axis label (1, 10, and 100) the axis is divided equally *in value*, although the marks are unequally spaced. So the eight axis markers between 10 and 100 represent 20, 30, 40, 50, 60, 70, 80, and 90. The point at one hour corresponds to about the third mark below 100, or 70%.

To find the amount of F-18 remaining, the original amount must be multiplied by 0.70. Therefore the correct answer is approximately 0.26 GBq.

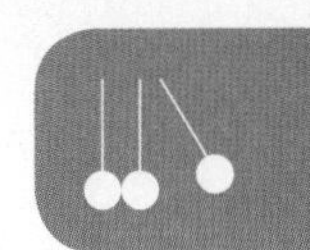

Interpreting Tables

Unlike with graphs, you should only take a brief moment to glance at the title of a table before approaching Test Day questions. Tables are more likely to contain disjointed information than either charts or graphs because they often contain categorical data or experimental results. Tables that do not have unusual data values (zeroes, outliers, changes in a trend, and so on) should be approached especially briefly.

When a table does contain significant organization (for example, listing results progressively), this structure is likely to be relevant while answering questions. For example, a trend that suddenly appears or disappears will often require an explanation.

Additionally, when provided with data in the form of a table, you should be able to convert it to a rough graph or to a linear equation. The MCAT may test on the interpretation of slope without actually providing a graph.

MCAT CONCEPT CHECK 12.6

Before you move on, assess your understanding of the material with these questions.

1. What type of data relationship is least likely to require transformation into a semilog or log–log plot?

____

2. Fill in the following table with the pros and cons of each type of visual data representation:

Type of Visual Aid	Pros	Cons
Pie Chart		
Bar Graph		
Box Plot		
Map		
Graph		
Table		

3. How do exponential and parabolic curves differ in shape?

- Exponential:

____

- Parabolic:

____

12.7 Applying Data

LEARNING OBJECTIVES

After Chapter 12.7, you will be able to:

- Distinguish between correlation and causation
- Relate the statistical results of a study to the impact of those findings on scientific knowledge and policy change

Finally, we have reached the discussion section of an academic paper, in which the data that we have gathered and interpreted is applied to the original problem. We can then begin drawing conclusions and creating new questions based on our results. Because much of this was covered in the discussion on experimental methods in Chapter 11 of *MCAT Physics and Math Review*, we will be terse in our review here.

Correlation and Causation

As discussed previously, we must be careful with our wording when discussing variable relationships. **Correlation** refers to a connection—direct relationship, inverse relationship, or otherwise—between data. If two variables trend together, that is as one increases so does the other, there is a positive correlation. If two variables trend in opposite directions (one increases as the other decreases) there is a negative correlation. These relationships can be quantified with a **correlation coefficient**, a number between -1 and $+1$ that represents the strength of the relationship. A correlation coefficient of $+1$ indicates a strong positive relationship, a value of -1 indicates a strong negative relationship, and a value of zero indicates no apparent relationship.

Correlation does not necessarily imply **causation**; we must avoid this assumption when there is insufficient evidence to draw such a conclusion. If an experiment cannot be performed, we must rely on Hill's criteria, discussed in Chapter 11 of *MCAT Physics and Math Review*. Remember that the only one of Hill's criteria that is uniformly necessary for causation is temporality.

In the Context of Scientific Knowledge

When interpreting data, it is important that we not only state the apparent relationships between data, but also begin to draw connections to other concepts in science and to our background knowledge. At a minimum, the impact of the new data on the existing hypothesis must be considered, although ideally the new data would be integrated into all future investigations on the topic. Additionally, we must develop a plausible rationale for the results. Finally, we must make decisions about our data's impact on the real world, and determine whether or not our evidence is substantial and impactful enough to necessitate changes in understanding or policy.

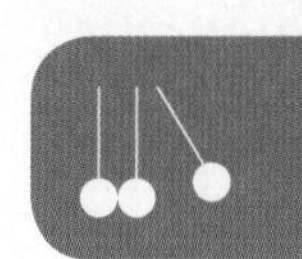

Example: A textbook publisher wanted to study the effectiveness of a new Advanced Placement (AP) study aide. From 2006 to 2015 the researchers recruited high schools across the country to participate, and randomly assigned each school to a group—those who received the study aide (Group A) and those who did not (Group B). Grades were tracked in AP Statistics over a 10 year time period. On which three years was there a statistically significant difference between the two groups, and a high likelihood of passing for those who used the study aide?

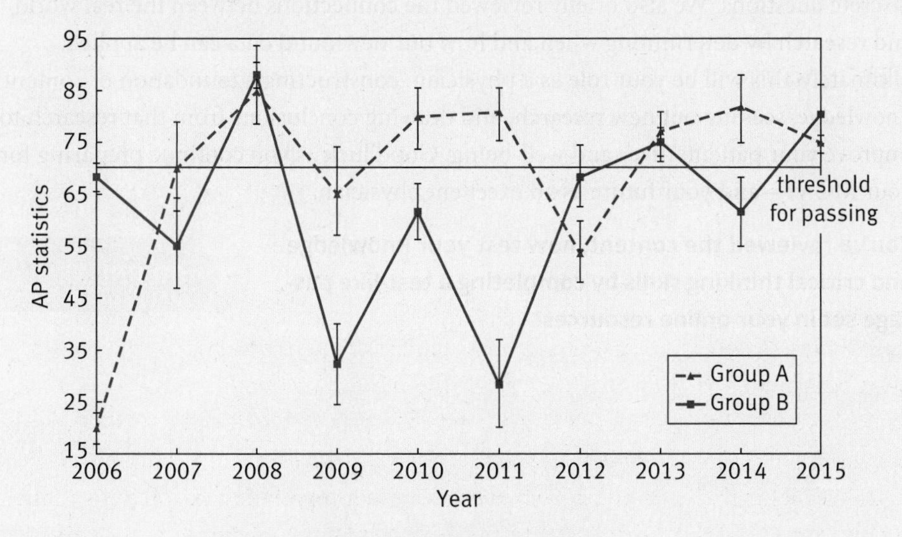

Average AP Grades by Group
Error bars show the 95% CI

Solution: In order to support the claim that the study aide helps student pass the AP exam, the evidence should show a statistically significant difference between those who received the aide and those who did not. Further, the true score (which should be within the confidence interval 95% of the time) of those who received the aide should be above passing and the true score of those who did not should be below passing. The data from 2010, 2011, and 2014 fit the description and provide the best support.

MCAT CONCEPT CHECK 12.7

Before you move on, assess your understanding of the material with these questions.

1. True or False: Statistical significance is sufficient criteria to enact policy change.

2. True or False: Two variables that are causally related will also be correlated with each other.

Conclusion

Congratulations on completing *MCAT Physics and Math Review*! While it has been a challenging journey, you are now equipped with all of the physics content knowledge and *Scientific Inquiry and Reasoning Skills* (SIRS) you need to perform well on Test Day. We completed our discussion of the MCAT SIRS by covering the transformation of raw data to actionable information. When taking the MCAT, these concepts may present themselves as the opportunity to use statistical methods and interpretation to draw conclusions, as well as the analysis of figures used as adjuncts to passages and discrete questions. We also briefly reviewed the connections between the real world and research by determining when and how our newfound data can be applied. Ultimately, this will be your role as a physician: constructing a foundation of content knowledge, seeking out new research, and drawing conclusions from that research to improve your patients' lives and well-being. Good luck as you continue preparing for your MCAT—and your future as an excellent physician.

You've reviewed the content, now test your knowledge and critical thinking skills by completing a test-like passage set in your online resources!

GO ONLINE

Concept Summary

Measures of Central Tendency

- **Measures of central tendency** provide a single value representation for the middle of a group of data.
- The **arithmetic mean** or **average** is a measure of central tendency that equally weighs all values; it is most affected by outliers.
- The **median** is the value that lies in the middle of the data set. Fifty percent of data points are above and below the median.
- The **mode** is the data point that appears most often; there may be multiple (or zero) modes in a data set.

Distributions

- Distributions have characteristic features that are exemplified by their shape. Distributions can be classified by measures of central tendency and measures of distribution.
- The **normal distribution** is symmetrical. The mean, median, and mode are all the same in the normal distribution.
  - The **standard distribution** is a normal distribution with a mean of zero and a standard deviation of one; it is used for most calculations.
  - 68% of data points occur within one standard deviation of the mean, 95% within two, and 99% within three.
- **Skewed distributions** have differences in their mean, median, and mode; the skew direction is the direction of the **tail** of the distribution.
- **Bimodal distributions** have multiple peaks, although not necessarily multiple modes, strictly speaking. It may be useful to perform data analysis on the two groups separately.

Measures of Distribution

- **Range** is the difference between the largest and smallest values in a data set.
- **Interquartile range** is the difference between the value of the **third quartile** and **first quartile**; interquartile range can be used to determine outliers.
- **Standard deviation** is a measurement of variability about the mean; standard deviation can also be used to determine outliers.
- **Outliers** may be a result of true population variability, measurement error, or a non normal distribution.
- Procedures for handling outliers should be formulated before the beginning of a study.

Probability

- The probability of **independent events** does not change based on the outcomes of other events.
- The probability of a **dependent event** changes depending on the outcomes of other events.
- **Mutually exclusive outcomes** cannot occur simultaneously.
- When a set of outcomes is **exhaustive**, there are no other possible outcomes.

Statistical Testing

- **Hypothesis tests** use a known distribution to determine whether a hypothesis of no difference (the **null hypothesis**) can be rejected.
- Whether or not a finding is statistically significant is determined by the comparison of a ***p*-value** to the selected **significance level (α)**. A significance level of 0.05 is commonly used.
- **Confidence intervals** are a range of values about a sample mean that are used to estimate the population mean. A wider interval is associated with a higher **confidence level** (95% is common).

Charts, Graphs, and Tables

- **Pie charts** (**circle charts**) and **bar charts** are both used to compare categorical data.
- **Histograms** and **box plots** (**box-and-whisker plots**) are both used to compare numerical data.
- Maps are used to compare up to two demographic indicators.
- **Linear**, **semilog**, and **log–log** plots can be distinguished by their axes.
- **Slope** can be calculated most easily from linear plots.
- Tables may contain related or unrelated categorical data.

Applying Data

- **Correlation** and **causation** are separate concepts that are linked by Hill's criteria.
- Data must be interpreted in the context of the current hypothesis and existing scientific knowledge.
- Statistical and practical significance are distinct.

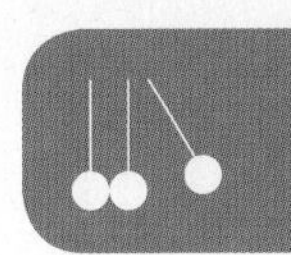

Answers to Concept Checks

12.1

1. The mean is the best measure of central tendency for a data set with a relatively normal distribution. The mean performs poorly in data sets with outliers.
2. Mean: $\frac{25+23+23+6+9+21+4+4+2}{9}=\frac{117}{9}=13$

   Median: The fifth position of 2, 4, 4, 6, 9, 21, 23, 23, 25 is 9

   Mode: There are two numbers that each appear twice: 4 and 23. These are both modes of this data set.

12.2

1. The mean of a right (positively) skewed distribution is to the right of the median, which is to the right of the mode.
2. Any distribution can be mathematically or procedurally transformed to follow a normal distribution by virtue of the central limit theorem, which is beyond the scope of the MCAT. Regardless, a distribution that is not normal may still be analyzed with these measures.
3. Bimodal distributions have two peaks, whereas normal or skewed distributions have only one.

12.3

1. Outliers can be defined as data points more than 1.5 × IQR below Q_1 or above Q_3. The can also be defined as data points more than 3σ above or below the mean. The cutoff values calculated through the two methods are likely to be different, and the selection of one method over the other is one of preference and study design. In general, the use of the standard deviation method is superior.
2. Where the data are not available, the range can be approximated as four times the standard deviation. For this data set, the relationship fails. The range is 9, which is only a little more than twice the standard deviation. This is because the data set does not fall in a normal distribution.
3. The average distance from the mean will always be zero. This is why, in calculations of standard deviation, we always square the distance from the mean and then take the square root at the end—it forces all of the values to be positive numbers, which will not cancel out to zero.

12.4

1. Simplify this question by rewording it as the probability of *not* having all girls. Having at least one boy and having all girls are mutually exclusive events, and no other possibilities can occur. Thus, the probability of having all girls is $(0.5)^{10}$ and the probability of having at least one boy is $1 - (0.5)^{10}$, or 99.90%.
2. Independence is a condition of events wherein the outcome of one event has no effect on the outcome of the other. Mutual exclusivity is a condition wherein two outcomes cannot occur simultaneously. When a set of outcomes is exhaustive, there are no other possible outcomes.

12.5

1. Hypothesis tests are used to validate or invalidate a claim that two populations are different, or that one population differs from a given parameter. In a hypothesis test, we calculate a *p*-value and compare it to a chosen significance level (α) to conclude if an observed difference between two populations (or between a population and the parameter) is significant or not. Confidence intervals are used to determine a potential range of values for the true mean of a population.
2. If the *p*-value is greater than α, then we fail to reject the null hypothesis.
3. After the test statistic is calculated, a computer program or table is consulted to determine the *p*-value of the statistic.
4. True. Power is the probability that the individual rejects the null hypothesis when the alternative hypothesis is true for the population.

12.6

1. Linear relationships can be analyzed without any data or axis transformation into semilog or log–log plots.
2.

Type of Visual Aid	Pros	Cons
Pie Chart	Easily constructed; useful for categorical data with a small number of categories.	Easily overwhelmed with multiple categories. Difficult to estimate values with circles.
Bar Graph	Multiple organization strategies. Good for large categorical data sets.	Axes are often misleading because of sizeable breaks.
Box Plot	Information-dense; can be useful for comparison.	May not highlight outliers or mean value of a data set. Only useful for numerical data.
Map	Provide relevant and integrated geographic and demographic information.	May only be used to represent at most two variables coherently.
Graph	Provide information about relationships. Useful for estimation.	Axis labels and logarithmic scales require careful interpretation.
Table	Categorical data can be presented without comparison. Does not require estimation for calculations.	Disorganized or unrelated data may be presented together.

3. Exponential and parabolic curves both have a steep component; however, exponential curves have horizontal asymptotes and become flat on one side while parabolic curves are symmetrical and have steep components on both sides of a center point.

12.7

1. False. As discussed in the last chapter, there must be practical (clinical), as well as statistical significance for a conclusion to be useful.
2. True. While two variables that are correlated are not necessarily causally related, all variables that are causally related must be correlated in some way (direct relationship, inverse relationship, or otherwise).

Science Mastery Assessment Explanations

1. **C**

The mean is computed by taking the sum of all of the data points and dividing by the total number of data points. If a very large data point was included, the sum would be significantly larger, inflating the mean. This observation justifies **(C)**. On the other hand, since the median is the middle number and the mode is the most common value, the inclusion of outliers would have minimal effect on these measures.

2. **B**

The mean is to the left of the median, which implies that the tail of the distribution is on the left side; therefore, this distribution is skewed left. It would be expected that there would be a low plateau on the left side of the distribution, which accounts for the shift in the mean.

3. **B**

The median is the central data point in an ordered list. Because this data set has seven numbers, the central point will be in the fourth position. Reordered, the list reads: 2, 4, 4, 7, 17, 23, 53. Thus, the median is 7. **(A)**, 4, is the mode while **(C)**, 15.7, is the mean.

4. **B**

Based on the population description in the question stem, we would expect to see a large number of finches with small beaks, a very small number of finches with intermediate beaks, and a large number of finches with large beaks. This pattern is consistent with a bimodal distribution, choice **(B)**.

5. **B**

Approximately 95% of values fall within two standard deviations ($\pm 2\sigma$) of the mean for a normal distribution. A confidence interval is constructed using the same values. Approximately 68% of the values are within one standard deviation, and 99% are within three standard deviations, eliminating the other answer choices.

6. **C**

When two events are independent of each other, the probability of them occurring simultaneously is equal to the product of the probability of each occurring separately. Thus, for weekly fast food consumption and prevalence of cardiovascular disease to be independent, the probability of both occurring must be (20%)(48%) = 9.6%, supporting **(C)**. While **(D)** is a tempting answer choice, a lack of cardiovascular disease among fast food consumers would not imply independence of these factors, rather it would imply that fast food consumption actually decreases or eliminates the likelihood of cardiovascular disease.

7. **A**

Because the error is in data transfer, the original source of data can be consulted to allow for the inclusion of the correct data point. An error in instrument calibration may introduce bias; while this should not affect the standard deviation of a sample, it would certainly affect the mean. The instrument would have to be recalibrated, and the relevant data points would have to be measured again to correct for this type of outlier, eliminating choice **(B)**. A skewed distribution is one that has a long tail. In this case, it may be more challenging to determine if a particular value is an outlier or simply a value in the long tail of the distribution. Repeated sampling or a large sample size is usually required to determine if a sample is truly skewed, eliminating choice **(C)**. An anomalous result is challenging to interpret, and how to correct for the result may be unclear. In some cases, the result should be inflated or weighted more heavily to reflect its significance; in other cases, it should be interpreted as a regular value. In still other cases, it is appropriate to drop the anomalous result. This decision should ideally be made before the study even begins, but this still certainly requires more consideration than simply checking a result from one's original data set, eliminating choice **(D)** and making choice **(A)** correct.

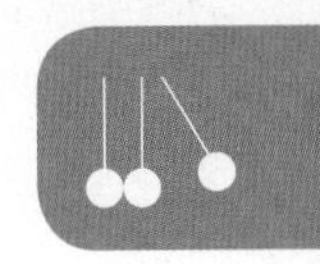

8. **A**

Because one parent is homozygous for both traits, we are only concerned with the other parent. This parent has a 50% chance of transmitting each independent trait, and thus a 25% chance of transmitting both $\left(\frac{1}{2} \times \frac{1}{2} = \frac{1}{4}\right)$. This probability is the same for both pregnancies because they are independent events; thus, the probability that both children exhibit both traits is: $\frac{1}{4} \times \frac{1}{4} = \frac{1}{16} = 0.0625 = 6.25\%$.

9. **C**

With data about percentages, we can only draw conclusions about percentages. Thus any information about number of people, as in **(A)**, is incorrect. This map shows us that a higher percentage of the residents in the middle of the country are elderly in comparison to other parts of the country. There are, of course, exceptions to this rule, including Florida, the Pacific Coast, and parts of Appalachia, which are all in the top category. Even so, there appears to be a clustering of counties with a high percentage of elderly individuals in the middle of the country. We also cannot say that most of the population is elderly in any place on this map because we are not given actual values for the percentages. There may be a plurality, but there is insufficient information to posit a majority, eliminating **(B)**. The map gives no indication of migration patterns, so we can also eliminate **(D)**.

10. **A**

To increase the confidence level, one must increase the size of the confidence interval to make it more likely that the true value of the mean is within the range. Therefore, the confidence interval must become wider.

11. **D**

Standard deviation is the most common measure of distribution. It is the most closely linked to the mean of a distribution and can be used to calculate *p*-values, which are probabilities (specifically, *p*-values are the probability that an observed difference between two populations is due to chance).

12. **C**

Outliers can be determined with respect to the interquartile range, Q3–Q1. The interquartile range for this box plot is 2280 – 2075, or 205. Values that are 1.5 times IQR below Q1 or above Q3 are considered outliers. 2075 – 1.5 times 205 is approximately 2075 – 300, or 1775 (actual = 1767.5). Therefore, 1575 is an outlier. 2280 + 1.5 times 205 is approximately 2580 (actual = 2587.5). Therefore, 2600 is also an outlier.

13. **D**

The attendant hypothesized that glucose in ice cream was the cause of the increase in violent behavior and designed an observational study. While he did see an increase in assaults with an increase in ice cream sales, this relationship is a correlation. Correlation does not prove causation. Thus, the conclusion that glucose was the cause is a significant flaw, making **(D)** the correct answer.

14. **D**

Because the histogram contains two peaks with a valley in between, it is a bimodal distribution. The color separation of two distinct populations provides evidence that there is a qualitative difference in the data between the two peaks, thus the data should be analyzed according to gender. There is indeed only one mode, at 5′6″. This is the measurement with the largest number of corresponding data points.

15. **C**

When considering whether the findings of a study merit policy change, both the statistical and clinical significance must be evaluated. In comparing the difference in both the systolic and diastolic blood pressure before and after treatment, the effect of the drug is minimal. While the results are statistically significant, it is very unlikely that this small reduction would lead to a significant clinical improvement. Thus, the doctor should not advocate, making **(C)** the correct answer.

Consult your online resources for additional practice. GO ONLINE

Equations to Remember

(12.1) **Arithmetic mean:** $\bar{x} = \frac{\sum_{i=1}^{n} x_i}{n}$

(12.2) **Median position:** median position $= \frac{(n+1)}{2}$

(12.3) **Range:** range $= x_{max} - x_{min}$

(12.4) **Interquartile range:** $\text{IQR} = Q_3 - Q_1$

(12.5) **Standard deviation:** $\sigma = \sqrt{\frac{\sum_{i=1}^{n}(x_i - \bar{x})^2}{n-1}}$

(12.6) **Probability of two independent events co-occurring:**
$P(\text{A} \cap \text{B}) = P(\text{A and B}) = P(\text{A}) \times P(\text{B})$

(12.7) **Probability of at least one event occurring:**
$P(\text{A} \cup \text{B}) = P(\text{A or B}) = P(\text{A}) + P(\text{B}) - P(\text{A and B})$

(12.8) **Slope:** $m = \frac{\text{rise}}{\text{run}} = \frac{\Delta y}{\Delta x}$

Shared Concepts

Behavioral Sciences Chapter 11
Social Structure and Demographics

Biology Chapter 12
Genetics and Evolution

General Chemistry Chapter 5
Chemical Kinetics

Physics and Math Chapter 1
Kinematics and Dynamics

Physics and Math Chapter 10
Mathematics

Physics and Math Chapter 11
Reasoning About the Design and Execution of Research

GLOSSARY

Aberration–Visual alterations as the result of an imperfect optical device; may be chromatic or spherical.

Absolute pressure–The actual pressure at a given depth in a fluid, including both ambient pressure at the surface and the pressure associated with increased depth in the fluid; also called hydrostatic pressure.

Absolute zero–The theoretically coldest temperature at which all atomic movements would halt (0 K).

Acceleration–The rate of change in the velocity of an object; related to force through mass and measured in $\frac{m}{s^2}$.

Accuracy–The tendency for data to represent the true answer; also known as validity.

Adhesion–The intermolecular force between molecules of a liquid and molecules of another substance.

Adiabatic–A thermodynamic process that occurs with no heat exchange.

Air resistance–The resistance that opposes the motion of a falling object.

Algebraic system–A method for determining the values of variables that are the same in two or more equations by relating them to each other.

α-particle–A helium nucleus $\left({}^{4}_{2}He\right)$ emitted during alpha decay.

Alternating current (AC)–In circuits, a pattern of current flow that changes direction periodically.

Ammeter–A device used to measure current within a circuit.

Amplitude–The maximum displacement from the equilibrium point during wave or oscillatory motion.

Antinode–A point of maximum displacement in a standing wave.

Archimedes' principle–States that a body immersed in a volume of fluid experiences a buoyant force equal to the weight of the displaced fluid.

Atomic absorption–Process in which an electron jumps from a lower to a higher energy orbit by absorbing a photon of light.

Atomic emission–Process by which an electron falls from a higher to a lower energy level and emits a photon of light.

Atomic number–The number of protons in the nucleus of a given element.

Attenuation–The loss of energy of a propagating wave as a result of nonconservative forces; also known as damping.

Autonomy–The ethical principle that states that individuals have the right to make decisions about their own healthcare.

Beat frequency–The difference between the frequencies of two interacting sound waves.

Beneficence–The ethical principle that states that practitioners should always act in their patients' best interests; in research ethics, also states that a research project should create a net positive change for both the study population and general population.

β-particle–An electron emitted during β^- decay, or a positron emitted during β^+ decay.

Bernoulli's equation–An equation that relates static and dynamic pressure for a fluid to the pressure exerted on the walls of a tube and the speed of the fluid.

Bias–A result of flaws in the data collection phase of an experimental or observational study that typically skews data within a study.

Bimodal distribution–A distribution of data with two peaks and a valley in between them.

Blackbody–An ideal absorber of all wavelengths of light.

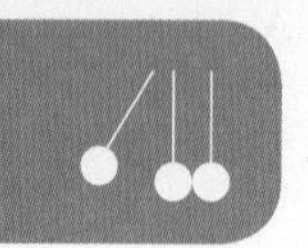

Blinding–Withholding information about a research subject's group assignment from the subject or evaluator to remove some potential bias from the results.

Boiling point–The temperature at which the vapor pressure of a liquid equals the ambient (incident) pressure, usually atmospheric pressure; the temperature at which the liquid boils.

Boundary layer–A region of laminar flow in an otherwise turbulent system that occurs at the very edges of the vessel.

Box-and-whisker plot–A visual representation of the range of data, quartiles, and the interquartile range; may contain outliers as separate points.

Buoyancy–The upward force that results from immersion in a fluid; described by Archimedes' principle.

Capacitance–A measure of the ability of a capacitor to store charge; the magnitude of the charge on one plate divided by the potential difference between the plates; measured in farads (F).

Capacitor–Two conducting surfaces that store charges of equal magnitude but opposite sign when connected to a voltage source.

Case-Control Study–An observational study that starts by identifying subjects with a given outcome, then looks for correlations to specific exposures within the group.

Causation–A relationship between two variables in which one (at least partially) depends on the other in order to occur.

Center of gravity–A point such that the entire force of gravity acting on an object can be thought of as acting at that point.

Center of mass–The point that acts as if the entire mass of an object was concentrated at that point.

Centripetal acceleration–The acceleration of an object that travels in a circle; it is always directed toward the center of the circle if the object is in uniform circular motion.

Centripetal force–The force responsible for centripetal acceleration; usually a result of gravity, tension, or a normal force.

Charges–Entities that can influence the environment through electrostatic forces or be influenced by electrostatic forces, measured in coulombs (C).

Chromatic aberration–A dispersive effect within a spherical lens.

Circular motion–A form of motion that occurs when forces cause an object to move in a circular pathway.

Cohesion–The intermolecular forces experienced between the molecules of a liquid.

Cohort study–An observational study in which subjects are sorted into groups based on different exposures, and then assessed at various intervals to determine outcome.

Concave–A surface that has a similar curvature to the interior of a sphere.

Condensation–The phase transition from a gas to a liquid.

Conductance–In the transfer of charge, the degree to which an object conducts electricity. Conductance can be metallic or electrolytic.

Conduction–In thermodynamics, the transfer of heat by physical motion of a fluid over a material.

Conduction pathway–a route for current to take through a resistor.

Conductor–A material that allows the free movement of electrical charge; one with very low or zero resistance.

Confidence–A statistical indicator of the likelihood that acquired results did not occur by random chance; equal to $1 - \alpha$.

Confounding–An error that results when a causal variable is associated with two other variables in a study but is not accounted for; may falsely indicate that the two variables are associated.

Conservative force–A force that does not cause energy to be dissipated from a system, such as gravity, electrostatic forces, and springs (approximately conservative); pathway independent and associated with a potential energy function.

Control–A set of experimental conditions meant to ensure that the results of the experimental group are a result of the intervention.

Convection–Heat transfer as a result of bulk flow of a fluid over an object.

Converging–The tendency to move parallel light rays toward one another; concave mirrors and convex lenses converge parallel light to a focal point.

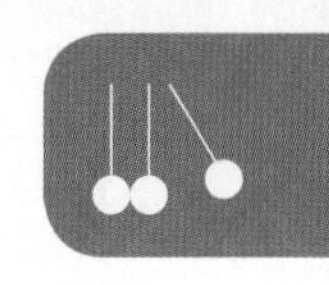

Convex–A surface that has a similar curvature to the exterior of a sphere.

Correlation–The degree to which two variables have a relationship with one another.

Correlation coefficient–A numerical value between -1 and $+1$ that indicates how strong a relationship is between two variables.

Coulomb's law–Relates the electrostatic force between two charged particles to their charges and the distance between them.

Critical angle–The angle above which any incident light will undergo total internal reflection; occurs when light is moving from a material with a higher refractive index to one with a lower refractive index.

Critical speed–The speed above which flow of a fluid will be turbulent.

Cross-sectional study–An observational study in which patients are categorized into different groups at the same point in time.

Current–The orderly movement of charge, often in a circuit; measured by convention as the direction that positive charge would flow within the circuit, and measured in ampères (A).

Decay constant–The proportionality constant between the rate at which radioactive nuclei decay and the number of radioactive nuclei remaining.

Density–A measure of mass per unit volume; useful for buoyancy calculations and usually measured in $\frac{g}{cm^3}$, $\frac{kg}{m^3}$, or $\frac{g}{mL}$.

Dependent variable–The measured or observed variable in an experiment that is affected by manipulations of the independent variable.

Detection bias–An error in data collection that results from the tendency to look more carefully for certain outcomes because a known association with that outcome exists.

Diamagnetic material–A material made of atoms with no unpaired electrons that have no net magnetic field.

Dielectric material–An insulating material used to increase capacitance.

Diffraction–The spreading or bending of light rays.

Dipole moment–In electrostatics calculations, the product of charge and separation distance.

Direct current (DC)–In circuits, a pattern of current in which charge flows in only one direction.

Direct relationships–A relationship in which increasing one variable proportionately increases the other.

Dispersion–The separation of light into its component wavelengths when passing through a medium, such as a prism.

Displacement–The vector representing the straight-line distance and direction from an initial point; not necessarily equal to total distance traveled, and measured in meters.

Diverging–The tendency to move parallel light rays away from one another; convex mirrors and concave lenses diverge parallel light rays from a focal point.

Doppler effect–Quantifies the perceived change in frequency of sound due to relative movement between the source and detector (observer).

Double-blind experiment–Experiment in which both the assessor and the subject do not know the subject's group.

Dynamics–In kinematics and dynamics, the study of forces and torques.

Efficiency–In machines, the ratio of useful work output compared to work input.

Electric dipole–A separation of equal and opposite charge by a small distance; can be seen in polar molecules.

Electric field–A region generated by an electric charge or multiple charges that can exert a force on another charge brought into the field; measured in $\frac{N}{C}$.

Electric meters–Devices used to measure circuit quantities like current, potential difference, or resistance.

Electric potential–A measure of electric potential energy per unit charge, given in volts (V); differences in electric potential (voltage) also drive current as the electromotive force in a circuit.

Electric potential energy–A form of potential energy that is dependent on the relative position of one charge with respect to another charge or to a collection of charges.

Electromagnetic radiation–A form of energy composed of oscillating electric and magnetic fields perpendicular to each other and perpendicular to the direction of propagation; includes visible light and other types of transverse waves, and can travel through a vacuum.

Electromagnetic spectrum–The full range of frequencies and wavelengths of electromagnetic waves.

Electromotive force–The difference in electric potential (voltage) that drives current in a circuit or battery.

Electron–A subatomic particle that remains outside the nucleus and carries a single negative charge.

Electron capture–A process in which an unstable atom absorbs an inner electron that combines with a proton to form a neutron, while releasing a neutrino.

Electrostatics–The study of stationary charges and the forces that are created by (and act upon) these charges.

Energy–The capacity to do work or transfer heat, measured in joules (J).

Entropy–A statistical measure of the distribution of unusable energy or heat; randomness introduced to a system, measured in $\frac{J}{g \cdot K}$.

Equilibrium–The state at which the net torque or net force is equal to zero, such that there is no acceleration.

Equipoise–The state of not knowing whether there is a difference between two interventions; ethically necessary for comparative study of the interventions.

Equipotential lines–Regions within an electric field with equal electric potential; movement from one point on these lines to another causes no change in the energy of the system.

Excited state–Describes an atom in which an electron occupies an energy state above the minimum energy (ground) state.

Exhaustive–Describes a set of outcomes that leave no room for other possible outcomes.

External validity–The ability to apply findings of a research study to other populations; also called generalizability.

Ferromagnetic material–A material made of atoms with unpaired electrons that become strongly magnetized when exposed to an external magnetic field.

Field line–A visual representation of the electric field; points to the direction a force would be exerted on a positive test charge in the electric field.

FINER method–A way to determine the usefulness of a research question on the basis of feasibility, interest, novelty, ethics, and relevance.

Fission–The splitting of a large nucleus into smaller nuclei with the release of energy.

Flow rate–The volume per unit time of a fluid in motion.

Fluid–A material that conforms to the shape of its container and that can flow.

Fluid dynamics–The study of fluids in motion.

Fluorescence–A process in which the electrons of certain substances are excited to higher energy levels by high-frequency photons, and then emit visible light as the energy is released in two or more steps back to the ground state.

Focal length–The distance from a mirror or lens to the focal point.

Focal point–The point at which rays of light parallel to the axis of a mirror or lens converge, or from which they appear to diverge when reflected by a mirror or refracted by a lens.

Force–A push or a pull, measured in newtons (N).

Free fall–A system in which the only force is gravity.

Freezing–The phase transition from liquid to solid; also called solidification.

Frequency–The rate at which a recurring event occurs; usually measured in hertz (Hz).

Friction–A nonconservative force that arises from the interactions between two surfaces in contact.

Fundamental frequency–The first harmonic of a pipe, string, or other standing wave.

Fusion–The merging of small nuclei into a larger nucleus with the release of energy.

γ-rays–High-energy photons released during gamma decay; part of the electromagnetic spectrum.

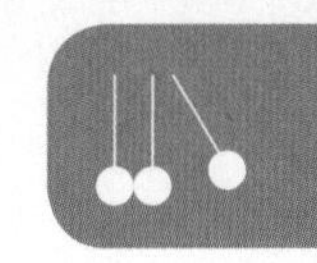

Gauge pressure–Pressure above and beyond atmospheric pressure.

Gravitational potential energy–A form of potential energy dependent on the relative position an object within a gravitational field.

Gravity–An attractive force between two objects that depends on their masses and the distance between them.

Ground–A means of returning charge to the earth.

Ground state–The lowest energy state of an atom.

Half-life–The amount of time it takes for one-half of a sample of radioactive nuclei to decay.

Harmonic series–The set of frequencies that can create standing waves in a given pipe or string.

Hawthorne effect–The tendency for research participants to change their behavior because they know they are being observed.

Heat–The transfer of thermal energy; measured in joules (J), calories (cal), or kilocalories (kcal or Cal).

Heat of transformation–The amount of heat necessary to cause a phase transition of a unit mass of a substance at the characteristic temperature and pressure of that phase transition; also called latent heat.

Hill's criteria–A systematized way of evaluating evidence for causality; only temporality is absolutely necessary to demonstrate causality.

Histogram–A visual representation for numerical data; related to a bar chart.

Hydraulic system–A simple machine that exerts mechanical advantage using an incompressible fluid; based on Pascal's principle and conservation of energy.

Hydrostatics–The study of fluid systems at rest.

Hyperopia–Farsightedness, or the ability to see distant objects while nearby objects are unfocused or blurry.

Hypothesis testing–A statistical method used to compare results between groups or to a theoretical value with a given level of confidence.

Image–The region where light rays converge or appear to converge after being reflected from a mirror or passing through a lens.

Independent variable–The manipulated variable in an experiment that affects measurements or observations of the dependent variable.

Index of refraction–A ratio of the speed of light in a vacuum to the speed of light in a given medium.

Inertia–An object's resistance to a change in its motion when a force is applied.

Informed consent–An ethical requirement for treatments or research, which requires that the patient or participant is able to understand the procedure and its consequences and alternatives; related to autonomy.

Infrared–A region of the electromagnetic spectrum that is not visible; may be perceived as heat.

Infrasonic–Sound that has a frequency that is lower than the range of human hearing.

Insulator–A material that resists the movement of charge because the electrons are tightly associated with their nuclei.

Intensity–The average rate of energy expenditure (power) per unit area, measured in $\frac{W}{m^2}$; in waves, intensity is related to the amplitude of the wave.

Interference–Interactions between waves traveling in the same space; may be constructive (waves adding together), destructive (waves cancelling each other), partially constructive, or partially destructive.

Internal validity–The ability to infer causality from a study or to replicate its results under the same conditions.

Interquartile range–A measure of distribution of a sample; outliers lie at least 1.5 interquartile ranges below Q_1 or above Q_3.

Inverse relationship–A relationship in which an increase in one variable is associated with a proportional decrease in the other.

Inverted–Describes an image that is upside down relative to the object; in single-mirror or single-lens systems, inverted images are always real.

Irreversible–A thermodynamic process that is extraordinarily unfavorable in reverse, usually as a result of changes in entropy.

Isobaric–A thermodynamic process that occurs under constant pressure.

Isothermal–A thermodynamic process that occurs under constant temperature.

Isotopes–Atoms of a given element with different numbers of neutrons and therefore different mass numbers.

Isovolumetric–A thermodynamic process that occurs under constant volume; also called isochoric.

Justice–The ethical principle that states that practitioners should fairly distribute healthcare resources, and which requires that differences in treatment choices between individuals are only due to morally relevant differences.

Kinetic energy–The energy of movement, which depends on both mass and speed; measured in joules (J).

Kinetic friction–The friction that exists between a sliding object and the surface over which the object slides.

Kirchhoff's laws–Rules that describe the conservation of charge and conservation of energy within an electric circuit; includes the junction rule and loop rule.

Laminar flow–Smooth flow within a fluid; characterized by streamlines that do not cross each other and an absence of backwards movement.

Lenses–Devices that act to create an image by refracting light; usually have spherical surfaces.

Logarithm–The inverse function of exponentiation; logarithmic scales are often used to mask large absolute differences between quantities by presenting them as small scale differences.

Longitudinal wave–A wave in which the oscillation of the material is parallel to the direction of propagation; sound is a classic example.

Lorentz force–The sum of the electrostatic and magnetic forces acting on a charge.

Loudness–Perceived intensity of a sound, which correlates with sound level; measured in decibels (dB).

Magnetic field–Field created by a moving charge.

Magnetic force–A force that is exerted when a charge moves in a magnetic field, provided the charge has a perpendicular component of velocity in comparison to the magnetic field vector.

Magnification–Apparent increase or decrease in size of an image as a result of forming the image with a converging or diverging system.

Mass–A measure of inertia or of the amount of "stuff" in an object; measured in kilograms.

Mass defect–The difference between the sum of the masses of unbound nucleons forming a nucleus and the mass of that nucleus in the bound state.

Mass number–The sum of the number of protons and neutrons in an atom; also called the atomic mass.

Mean–The average of a group of data; specifically, the arithmetic mean.

Measures of central tendency–Measures that describe the middle of a sample.

Mechanical advantage–The reduction in input force required to accomplish a desired amount of output work using a simple machine.

Median–The central value of a data set.

Melting–The phase transition from solid to liquid; also known as fusion.

Metric system–A system of measurements based on the powers of ten; most commonly used in scientific disciplines.

Microwaves–Long-wavelength electromagnetic radiation capable of inducing vibration in bonds.

Mode–The most common data point in a data set.

Monochromatic–Electromagnetic radiation wherein the wavelength is the same for all incident photons.

Mutually exclusive–Describes outcomes that cannot occur simultaneously.

Myopia–Nearsightedness, or the ability to see nearby objects while distant objects are unfocused or blurry.

Natural frequency–The frequency at which a system resonates; also called the resonant frequency.

Natural process–In thermodynamics, a process which would occur as expected in nature.

Newton's first law–The first law, also called the law of inertia, states that an object will remain at rest or move with a constant velocity if there is no net force on the object. This law thus accounts for the conservation of mechanical energy.

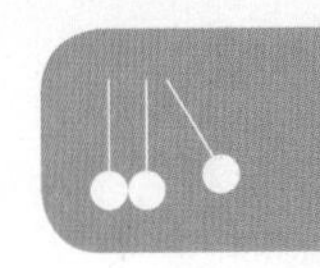

Newton's second law–This law states that any acceleration is the result of the sum of the forces acting on the object and its mass.

Newton's third law–This law states that any two objects interacting with one another experience equal and opposite forces as a result of their interaction.

Node–A point of zero displacement of a standing wave.

Nonconservative force–A force that causes energy to be dissipated from a system, such as friction, air resistance, and viscous drag; pathway dependent.

Nonmaleficence–The ethical principle that states that practitioners have an obligation to avoid treatments or interventions in which the potential for harm is greater than the potential for good.

Normal–A line perpendicular to the surface of interest.

Normal force–The force that two surfaces in contact exert on each other that is perpendicular to the plane of contact.

Nucleon–A proton or neutron.

Null hypothesis–The hypothesis of no difference; given enough statistical evidence, the null hypothesis may be rejected.

Observational study–A study that looks for the connection between exposures and outcomes without demonstrating causality.

Ohmmeter–A device used to measure resistance.

Ohm's law–Relates voltage, current, and resistance for a given circuit element.

Outlier–A data point that deviates significantly from the perceived pattern of distribution; depending on the context, an outlier may be disregarded, analyzed normally, or given disproportionate weight when calculating statistics.

Parallel–An arrangement of circuit elements in which the current can go through one element or the other, but not through both.

Paramagnetic material–A material made of atoms with unpaired electrons that will become weakly magnetized in the presence of an external magnetic field.

Parameter–A measure of population data.

Pascal's principle–States that pressure applied to a noncompressible fluid is distributed equally to all points within that fluid and the walls of the container.

Period–The amount of time it takes for a wave or oscillation to complete one cycle, measured in seconds; the inverse of frequency.

Phase difference–The difference in phase between waves with the same frequency, referenced at the same point.

Photoelectric effect–A phenomenon in which light of sufficiently high frequency incident on a metal in a vacuum causes the metal to emit electrons.

Pitch–A perception of sound that results from its frequency; as frequency increases, pitch gets higher.

Pitot tubes–Measurement devices for pressure or flow rate of a dynamic fluid system.

Plane mirrors–Reflecting surfaces with an infinite radius of curvature, which results in equal image and object distances.

Plane-polarized light–Electromagnetic radiation in which all of the electric field vectors are oriented parallel to one another.

Poiseuille's law–Relates viscosity, tube dimensions, and pressure differentials to the rate of flow between two points in a system.

Population–The group of all individuals who have certain desired characteristics.

Positron–Antiparticle of an electron; it has the same mass as an electron and the opposite charge (e^+ or β^+).

Potential difference–The difference of electric potential between two distinct points, measured in volts (V); also called voltage.

Potential energy–Energy associated with position, measured in joules (J); includes gravitational, elastic, chemical, and electrical forms.

Power–Rate at which work is accomplished, or energy expenditure per unit time; measured in watts (W). In statistics, the probability of correctly rejecting a false null hypothesis.

Precision–The tendency of measurements to agree with one another; also called reliability.

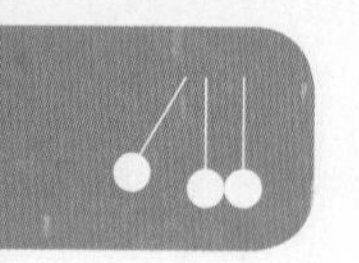

Pressure–The ratio of force to the area over which it is applied; measured in pascals (Pa), millimeters of mercury (mmHg) or torr, or atmospheres (atm).

Principle of superposition–When waves interact with each other, the displacement of the resultant wave at any point is the sum of the displacements of the two interacting waves.

Process functions–Physical quantities that depend on the path taken to get from one state to another; include work and heat.

Propagation–In waves and sound, the movement of a wave.

Proton–A subatomic particle that carries a single positive charge and has a mass slightly less than 1 amu.

Pythagorean theorem–Relationship between the three sides of a right triangle: the square of the hypotenuse is equal to the sum of the squares of the other two sides.

Quantum–A discrete bundle of energy such as the photon.

Quartiles–Values that separate data in ascending order into four evenly sized groups.

Radiation–A method of heat transfer that relies on electromagnetic waves; can occur in a vacuum.

Radioactive decay–A naturally occurring spontaneous decay of certain nuclei accompanied by the emission of specific particles.

Radio waves–Very long wavelength electromagnetic radiation.

Random error–In an experiment, error caused by natural variation in subjects and data points that can be reduced in impact by increasing sample size.

Randomization–A method of reducing bias and confounding during research in which participants are assigned to a group by a random number generator or similar method; participants or researchers cannot choose the groups.

Range–The difference between the smallest number in a data set and the largest.

Ray diagram–Visual representation of a geometrical optics system.

Real–Describes an image on the same side of a lens or mirror as the refracted or reflected light that can be projected on a screen; in single-mirror or single-lens systems, real images are always inverted.

Rectilinear propagation–A phenomenon in which light travels in a straight line when passing through a homogeneous medium.

Reflection–The return of light rays from a medium at an angle equal to the incident angle.

Refraction–The bending of light rays as a result of a change in the index of refraction between media.

Resistance–A measure of the opposition to current flow through a material, measured in ohms (Ω); the inverse of conductance.

Resistivity–A measure of the intrinsic resistance of a material independent of its shape or size; resistivity generally increases with temperature.

Resonance–Oscillation at maximum amplitude as the result of a periodically applied force at the natural (resonant) frequency of an object.

Respect for persons–A principle of research ethics that encompasses autonomy and informed consent.

Resultant–The sum, difference, or product of vector mathematics; also refers to the sum or difference of two waves.

Right-hand rule–A method for determining the direction of a vector that is the product of two vectors.

Rotation–The turning of an extended body about an axis or center.

Sample–A subset of a population that is used to make generalizations about the population as a whole.

Scalar–A mathematical quantity that lacks directionality.

Scientific method–A systematized way of evaluating data and investigating new hypotheses.

Scientific notation–A mathematical representation of quantities as multiples of powers of ten.

Selection bias–Occurs when research participants differ from the general population in a meaningful way.

Series–An arrangement of circuit elements in which the current must go through all of the elements.

Shear forces–Also called tangential forces, forces exerted on the surface of an object that are parallel to the surface of the object.

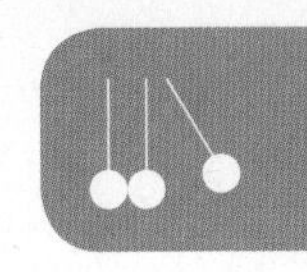

Shock wave–The buildup of wave fronts that occurs when the source is travelling at or above the speed of sound.

Significant figures–A tool for maintaining appropriate levels of precision when performing mathematical calculations.

Simple machine–A basic mechanical device for applying a force. The six simple machines include inclined planes, wedges, wheel and axle systems, levers, pulleys, and screws.

Single-blind experiment–An experiment in which only the subject or the assessor is blinded, but the other party is aware of the treatment the subject is receiving.

Snell's law–Relates the incident angle, refracted angle, and indices of refraction for two media.

Solid–A material with distinct boundaries and strong intermolecular forces capable of resisting shear forces.

Sound–The perception of longitudinal waves of pressure changes in air and other media.

Source charge (Q)–In electrostatics, the charge that creates the electric field.

Specific gravity–The ratio of an object's density to the density of water; unitless.

Specific heat–The relationship between thermal energy and temperature change per unit mass of a substance, measured in $\frac{\text{J}}{\text{g}\cdot\text{K}}$.

Speed–The ratio of distance traveled to time; at any given point, instantaneous speed is the magnitude of instantaneous velocity; measured in $\frac{\text{m}}{\text{s}}$.

Spherical aberration–A blurring of the periphery of an image as a result of inadequate reflection of parallel beams at the edge of a mirror or lens.

Spherical mirror–A mirror that causes convergence or divergence of light rays incident upon its surface.

Standard deviation–A measure of distribution of data from the mean of a sample; outliers lie at least three standard deviations above or below the mean.

Standing waves–Waveforms with steady nodes and antinodes formed from the interference of incident and reflected waves at a boundary.

State functions–Physical quantities that can be determined based on the state of an object, such as pressure, density, temperature, volume, enthalpy, internal energy, Gibbs free energy, and entropy; pathway independent.

Static friction–The friction that exists between a stationary object and the surface upon which it rests.

Statistic–A measure of sample data.

Streamlines–Visual representations of the movement of fluid during laminar flow.

Strong nuclear force–One of the four fundamental interactions; it is responsible for the binding of protons and neutrons together in the nucleus.

Surface tension–The result of the cohesive forces in a liquid creating a barrier at the interface between a liquid and the environment.

Surroundings–Everything that is not being measured as part of a given system.

System–The observed and quantified region of the universe of interest to the experimenter. Systems can be isolated (unable to exchange energy or matter with their surroundings), closed (able to exchange energy with their surroundings), or open (able to exchange matter and energy with their surroundings).

Systematic error–An error in an experiment that is typically caused by measuring instruments and creates a flaw in data that cannot be offset by increasing data pool.

Temperature–A measure of the average kinetic energy of particles in a substance; measured in degrees Fahrenheit (°F), degrees Celsius (°C), or kelvins (K).

Temporality–A necessary criterion for causality; the independent variable must occur before the dependent variable.

Terminal velocity–The velocity at which air resistance is equal to gravitational force and no acceleration occurs for an object in free fall.

Test charge (*q*)–In electrostatics, the charge placed in an electric field.

Thermal equilibrium–Equilibrium of thermal energy that exists when no heat flows between two objects in thermal contact.

Thermal expansion–An increase in length or volume of a substance as a result of an increase in temperature.

Threshold frequency–The minimum frequency of light that causes ejection of electrons.

Timbre–The quality of sound produced by an instrument.

Torque–The primary motivator for rotational movement that combines force, lever arm, and the angle between them; measured in N·m.

Total internal reflection–A phenomenon in which all the light incident on a boundary is reflected back into the original material.

Total mechanical energy–The sum of an object's potential and kinetic energies.

Translation–Motion through space without rotation.

Transverse wave–A wave that propagates in a direction perpendicular to the direction of oscillation.

Traveling wave–A wave that propagates through a medium with changes in the locations of crests and troughs.

Turbulent flow–Fluid movement that does not follow parallel streamlines; has backflow, eddies, and swirls.

Type I error–An error in conclusion in which the null hypothesis is incorrectly rejected.

Type II error–An error in conclusion in which the experimenter fails to correctly reject the null hypothesis.

Ultrasonic–Above the frequencies that humans can hear.

Ultrasound–A treatment and diagnostic modality using ultrasonic waves for medical purposes.

Ultraviolet–A region of the electromagnetic spectrum that is not visible; primarily responsible for the damaging effects of sunlight on skin.

Unnatural process–In thermodynamics, a process that does not undergo naturally predicted changes in heat/energy over time.

Upright–Describes an image that is the same orientation as the object; in single-mirror or single-lens systems, upright images are always virtual.

Vaporization–The phase transition from liquid to gas; also called boiling or evaporation.

Vector–A mathematical quantity that has both magnitude and direction.

Velocity–The rate of change in the displacement of an object; measured in $\frac{\text{m}}{\text{s}}$.

Venturi effect–Describes the relationship between the continuity equation and Bernoulli's equation; as cross-sectional area of a tube decreases, the speed of the fluid increases, and the pressure exerted on the walls of the tube decreases.

Virtual–Describes an image on the opposite side of a lens or mirror as the refracted or reflected light; in single-mirror or single-lens systems, virtual images are always upright.

Viscosity–A measure of the resistance to flow in a fluid.

Viscous drag–A nonconservative force exerted by fluids in a manner proportional to the viscosity of the fluid.

Visible region–The part of the electromagnetic spectrum visible to the human eye.

Voltmeter–A device used to measure voltage.

Wavelength–The distance between two corresponding points of successive cycles in a waveform, measured in meters.

Weak nuclear force–One of the four fundamental interactions; it is responsible for radioactive decay and contributes to nuclear stability.

Weight–The force of gravity acting on an object.

Work–A function of the applied force and the distance through which it is applied or the pressure and volume changes in a gas system; work is the use of energy to accomplish something and is measured in joules (J).

Work-energy theorem–States that net work is equal to the change in energy (usually kinetic energy) of an object.

X-rays–A type of electromagnetic radiation; primarily used for medical imaging.

Zeroth law of thermodynamics–If two thermodynamic systems are in thermal equilibrium with a third system, they are in thermal equilibrium with each other.

INDEX

Note: Material in figures or tables is indicated by italic *f* or *t* after the page number.

A

Aberration, 471
- chromatic, 304, 304*f*
- spherical, 303, 303*f*

Absolute pressure, 130–131, 471
Absolute zero, 93, 93*t*, 471
Absorption of light, 334–336, 334*f*, 335*f*, 348
Acceleration, 23
- *See also* Motion with constant acceleration
- centripetal, 31–32
- definition of, 23, 471
- due to gravity, 26
- force and (*See* Forces and acceleration)

Accuracy, 406, 407*f*, 471
Action and reaction, Newton's third law, 19, 25
Addition
- with exponents, 369
- with significant figures, 365
- vectors, 10–13, 11*f*–13*f*

Adhesion, 138, 471
Adiabatic processes, 103, 103*t*, 471
Aerodynamics, 144, 144*f*
Air resistance, 28, 61, 471
Algebraic systems, 379–382
- definition of, 471
- elimination, 380–381
- setting equations equal, 380
- substitution, 379–380

Alpha (α)-particle, 342, 471
Alpha decay, 342
Alternate explanations, as Hill's criteria, 410
Alternating current, 213, 471
Alternative hypothesis, 449
Ambient pressure, 131
Ammeters, 231, 471
Ampère, 8*t*, 213
Amplitude, of wave, 252, 252*f*, 261, 471
Analyses
- dimensional, 378
- regression, 409
- research data, 409
- unit, 378–379

and *vs.* or, 448
Ångströms, 9, 331
Angular frequency, 252
Antinodes in waves, 254, 264, 471
Archimedes' principle, 136–137, 471
Arithmetic, 364–367, 383
- *See also* Mathematics
- estimation, 366–367
- scientific notation, 364
- significant figures, 364–366

Arithmetic mean, 436–437, 476
Arrhenius equation for activation energy, 368*t*
Assessor, in experiment, 408
Atmosphere (atm), 129
Atmospheric pressure, 131
Atomic absorption, 334–336, 471
Atomic and nuclear phenomena, 323–356
- concept summary, 348–349
- equations for, 356
- light, absorption and emission of, 334–336
- mass defect, 336–338
- nuclear binding energy, 336–338
- nuclear reactions, 339–347
  - (*See also* Nuclear reactions)
- photoelectric effect, 330–332
  - (*See also* Photoelectric effect)

Atomic emission, 471
Atomic number, 339, 471
Attenuation, 255, 263, 471
Attractive forces, 168
Autonomy, 412, 471
Average, 436–437
Average acceleration, 23–24
Axes of linear graph, 455–457
Axis, 292
Axis ratio, 457

B

Backwards meniscus, 138, 138*f*
Bar charts, 453, 453*f*
Base-*e* logarithms, 371
Base-ten logarithms, 371
Base units, 9, 377*t*
- *See also* Units

Basic science research, 403–407, 419
- causality, 405–406
- controls, 403–404
- definition of, 403
- error sources, 406–407

Batteries, 218
Beat frequency, 264, 471
Bell curves, 440
Belmont Report, 413
Beneficence, 412, 414, 471
Bernoulli's equation, 143–147, 471
Beta decay, 342–343
Beta (β) particle, 342, 471
Bias
- definition of, 400, 471
- detection bias, 410–411
- error sources, 406, 410–411
- observation bias, 411
- selection bias, 410

Bimodal distributions, 441, 441*f*, 471
Binary variables, 409
Binding energy, 337
Blackbody, 288, 471
Blind experiments and blinding, 408, 472
Bohr model of atom, 334, 334*f*
Boiling and boiling points, 92, 93*t*, 102, 472
Bonds, metallic, 212
Boundaries, open *vs.* closed, 264
Boundary layers, 141, 472
Box and whisker plots, 453–454, 454*f*, 472
Box plots, 453–454
- *See also* Charts

British system, 8, 377
British thermal unit (BTU), 100
Bulk modulus, 257
Buoyancy, 136, 472
Buoyant force, 136–137

C

Calories, 100
Candela, 8*t*
Capacitance, 225–226, 234, 472
  due to dielectric material, 227
Capacitors, 225–230, 234
  definition of, 225, 472
  dielectric materials in, 226–228
  discharging of, 225
  in parallel, 229–230, 229*f*
  properties of, 225–226
  in series, 228–230, 229*f*
Case-control studies, 409, 472
Categorical variables, 409
Causality, in basic research, 405–406
Causation, 409–410, 460–461, 472
Cells (secondary batteries), 213, 218
Celsius scale, 92–93, 93*t*, 378
Center
  of curvature, 290
  of gravity, 22–23, 472
  of mass, 22–23, 22*f*, 472
Centimeters-grams-seconds (CGS, metric system), 8, 9*t*
Central tendency. *See* Measures of central tendency
Centripetal acceleration, 31–32, 472
Centripetal force, 31, 472
Charges, 168–169, 192
  attractive *vs.* repulsive forces, 168
  conductors, 169, 169*f*, 216
  definition of, 472
  insulators, 168, 169*f*, 216
  in magnetic field, 184–186
  and magnetic forces, 187–190
  source charge, 172, 173*f*
  static electricity, 168
  test charge, 172
Charts, 452–455, 464
  See also Visual data representations
  bar, 453, 454*f*
  box and whisker plots, 453–454, 454*f*
  circle, 452
  histograms, 453
  maps, 454–455, 455*f*
  pie, 452, 452*f*
Chromatic aberration, 304, 304*f*, 472
Circle charts, 452
Circuits, 203–240
  capacitance and capacitors, 225–230 (*See also* Capacitors)
  concept summary, 233–234
  current, 212–214 (*See also* Current)
  equations for, 240
  Kirchhoff's laws, 213–215
  meters, 231–232 (*See also* Meters)
  resistance, 216–224, in current (*See also* Resistance)
Circular motion, 31–32, 472
  See also Motion with constant acceleration
Circular polarization, 312, 312*f*
Circulatory system fluids, 148–149
Clinical significance/effect, 417
Closed boundaries, 264
Closed loop, in circulatory system, 148
Closed loop thermodynamic process, 104, 104*f*
Closed pipes, 266, 267, 267*f*
Closed systems, 97
Coefficient(s)
  of friction, 20–21
  of linear expansion, 94
  in scientific notation, 364
  of volumetric expansion, 95
Coherence, as Hill's criteria, 410
Cohesion, 137–138, 472
Cohort studies, 409, 472
Color, 287–288
Common logarithms, 371–372
Components of vector, 11, 11*f*
Compression, 251, 257
Concave
  definition of, 472
  lenses, 298, 299*f*
  meniscus, 138, 138*f*
  mirrors, 290–291, 292*f*
Concept summaries
  atomic and nuclear phenomena, 348–349
  circuits, 233–234
  data-based and statistical reasoning, 463–464
  electrostatics and magnetism, 192–194
  fluids, 151–152
  kinematics and dynamics, 39–41
  light and optics, 314–315
  mathematics, 383–384
  research design and execution, 419–421
  thermodynamics, 110–111
  waves and sound, 271–272
  work and energy, 76–77
Condensation, 102, 472
Conditions of equilibrium
  See also Mechanical equilibrium
  first, 35
  second, 37
Conductance, 212, 472
Conduction (heat transfer), 100, 472
Conduction pathways, 217, 472
Conductivity, 212–213
Conductors, 169, 169*f*, 216, 472
Confidence, 450, 472
Confidence intervals, 450–451
Confidentiality, 413
Confounding errors, 410, 411*f*, 472
Confounding variables, 409, 411
Consent, informed, 413, 475
Conservation
  of charge and energy in circuit, 213–224
  of mechanical energy, 59–62
  nucleon, 341–342
Conservative forces, 59–60, 60*f*, 472
Consistency, as Hill's criteria, 410
Constants
  Coulomb's, 170
  decay, 345
  dielectric, 226
  electrostatic, 170
  Faraday, 226
  G (universal gravitational), 19–20
  Planck's, 331
  spring, 59
Constructive interference, 253
Continuity equation, 143
Continuous variables, 409
Controls, in basic research, 403–404, 472
Convection (heat transfer), 100, 472
Converging
  definition of, 472
  lenses, 298, 299*f*
  mirrors, 290–291, 292*f*
Conversions, 376–378
  metric prefixes, 377*t*
  temperature scales, 93, 378
  between units, 377, 377*t*
Convex
  definition of, 473
  lenses, 298, 299*f*
  meniscus, 138, 138*f*
  mirrors, 290, 292, 293*f*
Correlation, 409–410, 460–461, 473
Cosine, 373, 375*t*
Coulomb, 168
Coulomb's constant, 170
Coulomb's law, 170–174, 192, 473
Crest of wave, 251–252, 252*f*
Critical angle, 297, 473
Critical speed, 141, 473
Cross product, 14–16
Cross-sectional area, as resistor property, 217
  See also Resistors
Cross-sectional studies, 409, 473
Current, 212–215, 233
  circuit laws, 213–215
  conductivity, 212–213
  definition of, 212, 213, 473
  direct *vs.* alternating, 213
  in magnetic fields, 185–186
  and magnetic forces, 187–190
  magnitude of, 213

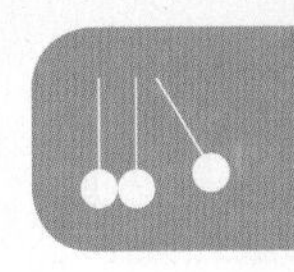

in photoelectric effect, 330
Curvature, center and radius of, 290, 300

D

Damping, 255, 263
Data analyses, in human subjects research, 409
Data-based and statistical reasoning, 427–470
central tendency measures, 436–439 (*See also* Measures of central tendency)
charts, graphs, and tables, 451–459 (*See also* Visual data representations)
concept summary, 463–464
correlation and causation, 409–410, 460–461
distribution measures, 442–446, measures of (*See also* Distributions)
distributions, 439–442 (*See also* Distributions)
equations for, 470
probability, 447–449 (*See also* Probability)
scientific knowledge contexts, 460–461
statistical testing, 449–451 (*See also* Statistical testing)
Datum, 57
Daughter nucleus, 341–342
Decay. *See* Radioactive decay
Decay constant, 345, 473
Deceleration, 23
Decibels, 261–262
Decompression, 251, 257
Defibrillators, 167, 225, 227
Density, 128, 473
of fluids and solids, 128–129
of ice, 136–137
as state function, 97
Dependence, 447
Dependent events, 447
Dependent variables, 405, 473
Derived units, 9, 9*t*
See also Units
Design in research. *See* Research design and execution
Destructive interference, 253
Detection bias, 410–411, 473
Diagrams, free body, 33–34
Diamagnetic materials, 184, 473
Dielectric constant, 226
Dielectric materials, 226–228, 473
Diffraction, 306–311, 315
definition of, 306, 473
gratings, 309–310
multiple slit, 307–310, 308*f*
single slit, 306, 306*f*
slit-lens system, 307, 307*f*
thin film interference, 309, 310*f*
x-ray, 310, 310*f*
Young's double-slit experiment, 307–309, 308*f*
Dimensional analyses, 378
Diopters, 301
Dipole moment, 180–184, 473
Dipoles
electric, 180–183, 180*f*, 183*f*
magnetic, 184–185
Direct current, 213, 473
Direct relationships, 376, 473
Directional alternative hypothesis, 449
Discharge of capacitors, 225, 227
Dispersion
chromatic aberration, 304, 304*f*
of energy, 106
of light, 303, 473
Displacement, 17, 39, 63, 473
of wave, 252
Distance, 17
Distributions, 439–442, 463
bell curve, 440
bimodal, 441, 441*f*
normal, 440, 440*f*
skewed, 440–441, 441*f*
standard, 440
Distributions, measures of, 442–446, 463
interquartile range, 443–444
outliers in, 436, 443, 446
range, 442
standard deviation, 444–445
Diverging
definition of, 473
lenses, 298, 299*f*
mirrors, 290, 292, 293*f*
Division
estimation, 367
with exponents, 369
with significant figures, 365
Doppler effect, 258–260, 259*f*, 473
Doppler ultrasound, 268
Dose-response relationships, 410
Dot product, 14
Double-blind experiments, 408, 473
Double-slit experiment, 307–309, 308*f*
Drag, viscous, 140
Drag force, 28
Dynamic pressure, 143
Dynamics, 33, 473
See also Fluid dynamics; Kinematics and dynamics
Dyne, 9*t*

E

Echolocation, 259
Eddies, 141, 141*f*
Efficiency, 71–73, 473
Effort, 71, 72
Effort distance, 71, 72
Elastic potential energy, 58–59
See also Potential energy
Electric dipoles, 180–183, 180*f*, 183*f*, 473
Electric fields, 172–174, 173*f*, 473
uniform, 226
Electric meters, 473
See also Meters
Electric potential, 176–178, 193, 473
Electric potential energy, 174–176, 192, 473
Electrolytic conductivity, 212–213
Electromagnetic radiation, 100, 286, 473–474
Electromagnetic spectrum, 286–288, 287*f*, 314
color and visible spectrum, 287–288
definition of, 286, 474
Electromagnetic waves, 286–288, 286*f*
See also Waves
Electromotive force, 213, 474
Electron-volts, 9
Electron(s)
capture, 344, 474
definition of, 168, 474
in electric charge, 168
kinetic energy of ejected, 331–333
in magnetic materials, 184–185
Electrostatic constant, 170
Electrostatics and magnetism, 159–201
charges, 168–169
concept summary, 192–194
Coulomb's law, 170–174
definition of, 167, 474
electric dipoles, 180–183, 180*f*, 183*f*
electric potential, 176–178
electric potential energy, 174–176
equations for, 201
equipotential lines, 179
magnetism, 184–190 (*See also* Magnetism)
Elimination, solving algebraic systems, 380–381
emf (electromotive force), 213, 474
Emission of light, 334–336, 334*f*, 335*f*, 348
Energy, 56–62, 76
See also Work and energy
definition of, 56, 474
equations for, 83
kinetic, 56–57
mechanical, 59–62 (*See also* Mechanical energy)
potential, 57–59 (*See also* Potential energy)
transfer by work or heat, 62–63, 98–100
work-energy theorem, 65–66
Energy density, 63, 144
Energy dispersion, 106
Enthalpy, 97
Entropy, 97, 106–108, 474
Equations
Bernoulli's, 143–147
common exponential and logarithmic, 368*t*
continuity, 143
lensmaker's, 299
solving linear systems, 379–381

Equations to remember
for atomic and nuclear phenomena, 356
for circuits, 240
for data-based and statistical reasoning, 470
for electrostatics and magnetism, 201
for fluids, 158
for kinematics and dynamics, 47–48
for light and optics, 321–322
for mathematics, 389
for thermodynamics, 117
for waves and sound, 278
for work and energy, 83
Equilibrium
condition of, first, 35
condition of, second, 37
definition of, 474
length, 58
mechanical (*See* Mechanical equilibrium)
position, of wave, 252
rotational, 36–37
thermal, 92
translational, 35–36, 70–71
Equipoise, 414, 474
Equipotential lines, 179, 193, 474
Equivalent resistance, 220
Equivalent units, 377*t*
See also Units
Erg, 9*t*
Erg per second, 9*t*
Error sources
in basic science research, 406–407
bias, 406, 410–411
confounding, 410, 411, 411*f*
in human subjects research, 410–411
Estimation
division, 367
logarithms, 372
multiplication, 366–367
square roots, 370, 370*t*
Ethics, 412–415, 420
autonomy, 412
beneficence, 412, 414
in FINER method, 401
justice, 413–414
nonmaleficence, 412
respect for persons, 413
Ethnographic studies, 409
Euler's number, 371
Evaporation, 102
Excited state, 334*f*, 474
Exclusivity, mutual, 447, 476
Execution of research. *See* Research design and execution
Exhaustive outcomes, 447, 474
Expansion
linear, coefficient of, 94
thermal, 94–96
volumetric, coefficient of, 95
Experimental approach, in human subjects research, 408–409
Experimentation, in scientific method, 400
Exponential decay, 345–346, 346*f*, 368*t*
See also Radioactive decay
Exponential linear graphs, 455, 456*f*
Exponents, 364, 368–370, 383
common equations using, 368*t*
definition of, 364
identities (arithmetic with), 369–370
square root estimations, 370, 370*t*
Exposures, in cohort studies, 409
External validity, 416, 474

F

Fahrenheit scale, 92–93, 93*t*, 378
Farad, 226
Faraday constant, 226
Feasibility, of research question, 401
Ferromagnetic materials, 184–185, 474
Field lines, 173, 474
Fields
electric, 172–174, 173*f*
magnetic, 184, 185–186
FINER method, 401, 474
First condition of equilibrium, 35
First law, Newton's, 24
First law of thermodynamics, 59, 66, 98–105, 99*t*, 110–111
Fission, 340–341, 474
Flow
laminar, 140–142, 140*f*
turbulent, 141–142, 141*f*
Flow rate, 142, 474
Fluid dynamics, 139–147, 152
Bernoulli's equation, 143–147
definition of, 139, 474
laminar flow, 140–142, 140*f*
Poiseuille's law, 141
streamlines, 142–143, 142*f*
turbulent flow, 141–142, 141*f*
Venturi flow meter, 144–145, 145*f*
viscosity, 140
Fluids, 119–158, 474
characteristics of, 128–132, 149–150
concept summary, 151–152
definition of, 128, 474
density, 128–129
dynamics (*See* Fluid dynamics)
equations for, 158
hydrostatics, 133–139 (*See also* Hydrostatics)
in physiology, 148–149
pressure, 129–132 (*See also* Pressure)
Fluorescence, 335, 474
Focal length, 290, 300, 474
Focal point, 290, 474
Foot (ft), 8
Foot-pound, 9*t*
Foot-pound per second, 9*t*
Foot-pound-second (FPS, Imperial system), 8, 9*t*
Force frequency, 255
Forced oscillation, 255
Force(s)
magnetic, 187–190 (*See also* Magnetic forces)
nuclear, weak *vs.* strong, 337
transference of energy by work, 62–63
Forces and acceleration, 19–24, 40
acceleration, 23
centripetal, 31–32
definition of, 19, 474
force, 19
friction, 20–21
gravitational, 19–20, 60, 171
gravity, 19–20
mass, 21–23, 22*f*
moment of force, 36
normal force, 20
weight, 21
Forces in mechanical energy
conservative, 59–60, 60*f*
nonconservative, 61
Free body diagrams, 33–34
Free fall, 26–27, 474
Free space
permeability of, 185
permittivity of, 170
Freezing and freezing points, 92, 93*t*, 102, 474
Frequency
angular, 252
definition of, 474
Doppler effect, 258–260, 259*f*
force, 255
fundamental, 265
of infrasonic waves, 258
natural (resonant), 254–255
threshold, 330–331
of ultrasonic waves, 258
of wavelength, 251–252
Friction, 20–21
coefficients of, 20–21
contact points, 21*f*
definition of, 20, 474
kinetic, 20–21
as nonconservative force, 61
static, 20–21
Fulcrum, 36
Fundamental frequency, 265, 474
Fundamental pitch, 255
Fusion (atomic), 339–340, 340*f*, 474
Fusion and heat of fusion, 102

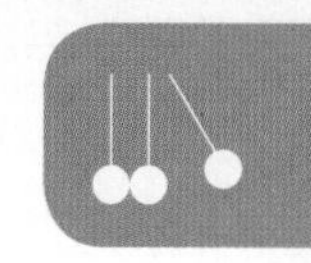

G

G (universal gravitational constant), 19–20
Galvanic (voltaic) cells, 213, 218
Gamma decay, 343–344
Gamma (γ) rays, 286, 287, 287*f*, 343–344, 474
Gas systems, pressure and volume, 63–65, 64*f*
Gauge pressure, 131–132, 475
Gauss, 184
Generalizability, 416
Geometrical optics, 288–305, 314–315
  definition of, 288
  dispersion, 303–305
  lenses, 298–303 (*See also* Lenses)
  rectilinear propagation, 288
  reflection, 289–294 (*See also* Mirrors; Reflection)
  refraction, 294–298 (*See also* Refraction)
Gibbs free energy, 97, 368*t*
Glossary, 471–480
Graphs, 455–458, 464
  See also Visual data representations
  axis ratio, 457
  linear graphs, 455–457, 456*f*
  log-log graph, 457–458
  pressure-volume, 63–64, 64*f*, 104
  semilog graphs, 457–458, 457*f*
  slope in, 456
Gravitational force, 19–20, 60, 171
Gravitational potential energy, 57–58, 475
  See also Potential energy
Gravity, 19–20, 475
  acceleration due to, 26
  center of, 22–23
  specific, 129
Ground
  charge, 168, 475
  state, 334*f*, 475
  zero potential energy, 57

H

Half-life, 344–345, 475
Hardy-Weinberg equilibrium, 447
Harmonic series, 265, 475
Harmonics
  of closed pipes, 267, 267*f*
  of open pipes, 266, 266*f*
  of strings, 264–265, 265*f*
Hawthorne effect, 411, 475
Hearing thresholds, 261–262, 262*t*
Heat, 99–103, 99*t*
  definition of, 92, 475
  energy transfer, 62–63, 98–100
  of fusion, 102
  latent, 102
  as process function, 97
  specific, 101
  of transformation, 101–103
  of vaporization, 102
Heat transfer, 100
  conduction, 100
  convection, 100
  radiation, 100
Henderson–Hasselbalch equation, 368*t*, 371
Hertz, 251, 255, 258
Hill's criteria, 409–410, 475
Histograms, 453, 475
  See also Charts
Human subjects research, 408–412, 419–420
  error sources in, 410–411
  experimental approach, 408–409
  observational approach, 409–410
Hydraulic systems, 133–135, 134*f*, 475
Hydrostatic pressure, 131
Hydrostatics, 133–139, 149–151
  Archimedes' principle, 136–137
  definition of, 133, 475
  molecular forces in liquids, 137–139, 138*f*
  Pascal's principle, 133–135, 134*f*
Hyperopia, 301, 475
Hypotheses, in scientific method, 400–401
Hypothesis testing, 449–450, 450*t*, 475
  See also Statistical testing

I

If-then statement, 400–401
Image(s)
  definition of, 475
  distance, 291
  inverted, 291, 300
  real, 289, 300
  in sign conventions, 293*t*, 300*t*
  upright, 291, 300
  virtual, 289, 300
Imperial System, 8
In phase *vs.* out of phase, 252, 253*f*
Incident pressure, 131
Inclined planes, 29–31, 68–69
Independence, 447
Independent events, 447
Independent variables, 405, 475
Index of refraction, 294, 294*t*, 295*f*, 303, 475
Inertia, 24, 475
Informed consent, 413, 475
Infrared, 286, 287, 287*f*, 475
Infrared spectroscopy, 334
Infrasonic waves, 258, 475
Instantaneous
  acceleration, 23
  speed, 17
  velocity, 17–18
Institutional review boards, 413
Insulation, 226
Insulators, 168, 169*f*, 216, 475
Intensity of sound, 261–264, 262*t*, 475
Interference, 307, 475
  constructive *vs.* destructive, 253
  in multiple slit diffraction, 307, 308*f*
Internal energy, 97, 99, 99*t*
Internal resistance, 218
Internal validity, 416, 475
International System of Units. *See* SI units
Interquartile ranges, 443–444, 475
Interventions support, 417
Inverse relationships, 376, 475
Inverse trigonometric functions, 374
Inverted images, 291, 300, 475
Inviscid fluids, 140
Irreversible process, 108, 475
Isobaric processes, 64, 103, 103*t*, 476
Isochoric processes, 64, 103
Isolated systems, 97
Isothermal processes, 103, 103*t*, 476
Isotope decay arithmetic, 341–342
Isotopes, 476
Isotopic notation, 339
Isovolumetric processes, 64, 103, 103*t*, 476

J

Joule, 9*t*, 56, 62, 100
Junction rule, Kirchhoff's, 214
Justice, 413–414, 476

K

Kelvin scale, 8*t*, 92–93, 93*t*, 378
Kilogram, 8–9, 8*t*, 21
Kinematics and dynamics, 5–48
  concept summary, 39–41
  displacement, 17 (*See also* Displacement)
  equations for, 47–48
  forces and acceleration, 19–24 (*See also* Forces and acceleration)
  mechanical equilibrium, 33–38 (*See also* Mechanical equilibrium)
  motion with constant acceleration, 26–32 (*See also* Motion with constant acceleration)
  Newton's laws, 19, 24–25
  units, 8–9, 377*t*
  vectors and scalars, 10–16 (*See also* Vectors)
  velocity, 17–18
Kinetic energy, 56–57, 476
  of ejected electrons, 331–333
Kinetic friction, 20–21, 476
Kirchhoff's laws, 213–215, 476
  junction rule, 214
  loop rule, 214–215

L

Laminar flow, 140–142, 140*f*, 476
Latent heat, 102
Laws
  Coulomb's, 170–174
  Kirchhoff's, 213–224
  Newton's, 19, 24–25 (*See also* Newton's laws)
  Ohm's, 217–219
  Poiseuille's, 141
  of reflection, 289, 289*f*
  Snell's, 294–296, 295*f*
  of thermodynamics, first, 59, 66, 98–105, 99*t*
  of thermodynamics, second, 99, 106–109
  of thermodynamics, third, 93
  of thermodynamics, zeroth, 92–96
Leading zeroes, 365
Length, as resistor property, 217
Lenses, 298–303
  See also Mirrors; Refraction
  aberrations of, 303, 303*f*, 304, 304*f*
  concave, 298, 299*f*
  convex, 298, 299*f*
  definition of, 298, 476
  multiple lens systems, 301–302
  power of, 301
  ray diagrams for, 299*f*
  real lenses, 299
  sign conventions for, 300, 300*t*
  slit-lens diffraction, 307, 307*f*
  thin spherical, 298, 299*f*
Lensmaker's equation, 299
Lever, as simple machine, 67
Lever arm, 36
Light
  absorption and emission of, 334–336, 334*f*, 335*f*
  color and visible spectrum, 287–288
  particle theory of, 332
  photoelectric effect, 330–332, 333*f*
  speed of, 287, 294
  visible, 286, 287, 287*f*
Light and optics, 279–322
  concept summary, 314–315
  diffraction, 306–311 (*See also* Diffraction)
  electromagnetic spectrum, 286–288 (*See also* Electromagnetic spectrum)
  equations for, 321–322
  geometrical optics, 288–305 (*See also* Geometrical optics)
  polarization, 311–313 (*See also* Polarization)
Lightning, 225, 227
Linear
  expansion, coefficient of, 94
  graphs, 455–457, 456*f* (*See also* Visual data representations)
  motion, 26–28
  solving systems of equations, 379–381
  speed, 143
Linearly polarized light, 311–312
Liquid molecular forces, 137–139, 138*f*
Lithotripsy, 269, 303
Load, 71, 72
Load distance, 71, 72
Log-log graphs, 457–458
Logarithms, 371–372, 383
  common equations using, 368*t*
  common *vs.* natural, 371
  definition of, 476
  estimation, 372
  linear graphs, logarithmic, 455, 456*f*
  rules of, 371
Longitudinal waves, 250–251, 251*f*, 257, 476
  See also Waves
Loop rule, Kirchhoff's, 214–215
Lorentz force, 187, 476
Loudness, 261–264, 262*t*, 476

M

Mach I, 260
Magnetic fields, 184, 185–186, 476
Magnetic forces, 187–190, 476
  force on current-carrying wire, 21
  force on moving charge, 187–188
  Lorentz force, 187
Magnetism, 184–190, 194
  concept summary, 194
  equations for, 201
  magnetic fields, 184, 185–186
  magnetic forces, 187–190
  material classification, 184–185
  right-hand rules, 185–188
Magnification, 291, 476
Mantissa, 364
Maps, 454–455, 455*f*
  See also Charts
Mass, 21–23, 22*f*, 476
Mass defect, 336–338, 348, 476
Mass number, 339, 476
Mathematics, 357–390
  arithmetic, 364–367 (*See also* Arithmetic)
  concept summary, 383–384
  equations, 389
  exponents and logarithms, 368–372 (*See also* Exponents; Logarithms)
  problem-solving, 376–382 (*See also* Problem-solving)
  trigonometry, 373–375 (*See also* Trigonometry)
Mean, 436–437, 476
Measurement units. *See* Units
Measures of central tendency, 436–437, 463
  definition of, 436, 476
  mean, 436–437
  median, 437–438
  mode, 438–439
Measures of distribution. *See* Distributions, measures of
Mechanical advantage, 67–74, 77
  definition of, 67, 476
  hydraulic systems, 133–135, 134*f*
  inclined planes, 68–69
  pulleys, 70–74, 70*f*–72*f*
  simple machines, 67
Mechanical energy, 59–62
  conservation of, 59–62
  conservative forces, 59–60, 60*f*
  nonconservative forces, 61
  total, 59
Mechanical equilibrium, 33–38, 41
  condition of, first, 35
  condition of, second, 37
  free body diagrams, 33–34
  rotational equilibrium, 36–37
  translational equilibrium, 35–36, 70–71
Median, 437–438, 476
Melting and melting points, 102, 476
Meniscus, 138, 138*f*
Metallic bond, 212
Metallic conductivity, 212
Meter, 8–9, 8*t*
Meters, 231–232, 234
  ammeters, 231
  electric, 473
  ohmmeters, 231
  Venturi flow, 144–145, 145*f*
  voltmeters, 231
Meters-kilograms-seconds (MKS, metric system), 8, 9*t*
Meters per second, 17
Meters per second squared, 23
Metric prefixes, 377*t*
Metric system, 8, 9*t*, 476
Microstates, 102
Microwaves, 286, 287, 287*f*, 476
Millimeters of mercury (mmHg), 129
Mirrors, 289–294
  See also Lenses; Reflection
  plane, 289, 290*f*
  sign conventions for, 291–294, 293*t*
  spherical, 290–293, 291*f*, 292*f*–293*f*
  spherical aberrations of, 303, 303*f*
Mode, 438–439, 476
Mole, 8*t*
Molecular forces in liquids, 137–139, 138*f*
Moment of force, 36
Monochromatic light, 307, 476
Morally relevant differences, 413–414
Motion, as kinetic energy, 56–57
Motion with constant acceleration, 26–32, 41

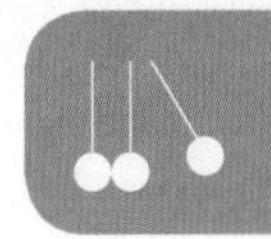

circular motion, 31–32
inclined planes, 29–31, 68–69
linear motion, 26–28
projectile motion, 28–29
Multiple lens systems, 301–302
Multiple slit diffraction, 307–310, 308*f*
Multiplication
estimation, 366–367 (*See also* Estimation)
with exponents, 369
with significant figures, 365
vectors by scalars, 13–14
vectors by vectors, 14–16
Musical sound, 257–258, 264–267
Mutually exclusive outcomes, 447, 476
Myopia, 301, 476

N

Nanometers, 9, 331
Natural
frequencies, 254–255, 476
logarithms, 371–372
phenomena, size of, 8*f*
process, 108, 476
Negative controls, 404
Neutrino, 342
Neutrons, 337, 339
Newton (N), 9, 9*t*, 19, 21
Newton per square meter, 129
Newton's laws, 24–25, 40, 476–477
first, 24
of gravitation, 19–20
second, 25
third, 19, 25
Nodes in waves, 254, 264, 477
Noise, 255
Noise-canceling headphones, 253
Nonconservative forces, 61, 477
Nondirectional alternative hypothesis, 449
Nonmaleficence, 412, 477
Normal, 289, 477
Normal distributions, 440, 440*f*
See also Distributions
Normal force, 20, 477
Notation
isotopic, 339
scientific, 364
Novelty, of research question, 401
Nuclear binding energy, 336–338, 348
Nuclear phenomena
See also Atomic and nuclear phenomena
mass defect, 336–338
nuclear forces, weak *vs.* strong, 337
Nuclear reactions, 339–347, 349
definition of, 339
fission, 340–341
fusion, 339–340, 340*f*
isotopic notation for, 339
radioactive decay, 341–347 (*See also* Radioactive decay)
Nucleon, 337, 477
conservation, 341–342
Null hypotheses, 449, 477

O

Observation, in scientific method, 400
Observation bias, 411
Observational approaches/studies, 409–410, 477
Ohm-meter, 216
Ohmmeters, 231, 477
Ohms, 217
Ohm's law, 217–219, 477
Open boundaries, 264
Open pipes, 266, 266*f*
Open systems, 97
Optics and light. *See* Geometrical optics; Light and optics or *vs.* and, 448
Oscillation of waves, 250–251, 251*f*, 286*f*
forced, 255
Out of phase, 252, 253*f*
Outcomes
in cohort studies, 409
exhaustive, 447
mutually exclusive, 447
Outliers, 436, 443, 446, 477
Overtones, 255, 265–267

P

p scales, equations for, 368*t*
p-value, 450
Parabolic linear graphs, 455, 456*f*
Parallel, 477
capacitors in, 229–230, 229*f*
resistors in, 221–224, 221*f*
Parallel plate capacitor, 225, 226
Paramagnetic materials, 184–185, 477
Parameters, 416, 477
Parent nucleus, 341–342
Partially constructive interference, 261
Partially destructive interference, 253
Particle theory of light, 332
Particles
alpha (α), 342
beta (β), 342
subatomic, charged, 168, 169
Pascal (Pa), 129
Pascal-second, 140
Pascal's principle, 133–135, 134*f*, 477
Peer reviews, 400
Period, of wave, 251, 477
Permeability of free space, 185
Permittivity of free space, 170
Perpendicular bisector of dipole, 182
Phacoemulsification, 269
Phase, of wave, 252, 253*f*
Phase changes, 92, 101–103
Phase difference of wave, 252, 253*f*, 477
Phenolphthalein indicator, 335
Photoelectric effect, 330–332, 333*f*
definition of, 330, 348, 477
kinetic energy of ejected electrons, 331–333
threshold frequency, 330–331
Photons, 331–332
Physics and math concepts
atomic and nuclear phenomena, 323–356
circuits, 203–240
data-based and statistical reasoning, 427–470
electrostatics and magnetism, 159–201
fluids, 119–158
glossary for, 471–480
kinematics and dynamics, 5–48
mathematics, 357–390
optics and light, 279–322
research design and execution, 391–426
thermodynamics, 85–117
waves and sound, 243–278
work and energy, 49–83
Physiology fluids, 152
circulatory system, 148–149
respiratory system, 149
Pie charts, 452, 452*f*
Pipes, and harmonics of
closed, 266, 267, 267*f*
open, 266, 266*f*
Pistons, 63
Pitch, 255, 258, 477
Pitot tubes, 144, 477
Placebo effect, 404, 408
Planck's constant, 331
Plane mirrors, 289, 290*f*, 477
Plane-polarized light, 311–312, 477
Planes, inclined, 29–31, 68–69
Plausibility, as Hill's criteria, 410
Plots, box and whisker, 453–454, 454*f*, 472
See also Charts
Poiseuille's law, 141, 477
Polarization, 311–313, 315
circular, 312, 312*f*
plane-polarized light, 311–312
Populations, in research, 416, 477
Positive controls, 404
Positron, 342, 477
Positron emission, 342
Potential difference, 177, 213, 477
Potential energy, 57–59
in capacitor, 226
definition of, 57, 477
elastic, 58–59
electric, 174–176

Potential energy (cont.)
  gravitational, 57–58
  zero (datum), 57
Pound (lb), 8
Power
  in energy transfer, 65, 477
  of lenses, 301
  measuring resistance, 218–219
  in probability of hypothesis, 450
Powers of ten, 364
Precision, 406, 407*f*, 477
Prefixes, metric, 377*t*
Pressure, 129–132
  absolute, 130–131
  atmospheric, 131
  definition of, 129, 478
  dynamic, 143
  energy density, 63, 144
  gauge, 131–132
  hydrostatic, 131
  P–V graphs, 63–64, 64*f*, 104*f*
  as state function, 97
  static, 144
Pressure–volume (P–V) curves, 63–64, 64*f*, 104*f*
  See also Graphs
Principle of superposition, 253, 478
Prism, 303, 304*f*
Probability, 447–449, 464
  calculations with and, or, 448, 448*f*
  exhaustiveness, 447
  independence, 447
  mutual exclusivity, 447
Problem-solving, 376–382, 384
  algebraic systems, 379–382 (*See also* Algebraic systems)
  conversions, 376–378 (*See also* Conversions)
  unit analyses, 378–379
  use of relationships, 376
Process functions, 97, 478
Processes, thermodynamic. *See* Thermodynamic processes
Projectile motion, 28–29
  See also Motion with constant acceleration
Propagation, 250–251, 478
  rectilinear, 288
Propagation speed, 251
Protons, 168, 337, 339, 478
Pulleys, 70–74, 70*f*–72*f*
Pulse, 148
Punnett squares, 447
P–V graphs, 63–64, 64*f*, 104*f*
Pythagorean theorem, 12, 12*f*, 478

Q

Quantum energy, 478
Quartiles, 443, 478

R

Radiation, 100, 286, 478
Radio waves, 286, 287, 287*f*, 478
Radioactive decay, 341–347
  alpha decay, 342
  beta decay, 342–343
  constant, 345
  definition of, 341, 478
  electron capture, 344
  exponential decay, 345–346, 346*f*
  gamma decay, 343–344
  half-life, 344–345
  isotope decay arithmetic, 341–342
  nucleon conservation, 341–342
Radius of curvature, 290, 300
Random error, 406, 478
Randomization, 408, 478
Range, 442, 478
  interquartile, 443–444
Rarefaction, 251, 257
Ray diagrams, 291–292, 478
  for lenses, 299*f*
  for mirrors, 291–292, 292*f*–293*f*
Rays
  gamma, 286, 287, 287*f*, 343–344
  x-rays, 286, 287, 287*f*, 310, 310*f*
Real images, 289, 300, 478
Real lenses, 299
Real world research, 416–417, 421
  generalizability, 416
  interventions support, 417
  populations *vs.* samples, 416
Reasoning. *See* Data-based and statistical reasoning; Research design and execution
Rectilinear propagation, 288, 478
Reflected waves, 254, 254*f*, 263, 268*f*
Reflection, 289–294, 289*f*
  definition of, 289, 478
  law of, 289, 289*f*
  plane mirrors, 289, 290*f*
  sign conventions for mirrors, 293–294, 293*t*
  spherical mirrors, 290–293, 291*f*, 292*f*–293*f*
  total internal, 297–298, 297*f*
Refraction, 294–298
  definition of, 294, 478
  index of, 294, 294*t*, 295*f*, 303
  Snell's law, 294–296, 295*f*
  total internal reflection, 297–298, 297*f*
Regression analyses, 409
Relationships, direct and inverse, 376
Relevancy of research question, 401
Reliability, 406, 407*f*
Repulsive forces, 168
Research design and execution, 391–426
  basic science research, 403–407 (*See also* Basic science research)
  concept summary, 419–421
  ethics in, 412–415 (*See also* Ethics)
  FINER method, 401
  human subjects research, 408–412 (*See also* Human subjects research)
  real world, 416–417 (*See also* Real world research)
  scientific method, 400–402
  studies, types of, 408–409 (*See also* Studies)
Resistance, air, 28, 61
Resistance, in current, 216–224, 233
  definition of, 216, 478
  internal, 218
  Ohm's law, 217–219
  power, 218–219
  resistors, 216–217, 219–224 (*See also* Resistors)
Resistivity, 216–217, 478
Resistors
  in parallel, 221–224, 221*f*
  properties of, 216–217
  in series, 219–220, 219*f*
Resonance, 254–255, 478
Resonant frequencies, 254–255
Resonating system, 255
Respect for persons, 413, 478
Respiratory system, 149
Resultant of vectors, 10–13, 478
Resultant resistance, 220
Reversible reaction, 108
Reynolds number, 142
Right-hand rules
  of magnetism, 185–188
  of vector multiplication, 14–16, 14*f*, 478
Right triangles and sides, 373, 373*f*, 374, 374*f*
Rotation, 478
Rotational equilibrium, 36–37
  See also Mechanical equilibrium
Rounding, 365

S

Samples, in research, 416, 478
Scalars, 10, 39, 478
  multiplication, vectors by scalars, 13–14
Scientific knowledge contexts, 460–461
Scientific method, 400–402, 419, 478
Scientific notation, 364, 478
Screw, as simple machine, 67
Second, 8–9, 8*t*
Second condition of equilibrium, 37
Second law, Newton's, 25
Second law of thermodynamics, 99, 106–109, 111
Secondary batteries, 218
Selection bias, 410, 478
Semilog graphs, 457–458, 457*f*
Series, 478

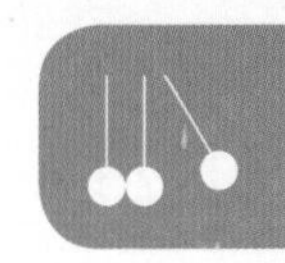

capacitors in, 228–230, 229*f*
resistors in, 219–220, 219*f*
Setting equations equal, solving algebraic systems, 380
Shear forces, 128, 478
Shock waves, 260, 479
SI units, 8, 8*t*, 9, 9*t*, 377, 479
ampère, 8*t*, 213
base units, 9
candela, 8*t*
coulomb, 168
farad, 226
gauss, 184
joule, 9*t*, 56, 62, 100
kelvin, 8*t*, 92–93
kilogram, 8–9, 8*t*, 21
meter, 8–9, 8*t*
meters per second, 17
meters per second squared, 23
mole, 8*t*
newton, 9, 9*t*, 19, 21
newton per square meter, 129
ohm-meter, 216
pascal, 129
pascal-second, 140
second, 8–9, 8*t*
siemens, 212
tesla, 184
watt, 9*t*, 65
watts per square meter, 261
Siemens (S), 212
Sign conventions
for Doppler effect, 258–259
for first law of thermodynamics, 99*t*
for lenses, 300, 300*t*
for mirrors, 291–294, 293*t*
Significance
clinical, 417
level, 450
statistical, 417
Significand, 364
Significant figures, 364–366, 383, 479
Simple machines, 67, 479
Sine, 373–374, 375*t*
Single-blind experiments, 408, 479
Single slit diffraction, 306, 306*f*
Sinusoidal waves, 250
Sisyphus myth, 55, 62–63, 67
Six-pulley system, 72–73, 72*f*
See also Pulleys
Skewed distributions, 440–441, 441*f*
Slit-lens system, diffraction, 307, 307*f*
Slope, 456
Slug, 8, 9*t*
Snell's law, 294–296, 295*f*, 479
Solidification, 102
Solids
See also Fluids
characteristics of, 128–132, 149–150
definition of, 128, 479
density, 128–129
pressure, 129–132
Sonic boom, 260
Sound, 257–270
See also Waves
attenuation, 255, 263
beat frequency, 264
closed pipes, 266, 267
concept summary, 272
definition of, 257, 479
Doppler effect, 258–260, 259*f*
equations for, 278
frequency, 258–260
harmonics (*See* Harmonics)
intensity of, 261–264, 262*t*
levels, 261–262, 262*t*, 368*t*
loudness of, 261–264, 262*t*
open pipes, 266
pitch, 255, 258–260
production of, 257–258
shock waves, 260
speed of, 257, 260
standing waves, 264–267 (*See also* Standing waves)
strings, 264–265
ultrasound, 268–269, 268*f*
Source charge, 172, 173*f*, 479
See also Charges
Special right triangles, 374, 374*f*
Specific gravity, 129, 479
Specific heat, 101, 479
Specific rotation, 312
Specificity, as Hill's criteria, 410
Spectroscopy, infrared *vs.* UV-Vis, 334
Spectrum, electromagnetic. *See* Electromagnetic spectrum
Speed
critical, 141
definition of, 17, 479
of light, 287, 294
linear, 143
propagation, of waves, 251
of sound, 257, 260
Spherical aberrations, 303, 303*f*, 479
Spherical lenses, 298, 299*f*
Spherical mirrors, 290–293, 291*f*, 292*f*–293*f*, 479
Spring, potential energy of, 58–59
See also Potential energy
Spring constant, 59
Square root estimation, 370, 370*t*
See also Estimation
Standard, in basic research, 403
Standard deviation, 444–445, 479
Standard distribution, 440
Standing waves, 254, 264–267, 479
See also Waves
closed pipes, 267
open pipes, 266
strings, 264–265
State functions, 97, 479
Static
charge buildup, 168
electricity, 168
friction, 20–21, 479
pressure, 144
Statistical significance/effect, 417
Statistical testing, 449–451, 464
See also Data-based and statistical reasoning
confidence intervals, 450–451
hypothesis testing, 449–450, 450*t*
Statistics, 416, 479
Streamlines, 142–143, 142*f*, 479
Strength, as Hill's criteria, 410
String harmonics, 264–265, 265*f*
Strong nuclear force, 337, 479
Studies
case-control, 409
cohort, 409
cross-sectional, 409
double-blind, 408
single-blind, 408
Subatomic particles, charged, 168, 169
Substitution, solving algebraic systems, 379–380
Subtraction
with exponents, 369
with significant figures, 365
vector, 13 (*See also* Vectors)
Superconductors, 218
Surface tension, 137, 479
Surroundings, 96, 479
Systematic error, 406, 479
Systems, 96–98, 110
closed, 97
definition of, 96, 479
isolated, 97
open, 97
pulleys, 70–74, 70*f*–72*f*
state functions, 97

T

t-score, 450–451
t-tests, 450
Tables, 459
See also Visual data representations
Tail, of skewed distribution, 440
Tangent, 374, 375*t*
Tangential forces, 128

Temperature, 92–93, 93*t*
conversions between scales, 93, 378
definition of, 479
as resistor property, 217 (*See also* Resistors)
as state function, 97
Temporality, 410, 479
Ten, powers of, 364
Tendency measures. *See* Measures of central tendency
Tension, surface, 137, 479
Terminal velocity, 26, 28, 479
Tesla (T), 184
Test charge, 176, 479
See also Charges
Test statistic, 450
Testable question, in scientific method, 400–401
Theorems
Pythagorean, 12, 12*f*
work-energy, 65–66
Thermal equilibrium, 92, 479
Thermal expansion, 94–96, 480
Thermodynamic processes, 103–105
adiabatic, 103, 103*t*
isobaric, 64, 103, 103*t*
isothermal, 103, 103*t*
isovolumetric, 64, 103, 103*t*
natural *vs.* unnatural, 108
reversible *vs.* irreversible, 108
Thermodynamics, 85–117
concept summary, 110–111
definition of, 91
energy dispersion, 106
entropy, 106–108
equations for, 117
first law of, 59, 66, 98–105, 99*t*
processes (*See* Thermodynamic processes)
second law of, 99, 106–109
systems, 96–98
third law of, 93
zeroth law of, 92–96 (*See also* Zeroth law of thermodynamics)
Thin film interference, 309, 310*f*
Thin spherical lenses, 298, 299*f*
Third law, Newton's, 19, 25
Third law of thermodynamics, 93
Threshold frequency, 330–331, 480
Threshold of hearing, 261–262, 262*t*
Timbre, 255, 480
Time's arrow, 107
Tip-to-tail method, 10, 11*f*
Torque, 36, 480
on electric dipole, 182–183, 183*f*
Torr, 129
Total internal reflection, 297–298, 297*f*, 480
Total mechanical energy, 59, 480
See also Mechanical energy
Trailing zeroes, 365
Transformation, heat of, 101–103, 475
Translational equilibrium, 35–36, 70–71, 480
See also Mechanical equilibrium
Transverse waves, 250–251, 251*f*, 480
See also Waves
Traveling waves, 254, 254*f*, 480
See also Waves
Trigonometry, 373–375, 384
common values, 374–375, 375*t*
cosine, 373, 375*t*
inverse functions, 374
right triangles, 373, 373*f*, 374, 374*f*
sine, 373–374, 375*t*
tangent, 374, 375*t*
Trough of wave, 252, 252*f*
Turbulence, 141–142
Turbulent flow, 141–142, 141*f*, 480
Tuskegee syphilis experiment, 413
Two-pulley system, 71, 71*f*
Type I error, 450, 480
Type II error, 450, 480

U

Ultrasonic frequency, 258, 480
Ultrasound, 268–269, 268*f*, 480
Ultraviolet, 286, 287, 287*f*, 480
fluorescence, 335
Uniform circular motion, 31, 31*f*
See also Motion with constant acceleration
Uniform electric field, 226
Unit analyses, 378–379
Units, 8–9, 39
base, 9, 377*t*
British, 8
derived, 9, 9*t*
equivalent, 377*t*
SI, 8, 8*t*, 9, 9*t* (*See also* SI units)
Universal gravitational constant (G), 19–20
Unnatural process, 108, 480
Upright images, 291, 300, 480
UV-Vis spectroscopy, 334

V

Validity, 406, 407*f*, 416
Vaporization, 102, 480
Variables
binary, 409
categorical, 409
confounding, 409, 411
continuous, 409
dependent *vs.* independent, 405
Vectors, 10–16, 39
addition, 10–13, 11*f*–13*f*
definition of, 10, 480
multiplication, vectors by scalars, 13–14
multiplication, vectors by vectors, 14–16
subtraction, 13
Velocity, 17–18, 39, 480
See also Acceleration
terminal, 26, 28
Ventricular fibrillation, 227
Venturi effect, 145, 480
Venturi flow meter, 144–145, 145*f*
Virtual images, 289, 300, 480
Viscosity, 140, 480
Viscous drag, 61, 140, 480
Visible light, 286, 287, 287*f*
Visible region, 287–288, 480
Visual data representations, 451–459, 464
charts, 452–455 (*See also* Charts)
graphs, 455–458 (*See also* Graphs)
tables, 459
Voltage, 177, 213
Voltaic (galvanic) cells, 213, 218
Voltmeters, 231, 480
Volts, 177
Volume
and pressure, in work, 63–65, 64*f*
of sound, 261
as state function, 97
Volumetric thermal expansion, 95
Vulnerable persons, 413

W

Watt, 9*t*, 65
Watts per square meter, 261
Wavelength, 251–252, 252*f*, 286*f*, 287, 480
Waves, 243–278
See also Sound
anatomy of, 251–252, 252*f*
concept summary, 271–272
electromagnetic, 286–288, 286*f*
equations for, 275
general characteristics of, 250–256
infrasonic waves, 258
longitudinal, 250–251, 251*f*, 257
phase and phase difference, 252
principle of superposition, 253
resonance, 254–255
shock waves, 260
standing waves, 254, 264–267
transverse waves, 250–251, 251*f*
traveling waves, 254, 254*f*
ultrasonic waves, 258
Weak nuclear force, 337, 480
Weight, 21–22, 480
Wheel and axle, 67
Work, 62–66, 76–77
definition of, 62, 480
energy transfer, 62–63, 99
equations for, 83

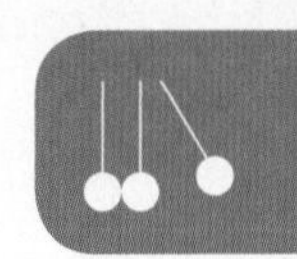

force and displacement, 63
power, 65
pressure and volume, 63–65, 64*f*
as process function, 97
sign convention, 99*t*
work-energy theorem, 65–66, 480
Work and energy, 49–83
concept summary, 76–77
energy (*See* Energy)
equations for, 83
mechanical advantage, 67–74 (*See also* Mechanical advantage)
work (*See* Work)
Work function, 331
Work input, 71–72
Work output, 72

X

X-component, 11, 11*f*
X-ray(s), 286, 287, 287*f*, 480
diffraction, 310, 310*f*

Y

Y-component, 11, 11*f*
Young's double-slit experiment, 307–309, 308*f*

Z

z-score, 450–451
z-tests, 450
Zero, absolute, 93, 93*t*, 471
Zero potential energy (datum), 57
Zeroes, leading and trailing, 365
Zeroth law of thermodynamics, 92–96, 110
definition of, 92, 480
temperature, 92–93, 93*t*
thermal expansion, 94–96

ART CREDITS

Figure 1.1—Image credited to Melissa Thomas. From The Great Cosmic Roller-Coaster Ride by Cliff Burgess and Fernando Quevado. Copyright © 2007 by Scientific American, Inc. All rights reserved.

Figure 1.7—Image credited to Jared Schneidman Designs. From Friction at the Atomic Scale by Jacqueline Krim. Copyright © 1996 by Scientific American, Inc. All rights reserved.

Chapter 4 Cover—Image credited to sakhorn. From Shutterstock.

Figure 4.5—Image credited to Kent Snodgrass/Precision Graphics. From Working Knowledge: Big Squeeze by Mark Fischetti. Copyright © 2006 by Scientific American, Inc. All rights reserved.

Chapter 5 Cover—Image credited to Piotr Krzeslak. From Shutterstock.

Chapter 6 Cover—Image credited to Jelena Aloskina. From Shutterstock.

Chapter 7 Cover—Image credited to Andreea Dragomir. From Shutterstock.

Sidebar, Chapter 7—Image credited to Samuel Velasco; Source: Bose Corporation. From Working Knowledge: Reducing a Roar by Mark Fischetti. Copyright © 2005 by Scientific American, Inc. All rights reserved.

Chapter 8 Cover—Image credited to Juan J. Jimenez. From Shutterstock.

Figure 8.9—Image credited to Melissa Thomas. From The Quest for the Superlens by John B. Pendry and David R. Smith. Copyright © 2006 by Scientific American, Inc. All rights reserved.

Figure 8.17—Image credited to Michael Goodman. From The Duality in Matter and Light by Berthold-Georg Englert, Marlan O. Scully, and Herbert Walther. Copyright © 1994 by Scientific American, Inc. All rights reserved.

Chapter 9 Cover—Image credited to Vaclav Volrab. From Shutterstock.

Figure 9.1—Image credited to Alfred T. Kamajian. From Everyday Einstein by Philip Yam. Copyright © 2004 by Scientific American, Inc. All rights reserved.

Figure 9.2—Image credited to George Retseck. From The Dark Ages of the Universe by Abraham Loeb. Copyright © 2006 by Scientific American, Inc. All rights reserved.

Chapter 10 Cover—Image credited to zphoto. From Shutterstock.

Chapter 11 Cover—Image credited to Serg Zastavkin. From Shutterstock.

Chapter 12 Cover—Image credited to Zorabc. From Shutterstock.

Figure 12.1—Image credited to User: Mwtoews. From Wikimedia Commons. Copyright © 2007. Used under license: CC-BY-2.5.

Notes

Notes

Notes

Notes